Diagrammatic Representation and Reasoning

Springer
London
Berlin
Heidelberg
New York
Barcelona
Hong Kong
Milan
Paris
Singapore
Tokyo

Michael Anderson, Bernd Meyer
and Patrick Olivier (Eds)

Diagrammatic Representation and Reasoning

Springer

Michael Anderson, MSc, PhD
Department of Computer and Information Sciences, Fordham University, 441 East Fordham Road, Bronx, NY 10458, USA

Bernd Meyer, Dr rer nat
School of Computer Science and Software Engineering, Monash University, PO Box 26, Clayton Campus, Victoria 3800, Australia

Patrick Olivier, MA, MSc, PhD
Department of Computer Science, University of York, Heslington, York, YO10 5DD, UK

British Library Cataloguing in Publication Data
Diagrammatic representation and reasoning
1.Artificial intelligence 2.Problem solving 3.Charts, diagrams, etc.
I.Anderson, Michael II.Meyer, Bernd III.Olivier, Patrick
006.3
ISBN 1852332425

Library of Congress Cataloging-in-Publication Data
Diagrammatic representation and reasoning / Michael Anderson, Bernd Meyer, and Patrick Olivier (eds.).
p. cm.
Includes bibliographical references and index.
ISBN 1-85233-242-5 (alk. paper)
1. Automatic theorem proving. 2. Artificial intelligence. 3. Image processing. 4. Problem solving. I. Anderson, Michael, 1965- II. Meyer, Bernd, 1951- III. Olivier, Patrick, 1966-
QA76.9.A96 D54 2000
006.3--dc21 00-063765

ISBN 1-85233-242-5 Springer-Verlag London Berlin Heidelberg
a member of BertelsmannSpringer Science+Business Media GmbH
http://www.springer.co.uk

Printed in Great Britain

Typesetting: Camera ready by editors
Printed and bound at the Athenæum Press Ltd., Gateshead, Tyne and Wear
34/3830-543210 Printed on acid-free paper SPIN 10748985

Preface

Diagrams are essential in most fields of human activity. There is substantial interest in diagrams and their use in many academic disciplines for the potential benefits they may confer on a wide range of tasks. Are we now in a position to claim that we have a science of diagrams—that is, a science which takes the nature of diagrams and their use as the central phenomena of interest? If we have a science of diagrams it is certainly constituted from multiple disciplines, including cognitive science, psychology, artificial intelligence, logic, mathematics, and others.

If there is a science of diagrams, then like other sciences there is an applications, or engineering, discipline that exists alongside the science. Applications and engineering provide tests of the theories and principles discovered by the science and extend the scope of the phenomena to be studied by generating new uses of diagrams, new media for presenting diagrams, or novel classes of diagram. This applications and engineering side of the science of diagrams also comprises multiple disciplines, including education, architecture, computer science, mathematics, human-computer interaction, knowledge acquisition, graphic design, engineering, history of science, statistics, medicine, biology, and others.

The chapters of this book reflect this diversity of interests in the nature and uses of diagrams and the synthesis of results presented at three recent events on diagrammatic representation and reasoning: the *American Association for Artificial Intelligence Fall Symposium on Diagrammatic Reasoning* held at MIT in November 1997; the *Thinking with Diagrams Workshop* held at the University of Wales, Aberystywth, in August 1998, and the *American Association for Artificial Intelligence Fall Symposium on Formalizing Reasoning with Visual and Diagrammatic Representations*, held in Orlando in October 1998. Without these events, this book would not have come about, and therefore we would like to thank the American Association for Aritificial Intelligence and the Engineering and Physical Sciences Research Council (UK) for generous funding in supporting these meetings.

Michael Anderson
Bernd Meyer
Patrick Olivier

April 2001

Preface

Contents

Contributors

Herman J. Adèr hj.ader.biostat@med.vu.nl
Vrije Universiteit, Clinical Epidemiology and Biostatistics, Van der Boechorststraat 7, 1081BT Amsterdam, The Netherlands.

Gerard Allwein gtall@cs.indiana.edu
Indiana University, Department of Computer Science, Lindley Hall 215, Bloomington, IN 47405, USA.

Michael Anderson anderson@cis.fordham.edu
Fordham University, Department of Computer and Information Sciences, 441 East Fordham Road, Bronx, NY 10458, USA.

Daniela M. Bailer-Jones daniela.bailer-jones@uni-bonn.de
University of Bonn, Department of Philosophy, LFB I, Am Hof 1, 53113 Bonn, Germany.

Sidney C. Bailin sbailin@waves.kevol.com
Knowledge Evolution Inc., 1050 17th Street NW, Suite 520, Washington DC, 20036, USA.

Dave Barker-Plummer dbp@csli.stanford.edu
Stanford University, Center for the Study of Language and Information, Ventura Hall, Stanford, California, 94305-4115, USA.

Alan F. Blackwell Alan.Blackwell@cl.cam.ac.uk
University of Cambridge, Computer Laboratory, Cambridge, UK.

Mark Blades m.blades@sheffield.ac.uk
University of Sheffield, Psychology Department, Western Bank, Sheffield S10 2TP, UK.

Alan Bundy A.Bundy@ed.ac.uk
University of Edinburgh, IRR, Division of Informatics, 80 South Bridge, Edinburgh, EH1 1HN, UK.

Jo Calder J.Calder@ed.ac.uk
University of Edinburgh, Division of Informatics, ICCS and LTG, University of Edinburgh, Division of Informatics, 2 Buccleuch Place, Edinburgh EH9 9LW, UK.

Peter C-H. Cheng peter.cheng@nottingham.ac.uk
University of Nottingham, ESRC Centre for Research in Development, Instruction and Training, School of Psychology, University Park, Nottingham NG7 2RD, UK.

Nathalie Cousin-Rittemard rittemar@maths.univ-rennes1.fr
Rennes 1 University, IRMAR, Equipe de Mecanique, Campus BEAULIEU, Université de Rennes 1, CS 74205 35042 Rennes Cedex, France.

Yuri Engelhardt yuriengelhardt@yahoo.com
University of Amsterdam, Department of Computational Linguistics, Palmgracht 35, 1015 HK Amsterdam, The Netherlands.

Dale E. Fish fish@engr.uconn.edu
University of Connecticut, Department of Computer Science and Engineering, School of Engineering, 261 Glenbrook Road, Storrs, CT 06269, USA.

Norman Foo norman@cse.unsw.edu.au
University of New South Wales, Department of Computer Science and Engineering, Sydney NSW 2052, Australia.

Jean-Louis Giavitto giavitto@lami.univ-evry.fr
CNRS, LaMI umr 8042, Université d'Evry Val d'Essonne, Boulevard F. Mitterrand, 91025 Evry Cedex, France.

Ian Green I.Green@ed.ac.uk
University of Edinburgh, IRR, Division of Informatics, 80 South Bridge, Edinburgh, EH1 1HN, UK.

Corin Gurr C.Gurr@ed.ac.uk
University of Edinburgh, Division of Informatics, 2 Buccleuch Place, Edinburgh EH8 9LW, UK.

Volker Haarslev haarslev@informatik.uni-hamburg.de
University of Hamburg, Computer Science Department, Vogt-Kölln-Str. 30, 22527 Hamburg, Germany.

Mary Hegarty hegarty@psych.ucsb.edu
University of California, Santa Barbara, Psychology Department, Santa Barbara, CA 93106-9660, USA.

Mateja Jamnik M.Jamnik@cs.bham.ac.uk
University of Birmingham, School of Computer Science, Birmingham B15 2TT, UK.

Peter Johnson pete@mimir.demon.co.uk
University of Newcastle, The Sowerby Center for Primary Health Care Informatics, Newcastle upon Tyne, NE4 6BE, UK.

Robert Kosara rkosara@ifs.tuwien.ac.at
Vienna University of Technology, Institute of Software Technology, Favoritenstraße 9–11/E 188, 1040 Vienna, Austria.

Maria Kozhevnikov mkozhevn@fas.harvard.edu
Harvard University, Graduate School of Education, Longfellow Hall, Appian Way, Cambridge MA 02138, USA.

Ellen Levy levy@psych.stanford.edu
Stanford University, Department of Psychology, Jordan Hall, Stanford University, Stanford, CA 94305-2130, USA.

Michael Lewis ml@sis.pitt.edu
University of Pittsburgh, Department of Information Science and Telecommunications, 135 North Bellefield Avenue, Pittsburgh, PA 15260, USA.

Robert K. Lindsay lindsay@umich.edu
University of Michigan, Mental Health Research Institute, 205 Zina Pitcher Place, Ann Arbor, Michigan 48109, USA.

Nadine Lucas Nadine.Lucas@info.unicaen.fr
Caen University, Groupe de Recherche en Informatique, Image, Instrumentation de Caen (GREYC), Campus II Universit de Caen BP 5186 F-14032 CAEN CEDEX, France.

Kim Marriott marriott@cs.monash.edu.au
Monash University, School of Computer Science and Software Engineering, PO Box 26, Clayton Campus, Victoria 3800, Australia.

Richard Mayer mayer@psych.ucsb.edu
University of California, Santa Barbara, Psychology Department, Santa Barbara, CA 93106-9660, USA.

Robert McCartney robert@cse.uconn.edu
University of Connecticut, Department of Computer Science and Engineering, School of Engineering, 261 Glenbrook Road, Storrs, CT 06269, USA.

Bernd Meyer bernd.meyer@acm.org
Monash University, School of Computer Science and Software Engineering, PO Box 26, Clayton Campus, Victoria 3800, Australia.

Silvia Miksch silvia@ifs.tuwien.ac.at
Vienna University of Technology, Institute of Software Technology, Favoritenstraße 9–11/E 188, 1040 Vienna, Austria.

Mark Minas Mark.Minas@informatik.uni-erlangen.de
Universität Erlangen-Nürnberg, Lehrstuhl für Programmiersprachen, Martensstr. 3, 91058 Erlangen, Germany.

Ralf Möller moeller@informatik.uni-hamburg.de
University of Hamburg, Computer Science Department, Vogt-Kölln-Str. 30, 22527 Hamburg, Germany.

Clive Richards C.Richards@coventry.ac.uk
Coventry University, Coventry School of Art and Design, Priory Street, Coventry CV1 5FB, UK.

Leonid G. Rozenblit leonid.rozenblit@yale.edu
Yale University, Department of Psychology, 2 Hillhouse Ave., New Haven, CT 06520-8205, USA.

Diane Schiano
At the time of writing the author was affiliated with Interval Research Corporation, Palo Alto, CA, USA.

Priti Shah priti@umich.edu
University of Michigan, Department of Psychology, 525 East University, Ann Arbor, MI 48109-1109, USA.

Yuval Shahar shahar@smi.stanford.edu
Stanford University, Stanford Medical Informatics, MSOB X-215, 251 Campus Drive, Stanford, CA 94 305 - 5479, USA.

Sun-Joo Shin shin.3@nd.edu
University of Notre Dame, Department of Philosophy, 336 O'Shaughnessy, Notre Dame, IN 46556, USA.

Aaron Sloman A.Sloman@cs.bham.ac.uk
The University of Birmingham, School of Computer Science, B15 2TT, Birmingham, UK.

Christopher Spencer c.p.spencer@sheffield.ac.uk
University of Sheffield, Psychology Department, Western Bank, Sheffield S10 2TP, UK.

Michael Spivey spivey@cornell.edu
Cornell University, Department of Psychology, Ithaca, NY 14853, USA.

Thomas F. Stahovich stahov@andrew.cmu.edu
Carnegie Mellon University, Mechanical Engineering Department, 415 Scaife Hall,Pittsburgh, PA 15213, USA.

Masaki Suwa suwa@sccs.chukyo-u.ac.jp
Chukyo University, School of Computer and Cognitive Sciences, 101 Tokodate, Kaizu-chou, Toyota, Aichi 470-0393, Japan.

Nik Swoboda swoboda@cs.indiana.edu
Indiana University, Computer Science Department, Lindley Hall, 150 S. Woodlawn Ave., Bloomington IN 47405-7104, USA.

Jozsef A. Toth jtoth@ida.org
Institute for Defense Analyses, 1801 North Beauregard Street, Alexandria, VA 22311-1772, USA.

Barbara Tversky bt@psych.stanford.edu
Stanford University, Department of Psychology, Building 420, Stanford, CA 94305, USA.

Simon Ungar ungar@lgu.ac.uk
London Guildhall University, Psychology Department, Calcutta House, Old Castle Street, London E1 7NT, UK.

Erika Valencia erika@limsi.fr
CNRS, LIMSI, Université Paris-XI, BBP 133, 91403 Orsay Cedex, France.

Michael Wessel mwessel@informatik.uni-hamburg.de
University of Hamburg, Computer Science Department, Vogt-Kölln-Str. 30, 22527 Hamburg, Germany.

Julie Wojslawowicz jwojslaw@wam.umd.edu
University of Maryland, Department of Human Development, 3304 Benjamin Building, College Park, MD 20740, USA.

Jeffrey M. Zacks jzacks@artsci.wustl.edu
Washington University, Psychology Department, 1 Brookings Drive, Saint Louis, MO 63130, USA.

Part I

Views of Diagrams

Introduction

There are a number of different perspectives on the nature of diagrams and this volume stands as a testament to the diversity of approaches adopted. Furthermore, harnessing the apparent power of diagrammatic reasoning is an essentially multidisciplinary endeavour. On the one hand, computer scientists posit the formal mechanisms that must be the basis for any truly automated diagrammatic reasoning capability, and on the other, cognitive scientists strive to understand the use of external representations and their role in reasoning, with the results of the psychological investigation not only shedding light on human cognition, but also serving to characterise the formalist's requirements. Of equal importance to these two enterprises is the work of diagram practitioners, the mathematicians, architects, scientists, designers, graphics artists and information display designers, whose work and wealth of experience are increasingly falling under the spotlight of academic interest.

This volume is organised into four (reasonably distinct) parts. This first part comprises a multidisciplinary first look at the nature of diagrams, and the flavour of these initial chapters is at the same time philosophical, historical, computational, psychological and aesthetic. The chapters of Part 2 are primarily concerned with cognitive issues of diagrams, Part 3 with formal aspects of diagrams, and the concluding part reports a number of approaches to applying and designing actual diagrams, at all times with an eye to cognitive and/or computational insights.

The author of Chapter 1, Aaron Sloman, has a special place in the debate on diagrammatic reasoning within the artificial intelligence community. Sloman's papers from the 1970s on the nature of analogical representations triggered the whole discussion as to the nature of possible alternatives to classical symbolic reasoning by intelligent systems. In his chapter Sloman sets the scene by posing a visualisation problem faced by Mr Bean; that is, how he can remove his underpants without taking his trousers off. The choice of an appropriate abstraction of Mr Bean's body, underpants and trousers is used to illustrate how it is an effective abstraction that facilitates effective reasoning. Indeed, at a purely topological level the underpants are in the same configuration at the start and end of the removal operation (other than being in contact with his body), so there is no problem. But the nature of the containment of the body by the underpants is in fact a functional one (i.e. not topological). The problem is in fact one of reasoning about the metric configuration, and as Sloman shows, some metric abstractions are more efficient than others.

So in the "underpants"domain visualisation is critical to what is best described as the pragmatics of reasoning, that is, the choice of an effective abstraction. Sloman goes on to make a case for the role of visualisation in reasoning about less tangible, non-finite, domains for which he posits the use of parallel representations, both symbolic and visual, each of which is invoked opportunistically. Interestingly, this mirrors a recent leaning in cognitive science away from a belief in core representations. Sloman's discussion sets us up nicely for both the chapters and parts that remain, concluding with a plethora of questions as to the nature of internal representations and the requirements and mechnanisms of an artificial diagrammatic reasoning machinery.

In Chapter 2, Lindsay takes up one of Sloman's challenges and considers the interaction between pictorial and linguistic reasoning. His discussion concerns the problem of geometry theorem proving and discovery, and takes as a starting point the quasi-formal proofs that are typical of classroom textbooks. In such cases, the linguistic argument is usually a highly stylised (and shorter) version of the complete formal proof, and is liberally augmented by diagrams illustrating one instance of a general proposition. Lindsay's insights cut to the very core of arguments surrounding diagrams and formal reasoning, and serve as the starting point for observations as to how diagrams, with all their limitation of specificity (correspondence to single instances), can be used to support more general reasoning about geometry (both theorem verification and discovery).

Research in diagrammatic reasoning is undoubtedly still in its infancy, or one might argue, at a pre-science stage. The multiplicity of characterisations of diagrams and terms for describing the nature of diagrams stands in evidence for the lack of core statements to which researchers (even within a single discipline) subscribe. Seemingly, each new research group in diagrammatic reasoning devises its own descriptive framework for diagrams and diagram use, and in this light Blackwell and Engelhardt's chapter (Chapter 3) is a welcome meta-analysis of such frameworks.

Through a survey of over fifty previous characterisations the chapter reports a meta-taxonomy of diagram research spanning disciplines as diverse as computer science, linguistics, history and philosophy of science and painting. The resulting framework is surprisingly coherent, with three principal groups of dimensions ("signs", "meaning" and "context-related aspects"). In particular, their chronological organisation of diagram research and analysis of discipline-specific research concerns (which has given rise to the diversity of frameworks) will be a useful reference for researchers in the science of diagrams, new and old alike.

Blackwell and Engelhardt's meta-taxonomy sets the scene for the final two chapters in this part: those by Bailer-Jones and Richards. Both are views from researchers outside the mainstream of computer science and cognitive psychology (in this case history and philosophy of science, and graphic design respectively); and, as the meta-taxonomy correctly describes, their concerns and methodologies are characteristically distinct. Echoing Sloman's chapter, Bailer-Jones emphasises the important distinction between internal and external representations in considering the use of diagrams in scientific theorising, retaining the notion of "visualisation" exclusively for internal (or "in the head") representations, particularly where the domain is non-finite, and referring to the two-dimensional sketches that scientists spontaneously produce as "mental reifications" of their scientific thought. Although she eludes to the use of diagrams in scientific discovery, her principal concern, and the specific sketches used, are from her own experience of being taught astrophysics. In this respect, in Blackwell and Engelhardt's terminology, her account is at its core concerned with the "social" elements of sketching.

To conclude this part Richards' view of diagrams is from the altogether fresh perspective of the graphic artist. In striving to develop a terminology for describing diagrams that is meaningful to the information designer he presents a taxonomic model of the fundamental design variables for "diagramming". Interestingly, his central notion of "modes of interpretation", comprising the organisational, depictive and correspondence modes, echo many of the existing taxonomies that have arisen from within the cognitive and computer sciences. This leaves us with the rather optimistic hope that not too far ahead there lies some characterisation of diagrams, that both theoreticians and practitioners alike can agree on and use to foster further collaboration.

1. Diagrams in the Mind?

Aaron Sloman

Clearly we can solve problems by thinking about them. Sometimes we have the impression that in doing so we use words, at other times diagrams or images. Often we use both. What is going on when we use mental diagrams or images? This question is addressed in relation to the more general multi-pronged question: what are representations, what are they for, how many different types are they, in how many different ways can they be used, and what difference does it make whether they are in the mind or on paper? The question is related to deep problems about how vision and spatial manipulation work. It is suggested that we are far from understanding what is going on. In particular we need to explain how people understand spatial structure and motion, and how we can think about objects in terms of a basic topological structure with more or less additional metrical information. I shall try to explain why this is a problem with hidden depths, since our grasp of spatial structure is inherently a grasp of a complex range of possibilities and their implications. Two classes of examples discussed at length illustrate requirements for human visualisation capabilities. One is the problem of removing undergarments without removing outer garments. The other is thinking about infinite discrete mathematical structures, such as infinite ordinals. More questions are asked than answered.

1.1 We can Think with Diagrams

Consider the trick performed by Mr Bean (actually the actor Rowan Atkinson): removing his (stretchable) underpants without removing his trousers. Is that really possible? Think about it if you haven't previously done so.[1]

Is it possible to remove the underpants without removing the trousers, leaving the waistband of the trousers constantly around the person's waist, allowing only continuous changes of shape of the body and the underpants

[1] I have previously given audiences the task of finding out how many possible numbers of intersection (or tangent) points there can be between a triangle and a circle in the same plane. It is easier than Mr Bean's problem, but many people miss out some cases unless prompted.

and trousers, e.g. stretching, bending, twisting, but with no separation of anything into disconnected parts, no creation of new holes, etc.? Does it matter whether the waistband of the trousers is tight or not?

Many people can answer this question by thinking about it and visualising the processes required, even if they have not seen Rowan Atkinson's performance. A harder question is: in how many significantly different ways can the underpants be removed?

1.2 Some Comments on the Underpants Problem

It is easier to consider the underpants being distorted, ignoring who does it and how, than trying to work out all the contortions of posture Mr Bean would have to go through to produce the appropriate sequence of changes.[2] If we abstract away from the problem of how the wearer makes the transformations happen we can suppose Mr Bean remains rigid and still and someone else pulls and stretches his underpants, perhaps using long thin tongs where necessary. (Is it obvious that this change makes no difference to the main problem? Why?)

Even with this abstraction there are several different ways of thinking about the underpants problem. Some use only topological relationships preserved under all continuous transformations, including those which change size, shape and distances. Some also use metrical relationships involving shape and size. We can also use topological relationships with structural features of under-specified metrical relationships.

Thinking purely topologically is quite hard to do, since it involves finding the most general way to characterise the relationship between Mr Bean and his garments in the initial and final states. From that point of view the start and end states are equivalent and there is no problem for Mr Bean to solve. So it cannot be the right way to think about the problem of how to do it. Most people do not think like that. They conceptualise the problem in a largely qualitative but partly metrical fashion, including various ways the underpants might stretch and fold. We shall see that it is useful to combine different abstractions.

[2] The first draft of this paper located Mr Bean in a launderette. Toby Smith corrected me, pointing out that the shy Mr Bean was on the beach, and wished to remove his underpants then put on his swimming trunks, both without removing his trousers. On 29th July 1995 I posted Mr Bean's problem as a followup to a discussion of achievements of AI in several Internet news groups (comp.ai, comp.ai.philosophy, sci.logic, sci.cognitive) and received a number of interesting and entertaining comments. Chris Malcolm pointed out the similarity with the bra and sweater problem, i.e. removing a bra without removing the sweater worn above it. Readers are invited to reinvent the jokes that were then posted, about which problem was easier for whom under which conditions. In particular, someone pointed out the distinction between difficulty due to unfamiliarity *vs* difficulty due to being distracted.

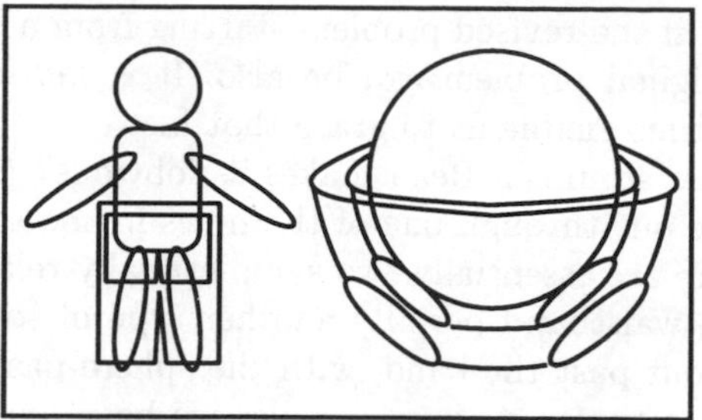

Figure 1.1. Mr Bean still wearing his trousers and underpants, before and after being continuously transformed into a sphere

1.2.1 How Many Distinct Solutions Are There?

Most people at first see only two symmetrically related solutions to the problem. One involves stretching the left side of the underpants down through the left trouser leg, over the foot and back up the left leg, leaving only the right leg through its hole. The underpants can then be slid down the right leg and out. A similar solution starts on the right side, with the underpants emerging through the left trouser leg.

If the waistband of the trousers is loose there are several more pairs of symmetrically related solutions, e.g. sliding one side of the underpants up over the head and down the other side and out through the leg, or sliding the central (leg-divider) part of the underpants down inside a leg then over the foot and up the same leg on the outside, then out past the waistband, over the head and down the other leg. It is easier to visualise than to describe! Another pair of solutions starts the same way, and ends with the underpants going off past the head. That is four pairs of solutions so far. But there is at least one still missing! (Or more, depending how solutions are counted.)

At first I saw only two solutions, and did not think of pulling the underpants over the head until a usenet poster mentioned the possibility. Then I looked for more solutions and noticed that the central part of the underpants could be moved first, leading to underpants around the waist. Eventually, after further abstraction, followed by some arithmetic, explained below, I found *nine* different solutions. Most people don't find them all.

1.3 A Spherical Bean

The solutions outlined above all used metrical notions including stretching and translation. We can de-emphasise metrical features (size, shape, distance, orientation, sizes of angles) and focus more on topology if we envisage the body shrinking to a sphere, or egg, as in Fig. 1.1, with the trousers and underpants following faithfully, so that each becomes a hemisphere with two holes, while their waistbands remain around the equator.

It is clear that if the revised problem starting from a spherical shape can be solved, the original problem can be also. It is *not* clear what kinds of cognitive mechanisms enable us to grasp that fact.

Considering the shrunken Bean makes it "obvious" (how?) that the underpants can slide out through one of the holes in the trousers. Since there are two holes there are essentially two symmetrically related solutions.

Loosening the waistband permits another type of solution in which the underpants slide out past the band, with the sphere passing through one of the leg holes. Since there are two leg holes we have another symmetrically related pair of solutions.

Another solution has the underpants sliding out past the waistband, without the sphere passing through the leg holes.

So with the trousers attached and impassable at the waist, there are two distinct solutions. Loosening the waistband enables several more distinct solutions. Have we found them all?

1.3.1 Holey Spheres

We can think of a two-holed hemisphere as a sphere with three holes! Then we can envisage underpants and trousers each as three-holed spherical sheets, concentric with each other and with the spherical Bean. The two sheets have their holes aligned, but we can ignore that.

What kind of cognitive process allows you to grasp the three-holed sphere view? I saw it like that only after I attended to the task of looking for more general characterisations of the problem and then saw that talking about the loose waistband was a distraction: it is just another hole in the trousers. Similarly there was all along just another hole in the underpants, at the waist.

What are the cognitive mechanisms that enable us to perform that sort of re-conceptualisation? How does the mechanism relate visual and non-visual information, e.g. about the nature of holes and waistbands? Why is the mechanism sometimes not invoked? What triggers its invocation?

There are two distinct but related re-conceptualisations. One involves noticing the similarity in structure and function between the big hole at the top and the two small holes at the bottom. Ignoring differences in size and location, they are similar in *function*: something inside (the underpants or trousers) can come out only by going through one of the three holes. Alternatively one can visualise a simple continuous deformation, i.e. stretching the garments up over the sphere, turning them into spheres with three similar holes.

That is we can discern the more abstract characterisation *either* by noting common aspects of the functional roles (causal powers) of the holes despite their difference in size (seeing affordances), *or* by visualising a deformation which makes them indistinguishable anyway (visualising structural changes and relationships). Different cognitive mechanisms and skills would be needed

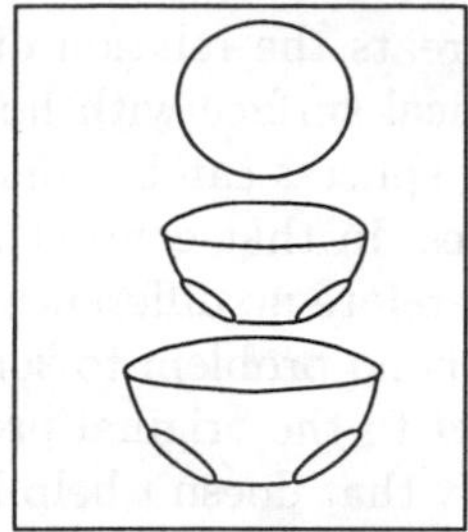

Figure 1.2. "Exploded" abstract representation of Mr Bean and his garments. The hemispheres can be continuously "flattened" into plates. Is it "obvious" that there is no topological difference between the original and final state?

for these two tasks. *How are these skills implemented? How do they develop? Which animals have them?* (Cf. [7])

Having noticed that Mr Bean with his lower garments is equivalent to a solid sphere surrounded by and concentric with two spherical rubber sheets each with three holes, we can also notice (*how?*) that removing the underpants involves two steps:

1. getting the sphere out of the underpants through one of the three holes in the inner sheet;
2. getting the underpants (the inner sheet) out of the trousers through one of the three holes in the outer sheet.

Suddenly it becomes clear that there are three ways of doing step (1) each consistent with three ways of doing step (2), so there must be $3 \times 3 = 9$ different solutions, covering all possible combinations at this level of abstraction, which ignores protrusions (e.g. legs) through holes. It is also possible to do step (2) before step (1), doubling the number of solutions!

It is worth noting that the type of abstraction identified here which enables us to reason about the combinations of steps does not require Mr Bean and the two garments to have any specific shape as long as the garments are approximately convex, or at least have a distinction between inside and outside and three communication ports between them. We can discuss the spheres and their changing relationships without assuming all the metrical properties of spheres, e.g. smoothness, constant curvature, fixed radius. This is what I meant by "structural features of under-specified metrical relationships".

1.3.2 Yet More Abstraction

Further abstractions are possible. The initial configuration is topologically equivalent (deformable by continuous changes) to one in which the three items are simply separated vertically, by moving the sphere up and the trousers down (as in Fig. 1.2). This treats the relation of being inside and the relation of being outside a spherical surface with holes as equivalent. Moreover The two stretchable enclosing spheres can be continuously deformed into two flat sheets each with two holes. In that context nothing is inside or outside anything else, and there is therefore no difference between the initial and the final state. Either way, there is no problem to solve!

Only mathematicians react to the original problem that way, concluding that it is trivial. Unfortunately that doesn't help Mr Bean get his underpants off. Even when a mathematically satisfying solution to a problem has been found at a high level of abstraction, there is still work to be done if detailed actions have to be specified.

When moving between different abstractions we need to know where to stop. For example, in analysing options for the removal process it is useful to go from the *fully metrical* initial specification, where the detailed shapes and sizes are relevant, to the *minimally metrical* nearly topological situation where only inside – outside relations are relevant (but still metrical because being "inside" an object with holes is a metrical property). Having enumerated possible strategies at the minimally metrical level (where each strategy involves use of one hole in the underpants and one in the trousers) we can then move to more detailed planning and evaluation in the fully metrical representation, where changes of shape and length are required, i.e. stretching of underpants over the head or down and under the foot. At that level there are far more options and the search space is much larger.

1.4 Coexisting Search Spaces

We found that there are nine different solutions when the problem is construed as involving three concentric spheres (or, to be more precise, three spheres totally ordered by an "encloses" relation). This discovery was not made by visualisation or simulation of the removal process, but by using the general information that for something to move from being inside a holed sphere to being outside it must go through one of the holes. (How does a child grasp *that* fact? Does a chimpanzee?)

Why was the full range of solutions not obvious with the original configuration? Not everyone spots the solution where both leg holes of the underpants are moved round to the top of Mr Bean's head, so that the underpants are upside down, and then pulled off upwards (i.e. Mr Bean exits the underpants through their waist hole while the underpants exit the trousers through the outer waist hole). There are different ways of doing this which are equivalent at a high level of abstraction, though they involve different contortions of Mr Bean and different locations where the underpants risk being torn.

At a fully metrical level the search space is far more complex: there are more detailed options, with more explosive combinatorics. At that level it is hard to see patterns among the routes, because the simpler structure got by grouping (almost) topologically equivalent options is not visible.

This is an illustration of the general fact that finding an abstract spatial representation and combining that with some abstract non-spatial (arithmetic or logical) reasoning can give a deeper insight into a problem than simply using very concrete spatial visualisation capabilities. Information about solutions at the abstract level can be transformed to lower-level solutions (e.g. with metrical information) by adding details, though generally there will not be a unique extension.

Having different views of a diagram or 3-D scenario involving different types of abstraction often helps in the process of solving a problem, e.g. planning a detailed sequence of actions. This is used by multi-level planners, which form meta-plans in one or more abstraction spaces (e.g. ABSTRIPS, NOAH) to control the search more effectively than a "flat" single-level planner can (e.g. STRIPS).

A related theme in the history of mathematics is the constant development of new forms of abstraction and techniques for relating and combining different abstractions. A similar theme can be found in work on child development, e.g. by Karmiloff-Smith.

In principle, given enough time to explore visually all the possible metrical transformations we could eventually discover instantiations of all nine possibilities described above, though we might not notice the partitioning into nine cases.

However, even given enough time, most people would not get around to considering all of the options because the more complex search space involves a more complex book-keeping task if the search is to be exhaustive. Human architectures do not cope well with deep stacks or long queues, though these are easy to implement on computers.

Our limitations may arise in part because different arrays of possibilities compete in parallel for attention. When considering any spatial structure there are indefinitely many changes of size, shape, orientation, colour, etc. that we can envisage if we think of them [18]. AI models of visual or spatial reasoning do not yet match this, though perhaps they will in the distant future.

Part of the price of such human flexibility is unmanageable combinatorics when searching for a sequence of changes to solve a problem. This can be alleviated by using more abstract patterns to control the search, though not everyone can do this equally well. Could this also explain the different achievements of the chimpanzees in Kohler's famous experiments?

Explaining how capabilities at different levels of abstraction are used and combined to control the search for a solution to a complex problem requires not only a specification for the representations and mechanisms used, but also the architecture which combines them and allows different processes to interact fruitfully [22].

1.4.1 What Makes Us Fail?

Why do people sometimes fail to visualise an action or change, or fail to draw an inference?

It may be due to (i) use of poor representations, (ii) use of inadequate mechanisms or algorithms for manipulating the representations, (iii) inadequate architecture for combining and integrating different sorts of representations and mechanisms (e.g. ability to construct only simple structures, limited possibilities for modifying structures, limited possibilities for analysing structures, limited short-term memory for storing sequences of modifications), (iv) wrong or incomplete stored information (e.g. about changes possible in a physical system, about consequences of changes), (v) inadequate mechanisms for monitoring effects of changes in order to infer consequences, (vi) lack of meta-level know-how and architectural support required for systematically exploring all the available information and all the available transformations, (vii) not using available know-how, e.g. because of an attention problem or a motivational problem or some kind of "fixation" on a different inadequate strategy.

The above points illustrate some of the requirements for a system able to explain or model human abilities. Some failures may involve transient dysfunctions, such as distracted attention, or forgetfulness. There may be others produced by brain damage, genetic brain malformations, drugs, chemical disorders, etc.[3] Some tasks may come too early for a developing architecture, in childhood.

1.5 External and Internal Diagrams

Our discussion shows that a diagram on paper is not necessarily a good model for what is grasped when someone visualises a spatial structure.

One person looking at the diagram may see only the more detailed, metrically specific configuration whereas another can see ("grasp"? "comprehend"?) in the same diagram a more abstract structure in which metrical relationships play a reduced role. The two views support different ways of seeing possible changes. So even if both perceivers had an internally inspectable 2-D diagram they might still view it quite differently. Simply having internal spatial structures cannot explain what it is to grasp or visualise a spatial aspect of a scene or problem. (Otherwise simply having a brain would suffice.)

Asking whether people can build internal diagrams is less important than asking how diagrams can be viewed, analysed, interpreted, and used, no matter whether they are internal or external. Introspective reports should be treated as highly ambiguous and incomplete descriptions, and certainly not as explanations.

[3] I conjectured in [14] that some autistics lack the perceptual ability to move up levels of abstraction in perception, also described in more recent papers [21, 22]

1.5.1 Representations and Transformations

All of the different ways of thinking about Mr Bean's problem require not only some way of representing the original configuration, but also a grasp of the possible *transformations* of that configuration, a capability discussed more fully in a discussion of "actual possibilities" in [18].

We have seen that different transformations are possible at different levels of abstraction. At one level there are many detailed changes of shape as Mr Bean pulls part of the underpants down his trouser leg, over the foot and then back up again. At the highest level of abstraction that is a non-operation: the sphere is still in the underpants, as if a protrusion from the sphere (the leg) has been squashed in, leaving the underpants free to rotate around the sphere.

So visual experiences of looking at the diagram at various levels of abstraction differ in (among other things) the *possibilities for change* that are seen. Mental visualisation without an external diagram must also involve assembling possibilities for change in thinking about a solution to the problem. Practice somehow develops fluency in doing this: *How?* I learnt a great deal by playing with Meccano sets, as a child. Different visualisation skills are developed by mathematical or other sorts of training. What changes during such learning?

Experienced software engineers gain facility in grasping very abstract configurations of data-structures along with procedures which transform them. Likewise, being a composer, painter, mechanical engineer, dressmaker, etc., involves acquiring specialised abilities to grasp structures along with classes of possible transformations of those structures and their consequences.

Different structures in the same general class can support very different numbers and types of transformations. A drawing with a few lines supports far fewer "immediately available" transformations than more complex line drawings with far more lines, junctions, regions etc. Thus as you visualise a structure changing, the requirements for grasping which further changes are possible may also be constantly changing. Often a change is made intentionally in order to allow new possibilities, e.g. visualising a mechanical link being shortened in order to allow it to rotate further before being stopped. *How do we grasp these second-order possibilities for change?*

1.6 Thinking with Qualia

All this is related to disputes about the nature of consciousness [2]. Are qualia simply unanalysable "givens" or are they best understood as crucial parts of the functioning of an information processing system (as I have argued in [20])? Our discussion shows that visual qualia (e.g. an experienced red patch) have rich "internal" differences depending on what sorts of possibilities for change the experiencer is capable of handling. Changes could include changes of shape, size, orientation, location, splitting into two or more patches, and many ways of acquiring new coloured sub-regions (e.g a blue patch in the middle or a green line traversing the red patch, and so on.)

Wittgenstein wrote: "The substratum of this experience is the mastery of a technique" [24](p208). A full account of visualisation (and thinking with diagrams or other spatial structures) would require us to analyse the huge variety of techniques implicit in even the simplest human experiences, thereby uncovering requirements for mechanisms able to support apparently simple qualia.

Other animals may have much simpler qualia, especially *precocial* species born or hatched with genetically formed visual mechanisms ready for use, e.g. chickens, deer, horses. *Altricial* species, e.g. birds of prey, hunting or tree-climbing animals and humans, start off more helpless and grow their brains while interacting with the environment. Perhaps this "bootstrapping" produces a much richer grasp of structure and motion than can easily be encoded in genes. (Contrast this with the popular opinion that humans are born so immature because their skulls would otherwise be too big to pass through a human pelvis. Elephants manage, so that can't be all there is to it.)

1.7 Visualising Infinite Structures

Some visualisation goes beyond what can be experienced in perception. How do we visualise infinite structures? The answer will depend on the type of infinite structure. When we visualise continuous objects or continuous changes this involves the possibility of "zooming in" to smaller and smaller portions of the object or motion, without limit. That is part of what is implied by being continuous. It also underlies some of Zeno's paradoxes.

Mr Bean's problem involves continuous change (stretching, bending, moving), but solving that problem does not deploy most of what we know about continuous motion. The difference between continuous change and a finite succession of discrete states would not make any difference to our previous discussion. In fact a useful way to tame a problem involving continuous change is to identify a small number of key states, and ignore intermediate states. That is how we found 9 or 18 distinct solutions.

We can also think about infinite discrete structures, like the set of integers or the set of proofs in some formalism. Clearly we cannot create something infinite inside our heads. So visualisation in this case (and probably in all the other cases too!) does not involve actual creation and inspection of the structure visualised. Something far more subtle happens: when you visualise a spatial structure or process there need not be any actual spatial structure or process that is inspected, nor anything isomorphic with the structure or process.

There might be only a *representation* of inspecting the structure or process. If done well, that could fool us into thinking we are doing something that we aren't. But being fooled doesn't matter as long as the process which produces the illusion is exactly what is needed to implement a powerful reasoner or problem solver: i.e. it is a good biological solution, like being fooled into thinking tables are smooth, solid, continuous and rigid, because they *look* and *feel* as if they are.

1.7.1 Infinite "Images" Involving Numbers

Let us consider some examples of infinite structures, such as the sequence **N** of natural numbers, 0, 1, 2, ... etc. This is easily visualised, going off into the distance away from us, or from left to right, for instance. **N** satisfies Peano's axioms for arithmetic. (i) There is an initial element. (ii) Every element has a unique successor. (iii) The initial element has no predecessor. (iv) Every non-initial element has a unique predecessor. (v) The axiom of induction: properties which are possessed by the initial element, and possessed by the successor of any possessor, are possessed by all the elements.

Any sequence satisfying those axioms, e.g. an infinite row of dots, or an infinite sequence of repeated actions, is a Peano structure. It is clear that there are many visualisable subsets of **N** which are Peano structures, e.g. the even numbers, 2, 4, 6, ..., or the numbers starting from 999 and continuing indefinitely: 999, 1000, 1001, ... It is also clear that Peano structures all have certain properties, some of which are easier to grasp than others.

Grasping the relationship between the axiomatic characterisation and the visualised structure is non-trivial. For hundreds (thousands?) of years before Peano came up with his axioms, people thought about and used numbers and were able to visualise the infinite sequence of numbers. Kant discussed some of the issues in 1781.

What cognitive mechanisms enabled Peano to find the axioms? Consider the different roles of the axioms in characterising the required set. Axioms (i) and (ii) guarantee that the set is not empty and that you can go on along the sequence forever, with no choice points (because of the word "unique"). Axiom (iii) prevents you going backwards beyond the initial element. Axiom (iv) implies that you can go back from any non-initial element, and again the word "unique" rules out choice points, thereby preventing the sequence doubling back and rejoining itself, as this one does: 0, 1, 2, 3, 4, 5, 6, 3, 4, 5, 6, 3, 4, 5, 6, ... That is, axiom (iv) prevents 3 having both 2 and 6 as predecessors. Axiom (v) is more subtle, and prevents sequences which go on forever, and then have more items beyond that, like **S1** defined below.

We can easily infer some properties of a visualised Peano structure. For example, given any two distinct elements in the structure, there must be a finite chain of successor elements starting with one of them and ending with the other. So the elements comprise a total ordering. Compare proving this from the axioms using logic. We can also see that every initial sequence of a

Peano structure is finite, and every alternate initial sequence can be arranged as a rectangular 2 by N block of items, where N is some number, and the intervening ones cannot.

1.7.2 More Complex Infinite Structures

We can also visualise structures violating Peano's axioms. For example, imagine the even and odd numbers separated out, into two sequences, 0, 2, 4, ... and 1, 3, 5, ... We can visualise these concatenated in a structure **S1** with all the even numbers going from left to right, followed by all the odd numbers going from left to right.

Then **S1** has a successor relation just as **N** did, but it is "obvious" that Peano's axioms are no longer satisfied in **S1**. First, not every non-initial number has a predecessor in the new configuration. (There is one exception.) Secondly the axiom of induction no longer holds: properties which are possessed by the initial number, and possessed by the successor of any possessor, are no longer possessed by all the integers in this new organisation. An example is *being even*.

We can visualise a different infinite series **S2** by reversing the odd numbers and adding them all *before* the even numbers. That produces a structure like the set of positive and negative integers which is infinite in both directions. There is no longer any item without a predecessor. **S2** has symmetry lacking in Peano structures.

Moreover, if we start from the fact that there are infinitely many prime numbers (which is provable algebraically, though not so easily proved visually), we can form infinitely many Peano structures and concatenate them. Starting from any prime number we can form a Peano structure consisting of all its powers, e.g. $2^1, 2^2, 2^3, ...3^1, 3^2, 3^3, ...5^1, 5^2, 5^3, ...$ It is then not hard to visualise *all* of these sequences concatenated to form **S3**, a totally ordered set of numbers, which has infinitely many elements violating axiom (iv) because they have no predecessor. This can either be proved formally from a logical specification of the construction of **S3**, or intuitively by visualising the process of construction and seeing that each time a new set of powers is added its first element has no predecessor.

1.7.3 Well-Ordered Structures

The original sequence **N** can be seen to be "well-ordered", i.e. every subset of **N** contains a "least" element, one which has no predecessor in the subset and which precedes all the others in the subset. This is connected with the fact that **N** is inherently asymmetric. It is built by starting with an initial element and going on indefinitely adding elements, one at a time, on one side only. Proving logically that every Peano structure is well-ordered is harder than *seeing* that it is.

Experienced mathematicians can also see that the structure **S3**, got by concatenating infinitely many Peano structures, is well-ordered.

This would not be true if we reversed some of the sub-sequences, e.g. if all the powers of 13 were included in reverse order. That would violate well-ordering since there would be a subset with no first element.

1.7.4 Justifying Peano's Axioms

Having noted that it is easy to visualise structures, like **S1**, **S2**, **S3**, which violate the axioms in different ways, we can see that one way to "justify" Peano's axioms is using them to rule out those structures. I have no idea if this is how Peano arrived at his axioms.

Whether those axioms suffice to determine uniquely the "intended" intuitive model is a controversial topic discussed more fully in my review [15] of Penrose.

A Peano structure whether specified axiomatically or visually is asymmetric. Moving along it in one direction always leads to the least element, whereas the other direction goes on forever, which we often represent by "..." Being "well-ordered" is another type of asymmetry: every subset has a first element, though not necessarily a last one.

1.7.5 How Do We Grasp an Infinite Ordered Sequence?

It may be that part of what makes the visualised infinite natural number sequence what it is rather than a non-Peano structure is an information-processing implementation of the asymmetry along with something closely related to the axiom of induction. I do not know how to make this precise.

Two aspects of such an implementation could be (1) a mechanism for expanding an incomplete sequence "on the right" as often as required, and (2) a reasoning mechanism that implicitly assumes that properties propagated to successors are propagated to *everything* further along. This sort of mechanism is not inherently connected with numbers.

Anyone who can visualise an infinite row of vertical dominoes going off to the right, and then visualise the wave of activation that occurs when the first domino falls over causing the second one to fall over, etc. and who finds it "obvious" that they will all (eventually) end up knocked over, is using the equivalent of the axiom of induction. *How is the ability to do this implemented in human brains?* It is probably part of a large suite of operations for manipulating finite and infinite discrete structures, which will be different in detail from those for continuous structures, but may have some overlap, e.g. the ability to concatenate structures, or to "move" something along a structure.

What makes something a visualisation of a Peano structure, rather than a different sort of structure such as **S1**, **S2**, or **S3**, depends on the applicability everywhere of this local property-transmitter. The infinite detail need never be constructed, as long as it is available when needed (as in lazily evaluated data-structures). This is partly analogous to whatever makes it possible indefinitely to zoom in to continuous structures. For Peano structures we use something like an ability indefinitely to "zoom to the right".

When and how do young children develop this ability? How did it evolve? Was it a side effect of other abilities?

1.7.6 Visualising Proofs and Refutations

It is easy to visualise counter-examples to the claim that all ordered structures are Peano structures, or that all ordered structures are all well-ordered. It is not so easy to use visualisation to prove generalisations, such as that *any* concatenation of a well-ordered set of well-ordered structures will also be well-ordered. For some people, and perhaps for all, that is much easier to prove by reasoning logically from definitions than to demonstrate by somehow visualising all possible concatenations of well-ordered sets. How would one do that?

In general it is easier to visualise a case that refutes a generalisation than to visualise all possible instances of a generalisation in a reliable way. Sometimes that can be done by visualising a sort of pattern or template which covers all the possibilities. Mateja Jamnik's work on verifying diagrammatic proofs, reported in this volume, includes the use of diagrams to reason over an infinite set of finite structures, e.g. in proving that for every N the sum of the first N odd numbers is N^2. This depends on a common pattern shared by all the structures, so that they can be visualised in a uniform way.

A much harder visualisation of an infinite structure (or process) is required to prove the Cantor-Bernstein theorem, which says that if there are two sets A and B each of which is in one-to-one correspondence with a subset of the other, then there is a one-to-one mapping between A and B. The proof involves constructing the new mapping from the two given ones, and it is helpful when thinking about this to visualise something like a pair of mirrors facing each other with rays bouncing back and forth indefinitely.

1.8 How Do We Do It?

What is going on when we visualise these infinite structures? We obviously don't construct infinite physical structures since our brains are finite. However, it may be accurate to say that infinite structures are constructed in some sort of virtual machine, like the familiar virtual machines that support sparse arrays or infinite lazily evaluated lists, constructable in some programming languages. It is not hard to create in a computer a sparse array with more locations than there are electrons in the universe, as long as we leave most locations containing the default value. Perhaps brains (or the virtual machines we call minds) use similar tricks for representing extremely large, or even infinite, structures.

It might be tempting to think that what we do when we visualise an infinite structure is construct a very large set and use that as an approximation to the infinite set, since after all a very very large visualised collection of dots, like a starry sky, might as well be infinite if we cannot take in the whole lot and see how many there.

But that won't do. If you visualise the structure **S1**, with *all* the even numbers followed by *all* the odd numbers, then no very large finite subset of the even numbers will do as an approximation to *all* of them. For example, the structure **S1** violates Peano's axioms, as explained above, whereas if there are only finitely many even numbers preceding the odd numbers then the axiom that every number has a unique predecessor will no longer be violated, for the first odd number will now have a predecessor, the last even number. Moreover the axiom of induction will again hold. That is, if we replace the infinite sequence of even numbers with a finite subset this will transform **S1** into a Peano structure. So a large finite row of even numbers cannot model the required infinite row in this context.

Something deep goes on when we visualise the two infinite sets as being concatenated. Perhaps the important point is that what we experience as pure visualisation is actually a combination of visualisation and unconscious but explicit specification of rules for indefinite expansion and rules for inference? (I think that sort of idea goes back to Immanuel Kant [5].) For example, we may have something like the previously mentioned mechanism for "continuing to the right" waiting in the wings to prevent any interpretation of the set of evens as a finite set, however large. This is like the "lazy evaluation" of an infinite list structure in a computer: the list has a "generator" procedure and looking beyond the already expanded portion of the list causes the generator procedure to be run, to produce previously unavailable list elements.

Using lazy evaluation is a fairly abstract and sophisticated kind of visualisation, on a par with the domino/induction mechanism that was previously waiting in the wings to propagate properties along all the natural number sequence.

How many other sorts of visualisations involve such a mixture of implicit rules or axioms or mechanisms along with something like a spatial structure? One of the requirements for a mechanism of the sort discussed here is that whether the visualised spatial structure is finite or infinite, discrete or continuous, the visualisation is possible only insofar as it implicitly involves the availability of a large number of *possible* changes in the structure, as previously discussed. What exactly is visualised depends on exactly which transformations are available.

1.9 Visualising is not Like Seeing

From the discussion so far, it is clear that whatever visualisation of a structure is, it *cannot* be something very similar to seeing even if it *feels* similar. That is because the kind of grasping of a spatial structure involved in visualising is *part* of what happens in seeing the structure. Hence if visualising involved seeing then visualisation would be part of visualising and we'd have an infinite regress.

Also we cannot *see* an infinite (discrete) structure but we can *visualise* one. And it is arguable that when we visualise the kind of abstract topological structure that we previously discussed, that cannot be like seeing because seeing always involves *specific* metrical or topological structures and relationships which are missing in the *abstract* visualisations.

We need a new way of thinking about the problem, other than proposing that the brain creates 2-D or 3-D arrays and then "looks at" or "inspects" them, for if the looking at or inspection involves understanding the spatial structure we are going round in circles chasing a non-existent homunculus. There must be a way of understanding spatial structure (or more generally) a way of understanding, which is not to be explained in terms of understanding another structure!

It must, however, be something like a type of information-rich control state, i.e. a state which affects what the system can or will do next. Elsewhere I have argued that we need to view minds as control systems and representations as control substates with syntax, pragmatics and in some cases semantics, e.g. [16,17,19].

What sort of control state? How does grasping some structure affect what you can do? Note that "what you can do" does not refer only to external behaviour. It includes the sorts of *internal* processing which become available when we grasp some structure. We need a theory of an architecture that can accommodate all these processes.

1.10 Other Problems Involving Visualisation

Mr Bean's task is just one of many problems which people seem to be able to solve by *visualising* transformations of a structure.

Some are much easier: e.g. if a penny with the "head" on top is turned over three times will the head or the tail be on top? That one is easy to do *either* by visualising the process (simulating it mentally) *or* by reasoning about it. If we modify the problem to one in which the penny is turned over three thousand and five times, it is much easier (and far more reliable) to reason about than to visualise [10].

Here the more sophisticated process, using meta-level knowledge about the nature of the less sophisticated process, is easier and faster to do than the less sophisticated process which blindly goes through the steps to get from the start state to the end state. Being able to discover new ways of solving old problems and being able to select between alternative approaches requires "meta-level" knowledge, i.e. the ability to reflect on and reason about knowledge and problem solving. One of the earliest interesting examples of this was Sussman's Hacker [23], which debugged itself by watching itself at work, though it dealt only with a tiny fragment of the problem, like most AI models so far.

Being able to understand the possibility of looking for and using "easy" short cuts requires a more sophisticated processing architecture than a typical problem solver or planner. It requires an architecture which supports mechanisms for observing, analysing, evaluating, and noticing patterns in internal processes [20].

However, having an architecture supporting such meta-level abilities does not guarantee general meta-level competence. It seems that humans have to learn to be reflective in different domains. For example, someone who is good at noticing opportunities for improving his software designs may fail to notice opportunities for improving communication and relationships with other people.

Much mathematical ability seems to depend on grasping patterns and structures in one's own thinking and reasoning processes, like noticing that the outcome of a counting process does not depend on the order in which items are counted, or noticing that a repetitive process can continue indefinitely. I suspect that our ability to visualise infinite structures is related to the ability to grasp and reflect on properties of repetitive processes, and our ability to manipulate them by performing operations like concatenation or reasoning about subsets depends on noticing analogies between infinite structures and finite structures.

Children don't seem to start off with these abilities, but, unless damaged by teachers (or parents?), they somehow manage to bootstrap the more sophisticated architecture and to apply it in different domains. (For some speculations about this in connection with learning about numbers, see [12] (Ch 8) and compare with [6].)

1.11 Some Questions

The examples discussed above raise a host of interesting questions, relevant both to understanding how human minds work and how to give intelligent machines the ability to reason spatially.

1. What sort of knowledge enables people to work out the answer? (This subsumes the deep question: what sort of knowledge enables them to understand the problem?)
2. How is that knowledge represented in their brains – both physically in chemical and neural structures and within the information-processing

virtual machines implemented in brains? How many different forms of representation do we have available for such knowledge? [3, 9, 13, 19]
3. Can the information used be expressed in predicate calculus? In first-order predicate calculus? In some other mathematical or logical notation?
4. What would the knowledge actually look like if expressed in some form of predicate calculus, or other logical system? (That is, which predicates, functions, etc. would be used? Which axioms? How would the initial state and desired end state be described? Would modal operators be needed, e.g. to express which transformations are *possible*? Would temporal operators be needed to express the notion of a *process* and the constraints on the process? How would the requirement that the waistband not be moved be expressed?)
5. What sorts of logic engines would be able to find the solution? What sort of search space is involved? How can such a search be controlled?
6. What alternatives are there to logical representations and manipulations? What are their advantages and disadvantages?
7. What sorts of reasoning mechanisms do people actually use for this sort of problem? Can they use logic? Do they ever use logic? What alternatives are available, for humans or intelligent machines?
8. Can some or all of the human competence be replicated on computer-based machines using a very different physical implementation?
9. Which of these abilities are shared by which other animals, e.g. a magpie building a nest in a treetop out of twigs of many shapes and sizes, a squirrel working out a route to the bag of nuts hung up for birds, a female orang-utang in a tree clutching her infant with one hand and using the other to weave a nest for the night, out of branches and leaves?

1.11.1 Has AI Made Much Progress on These Questions?

Like many others, I have been thinking (and writing) about such questions, and about how human and animal vision works, for many years (see the References) and have seen various ideas about this re-invented many times. But I remain deeply puzzled since nothing I have come across in AI, or in psychology or brain science, seems to come close to explaining human (and animal) visual and spatial reasoning abilities.

Often an implementation appears to be doing something like human visualisation, but on closer examination lacks the generality and power: give it a slightly different problem and it cannot cope. There are now many wonderful systems for generating stunningly realistic static or moving images on computer displays, yet such programs cannot perceive and understand such images. Programs which can reason by manipulating diagrams containing a few discrete structures cannot cope with continuous structures or continuous change. In general, programs which reason about images using 2-D arrays or networks do not have a grasp of space or time as continuous. Work by Hayes [4] and others related to the idea of "naive physics" helps to define some aspects of the problem of characterising our grasp of spatial structure, but does not as far as I know specify mechanisms that can solve the problem.

Psychological and neural theories do not answer the questions either. Neural theories tend to identify locations where low-level visual processes occur, but say little or nothing about higher-level capabilities or how visualisation mechanisms are used in problem solving. When attempts are made to formulate theories about how brains do visual reasoning I usually find that they do not describe anything that I can interpret as a workable design with explanatory power. For example, talking about mechanisms which "manipulate images" by rotating, or stretching or translating them explains nothing. It merely re-formulates what needs to be explained.

In order to add more detail to the specification of what needs to be explained, I have tried to show that visual reasoning covers a variety of different things, using two examples of what we can visualise: one a finite but deformable structure and one a discrete but infinite type of structure.

1.12 Spatial vs Logical: What's the Difference?

Introspectively, many people are convinced that there is a deep difference between solving problems by reasoning logically (or verbally) and solving them by visualising and transforming spatial structures. Whether such introspections are reliable is a matter of dispute.[4] However, it is not so commonly noticed that both sorts have much in common, and what they have in common is probably more important and harder to account for than the differences.[5]

Whenever we reason, whether with pictures, words, imagined movements, or anything else, processes occur in which structures are created and manipulated, usually in virtual machines. If you reason logically or algebraically using pencil and paper, you'll normally create a *sequence* of spatial structures, where the transition from one element of the sequence to the next corresponds to a step in the reasoning. (This is why visualisation of sequences plays such an important role in a lot of meta-mathematical reasoning.)

Problems in Euclidean geometry can often be solved without a spatial sequence: instead we modify a diagram *in situ* (see [8]). Modern interactive graphics technology supports this and also allows direct transformation of a single logical or algebraic structure presented on the screen without having to produce a sequence of spatially separate structures, as happens when we

[4] Some of the differences between "Fregean" (applicative) and "analogical" representations were analysed in [10]. The differences are often misdescribed.

[5] I have previously argued that there are not only two categories, but a wide range of significantly different types of representation, e.g. in [10, 11, 19]. Similar strictures apply to other alleged dichotomies, e.g. between implicit and explicit, computational and non-computational mechanisms, or procedural and declarative representations.

reason with sentences, equations, logical formulae. Perhaps brains got there first?

The collection of structure-manipulations possible in a class of structures defines a generalised notion of "syntax" for such structures. The kinds of parts that can be replaced and the kinds of features and relations that can be changed define the structural properties of the information medium, its syntax. We can also generalise a notion of "pragmatics" from linguistics, to refer to the functional roles of information structures in larger systems. In some cases there will also be "semantics" insofar as the structures are used to describe, summarise or plan other internal or external structures, actions or goals.

We need a better grasp of the types of structure manipulation mechanisms there are and the many ways in which different possibilities for further manipulation are actively made available by the current contents of a particular structure. This may enable us to come up with better theories of how brains or minds do all this. That would require, yet again, re-inventing ideas discovered long ago by evolution, and in the course of doing so we'll probably have to discard many of our cherished distinctions.

1.13 Conclusion

This paper draws attention to a collection of unexplained features of our frequently noted ability to think and to visualise. All such cases (whether diagrammatic or not) seem to involve the ability to create structures – not necessarily the structures we think we are visualising, and not necessarily physical structures, since they can be structures in virtual machines (the "physical symbol system hypothesis" taken literally is a huge red herring). They also involve the ability to have readily available a collection of mechanisms for manipulating those structures which somehow implement our grasp of the possibilities for change inherent in a structure. The possibilities for change determine how the structure is grasped or understood, and provide the basis for its pragmatic and semantic functions.

What constitutes a grasp of something spatial as opposed to algebraic, or continuous as opposed to discrete, or finite as opposed to infinite, or linear as opposed to tree structured, or planar as opposed to three-dimensional, etc. will depend in part on the collection of types of transformations and inferences available and ready to be applied to the structure.

In some cases the same structure may be viewed or understood in different ways by making different classes of transformations or inferences available, as in the difference between a metrical and a topological understanding of a spatial configuration.

Using such a grasp in solving a problem or making a plan involves somehow being able to orchestrate the collection of possible changes in such a way as to find collections of changes which satisfy some condition. When the situation represented is continuous, continuous changes can be visualised. Whether we can actually produce such changes or only convincing representations of them is not clear.

Being intelligent often involves simultaneously viewing something in two or more ways and relating the sets of possible changes in the different views. What does and does not work has to be learnt separately in the context of different classes of structures, different classes of manipulations and different classes of problems, which is why there is no such thing as totally general intelligence.

How all this can be implemented in brains or computers remains an open problem. If we study lots more special cases we may eventually understand what sorts of structures and mechanisms can implement such capabilities, and what sorts of general architecture can accommodate them all, along with closely related capabilities such as vision and motor control. I don't think this will be easy to do, not least because we still don't understand what the problem is.

Acknowledgements and Apologies

I have learnt much from reading papers by others who have written on these topics and from conversations I have since forgotten. My perspective was strongly influenced by reading Kant's views on the nature of mathematical knowledge [5], with which most philosophers and logicians disagree, wrongly in my view. I apologise for not providing a literature review. Useful sources can be found in [1, 3, 9]. For inspiration see the examples in [8]. Further information can be found at: http://www.cs.bham.ac.uk/~axs/ This research is partly funded by the Leverhulme Trust Grant Ref F/94/BW.

References

1. Brachman, R. and Levesque, H. (Eds) (1985). Readings in knowledge representation. Los Altos, CA: Morgan Kaufmann.
2. Chalmers, D.J. (1996). The conscious mind: In search of a fundamental theory. Oxford: Oxford University Press.
3. Glasgow, J., Narayanan, H. and Chandrasekaran (Eds) (1995). Diagrammatic reasoning: Computational and cognitive perspectives. Cambridge, MA: MIT Press.
4. Hayes, P. (1985). The second naive physics manifesto. In J.R. Hobbs and R.C. Moore (Eds), Formal theories of the commonsense world. Norwood, NJ: Ablex, pp. 1–36. Also in [1], pp. 468–485.
5. Kant, I. (1781). Critique of pure reason. London: Macmillan. Translated (1929) by N. Kemp Smith.

6. Karmiloff-Smith, A. (1996). Internal representations and external notations: A developmental perspective. In D.M. Peterson (Ed.), Forms of representation: An interdisciplinary theme for cognitive science. Exeter: Intellect Books, pp. 141–151.
7. Kohler, W. (1927). The mentality of apes (2nd edn). London: Routledge & Kegan Paul.
8. Nelsen, R. (1993). Proofs without words: Exercises in visual thinking. Washingon, DC: Mathematical Association of America.
9. Peterson, D.M. (Ed.) (1996). Forms of representation: An interdisciplinary theme for cognitive science. Exeter: Intellect Books.
10. Sloman, A. (1971). Interactions between philosophy and AI: The role of intuition and non-logical reasoning in intelligence. In Proceedings of the 2nd IJCAI, London. Reprinted in Artificial Intelligence 1971, 209–225, and in J.M. Nicholas (Ed.) Images, perception, and knowledge. Dordrecht: Reidel, 1977.
11. Sloman, A. (1975). Afterthoughts on analogical representation. In R. Schank and B. Nash-Webber (Eds), Theoretical issues in natural language processing (TINLAP). Cambridge, MA: MIT Press, pp. 431–439. Reprinted in [1].
12. Sloman, A. (1978). The computer revolution in philosophy: Philosophy, science and models of mind. Hassocks, UK: Harvester Press.
13. Sloman, A. (1985). Why we need many knowledge representation formalisms. In M. Bramer (Ed.), Research and development in expert systems. Cambridge, UK: Cambridge University Press, pp. 163–183.
14. Sloman, A. (1989). On designing a visual system (towards a Gibsonian computational model of vision). Journal of Experimental and Theoretical AI 1(4):289–337.
15. Sloman, A. (1992). The emperor's real mind. Artificial Intelligence 56:355–396. Review of Roger Penrose's The emperor's new mind: Concerning computers minds and the laws of physics.
16. Sloman, A. (1993a). The mind as a control system. In C. Hookway and D.M. Peterson (Eds), Philosophy and the cognitive sciences. Cambridge, UK: Cambridge University Press, pp. 69–110.
17. Sloman, A. (1993b). Varieties of formalisms for knowledge representation. Computational Intelligence 9(4):413–423 (special issue on computational imagery).
18. Sloman, A. (1996a). Actual possibilities. In L.C. Aiello and S.C. Shapiro (Eds), Principles of knowledge representation and reasoning: Proceedings of the fifth international conference (KR '96). Los Altos, CA: Morgan Kaufmann, pp. 627–638.
19. Sloman, A. (1996b). Towards a general theory of representations. In D.M. Peterson (Ed.), Forms of representation: An interdisciplinary theme for cognitive science. Exeter: Intellect Books, pp. 118–140.
20. Sloman, A. (2000a). Architectural requirements for human-like agents both natural and artificial. (What sorts of machines can love?). In K. Dautenhahn (Ed.), Human cognition and social agent technology. Advances in consciousness research. Amsterdam: John Benjamins, pp. 163–195.
21. Sloman, A. (2000b). Models of models of mind. In Proceedings of the symposium on how to design a functioning mind, AISB'00, Birmingham, UK.
22. Sloman, A. and Logan, B. (2000). Evolvable architectures for human-like minds. In G. Hatano, N. Okada and H. Tanabe (Eds), Affective minds. Amsterdam: Elsevier, pp. 169–181.
23. Sussman, G. (1975). A computational model of skill acquisition. New York: American Elsevier.
24. Wittgenstein, L. (1953). Philosophical investigations (2nd edn 1958). Oxford: Blackwell.

2. Knowing About Diagrams

Robert K. Lindsay

Understanding diagrams and using them in problem solving requires extensive knowledge about the properties of diagrams, what diagram elements denote, how their parts are distinguished and referenced, how they relate to linguistic statements, and so forth. This knowledge is most naturally represented linguistically. Nonetheless, diagrams or imaginal representations of them are used in substantive non-linguistic ways as part of the problem-solving process. The interaction of linguistic and diagrammatic representations must be understood in order to construct a theory of diagrammatic reasoning. In this chapter, an example is examined to illustrate some of the ways in which diagram manipulation may be used in geometric reasoning and to identify informally some of the knowledge necessary for such reasoning.

2.1 Introduction

Humans and other animals continually engage in interactions that require perceptual processes that enable complex cognitive and motor behaviour. For example, an experienced person can catch a thrown ball, cross a busy street, sight read a musical score, judge shape by touch, recognise a friend's face in a crowd, and so forth. Each of these activities requires complex and rapid computations, usually not accessible to conscious description. The neural mechanisms and processes that perform these tasks are general in the sense that they apply to an unbounded variety of similar situations. However, each application is specific to one situation, and explicit generalisation is not needed and usually not available to the person: a ballplayer cannot and need not describe precisely how he fields fly balls. Thus, for many perceptual reasoning abilities language is not required.

Other forms of human reasoning do involve explicit generalisations, often stated in a natural language or, on occasion, in a formal artificial language such as a predicate calculus. Reasoning with language about both special cases and generalisations is frequently, at least in part, done consciously.

Formal mathematical reasoning is usually seen as the quintessential form of linguistic reasoning. The in-principle mechanisability of proof verification and production by linguistic methods (viz., the manipulation of character strings), though qualified by certain undecidability issues, gives linguistic reasoning an aura of correctness, both as a model of correct reasoning and as a model of human cognitive processes as applied to mathematics. However, the lack of obvious procedures for the discovery of important new ideas or for inventing new mathematical tools, methods, and subfields of utility belies the completeness of this model as a description of how mathematicians understand and create mathematics. Further, the fact that mathematicians use both natural language explanations and non-linguistic notational devices in performing their work and communicating it to others suggests that a purely linguistic model of mathematical reasoning is at best incomplete.

Here I am focusing on a small piece of the extra-linguistic machinery of mathematical reasoning, namely the use of diagrams in understanding plane geometry.

One way to use diagrams for reasoning is to translate them into a formal predicate calculus and use deductive methods to make inferences. Recent work by several researchers has addressed the problem of translating simple diagrams to statements, but that research has not reduced the translation to a mechanical process for general cases. Note specifically work by Barker-Plummer and Bailin [3], Chou [4], Shin [13], and Wang [14].

Another approach is to use a non-linguistic representation of diagrams in conjunction with methods of inference that manipulate those representations directly. Diagrammatic reasoning in this sense employs perceptual reasoning in the service of establishing general statements by applying perception-like processing to diagrams (artifacts) or mental representations of them (mental images). For example, inference processes can be constructed from procedures that manipulate a diagram by adding new elements to it and moving other elements about, and then reading off the ensuing changes that are then interpreted as inferences. Viewed as a psychological model, the mind is assumed to run experiments, using the mind's eye and mind's hands to perform them. While this explanation appeals to many as more psychologically plausible than the manipulation of propositions, before the question of psychological validity can be addressed the perceptual models must be more precisely specified. The first step is to show they are plausible "first-order" theories, that is, that they are capable of reasoning at all. One way to formalise diagrammatic perceptual reasoning of the above sort is to devise an inventory of diagram manipulation processes and demonstrate that these can be used as a programming language for manipulating diagrams in productive ways. For example, Furnas [6] has devised the "bitpict" system, which allows the specification of processes that transform small pixel arrays into other pixel arrays, and has demonstrated how some "programs" composed of such operations can be written to solve problems. Anderson and McCartney [1, 2] defined a system of picture element manipulation processes that can be combined according to explicit rules into programs that solve problems strictly by manipulating pixel arrays. My work is similar to those approaches in that it also performs computations on pixel arrays. However, the picture manipulation processes are described at a higher level of aggregation, namely at the level of the manipulations of geometric objects, such as "construct a line between two points" and "determine if two lines are parallel". These processes are then available to write programs that create and manipulate diagrams. One class of such programs can follow diagram manipulation steps that demonstrate certain geometric propositions and verify that the results support the conclusion [10]. In the present paper I am concerned with how such processes might be employed strategically to devise programs appropriate to a given reasoning task.

I have emphasised elsewhere [9] that an account of conscious geometric reasoning with diagrams must employ propositional as well as pictorial representations and manipulations. This would be true if only because the mathematical statements proved, demonstrated, or understood are propositional in form, usually universally quantified assertions (e.g., "The sum of the interior angles of any plane triangle is equal in measure to a straight angle"), and some connection must be established between the perceptual manipulation and the conclusion. More fundamentally, the very strategy of using diagrams in a substantive way requires the constant interaction of diagram elements and propositions, notably in the form of constraints imposed on the diagram ("This angle is – always – to remain a right angle no matter what other changes are made"). Indeed it is undeniable that mental experiments can be run according to a variety of assumptions, for example that the objects are rigid, or that they are plastic. That is, diagrammatic reasoning is "cognitively penetrable" [11], where what is "cognitive" is described propositionally.

Propositional knowledge is used in many essential ways other than to state a conclusion and the premises underlying it. This essential knowledge includes generalisations about invariants, inventories of special cases, definitions of symmetry, knowledge of algebraic relations, generalisations of prior conclusions, and knowledge of problem-solving strategies and heuristics. Purely diagrammatic mathematical reasoning is an oxymoron, since mathematical reasoning always entails drawing explicit conclusions.

It is desirable to establish an inventory of the knowledge that, in addition to the knowledge of space built into the programming language of my system, is needed to devise methods that can achieve human-like reasoning about geometry. Once this is done, that knowledge must be represented in ways that can interact with diagrammatic representations and be used to verify demonstrations, create demonstrations, and ultimately perhaps to discover geometric propositions that were not explicitly conveyed to the program.

I will illustrate by example how a "programming language" must be augmented to support the invention of demonstrations and the discovery of geometric relations. I will use one example to illustrate a range of diagrammatic reasoning strategies and the knowledge underlying them.

2.2 The Example

The example concerns the Quadrilateral Theorem (QT), informally: "The figure formed by connecting the midpoints of the sides of any quadrilateral is a parallelogram." This theorem, perhaps initially surprising, is true for convex as well as concave quadrilaterals.

2.2.1 Understanding by Proof

One way to understand this theorem is to examine its fully formal proof, following the proof and checking its steps for validity one by one. Diagrams are not part of a fully formal proof. However, fully formal proofs are almost never actually constructed because they are extremely long, detailed, and opaque. Proofs that suffice for publication in mathematical journals and texts rely heavily on natural language descriptions. They must only be sufficiently detailed to convince professional mathematicians that the complete proof could be supplied. Only in the case of machine-generated proofs are the details actually fully produced.

For most of us, a better way to understand the QT is a "textbook proof". By textbook proof I mean the sort of quasi-formal argument favoured in textbooks and classroom teaching. They are almost invariably accompanied by a diagram whose connection to the propositional argument is not formalised. Typically the diagram accompanying a textbook proof illustrates only one specific instance of the general proposition. Textbook proofs usually omit proving many properties that are true of the diagram; the bulk of a fully formal proof typically addresses these omissions from the textbook proof. Thus textbook proofs are not rigorous, nor even semi-rigorous like a proof in a mathematical publication.

Understanding a textbook proof requires the ability to understand both natural language and diagrams. It also requires seeing how the conclusion of the theorem is based upon other theorems and lemmas that are previously understood, and so on back to the axioms of geometry that presumably are accepted as obvious, whatever might cause that. Devising a proof is a more difficult task than understanding one, and reflects an understanding of the theorem in an even deeper sense.

2.3 Reasoning by Diagram Manipulation

The following discussion illustrates a variety of ways that diagram observation and manipulation can be used to support reasoning. It will be understandable because the reader understands a number of basic assumptions that need to be captured if a program is to be able to follow or generate such arguments.

The QT applies to "any" quadrilateral; it is a universal statement. The starting point for using a diagram typically is to choose one particular illustration of the concept or proposition in question. For the QT this means selecting four points on a sheet of paper and connecting them to form a quadrilateral, or doing something of that sort mentally. However, a single diagram is notoriously deficient in its ability to represent universal propositions. One common defence against this deficiency is to choose a particular case that has no special properties that may be responsible for the generalisation. That is, one wants a representative instance. Care is taken so that the figure so selected is not one of a number of "special" quadrilaterals, those that have other properties, such as rhombuses or trapezoids. However, this presupposes knowledge about what properties are special in this sense. For the QT, each side and each angle of the chosen quadrilateral should be of different measure, and special angle measures, such as 0, 45, and 90 degrees, should also be avoided. No pairs of sides should be equal, parallel, or perpendicular. Draw or imagine a representative quadrilateral, find the midpoints of its sides and connect them pair-wise. Finally, examine the inscribed figure and note that it is a parallelogram. Since the original quadrilateral was chosen "arbitrarily" and had no properties known by previous experience to yield additional special features, one is encouraged to accept the generality of the proposition. As a further check one could try other quadrilaterals as starting points to further reduce the probability of having hit upon a special case. One could also attempt to construct a counterexample, since a single counterexample suffices to disprove a conjecture (another important item of knowledge).

2.4 Deeper Understanding

Although the observation of several appropriate special cases is compelling, there is a deeper and more important sense of understanding that is aided by diagrams. One would like to know "why" the theorem is true.

One could modify the original quadrilateral by dragging one vertex and noting that the constructed figure remains a parallelogram. This not only provides a sequence of several examples, but shows how altering one property alters others in an exactly compensatory way. Given that one has a specific instance of the appropriate diagram, say Fig. 2.1, one could ask if an arbitrary change will preserve the parallelogram property. A given quadrilateral can be transformed into any other quadrilateral by moving its vertices to the location of the new quadrilateral vertices. The order of making these alterations is immaterial; only the end result matters.

Furthermore, there is a "syntactic" symmetry to the situation [7,9] in the sense that, since the figure is arbitrary to begin with, if one could show that the movement of any one vertex preserved the property, one could show it true for the movement of any other vertex by a similar procedure. So we are

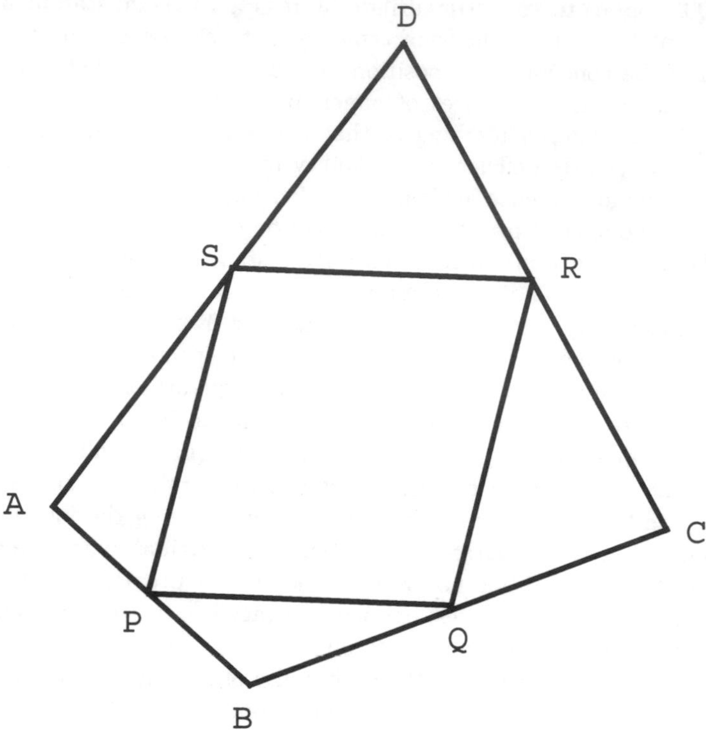

Figure 2.1.

encouraged to believe that no counter-example exists if we have one valid instance and can show that moving a single vertex preserves the midpoint-parallelogram property.

Note further that an arbitrary target movement of a vertex can be broken into any number of steps whose vector sum is the target movement. In particular, the target movement can be broken into just two steps in distinct directions that need not be orthogonal. If each of these sub-movements is property preserving, the total movement must be. This presumes path independence, that the relevant properties of a geometric figure are determined by its static configuration, independently of how it was constructed. This is a property of geometric figures, but not in general of physical situations involving energy transformations.

The next step might be to select two distinct directions such that movements in those directions can be easily seen to preserve the parallelogram property. The vertical and horizontal directions of the paper are not a priori interesting choices since geometry is orientation independent; the only relevant directions are those relative to objects in the figure. These include the directions of the quadrilateral sides and the sides of the parallelogram. Closer observation shows that the parallelogram sides are also the directions of the (unconstructed) diagonals of the quadrilateral. That observation, triggered by the search for significant directions, may be valuable (in particular, it helps to understand the textbook proof given later).

What happens if vertex C of Fig. 2.1 is moved in the direction of the orientation of segment PQ (from P toward Q, roughly "east"), directly away from its opposite vertex A? What happens if vertex C of Fig. 2.1 is moved in the direction of segment QR? We could actually make these movements and observe that the parallelogram property is preserved, but can we show in a deeper sense why it must be preserved?

Let's examine more closely the relation between quadrilaterals and inscribed parallelograms. Clearly it is possible to inscribe non-parallelograms by choosing appropriate points on the quadrilateral's sides. Is it possible to inscribe a parallelogram within a quadrilateral without connecting midpoints? If we connect two non-midpoints P and Q with a line (See Fig. 2.2) and attempt to construct a parallelogram with that as one side, the length and orientation of the opposite side of the inscribed parallelogram are determined. We can make a copy, P′Q′, of the first line PQ and move it parallel to itself until it contacts one of the other two quadrilateral sides (here P′ contacts AD), hold that end on the contacted side, maintaining orientation and length, and slide P′Q′ along the contacted side until it contacts the fourth side with P′ at S and Q′ at R. Completing the figure yields a parallelogram PQRS that does not connect the quadrilateral's midpoints. Thus the midpoints do not define the only inscribed parallelogram. R and S are uniquely determined once P and Q are chosen because there is at most one "slice" of the other half of the quadrilateral of that length and orientation because the sides converge to D. This is observed by diagram manipulation combined with an analysis of the available degrees of freedom.

If we had picked P and Q as midpoints, would this procedure force R and S to be midpoints? Experiment seems to confirm this but does not establish it in general. Here is another approach. Starting with an arbitrary parallelogram, let us circumscribe a quadrilateral whose sides are tangent to the parallelogram's vertices. Do the vertices of the parallelogram bisect the quadrilateral sides? This proves not to be true; an indefinite number of quadrilaterals can circumscribe a given parallelogram.

Next consider Fig. 2.3. Starting with a parallelogram PQRS and an arbitrary point A outside it, the directions of two circumscribing lines (1 and 2) are uniquely determined. This is because two points uniquely determine a line. However, one quadrilateral vertex on each of line 1 and line 2 may be freely chosen anywhere on the line subject to the constraint that they lie on the same side of the extended opposite parallelogram side QR as does A. Choice of these two points B and D determines three of the vertices of the quadrilateral. But placing B and D also determines the directions of the remaining two quadrilateral sides since they must pass through Q and R,

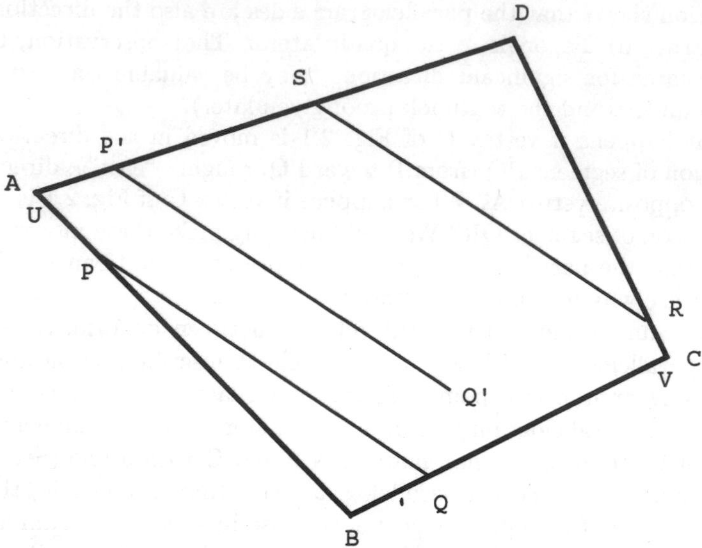

Figure 2.2.

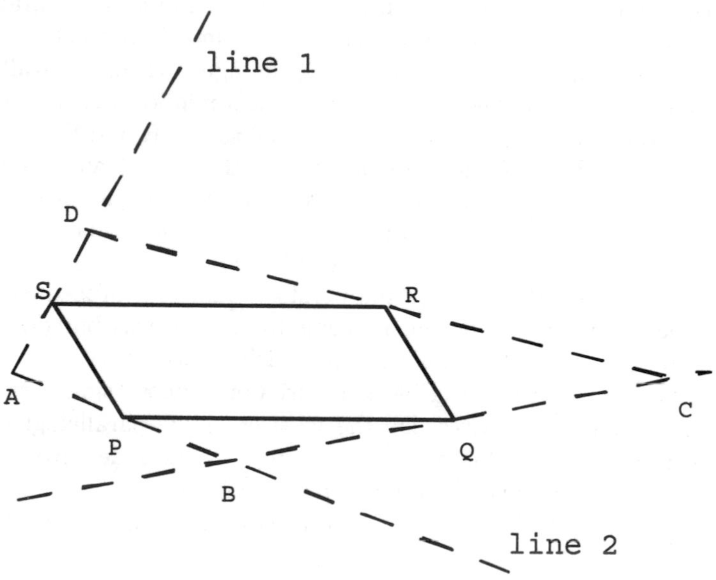

Figure 2.3.

respectively. Their intersection, the fourth vertex C, is thus also determined. Since points B and D may be altered subject only to the constraint above, we again see that the midpoint property is not necessary. What if we force it to hold by choosing B and D accordingly, as in Fig. 2.3? Then it will be observed that the quadrilateral is uniquely determined, and it satisfies the midpoint property. It is thus possible to construct a circumscribing quadrilateral obeying the parallelogram property around an arbitrary parallelogram. We have also learned that the situation has only a few degrees of freedom: selecting one quadrilateral vertex fixes the midpoint quadrilateral for a given parallelogram.

The foregoing experiments have not led to a complete understanding of why the QT should hold, but have provided information about the quadrilateral–parallelogram relationship.

Another useful strategy is to start with the simplest special case, and verify the proposition for it. This is often relatively easy because special properties enforce symmetries; that's what makes them special. We may then be able to show that departures from the special case do not alter the proposition. Here we would start with the simplest, most symmetric quadrilateral: a square. We see that the inscribed midpoint polygon is also a square. In the previous case, difficult measurements were needed to verify a potential relation. However, in this symmetric case the measurements are simple because they follow from the observation of bilateral symmetry, something human perception is good at.

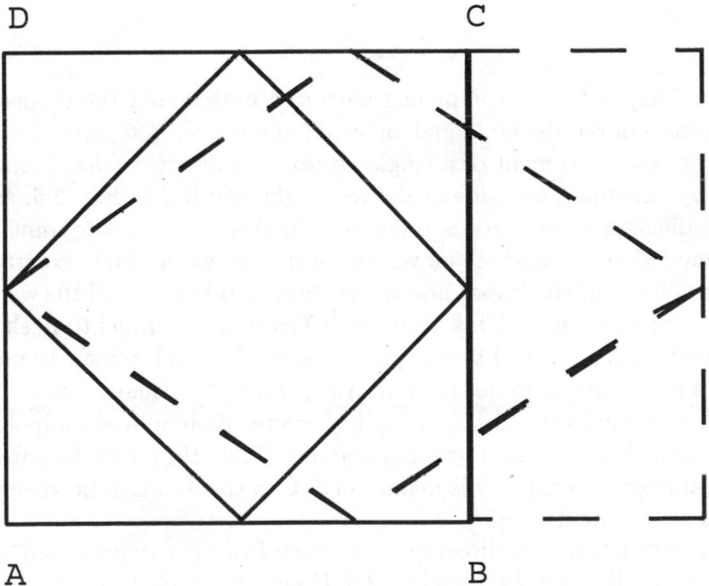

Figure 2.4.

Now what would happen if we make the square into a rectangle by lengthening a pair of opposite sides equally? See Fig. 2.4. It is readily seen that the inscribed square's vertices that lie on the elongated lines (AB and DC) move in the same direction and by the same amount: half of the elongation. The sides of the inscribed figure change orientation, but our bilateral symmetry detectors immediately see that opposite pairs change by the same amounts and the figure remains symmetric. We now have a parallelogram inscribed in a rectangle – the condition remains true. Again, syntactic symmetry detection shows us that the same will happen if we now stretch or shrink the rectangle in its other dimension.

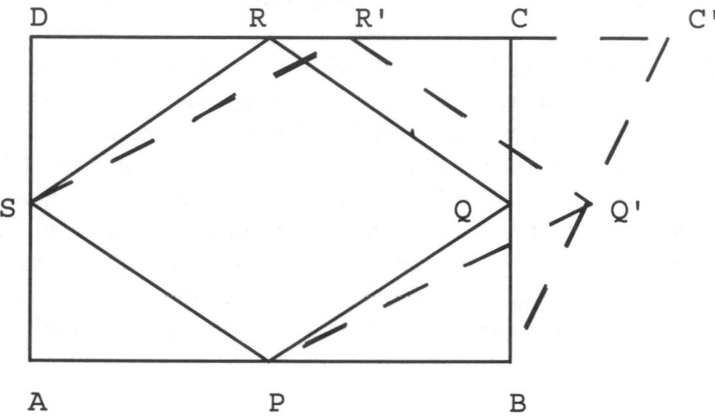

Figure 2.5.

Symmetry is thus a compelling means to understand the theorem in the special cases of rectilinear, equal, orthogonal stretching of pairs of sides. Consider next the movement of a single vertex in a direction that violates symmetry by extending one side of the rectangle, say DC in Fig. 2.5, and ask if the parallelogram property is preserved. In this case the midpoint Q of the right side BC moves half the movement of the vertex, which is exactly how far the midpoint R of the upper line moves. Seeing this is straightforward in this special, rectilinear case. This means that QR moves parallel to itself to Q′R′, and thus remains parallel to its opposite side SP, which remained unchanged since its endpoints were not altered. Also, while the orientations of the other two sides change, they change equally because their moved endpoints move in the same direction by the same amount. Thus they remain parallel, and the parallelogram property is maintained. Can this relation be generalised to arbitrary movements of C?

Move point C in any direction other than from D toward C, say in the direction of the diagonal PR. See Fig. 2.6. If we can establish that the midpoints of the altered segments move in the same direction by the same amount we know that the orientation of the line connecting them (R′Q′) has not changed. Examining a simpler case makes this easier to see. In Fig. 2.7 we consider a segment with one fixed endpoint E. Moving the other endpoint Y in a given direction to Y′ moves "every" point on the line in the same direction, but each point moves by a fraction of the endpoint movement proportional to its distance from the fixed point. Note that we are imagining the original segment and the new segment to have points in 1–1 correspondence. (Technically this is true, even though there is a non-denumerable set of points in the abstract idealisation of a line.) However, it is sufficient to imagine the line as composed of a finite number of points, with the line restricted to passing though each of them. Imagine the line as a rubber band that stretches. Applying this to Fig. 2.6 (EYY′ maps onto BCC′ and DCC′) means that R and Q each move by half the distance that C moves away from D and B, in each respective direction, and all three points move in the same net direction: the P to R direction. Thus both ends of QR move by the same amount in the same direction and the line must maintain its orientation and length.

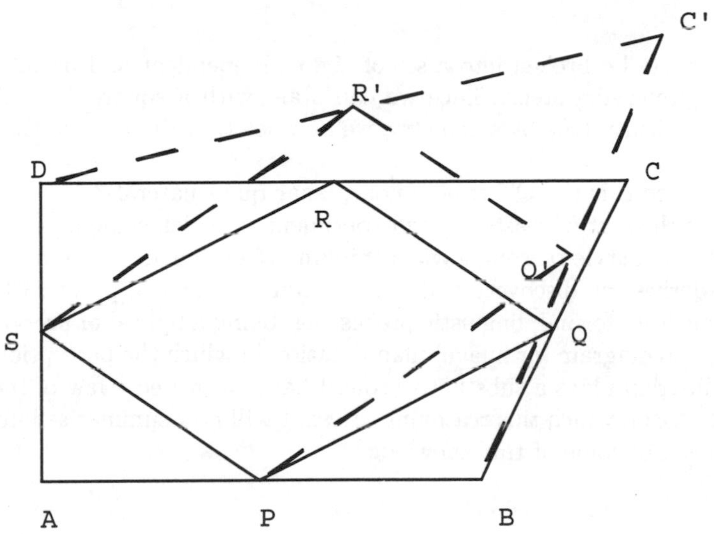

Figure 2.6.

Recalling path independence we now see that an arbitrarily chosen amount and direction of movement of one vertex of the quadrilateral can be broken into two components, one of which is parallel to one diagonal of the quadrilateral, the other of which is parallel to the other diagonal of the quadrilateral. Movements in diagonal directions are parallelogram property preserving, so any movement of the vertex is parallelogram property preserving. Finally this "same" argument applies to any vertex, mutatis mutandis, so any alteration of the quadrilateral is parallelogram property preserving

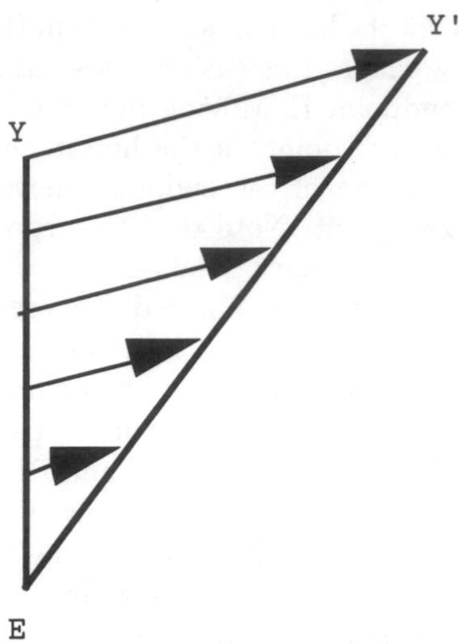

Figure 2.7.

because it can be broken into a set of eight independent and parallelogram property preserving steps. Since we can start with a square for which the theorem is clearly true by symmetry, we now see that the general theorem is true.

Extra credit: is the QT true of non-planar quadrilaterals?

This is the sort of reasoning by experiment in which some people engage in lieu of proof-style reasoning when thinking about geometry, and which for them underlies the discovery and understanding of formal proofs. Clearly it is different from formal, linguistic processing, being a hybrid of propositional thinking and diagram (or image) manipulation in which the perception of the altered diagram plays a substantive role. I have suggested a few of the items of knowledge on which this reasoning draws. I will now summarise informally an inventory of some of this knowledge.

2.5 Knowledge-Based Understanding

Some of the knowledge needed for diagrammatic reasoning about geometry resides in the ability to construct and manipulate a diagram according to propositional specifications. This includes knowledge implicit in the diagram (or a mental image or other representation of it) that enforces the essential properties of space. In addition there is knowledge of basic concepts such as line segment, midpoint, quadrilateral, and so forth. All of this knowledge is already represented in my programmed system in ways that permit it to be employed appropriately. Thus the system can be told what experiments to perform, and then do them. In addition to performing constructions and constrained manipulations it can observe the effects of the changes and record them in its inventory of facts about the diagram.

The ability to propose the experiments of a demonstration must employ additional knowledge, such as the following.

Some of the properties of relevance to the theorems of plane geometry depend upon relative values and are independent of absolute values. For example, the shape of an object is independent of its absolute location in space. Similarly, some of the properties of relevance, such as shape again, are independent of orientation. Other properties are independent of scale as well, while some are not. The proficient geometer knows which invariances apply to which properties.

Many propositions of plane geometry are implicitly about rigid figures. In general, the properties of rigid figures are static properties in the sense that they can be determined by examination of the static configuration without knowledge of how the figure was constructed, e.g., in which order the components were drawn. This was referred to earlier as the assumption of path independence.

In the case of diagrams, one knows that the object of study is not the diagram itself, but the abstract idealisation that it represents, where lines have no thickness and are meant to be perfectly straight. It is known that the imperfections of an actual diagram may yield incorrect results and one must know how to avoid such traps.

Other knowledge is provided by definitions. For example, the definition of a parallelogram as a four-sided figure with opposite sides parallel must be known in order to understand statements about parallelograms. Furthermore there is knowledge about the relations of classes of objects, such as the fact that all squares are rectangles. This knowledge may be used in conjunction with knowledge of the inheritance of properties concept to conclude that a property shown to be true of every rectangle is also true of every square, for example.

Overlying all of the foregoing knowledge is the knowledge of when each is true; for example, what properties are independent of scale or which ones are not. This knowledge perhaps can only be represented as explicit lists of facts.

There is also knowledge of logic. For example, a universal statement can be disproved by a single counterexample but generally cannot be proved by even a large number of positive instances. This knowledge is tempered in informal problem solving by other beliefs, some based on the concept of probability. For example, a larger number of positive examples increases support for a conclusion. If the possibilities can be factored into a finite number of cases, the generality can be concluded if it is shown to be true of each case.

Belief in a generalisation is increased if it can be shown that it is true of any arbitrarily selected member of the class to which it applies. The definition of arbitrary selection is difficult to capture. Usually it means that no constraints are placed on the example to make it a "special" case. The constraints are those that apply to the properties of the figure class in question. Thus "any quadrilateral" should exclude special cases, such as degenerate cases (where two or more vertices are identical, for example). A special case is where any subset of elements of the figure are related in "special ways".

"Special ways" is an inventory of properties that are known, presumably from prior learning, to be important. For experienced geometers this includes properties of parallelness, perpendicularity, equal length, and equal angle measure. Special angle measures are 0, 90, 45, 30, and 60 degrees and their integer multiples. Although absolute lengths are arbitrary, relative lengths are not; special cases of length are those where one length is an integer multiple of another.

Many problems also require knowledge about how to reason about some equality and ordering relations. For example, the equality (of area, of length, etc.) relation is symmetric, transitive, and reflexive, the greater-than relation is anti-symmetric, transitive, and irreflexive. A geometer must know what these properties mean and how to use them to draw new conclusions from established facts.

Other knowledge takes the form of previous conclusions, for example previously proved theorems in the case of formal geometry. Having this knowledge available in turn requires knowledge of how to represent and recall generalisations and how to apply them to new cases. Thus to demonstrate that two specific triangles are congruent, one might rotate and translate them into superposition (ignoring mirror images which require flipping in 3-space). One might then construct or examine a demonstration of the side–angle–side congruency theorem. The demonstration might take the form of showing that specifying these properties of a triangle leads to a rigid figure, that is, one that cannot be altered in shape if those values are fixed [9]. Having accepted this generalisation it would be stored. It could then be used in the future to establish congruence without the perceptual reasoning steps of rotation and translation. Other knowledge involves knowing strategies for problem solving, including knowledge of when they are useful. The use of symmetry is one example. If bilateral symmetry is detected, certain conclusions can be immediately drawn about the sameness of the relations among objects on opposite sides of the axis of symmetry. Syntactic symmetry is a different method. This involves detecting the similarity of procedures and discovering a mapping of variables that preserves the form and permits a similar conclusion. Degree-of-freedom analysis is another useful method for systematically applying a set of constraints to sequentially limit the possible properties of an object [8]. When the object is a point and the property is its location, this is the method of loci. Beginning with a simple, highly constrained example and relaxing the constraints is another frequently useful strategy. Another strategy is the analysis of a movement into a series of steps whose effects are easier to see. This was used in the QT example, and is of wide use.

2.6 Beyond Verification: Discovering Demonstrations

Most of the above knowledge is propositional, although it refers to diagrammatic properties and is only useful when it becomes procedural. Knowledge must be related to the available inventory of diagram-processing functions. The above inventory is only a sample of the relevant knowledge needed, even if we restrict the topic to plane geometry. To invent demonstrations requires the selective use of this knowledge, in appropriate sequence, so that the demonstration will be relevant to the theorem.

One way to employ this additional strategic knowledge is to encode it in a representation analogous to a "script" [12]. For example, "When attempting to demonstrate a claim about an entire class of figures, select one according to its rules of non-arbitrariness, examine it to see that the alleged property holds; repeat with other cases." And "After demonstrating the property for one arbitrary instance of the class, alter that instance by altering in turn each of its components in ways that are arbitrary for that component." And "To demonstrate congruence of figures, attempt to rotate and translate the figures into coincidence." This model of demonstration construction would yield a set of special cases (scripts), rather than a general procedure. The system would find the relevant script, if it knows one, and apply it, instantiated with the appropriate definitions of its parameters (e.g., "arbitrary" is interpreted in the context of the particular component). New models for other classes of demonstrations could be constructed ad hoc, or could perhaps be induced from a set of cases.

2.6.1 Understanding the Textbook Proof

A textbook proof of the QT is given in the Table 2.1 (paraphrased from Fogiel, pp. 66–67 [5]). Note that several steps of the QT textbook proof call upon previous definitions and theorems. The proof establishes that one pair of sides must be both parallel and of equal length, and the result follows by a previous theorem. The key idea is the construction of a quadrilateral diagonal and the use of theorems about similar triangles. See Fig. 2.8.

Notice that the textbook proof is no more rigorous than the perceptual proof. It is defined in terms of a specific diagram, and makes no explicit effort to show that this diagram is not a special case. It uses informal reference ("the third side", "each other" and so forth) and alludes to "reasons" whose application is vague ("by definition", "transitivity property" for example). Thus while the textbook proof has a surface appearance of succinctness, that is bought at the price of vagueness and incompleteness. Furthermore, the textbook proof relies on the very same knowledge underlying perceptual reasoning, but in an impoverished way, and in many ways is inferior to perceptual reasoning as a means of understanding. Rather than observing how components and properties of the figure interact, the proof examines the static figure. In contrast, the manipulation approach addresses the problem from a

Table 2.1.

Num.	Statements	Reasons
1	P, Q, R, and S are the respective midpoints of sides AB, BC, CD, and AD of quadrilateral ABCD	Given
2	SP is a midline of triangle ABD	A midline of a triangle is the line segment joining the midpoints of two sides of the triangle
3	SP parallel to DB	The midline of a triangle is parallel to the third side
4	SP = 1/2 DB	The midline of a triangle is half as long as the third side of the triangle
5	QR is a midline of triangle CDB	Definition of midline
6	QR parallel to DB	The midline of a triangle is parallel to the third side of the triangle
7	QR = 1/2 DB	The midline of a triangle is half as long a as the third side of the triangle
8	SP parallel to QR	If each of the two lines is parallel to a third line, then they are parallel to each other
9	SP = QR	Transitivity property
10	Quadrilateral PQRS is a parallelogram.	A quadrilateral is a parallelogram if two of its sides are both congruent and parallel

variety of ways, playing with different figures and noting how movements reveal the connections among properties in a perceptually related way. Clearly the two approaches are not exclusive alternatives, but are complementary and synergistic.

Neither the textbook proof nor the perceptual argument cited above is formal, and yet both are revealing and perhaps convincing. The reason in each case is that the result is related, through intermediate stages, to underlying properties that may be observed in a particular diagram that is accepted as representative. Ultimately, the credibility of each depends on the reader's acceptance of the validity of the arguments because they coincide with his perception and understanding of visually processed information. Each also requires knowledge that can only be understood propositionally, including knowledge of how the perceptual and the propositional are related.

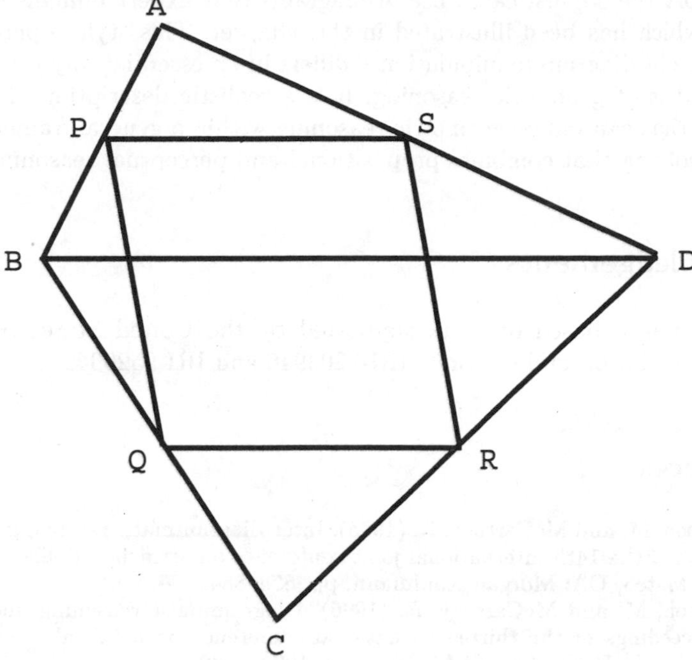

Figure 2.8.

2.7 Conclusion

Human reasoning about geometry uses diagrams in non-linguistic ways, but also uses linguistic reasoning in essential ways. Geometric reasoning requires a wide variety of propositional knowledge, some of which is specific to geometry but much of which is generic to problem solving, including knowledge of logic and of strategy heuristics.

In general artificial intelligence models are programming languages that implement certain reasoning strategies and styles but which rely on the "programmer" to put these together to achieve a goal since machines are not as yet autonomous and goal-directed in and of themselves. Thus AI models, including logic-based models, finesse the most fundamental problems of cognition. To move toward more fully comprehensive models will require the incorporation of a diverse and fluid knowledge system that is accessed in a variety of extremely complex ways. This chapter illustrates in the context of geometry some of this diversity and fluidity. A satisfactory model of human reasoning, even though restricted to a particular narrow topic such as Euclidean plane geometry, cannot be based on a single homogeneous representation and process, whether it is predicate calculus or pixel computations. The model I have developed, while not automating this knowledge-based mechanism, does provide an explicit model of diagram representation and manipulation that can support the sophisticated use of diagrams that expert humans employ, some of which has been illustrated in this chapter. This style – perceptual reasoning and diagram manipulation – differs in an essential way from logic-based models of geometric reasoning. It is a realistic description of human cognition that can imbed geometric reasoning within a general framework of problem solving that combines propositional and perceptual reasoning.

Acknowledgements

This material is based on work supported by the United States National Science Foundation under Grants IRI-9203946 and IRI-9526942.

References

1. Anderson, M. and McCartney, R. (1995). Inter-diagrammatic reasoning. In Proceedings of the 14th international joint conference on artificial intelligence, Vol. 1. San Mateo, CA: Morgan Kaufmann, pp. 878–884.
2. Anderson, M. and McCartney, R. (1996). Diagrammatic reasoning and cases. In Proceedings of the thirteenth national conference on artificial intelligence, Vol. 2. Menlo Park, CA: AAAI Press, pp. 1004–1009.
3. Barker-Plummer, D. and Bailin, S.C. (1992). Proofs and pictures: Proving the diamond lemma with the GROVER theorem proving system. In Reasoning with diagrammatic representations. Technical report SS-92-02. Menlo Park, CA: American Association for Artificial Intelligence, pp. 102–107.
4. Chou, S.-C. (1988). Mechanical geometry theorem proving. Dordrecht:Reidel.
5. Fogiel, M. (Ed.) (1994). The high school geometry tutor. (2nd edn.) Piscatawa, NJ: Research and Education Association.
6. Furnas, G.W. (1992). Reasoning with diagrams only. In Reasoning with diagrammatic representations. Technical report SS-92-02. Menlo Park, CA: American Association for Artificial Intelligence, pp. 118-123.
7. Gelernter, H. (1959). A note on syntactic symmetry and the manipulation of formal systems by machine. Information and Control 2:80–89.
8. Kramer, G.A. (1992). Solving geometric constraint systems: A case study in kinematics. Cambridge, MA: MIT Press.
9. Lindsay, R.K. (1996). Generalizing from diagrams. In Cognitive and computational models of spatial reasoning. Menlo Park, CA: American Association for Artificial Intelligence, pp. 51–55.
10. Lindsay, R.K. (1998). Using diagrams to understand geometry. Computational Intelligence 14:228–256.
11. Pylyshyn, Z.W. (1984). Computation and cognition: Toward a foundation for cognitive science. Cambridge, MA: MIT Press.
12. Schank, R. and Abelson, R. (1977). Scripts, plans, goals, and understanding: An inquiry into human knowledge structures. Hillsdale, NJ: Erlbaum.
13. Shin, S.-J. (1995). The logical status of diagrams. Cambridge, UK: Cambridge University Press.
14. Wang, D. (1995). Studies on the formal semantics of pictures. Ph.D. dissertation, Institute for Logic, Language, and Computation, University of Amsterdam.

3. A Meta-Taxonomy for Diagram Research

Alan Blackwell
Yuri Engelhardt

What is the common ground for a science of diagrams? A simple definition of which notations qualify as diagrams, if it were possible to achieve one, is likely to exclude valuable insights. As an alternative we suggest that common ground should be established on a taxonomic basis. A wide range of candidate taxonomies has already been described in several different academic fields. When taxonomies are needed, we propose that the taxonomic precedents should be treated more analytically than simply selecting the most inclusive or rigorous to be extended as necessary.

3.1 Introduction

We do not have space here to describe individual taxonomies in any detail, nor to assess their relative advantages. We recommend the original sources listed in the bibliography. Most of the taxonomies have been proposed from within a small range of academic contexts. An example is that of software engineering notations, including both diagrams used for system design [40,43] and visual programming languages [48]. The ergonomic implications of these diagrams have been categorised in the Cognitive Dimensions of Notations [23]. Further examples include the selection of representations for educational contexts [13,14,21], or in cartography, typography, and graphic design [6,18, 20,37,50,61,62].

The study of aesthetics and representation often generates taxonomic distinctions [22,28], as do theories of language [54,57,64]. Psychologists investigate the notational factors underlying cognitive performance both to gain insight into performance [2,4,25], but also as a basis for more general theories of cognition [34,55,69]. Some recent reviews have attempted to place previous research into diagram use within a framework of these cognitive questions [7,59].

3.2 Nine Aspects of Diagrams and Diagram Use

The taxonomies that we have studied all propose distinctions concerning some aspect of diagrams or diagram use. Some taxonomic systems make distinctions concerning more than one aspect. In our meta-taxonomy we are classifying these aspects along which the existing taxonomies make their distinctions. This set of taxonomic aspects can be used for comparative study of previous taxonomies, as the basis for creating new taxonomies of diagrams, and also to compare and contrast the research priorities of different fields contributing to the science of diagrams.

Signs or components of a diagram:

1. Basic graphic vocabulary
2. Types of tokens
3. Pictorial abstraction

Graphic structure of a diagram:

4. Graphic structure

Meaning of the diagram:

5. Mode of correspondence
6. The represented information

Context related aspects:

7. Task and interaction
8. Cognitive processes
9. Social context

Note that, although we refer to individual aspects examined in each taxonomy, these are sometimes combined into a matrix or some other multi-dimensional scheme by the original authors.

Our list of aspects was originally derived from our characterisations of the different research interests listed at the beginning of this chapter. We give a more complete description of the research fields motivating research taxonomies in Section 3.4. We should note that, although we proposed this exercise as an alternative to strict definition of the concept of diagram, we do of course make an implicit definition in our selection of research fields. We include disciplines such as cartography and typography whose subject matters (maps and words) might not be considered diagrams. Any visual representation that is not purely textual or purely pictorial can usefully be analysed to discover its diagrammatic content, whether or not it should formally be defined as a "diagram". This has provided us with further breadth of taxonomic perspective, for example from Kandinsky's classification of the fundamental elements of painting.

We can divide the nine aspects into representation-related aspects (1–6) and context-related aspects (7–9). The complete context is illustrated structurally in Fig. 3.1, and the structural relationship between those aspects that relate specifically to the representation are illustrated in Fig. 3.2.

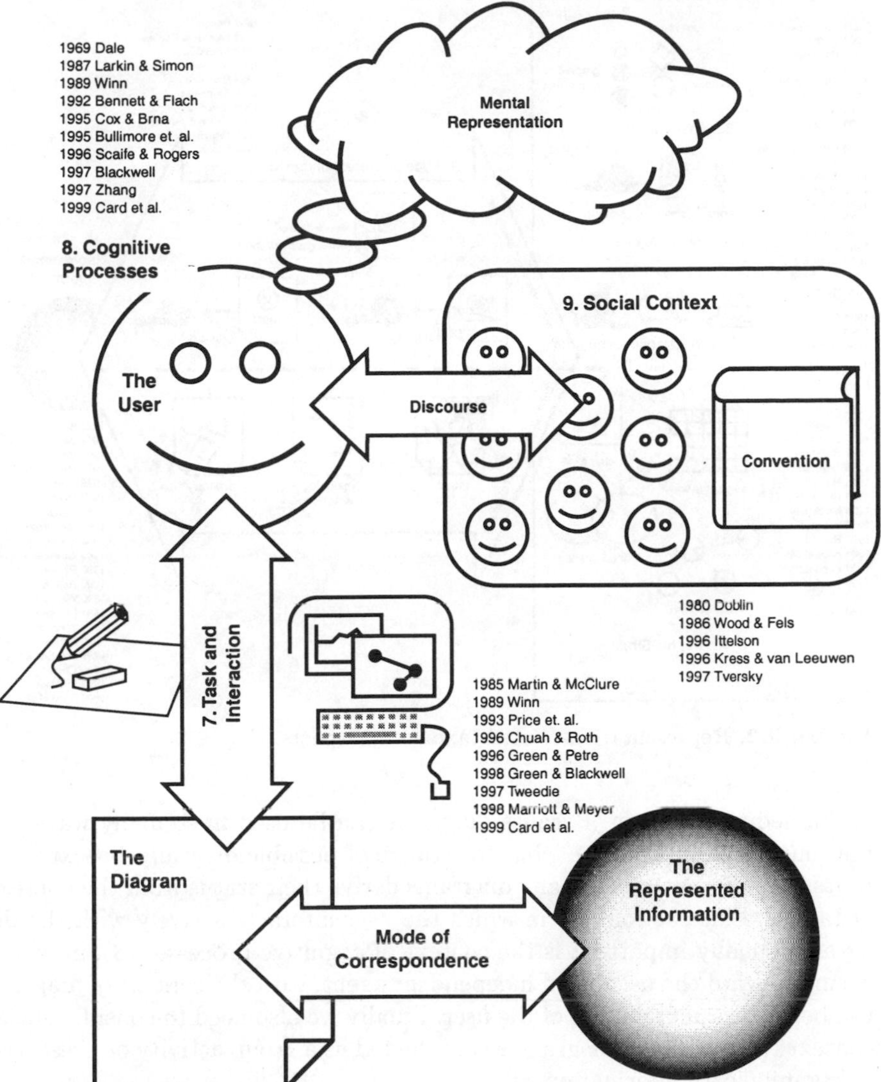

Figure 3.1. Contextual taxonomic aspects.

Representation-related aspects are related to the semiotic dyad of Saussure, which connects a representation to its meaning. Peirce's semiotic triangle [47] introduces the relevance of the interpretant – the intended result of the communication. In the case of diagrams, the potential interpretants include a great variety of possible contexts: the diagram may be used to communicate information to an audience (which is close to the conventional semiotic triangle), or it may be a sketch manipulated by a person who never intends showing it to anyone or keeping it for any longer than it takes to solve

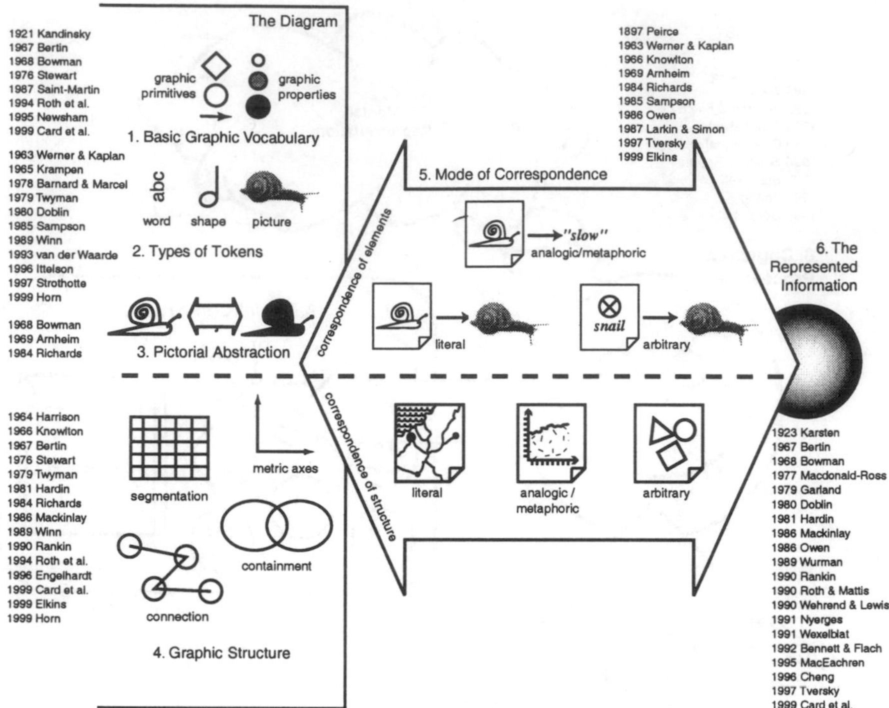

Figure 3.2. Representation-related taxonomic aspects.

an immediate problem. Because diagrams can be used in so many ways, we refer not to an interpretant, but to a range of possible diagram contexts (aspects 7–9). For example, many diagrams derive their status from the context of task and interaction (7), in which the user interacts actively with the diagram. Equally important is the context of cognitive processes (8), in which we may regard the user as an independent agent, with the context of diagram use being the mental state of the user. Finally we also need to consider social context (9), where the diagram is constructed as a group activity or presented to communicate information.

Representation-related aspects (1–6) deal with those representational properties of diagrams that apply in many different contexts. This relative context independence results in a potentially formalisable definition of form and meaning. In an earlier version of this chapter we called these aspects context-free, but clearly they do occur in a context, and that context guides the interpretation. However, many taxonomies attempt to describe diagrams in terms of some absolute properties (as if they could be pure, mathematical or context-free), and we adopt this perspective for the purpose of this part of the analysis (aspects 1–6).

Representation-related aspects relate either to the diagram itself (1–4), or to its meaning (5–6). The aspects regarding the diagram itself are concerned with either the signs that are the components of the diagram (1–3), or with the graphic structure of the diagram (4). The aspects regarding meaning are concernd with either mode of correspondence (5) or with the represented information (6). Aspects 1–5 and the many taxonomies that analyse these aspects are discussed in detail in Engelhardt's forthcoming Ph.D. thesis.

3.2.1 Basic Graphic Vocabulary (1)

The basic graphic vocabulary consists of the graphic primitive elements and the graphic properties that graphic representations are composed of. Example distinctions concerning this aspect are "point, line, area" and "colour, size, shape".

3.2.2 Types of Tokens (2)

A common taxonomic distinction is between words, shapes and pictures. Where these types of tokens appear within diagrams, they generally represent meaning that is borrowed from another symbolic convention – that of spoken language, for example. The simple distinction between words and pictures includes some element of our *mode of correspondence* (5), while that between shapes and pictures includes our *pictorial abstraction* (3). These nested symbolic conventions are also subject to taxonomic analysis: Werner and Kaplan [64] and Sampson [54] have both proposed classifications of words and symbols according to their mode of representation. Some words and symbols are apparently constructed from arbitrary smaller elements, while other words and symbols are constructed from smaller elements with meaning. Although this distinction is lost in taxonomies that refer to words, symbols and pictures as atomic elements, these tokens can be reconsidered in terms of our other taxonomic aspects if the taxonomy is applied recursively to analyse diagram tokens, even those that are at first sight purely conventional. An alternative, even further simplified version of the division between types of tokens such as "word, shape, picture" is the dichotomy "abstract vs. pictorial".

3.2.3 Pictorial Abstraction (3)

Concerning the depiction of physical objects or scenes, a *continuum* of pictorial abstraction can be observed, from the very realistic via the schematic to the completely abstract.

3.2.4 Graphic Structure (4)

Graphic structure, also referred to as "configuration", is concerned with the organisational principles according to which individual signs are combined into a diagram. Example distinctions concerning this aspect are "linear sequence, two-axis chart, table, tree structure".

3.2.5 Mode of Correspondence (5)

Mode of correspondence is about the kind of relationship between a representation and its meaning. Example distinctions concerning this aspect are "literal vs. metaphorical", "direct vs. indirect", and "iconic vs. symbolic". As both Arnheim [1, (pp. 135–136)] and Eco [16, (pp. 177–178)] have noted, these kinds of distinctions do not concern types of *signs*, but rather types of sign *functioning*. In different contexts, the same sign may function in different ways, and therefore mean different things. For example, depending on the context, a drawing of a wine glass may stand for a wine glass (literal correspondence), for "bar" (metonymic correspondence), or for "fragile" (metaphorical correspondence).

3.2.6 The Represented Information (6)

The information represented by the diagram has also been classified by various researchers. This includes classifications of information domains and classifications of relational properties. An example distinction concerning information domains is "space, time, other". An example distinction concerning relational properties is "nominal, ordinal, quantitative".

3.2.7 Task and Interaction (7)

The activity of a person interacting with a diagram, the structure of the task, and the tools that are used to complete that task, are also subject to taxonomic classification. This aspect includes taxonomic elements related to computational tools such as diagram parsers and editors, as well as task classifications (e.g. drawing, sketching, transcribing, restructuring). Although these considerations normally concentrate on the style of interaction where a user is creating or modifying the diagram, complex diagrams may require physical interaction even to read them. This may involve some computer program, the user moving a finger to track long paths on a piece of paper, or even the process of directing one's gaze along a locus of visual attention.

3.2.8 Cognitive Processes (8)

Many characteristics of diagram function are determined by the diagram user rather than by the representation, and these are reflected in taxonomic considerations of perceptual characteristics and support for cognitive function. The cognitive status of diagrammatic representations has led to classification of mental representations of diagrams, especially the contrast between hypothetical image-like mental representations and propositional representations. This aspect also includes the cognitive implications of diagram properties related to perception, interpretation and problem solving, as well as individual differences in ability, expertise or strategy.

3.2.9 Social Context (9)

Diagram users are not self-sufficient. Despite occasional naive claims regarding the inherently intuitive nature of graphics, the way that we interpret any representational conventions depends on cultural context as well as the conventions of particular media types. Analysis of a diagram must consider which information is present in the diagram, and which information comes from other sources. Furthermore, the content of a diagram must be considered in terms of its context in discourse.

3.3 Example: The London Underground Diagram

This section provides an illustration of the proposed taxonomic aspects of diagrams and diagram use. For each aspect, we will try to point out some specific distinctions that have been proposed in an actual taxonomy, using the London Underground diagram as a familiar example.

Regarding the *basic graphic vocabulary* (1), Bertin's [5] analysis of the graphic domain would suggest that the Underground diagram uses two "implantations": "points" (the stations) and "lines" (the connections). These encode information through two of Bertin's "visual variables": "shape" (types of stations), and "colour" (different lines). Bowman [8] would identify the same "vocabulary of form" in the Underground diagram as Bertin: "point, line, shape, colour".

Regarding *types of tokens* (2), Twyman's [61] analysis of "mode of symbolisation" suggests that the Underground diagram contains both "schematic" tokens (shapes, such as lines and marks) and "verbal" tokens (words, such as the station names).

Regarding *pictorial abstraction* (3), Richards [50] describes the "mode of depiction" of the Underground diagram as "non-figurative", since it hardly contains pictorial signs, except maybe for the River Thames.

Regarding *graphic structure* (4), Richards [50] points out "organisation" by "linking" (the lines) in the Underground diagram. Twyman [61], in his spectrum from linear to non-linear "configuration", regards the Underground diagram as a "non-linear" configuration with "directed viewing". Engelhardt [18], analysing "meaningful space", would point out the combination here of structuring by both "links" (the lines) as well as by crude geographic topology (the positions). In Lohse et al.'s "classification of visual representations", the assignment of the Underground diagram is not quite clear to us – it could be referred to either as a "map" or as a "structure diagram" or as a "process diagram" or as as a "network chart" in their classification system. Lohse et al.'s distinctions seem to consider both graphic structure (4) and the nature of the represented information (6).

Regarding *mode of correspondence* (5), Richards classifies the Underground diagram as "semi-literal". Eco [16, (pp. 178–179)] points out that the Underground diagram is both "iconic" (in its reference to the layout of tracks through the city) and "symbolic" (in its use of plain circles for stations and straight unidimensional lines for the fragmented routes).

Regarding *the represented information* (6), the Underground diagram may be considered to be a representation of "spatial" and "ordinal" information, or it may be taken to represent a "sequence of actions" required to reach a particular destination.

Concerning the context-related aspects of the Underground diagram, we note that there has been less empirical investigation of this diagram than there has been semiotic analysis. However, our list of aspects suggest several areas of investigation that could be pursued.

Regarding *task and interaction* with the Underground diagram (7), we can conjecture about interaction with it on the basis of the observation that maps in the underground are worn out in a patch near the current station. Users presumably put their finger on that patch, then trace a route to where they want to go. The finger seems to be an essential tool for interacting with such representations.

Regarding *cognitive processes* (8), we note that perceptual attributes of the diagram, including line weights, colour discrimination, font legibility and so on are a prerequisite to its usability. The Underground diagram has also affected mental representations: according to Garland, it has changed people's mental map of distances across London. No doubt the many versions also accommodate interpersonal variation – versions for use by the visually impaired, for example.

Finally, regarding *social context* (9), the Underground diagram certainly has a complex cultural and communicative context. When the diagram is printed on a T-shirt, what is its diagrammatic function? The Underground diagram can also be used as a pragmatic substrate for other messages. For example, shops in London often use a customised version of the diagram in their advertisements, to highlight their location.

3.3.1 Chronological Overview of Taxonomies

Figure 3.3 is a chronological listing of the taxonomies that we have examined. For each taxonomy we show which of the nine aspects it analyses. The chart can be used in two main ways. One way of using it is to look for researchers that have analysed a specific aspect. The other way of using it is to compare whether specific taxonomies analyse the same aspect(s) or not.

	EXAMINED ASPECT								
	Diagram				Meaning		Context		
RESEARCHER	Vocab	Tok	Abstr	Struc	Corr	Inform	Ta&In	Cogni	Social
1897 Peirce					5				
1921 Kandinsky	1								
1923 Karsten						6			
1963 Werner & Kaplan		2			5				
1964 Harrison				4					
1965 Barthes					5				
1965 Krampen		2							
1966 Knowlton				4	5				
1967 Bertin	1			4		6			
1968 Bowman	1		3			6			
1969 Arnheim			3		5				
1969 Dale								8	
1976 Stewart	1			4					
1977 Macdonald-Ross						6			
1978 Barnard & Marcel		2							
1979 Garland						6			
1979 Twyman		2		4					
1980 Doblin		2				6			9
1981 Hardin				4		6			
1984 Richards			3	4	5				
1985 Martin & McClure							7		
1985 Sampson		2			5				
1986 Mackinlay				4		6			
1986 Owen					5	6			
1986 Wood & Fels									9
1987 Larkin & Simon					5			8	
1987 Saint-Martin	1								
1989 Winn		2		4			7	8	
1989 Wurman						6			
1990 Rankin				4		6			
1990 Roth & Mattis						6			
1990 Wehrend & Lewis						6			
1991 Nyerges						6			
1991 Wexelblat						6			
1992 Bennett & Flach						6		8	
1993 Price et. al.							7		
1993 van der Waarde		2							
1994 Roth et al.	1			4					
1995 Bullimore et. al.								8	
1995 Cox & Brna								8	
1995 MacEachren						6			
1995 Newsham	1								
1996 Cheng						6			
1996 Chuah & Roth							7		
1996 Engelhardt				4					
1996 Green & Petre							7		
1996 Ittelson	1	2							9
1996 Kress & van Leeuwen									9
1996 Scaife & Rogers								8	
1997 Blackwell								8	
1997 Strothotte		2							
1997 Tversky					5	6			9
1997 Tweedie							7		
1997 Zhang								8	
1998 Green & Blackwell							7		
1998 Marriott & Meyer							7		
1999 Card et al.	1			4		6	7	8	
1999 Elkins				4	5				
1999 Horn		2		4					
	Vocab	Tok	Abstr	Struc	Corr	Inform	Ta&In	Cogni	Social

Figure 3.3. Chronological check-list of taxonomies addressing different aspects.

3.4 Characterisation of Research Interests

This section summarises research interests and objectives of the different academic fields that were identified as the basis for our meta-taxonomy in the introduction to this chapter.

3.4.1 Applied Psychology

How do different diagram types and diagram features affect human problem solving? It is obvious that a wide range of cognitive tasks is involved in constructing and interpreting diagrams, but we are not necessarily able to enumerate them or characterise them. If we could do so, we would be able to propose ways to make the use of diagrams faster, more accurate or easier to learn. These applied objectives have further scientific implications, however. Investigating the use of diagrams can teach us about the nature of cognition; this is central to the traditional concerns of cognitive psychology with perception, memory and problem solving.

Cognitive science is already focused on definite descriptions of diagrammatic reasoning tasks. As with all cognitive science, the use of computer models allows researchers to propose and investigate systematic (and potentially formalisable) models of reasoning. A primary focus of research into reasoning with diagrammatic representations is to investigate how these models can accommodate analogue representations rather than symbolic logic. Cognitive scientists carry out formal analyses of these different types of representations, but are also concerned with the nature of "internal representations" in human reasoning – the long-standing debate over mental imagery, for example, is regularly informed by research into diagrams. A further product of cognitive science research is that of artificial intelligence – what can we discover about computational questions or engineering solutions by building computer systems that use analogue representations?

3.4.2 Linguistics

How can the syntax, semantics, and pragmatics of diagrams be analysed? Modern critical theory has given us all a passing familiarity with this linguistic trichotomy; this is evident from the similarities between the taxonomies that we compared. Computational linguistics is concerned with properties of diagrams that enable research analogous to the research on verbal language. How can we formulate grammars that will allow automatic parsing and generation of diagrams? How can we write and test those grammars, or use them to analyse the structure of discourse? Situational semantics considers communication in a context that includes artefacts such as diagrams, and allows statements to be made about the informational status of the whole situation. How can we apply theories of conversational implicature to diagrams? What does the viewer expect a diagram to mean, and how can the producer exploit that expectation?

3.4.3 Visual Programming

Diagrams have held promise as a means of programming computers for many years. The objective is egalitarian – will diagrams make programming accessible to more people? This may be social – encouraging communication between programmers and their managers; educational – providing notations usable by children; or democratising – making programming accessible to novices rather than experts. How can these be achieved? Early work was unreasonably optimistic about the value of diagrams, and much research is now focused on using Green's Cognitive Dimensions of Notations to assess suitability for different tasks. Educational use of visual programming implies that skills learned from diagrams can be transferred to other notations – is this true? We need to understand what the user is learning – is it a virtual machine, mathematical properties of an algorithm, or just execution statistics? Finally, the traditional concerns of computer science intersect with these properties of diagrams. How can they be parsed and executed efficiently, and will they "scale up" to large and complex problems?

3.4.4 Data Visualisation

Most published diagrams are created by a person with a communicative intent and an understanding of the expected reader. How can data be characterised and design rules formalised in order to automate the creation of visual representations? This might involve choosing an appropriate interpretation of higher-dimensional data in two or three dimensions – if three, how can we deal with the occlusion problem on a visual display screen? How can such a system choose to allocate dimensions to isomorphic representation of physical spaces rather than symbolic information? What is the space of possible visual interactions between the user and the machine? Will we facilitate deeper understanding by making diagrams interactive or immersive? In all of these considerations, we need to know how the choice of an appropriate visualisation depends on the user's information-seeking goals.

3.4.5 Graphic Design

The task of a graphic designer is to start with some set of information, and prepare a way of communicating that information effectively. In order to do this, they must have access to some set of possible design solutions. How can this set be systematised? What is the visual vocabulary available to the designer? These are the topics of research in graphic design. The visual vocabulary must be extended by a "space" of possible spatial organisation (set-up, layout) for a given design problem. Graphic design also takes place within a social context: how does the genre (instructions, signage, newspaper graphics, textbook illustration, forms, etc.) affect design decisions beyond the simple limitations of media type and the vagaries of graphical fashion?

3.4.6 Education

Which types of diagrams are appropriate for what teaching goals? The field of education considers two separate questions: first, the need to educate children to be graphically "literate". What are the requirements of graphical literacy, how can it be assessed, and at what age should it be taught? Does it belong within a specific subject (e.g. mathematics), or should it span the curriculum? The second objective in education research is to understand how use of diagrams can facilitate education in all subjects. When and how and to what extent should we use diagrams in a textbook or a classroom presentation? Are they likely to enliven dull material, or will students ignore them as overly technical? If school material is learned diagrammatically, might this inhibit the development of abstract thought?

3.4.7 History and Philosophy of Science

Scientific discovery has often been associated with novel uses of representations. Some of these involve representations that are now widespread, such as algebra or Cartesian coordinates, while others were completely personal (Einstein's thought experiments, or Kekule's insight into the structure of the benzene ring). Are representations essential to science? Classicists such as Netz [42] are doing "cognitive history" by investigating the diagrams on which past discoveries have been founded. The narratives of creative discovery from scientists bear great resemblance to architectural theories of creative sketching. Is such creativity restricted to intellectual giants, or can ordinary people exploit these diagrammatic strategies in the course of problem solving?

3.4.8 Architecture

Architects spend much of their time interacting with visual representations. Some of these are described by the architects themselves as diagrams, although their most common representation is the sketch. The uses of sketches are closely related to the uses of diagrams, however. How can a broader definition of diagram types support architectural problem solving? Are there some representations which constrain the possible design solutions? This is an issue of major concern whenever architects use computer-aided design tools. Alternatively, what sort of diagrams facilitate the creativity that architects experience when sketching? Research into architectural sketching provides a radically different perspective on the cognitive function of diagrams – one that has also been discovered in studies of software engineers.

3.4.9 Cartography

Maps are seldom classed as diagrams in common usages of the term, but they share many of the interesting characteristics that have been covered in the taxonomies here. Furthermore maps are among the oldest visual representations that confronted the need to stylise and schematise from visual observations. Modern cartographic analysis also emphasises very challenging issues of social context, such as the implicit representation of political power in choice of representational conventions. It is very likely that these issues will eventually be noted as equally relevant in other areas of diagram use.

3.4.10 Decision Support

Although we are aware that researchers in decision support are interested in many of the issues raised here, neither of us feel sufficiently qualified to make any statements about their research objectives. We have therefore noted the existence of this topic, but will consider it no further here.

3.5 Discussion

Several other researchers have proposed classifications of existing diagram-related taxonomies. A common distinction [35,49] is between functional taxonomies and structural taxonomies. Functional taxonomies involve our aspects "task and interaction" (7), and "social context" (9). Structural taxonomies involve our aspects "basic graphic vocabulary" (1) and "graphic structure" (4). However, some of the aspects discussed in this paper, like "cognitive processes" (8) and "mode of correspondence" (5), fall outside the functional/structural distinction.

Goldsmith's [21] work is similar to ours in the sense that she is also trying to create a high-level reference frame for positioning and discussing existing research on visual representation. However, her work is quite different to ours in the sense that the existing research that she is looking at is not about diagrammatic representations and their use, but about depictive illustration, including issues of object identification and perception of pictorial depth. The aspects that she examines are geared towards those issues. In addition, the existing research work that she is studying and structuring is not taxonomic in nature, as in our case, but empirical.

Narayanan [41] offers a taxonomy of the research that has been conducted into diagrammatic communication. His taxonomy of diagram research includes a review of previous taxonomies. Narayanan proposes a detailed theoretical definition of diagrammaticity (a more recent theoretical discussion of definition can be found in Shimojima [56]), and summarises Bertin [6], Goodman [22], Lohse et al. [35], Engelhardt et al. [19], and many others. Narayanan's review of psychological research is particularly valuable, and provides a coherent taxonomy of mental representation, perceptual and interaction processes.

Lohse et al. [35], have produced a widely cited taxonomy of visual representations. This taxonomy is not based on an academic analysis of the representations themselves, but on similarity and other assessments made by experimental subjects for a sample of typical visual representations. This is a valuable exercise, but it is not a substitute for principled analysis. If anything, the whole project falls within our aspect of social context – it was a study of people's attitudes toward representations rather than their intrinsic characteristics.

In our survey, it is obvious that the majority of the taxonomies developed so far in diagram research concern our representation-related aspects. These aspects regard formalisable structure, and the attributes of diagrams that are most apparent by inspection. The representation-related aspects also fit most easily within the framework of semiotics, although semiotics does also include pragmatic context. Our characterisation of outstanding research issues, however, is generally grouped toward the later aspects. These aspects concern questions of performance, interpretation and cognition.

The reasons why taxonomies have tended to ignore these later issues is also clear – they are less easily formalised. We must ask ourselves, though, whether this neglect is either desirable or necessary. As Ittelson points out, diagrams have no meaning without an interpreter and a communicative intent. There are many academic disciplines which quite reasonably give separate consideration to those questions more amenable to formal analysis. Contextual issues require different research methods, even different academic disciplines.

An interdisciplinary science of thinking with diagrams cannot afford to concentrate only on formal analyses without context. A lack of context-related approaches is evident from our inventory of taxonomies. While the meta-taxonomic framework that we have proposed here is certainly still work in progress, we do hope that it can provide a reference frame for the future development of diagram-related taxonomies.

Acknowledgements

We would like to thank Remko Scha, Karel van der Waarde, Ingrid von Engelhardt, Matthias Mayer, Alexander Klippel, Peter van Emde Boas, Theo Janssen, and Alan Swanson for their helpful comments on earlier drafts of this paper. Alan Blackwell's research was funded by the Engineering and Physical Sciences Research Council, under grant GR/M16924 "New paradigms for visual interaction".

References

1. Arnheim, R. (1969). Visual thinking. Berkeley, CA: University of California Press.
2. Barnard, P. and Marcel, T. (1978). Representation and understanding in the use of symbols and pictograms. In R. Easterby and H. Zwaga (Eds), Information design. Chichester: Wiley, pp. 37–75.

3. Barthes, R. (1965). Elements de semiologie. Translated by A. Lavers and C. Smith, Elements of semiology. New York: Hill and Wang (1975).
4. Bennett, K.B. and Flach, J.M. (1992). Graphical displays: Implications for divided attention, focused attention and problem solving. Human Factors 34(5):513–533.
5. Bertin, J. (1967). Semiologie graphique: les diagrammes, les reseaux, les cartes. The Hague/Paris: Mouton/Gauthiers-Villars.
6. Bertin, J. (1977). La graphique et le traitement graphique de l'information. Paris: Flammarion.
7. Blackwell, A.F. (1997). Diagrams about thinking about thinking about diagrams. In M. Anderson (Ed.), Reasoning with diagrammatic representations II: Papers from the AAAI 1997 fall symposium. Technical report FS-97-02. Menlo Park, CA: AAAI Press, pp. 77–84.
8. Bowman, W.J. (1968). Graphic communication. New York: Wiley.
9. Bullimore, M.A., Howarth, P.A. and Fulton, E.J. (1995). Assessment of visual performance. In J.R. Wilson and E.N. Corlett (Eds), Evaluation of human work (2nd edn). London: Taylor and Francis, pp. 804–839.
10. Card, S., Mackinlay J., and Shneiderman, B. (1999). Readings in information visualization: Using vision to think. San Francisco: Morgan Kaufmann, pp. 1–34.
11. Cheng, P.C.-H. (1996). Functional roles for the cognitive analysis of diagrams in problem solving. In Proceedings of 18th annual conference of the Cognitive Science Society. Hillsdale, NJ: Erlbaum, pp. 207–212.
12. Chuah, M.C., and Roth, S.F. (1996). On the semantics of interactive visualizations. In Proceedings of information visualization, IEEE, San Francisco, October 1996, pp. 29–36.
13. Cox, R. and Brna, P. (1995). Supporting the use of external representations in problem solving: The need for flexible learning environments. Journal of Artificial Intelligence in Education 6(2):239–302.
14. Dale, E. (1969). Audiovisual methods in teaching (3rd edn). New York, Holt, Rhinehart & Winston.
15. Doblin, J. (1980). A structure for nontextual communications. In P.A. Kolers, M.E. Wrolstad and H. Bouma (Eds), Processing of visible language 2. New York: Plenum Press, pp. 89–111.
16. Eco, U. (1985). Producing signs. In M. Blonsky (Ed.), On signs. Baltimore: John Hopkins University Press, pp. 176–183.
17. Elkins, J. (1999). The domain of images. Ithaca, NY: Cornell University Press.
18. Engelhardt, Y. (1998). Meaningful space: How graphics use space to convey information. In Proceedings of vision plus 4, School of Design, Carnegie Mellon University, Pittsburgh, pp. 108–126.
19. Engelhardt, Y., Bruin, J., Janssen, T. and Scha, R. (1996). The visual grammar of information graphics. In N.H. Narayanan and J. Damski (Eds), Proceedings of AID '96 workshop on visual representation, reasoning and interaction in design, Key Centre for Design Computing, University of Sydney.
20. Garland, K. (1979). Some general characteristics present in diagrams denoting activity, event and relationship. Information Design Journal 1(1):15–22.
21. Goldsmith, E. (1984). Research into illustration: An approach and a review. Cambridge, UK: Cambridge University Press.
22. Goodman, N. (1969). Languages of art: An approach to a theory of symbols. London: Oxford University Press.
23. Green T.R.G. and Petre M. (1996). Usability analysis of visual programming environments: A "cognitive dimensions" approach. Journal of Visual Languages and Computing 7:131–174.

24. Green, T.R.G. and Blackwell, A.F. (1998). Design for usability using cognitive dimensions. Tutorial presented at BCS conference on human-computer interaction HCI'98, Sheffield, UK.
25. Hardin, P. (1981). Representational characteristics in diagrams of statements of relationships. Unpublished PhD thesis, University of Iowa, UM 812 8401.
26. Harrison, R.P. (1964). Pictic analysis: Toward a vocabulary and syntax for the pictorial code; with research on facial expression. Unpublished PhD thesis, Michigan State University.
27. Horn, R.E. (1999). Visual language: Global communication for the 21st century. Bainbridge Island, WA: MacroVU, Inc.
28. Ittelson, W.H. (1996). Visual perception of markings. Psychonomic Bulletin and Review 3:171–187.
29. Kandinsky, W. (1921). Fundamental elements of painting [in Russian]. In a report to the People's Commissariat for Public Education, Moscow. English translation published in Languages in Design 1(3, 1993):267–271.
30. Karsten, K.G. (1923). Charts and graphs. New York: Prentice-Hall.
31. Knowlton, J.Q. (1966). On the definition of "picture". AV Communication Review 14:157–183.
32. Krampen, M. (1965). Signs and symbols in graphic communication. Design Quarterly 62:1–31.
33. Kress, G. and van Leeuwen, T. (1996). Reading images: The grammar of visual design. London: Routledge.
34. Larkin, J.H. and Simon, H.A. (1987). Why a diagram is (sometimes) worth ten thousand words. Cognitive Science 11:65–99.
35. Lohse, G.L., Biolisi, K., Walker, N. and Rueter, H.H. (1994). A classification of visual representations. Communications of the ACM 37(12):36–49.
36. MacEachren, A.M. (1995). How maps work: Representation, visualization, and design. New York: Guilford Press.
37. Macdonald-Ross, M. (1977). Graphics in texts. In L.S. Shulman (Ed.), Review of research in education, Vol. 5. Itasca, IL: Peacock.
38. Mackinlay, J. (1986). Automating the design of graphical presentations of relational information. ACM Transactions on Graphics 5(2):110–141.
39. Marriott, K. and Meyer, B. (1998). The CCMG visual language hierarchy. In K. Marriott and B. Meyer (Eds), Visual language theory. Berlin: Springer, pp. 129–170.
40. Martin, J. and McClure, C. (1985). Diagramming techniques for analysts and programmers. Englewood Cliffs, NJ: Prentice–Hall.
41. Narayanan, N.H. (1997). Diagrammatic communication: A taxonomic overview. In N. Kokinov (Ed.), Perspectives on cognitive science, Vol. 3. Sofia: New Bulgarian University Press.
42. Netz, R. (1999). The shaping of deduction in Greek mathematics: A study in cognitive history. Cambridge, UK: Cambridge University Press.
43. Newsham, R. (1995). Symbolic representation in object-oriented methodologies: Modeling the essence of the computer system. Unpublished Master's thesis, Department of Computer Science, Nottingham Trent University.
44. Nyerges, T.L. (1991a). Geographic information abstractions: Conceptual clarity for geographic modeling. Environment and Planning A 23:1483–1499.
45. Nyerges, T.L. (1991b). Representing geographical meaning. In B.P. Buttenfield and R.B. McMaster (Eds), Map generalization: Making rules for knowledge representation. Essex, UK: Longman, pp. 59–85.
46. Owen, C.L. (1986). Technology, literacy, and graphic systems. In M.E. Wrolstad and D.F. Fisher (Eds), Towards a new understanding of literacy. In Proceedings

of the third conference on processing of visual language, 31 May–3 June 1982, Airlie House, Airlie, VA.
47. Peirce, C.S. (written around 1897, republished in 1932). Elements of logic. In C. Hartshorne and P. Weiss (Eds), The collected papers of C.S. Peirce. Cambridge, MA: Harvard University Press.
48. Price, B.A., Baecker, R.M. and Small, I.S. (1993). A principled taxonomy of software visualization. Journal of Visual Languages and Computing 4(3):211–266.
49. Rankin, R. (1990). A taxonomy of graph types. Information Design Journal 6(2):147–159.
50. Richards, C.J. (1984). Diagrammatics. PhD thesis, Royal College of Art, London.
51. Roth, S.F. and Mattis, J. (1990). Data characterization for intelligent graphics presentation. In Proceedings of the conference on human factors in computing systems (SIGCHI '90), Seattle, WA, April 1990, pp. 193–200.
52. Roth, S.F., Kolojejchick, J., Mattis, J. and Goldstein, J. (1994). Interactive graphic design using automatic presentation knowledge. In Proceedings of the CHI'94 conference on human factors in computing systems. New York: ACM Press, pp. 112–117.
53. Saint-Martin, F. (1987). Semiotics of visual language. Bloomington, IN: Indiana University Press.
54. Sampson, G. (1985). Writing systems: A linguistic introduction. London: Hutchinson.
55. Scaife, M. and Rogers, Y. (1996). External cognition: How do graphical representations work? International Journal of Human Computer Studies 45:185–214.
56. Shimojima, A. (1999). The graphic–linguistic distinction – exploring alternatives. Artificial Intelligence Review 13:313–335.
57. Stewart, A.H. (1976). Graphic representation of models in linguistic theory. Bloomington, IN: Indiana University Press.
58. Strothotte, C. and Strothotte, T. (1997). Seeing between the pixels. Berlin: Springer Verlag.
59. Tversky, B. (1997). Cognitive principles of graphic displays. In M. Anderson (Ed.), Reasoning with diagrammatic representations II: Papers from the AAAI 1997 fall symposium. Technical report FS-97-02. Menlo Park, CA: AAAI Press, pp. 116–124.
60. Tweedie, L. (1997). Characterizing interactive externalizations. In Proceedings of the CHI'97 conference on human factors in computing systems. New York: ACM Press, pp. 375–382.
61. Twyman, M. (1979). A schema for the study of graphic language. In P.A. Kolers, M.E. Wrolstad and H. Bouma (Eds), Processing of visible language, Vol. 1. New York: Plenum Press, pp. 117–150.
62. van der Waarde, K. (1993). An investigation into the suitability of the graphic presentation of patient package inserts. Unpublished PhD thesis, Department of Typography and Graphic Communication, University of Reading, UK.
63. Wehrend, R. and Lewis, C. (1990). A problem-oriented classification of visualization techniques. In Proceedings of the first IEEE conference on visualization: Visualization 90, October 1990. Los Alamitos, CA: IEEE, pp. 139–143.
64. Werner, H. and Kaplan, B. (1963). Symbol formation: An organismic-developmental approach to language and the expression of thought. New York: Wiley.
65. Winn, W. (1989). The design and use of instructional graphics. In H. Mandl and J.R. Levin (Eds), Knowledge acquisition from text and pictures. North-Holland: Elsevier.

66. Wexelblat, A. (1991). Giving meaning to place: semantic spaces. In M. Benedikt (Ed.), Cyberspace: First steps. Cambridge, MA: MIT Press, pp. 255–271.
67. Wood, D. and Fels, J. (1986). Designs on signs: Myth and meaning in maps. Cartographica 23:54–103.
68. Wurman, R.S. (1991). Information anxiety. New York: Doubleday.
69. Zhang, J. (1997). The nature of external representations in problem solving. Cognitive Science 21:179–217.

4. Sketches as Mental Reifications of Theoretical Scientific Treatment

Daniela M. Bailer-Jones

Sketches are used abundantly in the practice of science. The purposes for which they are used are diverse, such as illustration, exemplification or data display. Moreover, sketches in science can be of very different things – from concrete objects to abstract conceptions. The graph of a mathematical function is very different in what it shows from a drawing of an experimental set-up or from an X-ray picture. My focus, in this paper, is on a quite specific use of sketches: sketches that *stand for* the theoretical treatment of empirical phenomena;[1] they function as substitutes for the physical processes (and their theoretical description) that are thought to produce these phenomena. Examples for such sketches will be given below. Relying on conventions old and new, the point is that theoretical information – abstract theories from physics as they are applied to an empirical phenomenon – is put into pictures. This means that, in effect, abstract and theoretical conceptions about the empirical world, mostly expressed in mathematical formulae, are *reified*, i.e. "made into a thing".[2] By being put into a picture, scientific treatment dominated by theoretical conceptions is effectively made into two-dimensional objects.

4.1 Introduction

How can theoretical information be put into pictures? First of all, one would expect such a transformation to serve a specific purpose in the process of scientific theorising – more on this in Section 4.3.1. Moreover, there is no known set of rules that guides the transformation. This implies that putting theoretical information into pictures is a creative mental process, not one that obeys strict rules, but one guided by intuition, conventions and convenience. One might think of "visualisation" in this context, but I reserve this term

[1] Such theoretical treatment of empirical phenomena is commonly formulated in terms of scientific models, and developing scientific models is seen as a form of problem solving (cf. [4]). The problem in question is how to explain a certain aspect of the empirical world.

[2] Nersessian [13, (p. 24)] talks about "physical embodiment" instead.

for something else and therefore exclusively talk of (mental) reification when I want to imply that theoretical information gets transformed into a sketch or diagram. I call it mental reification because making something abstract and theoretical into something concrete and two-dimensional, like a picture, is a *mental process* depending on the imagination of those creating and using these pictures. For visualisation to mean that one imagines a sketch or a picture in the mind is the use of "visualisation" which I employ in the current study. Any picture we can see ordinarily, e.g. drawn on a piece of paper, is *visualized* when it is reproduced "before our mental eye". Accordingly, mental reifications can be visualised because, besides being drawn on paper, they can be "held in the mind". I employ this narrow understanding of visualisation because it is the only one required for my purposes. This understanding of visualisation excludes the possible notion of visualising things that cannot be drawn in an ordinary picture, such as, for instance, an infinite row of houses. While one may, in a certain sense, claim to be able to "visualise" such infinity, one will certainly not be able to capture it in an ordinary (non-metaphoric) picture. For my current purposes, I am not concerned with any meanings of visualisation that go beyond "holding a picture in the mind". Also, I cannot begin to offer speculations about what it is like to "hold a picture in one mind" or to have a mental image, and how this compares to ordinary seeing. Most people, however, seem to share the experience of visualising a picture in the mind in a very similar way in which that picture could be drawn on paper. As a result, the visual image of a piece of abstract and theoretical information can be either on paper or in the mind.

Accepting this notion of visualisation as recreating and holding a picture in one's mind, I intend to examine further the implications of mental reification. My aim is to show that the latter centrally serves scientific exploration and theorising because sketches, or visualisations thereof, can function as useful representations of detailed theoretical treatment. They are symbols and memory aids that can be contrasted to other, more direct cases of visual problem solving, as discussed in Section 4.2. Later in this chapter, in Section 4.3.2, I shall examine why mental reification, as a "less direct" form of visual problem solving, might nonetheless play an important role in scientific theorising. Before I can do so, however, I shall, in Section 4.3.1, provide evidence that there is indeed reason to think that theoretical information about an empirical phenomenon is imported into simple sketches. The example I use to illustrate my points about mental reification and visualisation is research work on the phenomenon of extended extragalactic radio sources (EERSs) in astrophysics. To support my claims about the transformation from theoretical treatment to pictures I use sketches on EERSs drawn by a physicist in a research context.

4.2 Using Sketches

Mental reification is not a phenomenon that occurs in all problem solving tasks. Some problems and/or their solutions are formulated in terms of a visual expression in the first place, which is why they neither require nor allow for reification. In such visual problem solving tasks, problems are solved *through* diagrams, e.g. *through* Feynman diagrams (cf. [12], pp. 170–172) or *through* Peter Cheng's AVOW diagrams (see his contribution to this volume). Another example of a problem that is intrinsically visual is the match task discussed below. In that case, the special insight to be promoted by the picture is intimately tied to the picture and could hardly be communicated without the specific picture. I use this example to contrast it with the case of mental reification and to emphasise how the visual aspect of mental reification is important, but quite differently from "classical" visual problem solving tasks. Thus, let me now examine the case of an intrinsically *visual* problem solving task.

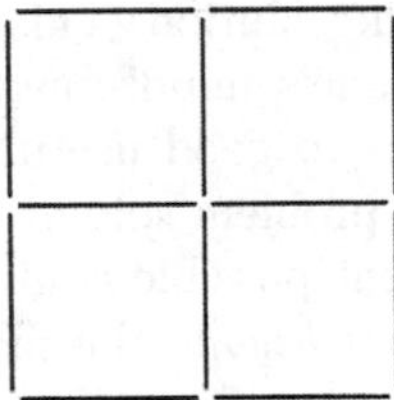

Figure 4.1. Configuration of matches.

Take a match task, such as the following: consider the configuration of matches, shown in Fig. 4.1, consisting of 12 matches and forming three squares. The problem to solve is: How can one produce five squares by moving three matches?[3] This kind of problem to be solved is formulated in terms of real, physical matches, in terms of a sketch of the configuration of matches, as used in Fig. 4.1 for purposes of illustration. Abstract notions, such as squares and numbers, play a role, but because the problem is already posed in terms of matches or of a sketch thereof there is no room for reification.

The easiest way for most people to solve the puzzle is to move around a few matches and try out alternative configurations in order to approach a solution. Thus they include the "physical", thing-like representation of the problem in their problem solving. Needless to say, it is also possible to find the solution without touching a single match – by trying out different moves and configurations in one's mind, i.e. by visualising the images of different

[3] This and other match tasks are discussed in [9, (pp. 58–80)].

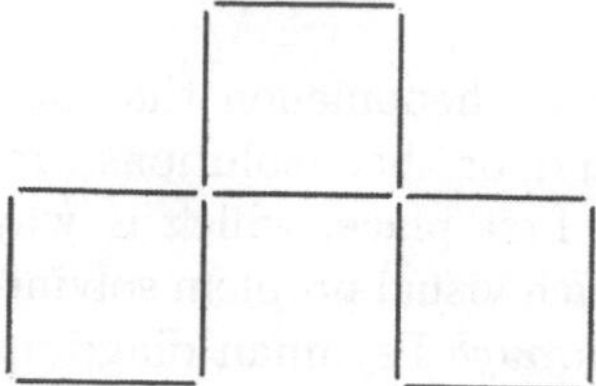

Figure 4.2. The solution to the puzzle with the matches – producing five squares by moving three matches.

configurations. The whole puzzle is based on a picture, a two-dimensional representation of the problem, and the solution to the puzzle is found *in* the picture. It is notable, however, that the successful solution, as shown in Fig. 4.2, requires a bit of "mental flexibility". If one expects all the squares to be the same size, as the original configuration suggests, it is not possible to solve the puzzle. Whoever thinks only in terms of squares of one size cannot produce and grasp the solution. This consideration goes beyond the mere "holding a picture in one's mind", even though it remains true that such visualisation can be put to good use in problem solving, but it is not problem solving. The actual problem solving of the match task has not just to do with computing different possible configurations that emerge through moving three matches. It goes *beyond* the picture and has a lot to do with "seeing" the large square in the configuration displayed in Fig. 4.2,[4] although the solution can be found *in* the picture because the task as such is deeply visual. In contrast to this, where do we encounter pictures that are the result of mental reification? I suggest that many pictures employed in the context of scientific problem solving can be interpreted in just this way. The usefulness of sketches and diagrams alongside scientific texts cannot be denied and is striking in comparison to humanity subjects that are much less prone to employ diagrams etc. Diagrams are widely used in all areas of science, with the depicted ranging from abstract – often mathematical – to concrete. There is also plenty of evidence that the comprehension of texts is aided by diagrams and pictures (e.g. [7]). Unlike the above match task, the problems that need solving in scientific practice are only in exceptional cases problems that lend themselves to a purely visual strategy of tackling them. After introducing visual aids that can be interpreted as mental reifications, I shall show how these are important in dealing with – often very abstract and often very complex – scientific problem solving tasks.

[4] For a similar kind of example, see [15, (pp. 112–113)]. The problem to solve there is: Can you intersect a cube with a plane in such a way as to make a regular hexagon? The point there is that this problem can hardly be solved without visualisation.

4.3 Sketches in the Context of EERSs

The research context from which I draw my examples is that of extended extragalactic radio sources (EERSs). The first radio source was discovered in the 1940s [8] and a great deal of research effort has gone into radio sources since. They distinguish themselves from other celestial bodies by predominantly radiating at radio wavelengths, being exceedingly powerful and having large redshifts. They are outside our galaxy and tend to be very far away. The processes thought to lie at the root of these phenomena are mostly described with the means of high-energy astrophysics, e.g. the synchrotron process, and plasma physics. I shall omit here the discussion of observational evidence and the physics involved in radio sources because they can be found elsewhere [2,4] and are not necessary for the discussion to follow.

Although EERSs are very far away and rarely visible at optical wavelengths, it is possible to display observational information about them in the form of a *radio map*. This is due to their enormous size and brightness, and the map displays the brightness distribution of the source in question. In the same way as contour lines mark the different heights and slopes of a mountain range on a topographical map, the lines on a radio map show the variation in intensity of radiation within a radio source. Figure 4.3 is an example of such a contour map. Guided by such a map – which is a systematic display of data using a known imaging technique – one can now *draw a sketch*, as in Fig. 4.4, and outline the phenomenon of EERSs with its help, mostly avoiding technical detail and the corresponding empirical evidence. Notice that this sketch is a different form of pictorial display from the map. It resembles the map in a phenomenological sense, but is not a systematic display of data.

Let me now introduce some phenomenological features of EERSs, illustrated by the sketch in Fig. 4.4. In the centre of the radio source there appears a zone of high brightness at radio wavelengths, the *core*. This is thought to be the "powerhouse", the area where the enormous energies are produced not only to make EERSs visible over huge distances, but also to make them expand to large sizes into space at mildly relativistic speeds. From the core, two *lobes* extend out in opposite directions. The lobes seem to be formed by two *jets* going out from the core and pointing towards the outermost edges of the lobes. The jets consist of particle streams, a continuous outflow of mass from the centre. Where the jets of the expanding source hit the intergalactic medium one finds more zones of enhanced brightness, the *hotspots*. Because the pressure of the ambient matter and the pressure caused by the source's expansion resist the jets, the particles from the jets are pushed towards the side and backwards where they encounter the intergalactic medium. Thus the plasma stream that originated from the core and moved outwards is diverted, pushed back and finally forms the lobes. The lobes are areas of reduced brightness that seem to contain older material.

The rough sketch I have drawn in Fig. 4.4 is a bit like the pictures radio astronomers may draw when they talk about EERSs, as I will illustrate in a

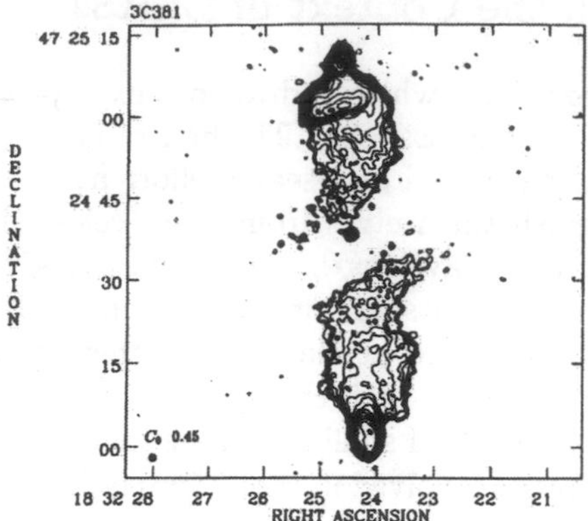

Figure 4.3. Map of 3C 381 (Leahy and Perley ©1991. Reprinted from *VLA images of 23 extragalactic radio sources*, Patrick Leahy and Rick Perley, Astronomical Journal, volume 102, pp. 537–561, American Astronomical Society, with kind permission.) Declination and right ascension are the coordinates of the object in the sky. The brightness of the object increases as one goes across the contours from the "outside" to the "inside".

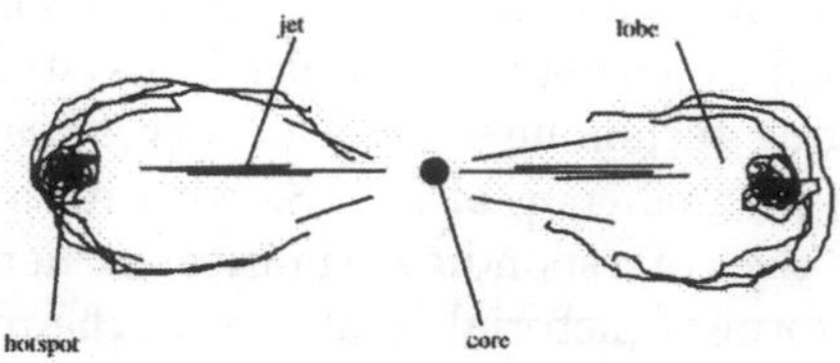

Figure 4.4. Simple sketch of the model of an extended radio source.

moment. Note that the natural phenomena transformed into a picture here do not look anything like the picture – nor the map, for that matter. Many of the radio sources are not visible to the naked eye, because they hardly radiate at optical wavelengths and are too far away. Moreover, neither sketches nor maps reflect anything of the enormous sizes of EERSs. Lastly, the radio maps display brightness distributions at *one* frequency, so the same object can appear quite differently at a different frequency. Both forms of display are somewhat abstracted, the map with regard to the choice of data displayed and the sketch with regard to the features of the phenomenon that are highlighted and displayed. Incidentally, it is not just EERSs as a whole that are drawn. Different theoretical aspects, all part of the overall phenomenon, can be drawn in a symbolic, representative manner; the resulting sketches pick out a specific aspect, or a specific theoretical perspective, as if with a magnifying glass. I call such theoretical treatments of the phenomenon that only refer to aspects of the empirical phenomenon *sub-models* [4]. In either case, that of sub-models or that of the overall model, the content of the pictures is not as concrete as it appears, because theoretical information is encoded in the pictures, employing a diverse range of symbols and conventions.

It is now time to turn to some evidence for my proposal that theoretical considerations about EERSs are *imported* into sketches of EERSs and that theoretical information is *transformed*, *compressed* and *reified* for easier handling. My evidence stems from sketches which my physics supervisor, Professor Malcolm Longair, drew during supervisions while I was doing research for an M.Phil. at the Cavendish Laboratory in the University of Cambridge [1].[5] The sketches served educational purposes, teaching me about EERSs, but also provided visual aids in discussions of theoretical issues concerning my research. The important point is that the sketches were drawn in a physics and a research context providing some evidence of the practice of thinking about EERSs. I shall now describe and analyse four sheets of such sketches, (A), (B), (C) and (D), which serve as case studies of the use of diagrams and pictures in teaching and research in contemporary astrophysics.

4.3.1 The Case Studies

Sheet A in Fig. 4.5 accompanied a discussion of the empirical finding of EERSs that sources of lower radio luminosity are edge-darkened, i.e. have their hotspots towards the core rather than the edge of the lobes. For some unknown reason, this class of sources generally occurs for radio luminosities with power at $P < 5\times10^{25}$ W Hz $^{-1}$ at 178 MHz. This class of sources is called FR I, after the finders of this correlation, Fanaroff and Riley (1974), while the edge-brightened sources with luminosities above the mentioned limit are called FR II. In sheet A, such an FR I source is sketched on the left towards the bottom. Above it, at the top of the page, there is a sketch of an edge-brightened FR II source. The diagram at the bottom of the page shows a distribution of EERSs according to their radio luminosities. Arrows from the sketches of the two types of sources point to the diagram at the bottom to show which part of the diagram, the luminosity-dependent distribution, is relevant for which type of source. The rest of the page is filled with the formulae for flux density and luminosity.

It should be noted here that two types of diagrammatic representation, phenomenological sketches of EERSs and a diagram of distribution of number per luminosity, are employed together. This indicates that the different forms of display, at different levels of abstraction, complement each other. Both types of pictures are beneficial, if not indispensable, for the understanding of findings about EERSs. The drawing testifies to the fact that these different types of pictures need to be considered *together* and in their

[5] I most gratefully acknowledge Malcolm Longair's permission to use his sketches for this paper. By chance I rediscovered them in a folder among my own physics notes and found, perhaps not surprisingly, how much they bear evidence of what my own experience is and what I argue for in this paper.

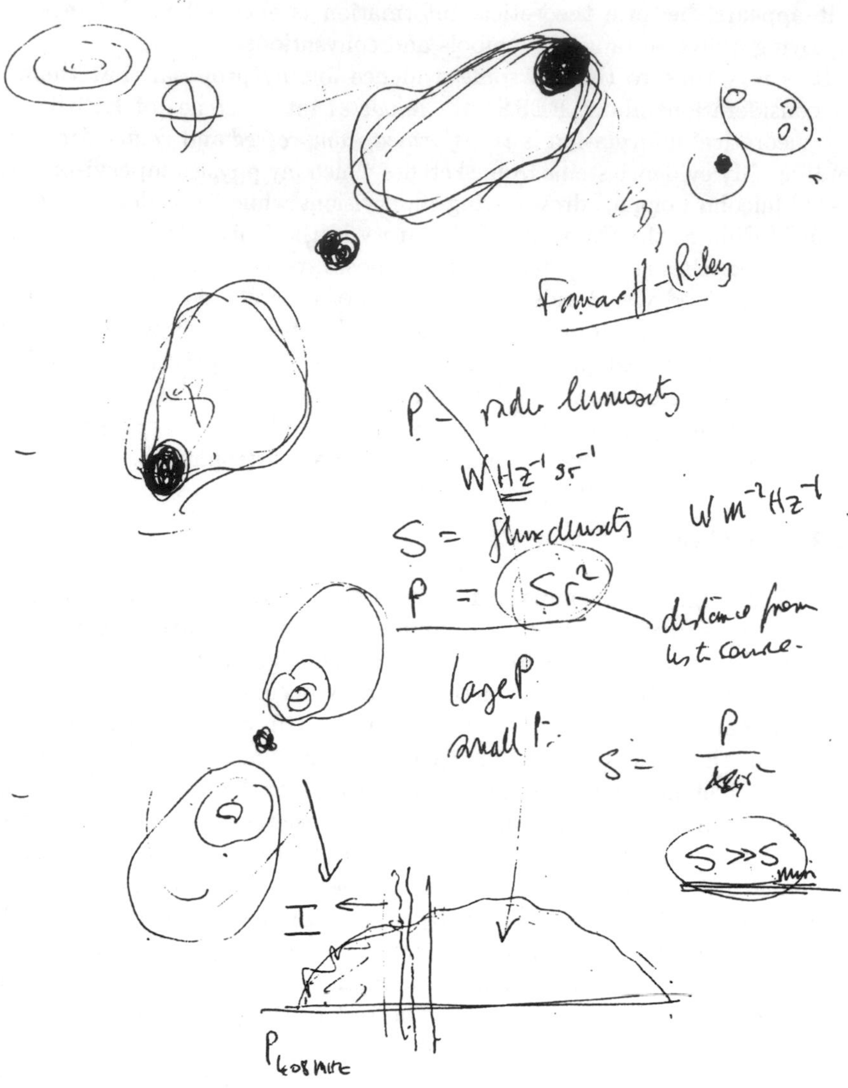

Figure 4.5. Sheet A: two types of EERSs, Fanaroff-Riley class I and II (©Malcolm Longair. Reprinted with kind permission).

relation to each other: the arrows on sheet A express this very explicitly. They indicate that the information of one form of display (the phenomenological) can be *imported* into the other (the number distribution). In other words, additional information is contained in the fact that one picture is displayed as importable into the other. While the main relationships displayed are between the various forms of pictorial display, the formulae and quantities that accompany the pictures closely function as a theoretical "commentary" of what the pictures are about.

Sheet B in Fig. 4.6 tells the story of the synchrotron process and the power-law spectrum of EERSs. EERSs have a spectrum that differs in shape significantly from spectra of many other celestial objects, such as stars. The cause for the power-law spectra, as they are observed in EERSs, is thought to be synchrotron radiation produced by electrons moving at relativistic speeds in a magnetic field. The various diagrams of sheet B belong in this theoretical context. Accelerated charged particles radiate perpendicularly to the direction of the acceleration. This phenomenon is well known as dipole radiation, as it is found in any ordinary antenna. Reading the sheet from top to bottom, the story begins with the general case of radiation of a non-relativistic, but accelerated electron. This is sketched in the top left-hand corner of sheet B in Fig. 4.6. The middle of the sheet shows a charged particle moving on a curved path. The significance of the curved path is that acceleration is required to change the direction of a particle along the path; this acceleration is perpendicular to the direction in which the particle moves. Thus, any charged particle that moves on a curved path radiates. This is shown in the middle of the sheet where the dipole from the top of the page reappears on a bending line that represents the curved path. The sketch at the bottom of sheet B concerns the relativistic case when electrons, as in the case of EERSs, are accelerated to relativistic speeds whereby the solid angle decreases into which any individual electron radiates. This is due to length contraction which distorts, or rather narrows, the angles at which a charged particle radiates. The radiation is "beamed" into a cone in the direction in which the electron moves.

The three consecutive pictures of sheet B represent and explain the theoretical details relevant for illustrating the synchrotron process step by step. Gradually, more complicating factors are added, from an accelerated electron acting as a dipole to a moving electron being accelerated on a curved path and finally to the case of a relativistically accelerated electron. In the same way that relativistic terms can be inserted into classical equations, one picture is here inserted into the next. In a way, the various theoretical steps on the way to the appropriate synchrotron equation are mimicked by a sequence of sketches on the sheet – a sequence that has its continuation on the next sheet to be discussed, sheet C. As previously, the pictorial illustration of the physical process is accompanied, in the centre of the sheet, by the mathematical tools that concern the effect of relativistic beaming of radiation.

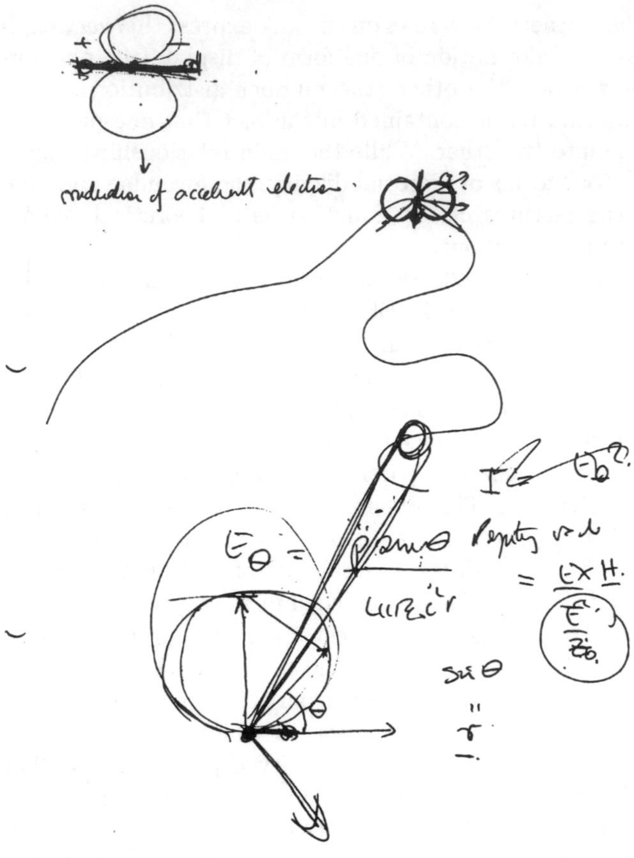

Figure 4.6. Sheet B: radiation of an accelerated electron (©Malcolm Longair. Reprinted with kind permission).

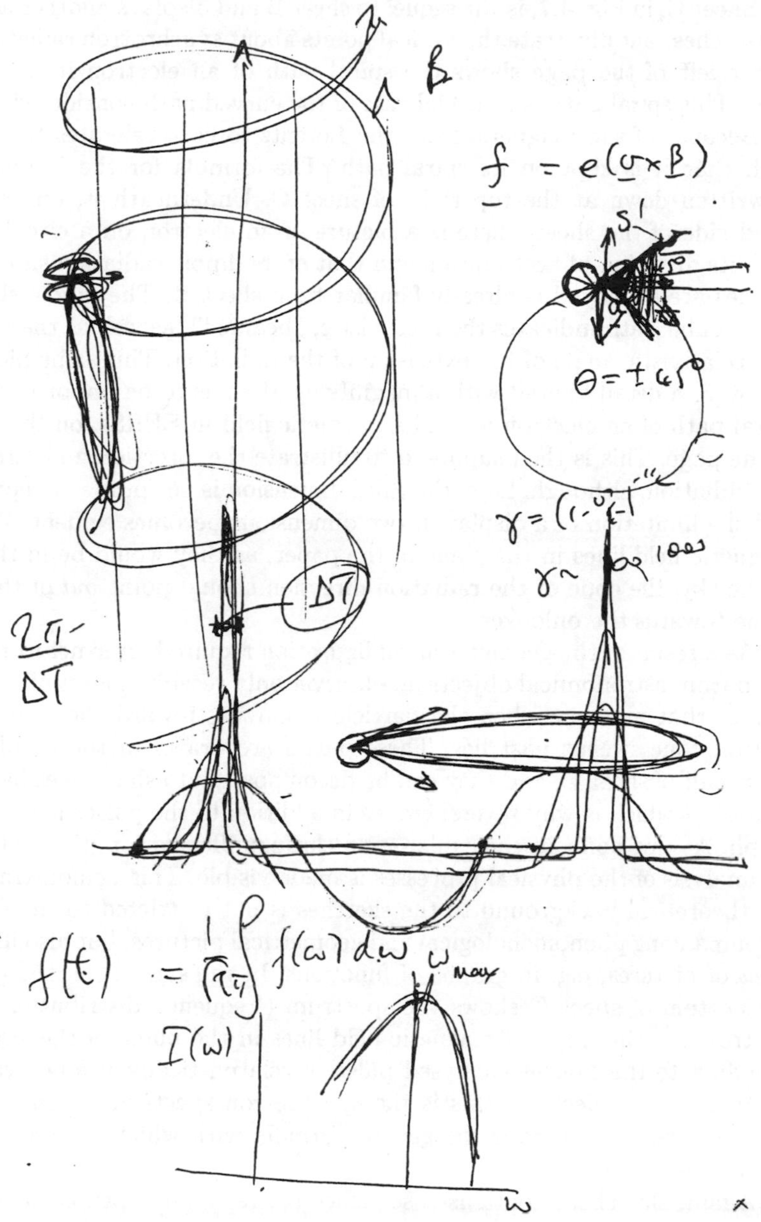

Figure 4.7. Sheet C: synchrotron radiation (©Malcolm Longair. Reprinted with kind permission).

Sheet C, in Fig. 4.7, is the sequel to sheet B and displays another assembly of sketches that illustrate theoretical points about synchrotron radiation. The upper left of the page shows the spiral path of an electron in a magnetic field. This spiral path is a special case of the curved path considered on sheet B. Because of the magnetic field, the Lorentz force accelerates the particle such that it is kept on its spiral path. The formula for the Lorentz force is written down at the top right of sheet C. Underneath it, on the right-hand side of the sheet, there is a picture of an electron on a circular path, with its direction of acceleration and that of its dipole radiation shown. How dipoles are displayed is already familiar from sheet B. The dipole sketch on the circular path indicates the relativistic, "beamed", as well as the ordinary, non-relativistic angle of the extension of the radiation. This same picture, as if it were a detail viewed with a magnifying glass, is to be imported into the spiral path of an electron around a magnetic field in EERSs, on the left side of the page. This is then supposed to illustrate the direction and strength of the radiation, although, here, the third dimension is not properly represented and the limitation of a display in two dimensions becomes evident. With the magnetic field lines in the plane of the paper, as they would be in the plane of the sky, the cone of the radiation direction should point *out* of the paper plane towards the onlooker.

As a result of the geometrical configuration required for synchrotron radiation from astronomical objects an observer only receives *pulses* of radiation, just in that moment when the particle is moving towards her and the radiation cone sweeps past her. These pulses are drawn in the graph in the lower half of sheet C, and they can be decomposed into sine waves by Fourier analysis – which is why several curves in addition to the pulses feature in the graph. At this point in the explanation through illustration, the mathematical analysis of the physical processes is made visible. This demonstrates that the theoretical background to the sketches is not restricted to the formulae accompanying phenomenological and geometrical pictures, but also in certain *types* of pictures, e.g. in graphs of functions. In the same vein, the graph at the bottom of sheet C shows the spectrum (frequency distribution) of *one* electron gyrating around magnetic field lines in the plane of the sky, corresponding to the Fourier-analysed pulse of relativistically beamed radiation emitted by one electron; this is the synchrotron spectrum of *one* electron. Directly above it, there is the general formula with which such a spectrum can be calculated.

Again, this sheet contains a sequence of steps, all captured in pictorial sketches, which tells a story beginning at the top right with the radiation mechanism, continuing with the geometrical configuration at the top left and moving down the page to the display of the spectrum based on the mathematical analysis of the pulses of emission resulting from the geometrical configuration.

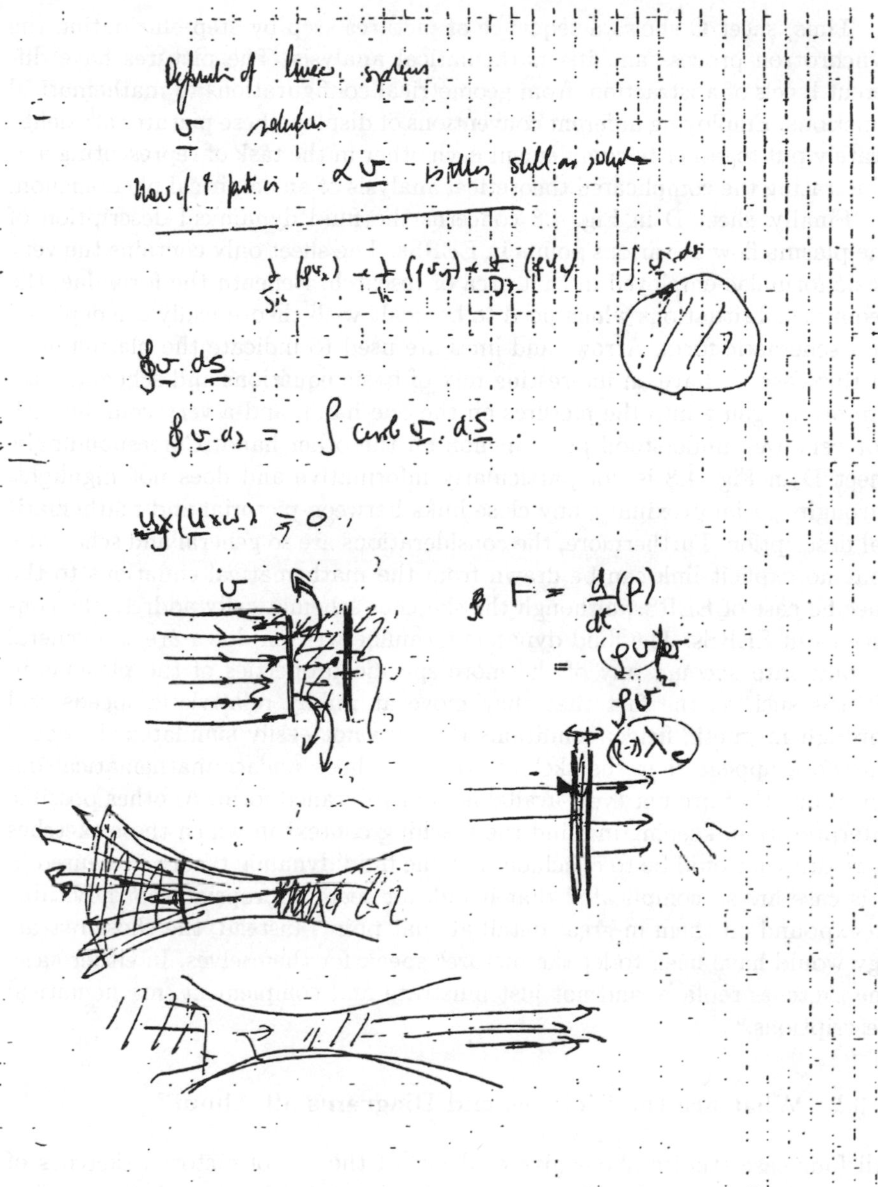

Figure 4.8. Sheet D: fluid dynamical issues in EERSs (©Malcolm Longair. Reprinted with kind permission).

Thus, sheet C shows a sequence of pictures step by step elucidating the synchrotron process and its mathematical analysis. The pictures have different levels of abstraction, from geometrical configurations to mathematical functions. Employing different conventions of display, these pictures are deliberately put together to complement each other in the task of representing and illustrating the complicated theoretical analysis of an empirical phenomenon.

Finally, sheet D in Fig. 4.8 concerns the fluid dynamical description of the plasma flow at various points in EERSs. The sheet only contains the very basic formulae employed in that area of research. Beneath the formulae, the geometrical situations which need to be dealt with theoretically are depicted in a schematic form. Arrows and lines are used to indicate the plasma flow. In this case we have an interesting mix of basic equations and schematic assumptions going into the pictures on the one hand, and a very complicated, not very well understood phenomenon on the other hand. Correspondingly, sheet D in Fig. 4.8 is not particularly informative and does not highlight, through spatial proximity, any close links between pictorial and mathematical description. Furthermore, the considerations are so general and schematic that no explicit link can be drawn from the mathematical equations to the specific case of EERSs, although the sketches schematically address the conditions in EERSs. The fluid dynamic formulae written down are too general to take into account any of the more specific properties of the plasmas in EERSs such as the fact that they move at mildly relativistic speeds and through magnetic fields, conditions that are not easily simulated. It would therefore appear that the sketches on this sheet *replace* mathematical descriptions that are not (yet) available in any detailed form. Another possible interpretation, keeping in mind the teaching context in which these sketches were drawn, would be to conclude that the fluid dynamic theories required in this case are so complicated that it did not seem appropriate nor advisable to expound on them in great detail at that point. Instead, the chosen strategy would have been to let the pictures speak for themselves. In either case, the sketches replace, and not just illustrate and complement, mathematical descriptions.[6]

4.3.2 What are the Pictures and Diagrams all About?

All four case studies above give evidence of the use of pictorial sketches of various degrees of abstraction that are accompanied by the relevant mathematical description of the issue at stake. Sometimes the connection between the pictures and the accompanying formulae is quite loose, e.g. on sheet D, because the theoretical treatment (of the fluid dynamic processes) is rather general and hand-waving. In other cases, such as on sheet C, the mathematical

[6] This can be compared to the case of adaptive landscapes in evolutionary biology for which Michael Ruse [14] argues that pictures allow evolutionary biologists to avoid the complicated mathematics which very few of them could understand.

expressions are closely intertwined with the pictures because the theoretical treatment (of the synchrotron process) is established and worked out in a considerable amount of detail. The close proximity of sketches and mathematical expressions indicates that they may unfold their meaning better jointly than in isolation. This implies that the pictures and diagrams support the theoretical treatment in such a way that they are not dispensable. It seems that it is the combination of pictures and theoretical treatment that works well when scientists have to think about and communicate about empirical phenomena.

The sheets of sketches also show that the pictorial displays have various degrees of abstraction which *together*, and despite their differences in type and abstraction, contribute to the treatment and understanding of a phenomenon. Frequently information from one form of display is *imported* into another. For this, arrows are used, as on sheet A, or the various pictures are arranged in a sequence which gradually introduces more detail or more complexity of the scientific treatment, as on sheets B and C.

On occasion – like on sheet D – the fluid dynamic example, sketches and pictures become important in cases where the theoretical description has not been worked out for the specific, concrete empirical case. Then pictures can be seen to practically replace theoretical treatment; pictures then do not just *stand for* theoretical treatment, but more or less *substitute* it. A fully-fledged theoretical treatment may not exist, or pictures may be preferred due to their convenience in representing theoretical treatment. In either case, sketches function as an abbreviation in visualised form of a whole range of considerations and theoretical information (known or unknown) concerning a phenomenon. This kind of abbreviation is not restricted to being used when things are written down, e.g. illustrating a scientific text or accompanying a scientific discussion on a blackboard or a piece of paper. Scientists can hold these abbreviating pictures in their minds – in the sense of visualisation discussed in the introduction – when they think about issues related to the scientific treatment of a phenomenon.

To create various types of pictures, geometrical (e.g. on sheets C and D), phenomenological (e.g. in Fig. 4 and on sheet A), or diagrammatic (e.g. on sheets B and C), scientists are left to their creative intuition and their shared use of conventions and symbols. Important is not what specific pictures are chosen in individual cases of scientific treatment, but *that* theoretical treatment is put into pictures, i.e. is reified. Clearly, no picture can explicitly contain every detail of information about the phenomenon for which it stands. Often, using pictures enables people to get around having to deal with mathematical complexities, at least temporarily. Pictures are concrete and therefore easy-to-use placeholders for more complicated and abstract considerations. Such enhanced accessibility gained through pictorial display may benefit various groups of people. One group are those who are unable to understand the complex mathematics behind some aspects of the treatment of EERSs. For them, the symbols and conventions used in the picture will aid the communication about basic ideas concerning the theoretical treatment of EERSs. Another group of people are those who are, in principle, capable of grasping the formalism, but may not bother in the specific situation, in order to save time and effort. Consequently, some of the pictures introduced here can be useful to researchers who conduct research on only a small aspect of the overall phenomenon of EERSs. Aspects of the phenomenon that are only marginal to their research interest can be considered in the accessible form of a picture that stands for a corresponding theoretical treatment. This grants the possibility to acquire a *working picture* of EERSs through sketches, diagrams and pictures, without having to grasp the mathematical subtleties of all aspects of scientific treatment of EERSs in full detail.

This means, from a pragmatic point of view, that mental reifications with their easy accessibility facilitate specialised research efforts, because researchers can concentrate on individual issues while having the pictures, as a form of compressed information, to guide them and inform them about the wider context into which their specialised research belongs. Thus, researchers of different fields can collaborate beyond their individual areas of specialisation. It then becomes possible to share out the research task between specialists in different areas and yet to maintain connections. Pictorial sketches, as I have discussed them in this paper, provide researchers with channels of communication through which they can "see" that (and how) they are all working on the same project: on different aspects of a single empirical phenomenon. Such a strategy, underlying empirical research and facilitated by the use of pictures and diagrams, is economical and guarantees an efficient use of people's skills and capabilities. Viewed in this manner, one can easily argue for a *practical* or *empirical necessity* of pictures for the research process (cf. [14], p. 73). Pictorial sketches, as mental reifications, save time, because they are something about the phenomena and processes for which they stand that can be grasped easily.

The idea of sharing out the research tasks can also be applied to an individual researcher. One could hypothesise that one researcher keeps various theoretical considerations in mind in the form of pictures, while she concentrates on some other, very specific theoretical problem in much detail. Although she may be in full possession of the theoretical knowledge concerning the various aspects represented by the pictures, the use of such pictorial "abbreviations" may save mental capacity momentarily (cf. [13], pp. 24–25). Whatever the precise level of knowledge about theoretical treatment, the practice of mental reification means that the mind finds itself shortcuts: in some cases to avoid complexity altogether, in other cases to suspend it temporarily. Through the procedure of employing a visualised representation – a mental reification of something abstract and complicated – mental capacity – is freed for dealing with a specific issue at stake, while more distantly related problems can be kept in mind, in the background and with little effort.

4.4 Conclusions

A problem, such as the match task in Section 4.2, could not be solved without some form of visual representation because of its very nature: the problem is itself formulated in terms of something physical, i.e. matches, or at least a picture of matches. Things are not as clear-cut for problem solving in the scientific context of EERSs. There, the visual pictures, as reifications of theoretical treatment, seem added and arbitrary in the sense that they rely on conventions and accepted symbols. It is therefore not self- evident that such pictures are necessarily required for problem solving, i.e. for scientific research concerning EERSs. Nonetheless, I suggest that the problem solving process is substantially aided by these pictures, taking my evidence from the research practice of science. Wanting to draw an analogy to the need for visualisation and pictures in the match task, one could say that the nature of theorising about a very complex phenomenon appears to be such that in practice it cannot be done without pictures and sketches that abbreviate aspects of the theoretical treatment.

That this is so is hard to prove, even if it seems plausible. There is one kind of problem in scientific theorising, however, that can be solved well through pictures, and this is to bring together information from a range of specialised areas of research and from different levels of abstractness of treatment. In other words, if all this information is depicted in terms of various types of two-dimensional pictures, different pieces of information can be *imported* into others, as found, for instance, on sheet A. Pictures can be brought to overlap [3] in order to illustrate how they all belong to one and the same phenomenon, or they can be inserted into others, like on sheet C. Unifying, or at least bringing together the diversity of research into different aspects of an empirical phenomenon, is the kind of task that can be achieved well by means of pictures and diagrams that are mental reifications of theoretical treatment. So, while the research task in scientific theorising may not be visual in the first instance, pictorial and diagrammatic forms of representation are developed commonly to assist the organisation of the task in the human mind. As in the solution to the match task discussed in Section 4.2, if the pictures are to bear fruit, those who use them need to do so creatively to see more in them than meets the eye. In the case of modelling aspects of EERSs, this is their unity and "importability".

Whenever the use of pictures and diagrams in science is discussed, an underlying issue seems to be whether or not diagrams are crucial and indispensable for doing science (e.g. [6]), in the spirit of Larkin's and Simon's [10] catchy title "Why a diagram is (sometimes) worth ten thousand words". Of course, there is a pragmatic argument: the fact that scientists use sketches and diagrams frequently and without hesitation suggests their importance for doing science. On the other hand, pictures and diagrams tend not to be used in isolation. Importantly, on the sheets of my case studies, pictorial sketches and the mathematical tools of astrophysics are put on paper in close spatial proximity. They are closely connected with each other in written-down material, such as the notes and illustrations accompanying oral discussions which I analysed. It seems therefore reasonable to conclude that *together* mathematical tools and pictures thereof are centrally relevant to the theoretical treatment of scientific phenomena, such as EERSs. Helpfully, in certain situations, pictures and the theoretical treatment for which they stand are interchangeable. As an abbreviation to theoretical details, sketches are likely to make it easier to grasp a theoretical idea because, aided by a picture, the theoretical information depicted begins to appear more concrete and more familiar. Perhaps, such a link to our ordinary experience is rather tenuous, but it would explain why reification of theoretical treatment is so important for scientific theorising and therefore so widespread.

References

1. Bailer, D.M. (1993). The dynamics of extragalactic radio sources. M.Phil. Dissertation, University of Cambridge.
2. Bailer-Jones, D.M. (1997). Scientific models: A cognitive approach with an application in astrophysics. Ph.D. thesis, University of Cambridge.
3. Bailer-Jones, D.M. (1999). Creative strategies employed in modelling: A case study. Foundations of Science 4:375–388.
4. Bailer-Jones, D.M. (2000). Modelling extended extragalactic radio sources. Studies in History and Philosophy of Modern Physics 31B:49–74.
5. Best, P.N., Bailer, D.M., Longair, M.S. and Riley, J.M. (1995). Radio source asymmetries and unified schemes. Monthly Notices of the Royal Astronomical Society 275:1171–1184.
6. Giere, R. (1996). Visual models and scientific judgment. In: B.S. Baigrie (Ed.), Scientific illustration: Historical and philosophical problems concerning the interaction between art and science. Toronto: University of Toronto Press, pp. 269–302.
7. Glenberg, A.M. and Langston, W.E. (1992). Comprehension of illustrated text: Pictures help to build mental models. Journal of Memory and Language 31:129–151.
8. Hey, J.S., Parson, S.J. and Phillips, J.W. (1946). Fluctuations in cosmic radiation at radio frequencies. Nature 158:234.
9. Katona, G. (1940). Organizing and memorizing: Studies in the psychology of learning and teaching. New York: Columbia University Press.
10. Larkin, J.H. and Simon, H.A. (1987). Why a diagram is (sometimes) worth ten thousand words. Cognitive Science 11:65–99.
11. Leahy, J.P. and Perley, R.A. (1991). VLA images of 23 extragalactic radio sources. Astronomical Journal 102:537–561.
12. Miller, A. (1986). Imagery in scientific thought. Cambridge, MA: MIT Press.
13. Nersessian, N.J. (1992). How do scientists think? Capturing the dynamics of conceptual change in science. In R. Giere (Ed.), Cognitive models of science. Minnesota studies in the philosophy of science. Vol. XV. Minneapolis: University of Minnesota Press, pp. 3–44.
14. Ruse, M. (1991). Are pictures really necessary? The case of Sewell Wright's "adaptive landscapes". In A. Fine, M. Forbes and L. Wessels (Eds), PSA 1990, Vol. 2. East Lansing, MI: Philosophy of Science Association, pp. 63–77.

15. Wimsatt, W.C. (1991). Taming the dimensions: Visualizations in science. In A. Fine, M. Forbes and L. Wessels (Eds), PSA 1990. East Lansing, MI: Philosophy of Science Association, pp. 111–135.

5. The Fundamental Design Variables of Diagramming

Clive Richards

A terminology for discussing diagrams is offered. In part this terminology arises from the use of a grammatical analogy to identify the key features of graphic displays which express relationships. In this chapter a three-dimensional taxonomic model is presented which shows the relationship between the fundamental design variables available for diagramming. This spatial model may be used as a conceptual tool for diagram designers.

5.1 Introduction

Diagrams are a very ancient form of visual expression. They are used as tools for thought [1, 9], aids to memory [19] and as a means of instruction [8]. As pointed out elsewhere [14], diagrams, in the broadest sense, are going to play an increasingly important role in the so-called information revolution, providing valuable information resources for both learners and experts alike. This is especially so now that diagrammatic presentations can be interactive [15], instantly updated and potentially available across networks to users anywhere [12].

These technological developments have recently stimulated an interest in diagramming, a topic which previously has been somewhat neglected by researchers.[1] This chapter describes an analytical framework developed within the context of the design of diagrammatic materials for the printed page. However, it may well provide a useful basis for further work in the newer digital arena. One can expect some principles of presentation to be selectively transferable from the old to the new domain, for even though this new domain is in many respects quite different there are similarities. One obvious example is that despite being time-based and interactive multimedia make much use of static imagery.

Consider the proposition that pictorial illustrations show physical appearances, symbols indicate a presence or act as pointers and diagrams exhibit

[1] The lack of, and need for research into diagramming was noted by McDonald-Ross and Smith [10] in the section on "Scientific and mathematical diagrams" of their monograph (p. 30).

relationships.[2] Even accepting this broad categorisation distinctions between diagrams and non-diagrams are not always easy to maintain. One often finds hybrid forms which apparently function in part as illustrations, in part as diagrams, or even in part as symbols. As well as having symbols embedded within them, whole diagrams can sometimes function as single symbols. For example, when the Tube map is printed on a plastic carrier bag, or teeshirt, it becomes a symbol for tourist London. Notwithstanding the assertion that there can be no clear-cut boundary to the category we call "diagram", this chapter will nevertheless:

1. offer a terminology for discussing diagrams (these terms are identified in the text by the use of *italics* at the point where they are explained);[3]
2. present a taxonomic model of the fundamental design variables available for diagramming.[4]

Developing this taxonomy has involved the examination of a large sample of graphic displays from diverse application areas and the detailed analysis of selected specimens [13].

5.2 Grammatical Analogy

The terminology proposed here in part arises from the use of a grammatical analogy to identify the key features of diagrams. This approach has the merit of being applicable to some kinds of figurative illustrations as well as highly

[2] From the etymology of 'diagram' given in the Universal English Dictionary (Wyld, n.d.) we see it has Greek roots. Dia is a prefix meaning "through, throughout (of place and time); through the agency of" and gram is a suffix meaning "something written, a letter". Dia in this context presumably means one thing standing for another (through the agency of), and the suffix gram implies marks on a surface.

A useful definition of "diagram" by James Clark Maxwell may be found in the 1910 edition of the Encyclopedia Britannica: "... to mark out by lines, a figure drawn in such a manner that the geometrical relations between the parts of the figure illustrate relations between other objects. They may be classed according to the manner in which they are intended to be used, and also according to the kind of analogy which we recognise between the diagram and the thing represented..."

The following Maxwell-inspired definition is offered: "A diagram is primarily a graphic display which depicts spatial relations. The spatial relations depicted in a diagram may represent other spatial relations in some literal way, or they may represent non-spatial relations by means of graphic metaphor" [15, (pp. 10–17)].

[3] So as not to interrupt the flow of presentation these terms are not always explained on their first use as it is assumed that in the context in which they initially appear they will not demand the level of specification given later.

[4] Aspects of this model were first introduced in [16], more fully developed in [13] and partly outlined in [14]. It has informed work on cinegrammatics [4] and recent work on diagram theory [3].

schematised diagrams. In any case it has already been suggested that diagrams should not be thought of as a completely distinct species. It is perhaps more appropriate to speak of diagrammatic tendencies in graphic displays, and of diagramming as a form of picture making for a particular class of exposition. It can be argued that the only tenable determinant for establishing the diagrammatic nature of a graphic display is its function; that is, its use in displaying relations. Its formal qualities, such as, say, being composed only of lines, are in a sense incidental. A high level of schematisation may serve to emphasise what is relevant but is not an essential quality.

What makes a diagram a diagram is the ability of users to recognise in it spatial relations which in some way correspond to the relationships represented.

The relational meaning of a diagram is taken from the arrangement of its elements, and in this respect it is akin to a sentence or text.[5] By varying the sequence of a sentence, but retaining the same words, we may alter its meaning. In the same way we can change the meaning of a diagram by varying the way its elements are arranged.

The possibility that diagrams can provide evidence of a link between spatial intuition and language competence has been noted by others (cf. [7] p. 8). Whether this is so or not, it does seem that the meaning of certain types of graphic display is governed by a form of visual grammar.[6] Such displays allow distinctions to be made between the various, apparently discontinuous, spaces which they depict. If a graphic display shows a single homogeneous space we may have what is normally called a pictorial illustration, for example a perspective view showing a landscape at a single moment in time. The more that discontinuous spaces seem to be detectable, the more likely we are to have a graphic display which tends to have a diagrammatic function. Fig. 5.1 appears to show multiple spaces representing various points in time.

The depictions of machines each occupy their own discrete perspective fields, while the tapering flowlines occupy another "flat" world. Together the components in these two kinds of space tell us "this machine precedes that machine" and "these machines are related".

The depictions of machines can be thought as having a noun-like function, while the tapering lines operate in a verb-like fashion. When reading the diagram we first have to work out the visual syntax. Although strongly mediated by conventions, it may be that this visual syntax, this (dia)grammar, is rooted in the way we perceive and picture the physical world of everyday experience.

---

[5] The notion of picturing as some form of language is common in the literature, cf. Twyman [18], who offers a useful "... schema for the study of graphic language", which distinguishes between "verbal graphic language" and "pictorial" or "schematic graphic language".

[6] Parallels between diagrammatic and linguistic forms even extend to the notion of rhetoric. The equivalents of hyperbole, metonymy and synecdoche, for example, are often used in diagramming.

Now it is important to keep clear the distinction between the situation apparently depicted by a diagram and what it is that situation is supposed to stand for.

For the situation depicted the term *content model* is used to distinguish it from the content proper, that is, the meaning we attach to the diagram.[7] For example, in Fig. 5.2 the diagram depicts a content model we take to be a tree. However, the content itself is concerned with some form of family relations. A content model is what we recognise in the marks on the paper, or other image on a plane surface, such as a computer screen. The term *graphic display* is used to refer to the physical presentation, if a distinction between it and the depicted model is needed. The content model is made up of various *significant elements*, the smallest signifying parts, which in the case of the tree example are its labels and branches.[8] Their depictions are the basic graphic components out of which the diagram is constructed.

A single significant element may contain several formal characteristics, each capable of having different relational meanings ascribed to it. These characteristics are termed *relational features*. For example, a group of significant elements may be connected by a line which changes width as it progresses from element to element. An increase in width might be interpreted as indicating some increase in status of the connected elements. In this case the line would have two *relational features*. One is the property of linking between the elements, and the other is some gradual change in importance. Following the grammatical analogy, such modifications of one feature by another can be thought of as adverbial or adjectival.

So significant elements may have a mixture of organisational modes through having more than one relational feature.

If, in the example just given, the line exhibited no variation along its length, it would have only the linking mode of organisation. It is also possible for a significant element to have two features which have essentially the same meaning and are thus complementary. This is sometimes described as

[7] The term "content model" is also used by Umberto Eco (cf. [2] p. 200).

[8] It is acknowledged that the term "significant element" represents a somewhat imprecise concept. It of course depends on the intentions of the designer and the compliance of the viewer as to whether a particular collection of marks represents a single or multiple significant elements. At this level the parallel between spoken and written language starts to break down. For picturing there is no single limited set of basic graphic marks, in the same way as there is a limited set of phonemes for speaking and a single set of alphabetic letters for writing.

In one picture a representation of a texture may be made up of many tiny ticks, and the presence or absence of one tick will be of no significance. In another picture the same tick may represent a bird in the distant sky and be a key feature of the pictorial story.

Notwithstanding this, others have proposed fully formalised picturing systems with such components as "pictemes" which are supposed to be broadly equivalent to phonemes.

redundant recoding.[9] For instance, as the bars of a histogram increase in length so they might increase in, say, intensity of redness. Here two sorts of variation reinforce each other.

This example also illustrates another useful distinction, that is, the one which may be drawn between ***intrinsic*** and ***extrinsic*** features.

The increasing intensity in redness is a feature intrinsic to the significant elements which are the bars of the histogram. On the other hand, extrinsic features are those which are, as it were, not an identifiable characteristic of individual elements, but are properties of groups. This includes all spatial arrangements such as grouping by proximity, or rankings established by variations in size, as in the histogram example.

So the significant elements functioning like nouns are those that represent the things about which the diagram purports to say something. Relational meaning can be inferred from both their intrinsic and extrinsic features. Such significant elements may also be organised by other significant elements functioning like verbs. The intrinsic and extrinsic features of these verb-like elements take their meaning from the relations they impose on the noun-like elements. Without the noun-like elements we do not have anything the diagram can be about. Without the relational features with verb-like functions no relationships are exhibited.

A graphic display then depicts a content model that may be analysed into various significant elements, which in turn may have one or more relational features that can be intrinsic or extrinsic. This is summarised in Fig. 5.3.

5.3 Design Variables

The arrangement of significant elements within a given content model determines the relational meaning it displays, and picturing relationships is what distinguishes diagrams from other forms of graphic display. These relationships may be spatial, in which case a diagram can represent them in a more-or-less direct manner. When the relationships are non-spatial, the diagram performs as a kind of metaphoric picture.

Fig. 5.4 shows two network diagrams, but each deals with entirely different domains. Fig. 5.4(a) is a wiring diagram and is essentially literal in its mode of correspondence to what it stands for. While not showing exactly what the wiring looks like, the topology is preserved. Fig. 5.4(b) on the other hand bears no physical resemblance at all to what it represents. Here the connecting lines are a graphic metaphor for ownership, showing which companies have a financial stake in each other.

Fig. 5.5(a) is literally about the heights at which aeroplanes have flown, while Fig. 5.5(b), using very similar signifying elements, is dealing with rises in cost of living, shown metaphorically by means of altitudes.

[9] The value of redundant recoding is discussed by Fitter and Green [5, (p. 252)].

As can be seen from the examples given, the spatial organisation within diagrams does not necessarily fall into clear-cut categories of literal or metaphoric (or non-literal). While the altitudes in Fig. 5.5(a) correspond proportionately to the heights actually flown, the horizontal axis has been rearranged to suit the confines of the diagram. Time has been transposed metaphorically into distance and runs from left to right.

One may think there is no literal correspondence between Fig. 5.6 and its content, nevertheless there does seem to be a sense in which the vertical sequence of captions as shown is more appropriate than any other arrangement might be. As in spoken language, there are degrees of metaphoricity: it can range from strong contrast to near similitude, where it is hardly a metaphor at all.[10] Often diagrams present mixtures of the literal and metaphoric.

The degree to which the content models of diagrams correspond to what they represent can be thought of as a continuum ranging from the *literal*, through the *semi-literal* to the *non-literal*, the semi- and non-literal employing graphic metaphor in varying degrees. This useful variable should not be confused with another important continuum available to the diagram designer: that of the degree to which elements may be depicted in a more-or-less figurative manner.[11] For example, neither of the structures shown in Fig. 5.7 has anything to do with trees. The graphic metaphor of branching is employed in both diagrams to present family relationships within linguistics. However, while Fig. 5.7(b) is highly schematised in its presentation, Fig. 5.7(a) is far more figurative, and like the tree in Fig. 5.2 shows bark and even leaves.

Figures 5.8(a) and 5.8(b) each show the same factory layout, both corresponding in a quite literal manner to the spatial layout of the actual machines. However, Fig. 5.8(a) is depicted in a highly figurative manner while Fig. 5.8(b) is not.

It may be noted that even in very highly schematised diagrams there may still be faintly figurative elements. For example, in Fig. 5.9, the curved flowlines may be perhaps more appropriate than if they had been angular, although both styles would express the idea of linking in some way. The curves suggest a pouring in, curves being a characteristic of fluids tipped from one vessel to another. Sometimes vestiges of resemblance can be difficult to completely remove, and indeed their deliberate use can often be helpful in prompting appropriate interpretations by readers.

When diagrams contain non-literal features words are normally needed to ensure that the reader is tuned in to the graphic metaphor being employed. Sometimes a diagram can be largely pictorial with supporting annotations. Sometimes it can be more like a text, with various routes through it being, as it were, signposted by graphical indicators. An example of the latter category

[10] An interesting analysis which uses a spatial model of aptness in metaphor is presented by Tourangeau and Sternberg [17].

[11] "... all representations can be arranged along a scale which extends from the schematic to the impressionistic" [6, (p. 247)].

is the non-figurative flowchart with captions used for the expression of rules. Diagrams occupy that hinterland between written text and the purely graphical. This is their great strength, enabling, often through the use of graphic metaphor, the visual presentation of the otherwise invisible.

There are then two key and independent design variables available to the diagram designer: the *mode of correspondence*, ranging from the literal to the non-literal, and the *mode of depiction*, ranging from the *figurative* to the *non-figurative*. These two variables may be represented as diametrically opposing scales as shown here in Fig. 5.10.

Notionally content models can be located within the two-dimensional space defined by this graph, depending on the modes of correspondence and depiction used. One finding of the analysis previously mentioned [13] was that the less literal and the less figurative diagrams are (i.e. the more they cluster in the top right-hand corner of the graph shown in Fig. 5.10) the more evident become the three fundamental organisational modes available for diagramming. These *modes of organisation* are:

- *grouping*; where elements imply belonging, e.g. by sharing a common colour, shape or enclosing boundaries
- *linking*; where elements display connectivity, e.g. by a means of a joining line
- *variation*; where elements suggest value, e.g. by changes in size, distance from a baseline, or intensities in colour.

These are exemplified in Fig. 5.11. Typifying examples of diagrams which deploy these modes of organisation in a more-or-less pure form are indicated in Table 5.1.

Table 5.1. Typifying examples of diagrams

Mode of graphic organisation	Usual diagrammatic interpretation	Some characteristic application	Typical diagram
Grouping	Association	Classification	Venn
Linking	Sequence	Process	Flow chart
Variation	Value	Ranking	Bar chart

Used in combination this apparently limited set of possibilities provides an extremely rich graphic vocabulary. They can be present in strong, weak or hybrid forms. In Fig. 5.12 all three modes of organisation are clearly in evidence. It is a combination of time chart, flow chart and Venn diagram (where the variation in size of the circles also has meaning). To summarise, content models will have their own modes of correspondence and depiction and the relational features of their significant elements will have various modes of organisation. Together the modes of correspondence, depiction and organisation may be referred to as the *modes of interpretation.*[12]

The possible relationships between these modes of interpretation are presented in the taxonomic model shown in Fig. 5.13.

The various categories displayed in this model should not be regarded as being discrete. Each should be thought of as merging into the next. It is only a matter of presentational convenience that this model is shown as three disks. A better way to think of it is as a column of some coloured medium which changes tone, from dark to light, in an upward direction with its hues becoming increasingly saturated towards its periphery. Three primary colours could represent the three modes of organisation, and all the mixtures this produces would represent all the possible relations between the various modes of interpretation. Governed by their classification, significant elements in a given content model will have particular locations (and consequently colours) in the classificatory space defined by the column.

Looking at Fig. 5.13 we see that, depending on the direction of schematisation, the man, the woman and the dog may be represented in terms of their groupings (one outline grouping the two on the dais, and one outline enclosing all three), their linking (between the woman and the dog), and their variation (in relative sizes). The depictions of the man, woman and dog stand on the literal plane and should, therefore, be taken literally. This same illustration, and its non-figurative derivatives, could have been on the bottom plane, in which case it would not literally represent a man, a woman and a dog. It might, for example, represent the economic or political relations between two countries and some satellite state.

5.4 Conclusion

This taxonomic model evolved through an iterative process involving audits of extant diagrams conducted as part of the research already referred to here [13]. As this taxonomic model shows the range of fundamental design variables available when designing diagrams it has also been proposed [13, (pp. 10–17)] that it may be useful to graphic designers. It could be used as a kind of check-list of the theoretical possibilities for generating a series of potential content models for a particular diagramming task.

Diagrams, as "pictures" of relationships, have features which in some sense seem to be grammatically organised. Thinking of diagrams in this way has led the development of a terminology for discussing diagrammatic characteristics and the production of a taxonomic model which may provide a useful conceptual tool for diagram designers and researchers.

[12] The categorisation of the modes of interpretation into organisation, depiction and correspondence has some equivalence to the semiotic rules of syntactics, semantics and pragmatics [11, (p. 35)].

Summary of Special Terms

Content model the situation depicted as distinct from its diagrammatic meaning.
Figurative high in pictorial detail.
Extrinsic feature formal characteristic belonging to a group rather than individual significant elements.
Graphic display the physical display as distinct from its content model.
Graphic metaphor the mechanism by which a depicted spatial arrangement stands for some non-spatial situation.
Grouping the mode of organisation which displays association.
Intrinsic feature formal characteristic identifiable within a significant element.
Linking the mode of organisation which displays connection.
Literal corresponding to what is represented in a direct spatial manner.
Modes of interpretation the modes of organisation, depiction and correspondence.
Modes of correspondence collectively the literal, semi-literal and non-literal modes.
Modes of depiction can range from the figurative to non-figurative.
Modes of organisation collectively linking, grouping and variation.
Non-figurative highly schematised image.
Non-literal corresponds to what is represented via graphic metaphor.
Semi-figurative between figurative and non-figurative.
Semi-literal between literal and non-literal.
Significant element the smallest meaningful component.
Relational feature any formal characteristic of a significant element which can imply a relationship.
Variation the mode of organisation which displays various values.

References

1. Albarn, K. and Smith, J.M. (1977). Diagram: An instrument of thought. London: Thames & Hudson.
2. Eco, U. (1977). A theory of semiotics. London: MacMillan.
3. Engelhardt, Y. (1998). Meaningful space: How graphics use space to convey information. In The Republic of Information proceedings of the Vision Plus 4 symposium, Carnegie Mellon University, pp. 108–126.
4. Fischer, D. (1997). A theory of presentation and its implications for the design of on-line technical documentation. PhD thesis, Coventry University.
5. Fitter, M. and Green, T.R.G. (1979). When do diagrams make good computer languages? International Journal of Man–Machine Studies 11(2):235–261.
6. Gombrich, E.H. (1977). Art and illusion. Oxford: Phaidon.
7. Hardin, P. (1981). Representational characteristics in diagrams of statements of relationships. PhD thesis, University of Iowa.
8. Lowe, R. (1993). Successful instructional diagrams. London: Kogan Page.

9. Macdonald-Ross, M. (1979). Scientific diagrams and the generation of plausible hypotheses: An essay in the history of ideas. Instructional Science 8:223–234.
10. Macdonald-Ross, M. and Smith, E. (1977). Graphics in text: A bibliography. IET monograph no. 6. Milton Keynes: Open University.
11. Morris, C. (1938). Foundation of the theory of signs. International encyclopedia of unified science, Vol. 1, No. 2. Chicago: University of Chicago Press.
12. Newman, R., Richards, C.J. and Fischer, D. (1997). On-line multimedia information for maintenance and operation (OMIMO), D06.3 final report, project IE2054, Coventry University.
13. Richards, C.J. (1984). Diagrammatics: An investigation aimed at providing a theoretical framework for studying diagrams and for establishing a taxonomy of their fundamental modes of graphic organisation. PhD thesis, Royal College of Art, London.
14. Richards, C.J. (1997). Getting the picture: Diagram design and the information revolution. Professorial Lectures 16, Coventry University.
15. Richards, C.J. and Fischer, D. (1994). Cinegrams: Interactive animated systems diagrams for technical documentation. In The use of IT in art and design. Technical report 26. Loughborough: Advisory Group on Computer Graphics (AGOCG).
16. Richards, C.J. and Smith, R.H. (1983). La diagrammaire (Cleve [sic] Richards, présentée par Roger Smith), Recontres Internationales de Lure: L'image Schematique, August 1983, Lurs-en-Provence, France, 26 pp.
17. Tourangeau, R. and Sternberg, R.J. (1981). Aptness in metaphor. Cognitive Psychology 13:27–55.
18. Twyman, M. (1982). The graphic presentation of language. Information Design Journal 3(1):2–22.
19. Yates, F.A. (1986). The art of memory. London: Ark.

Appendix

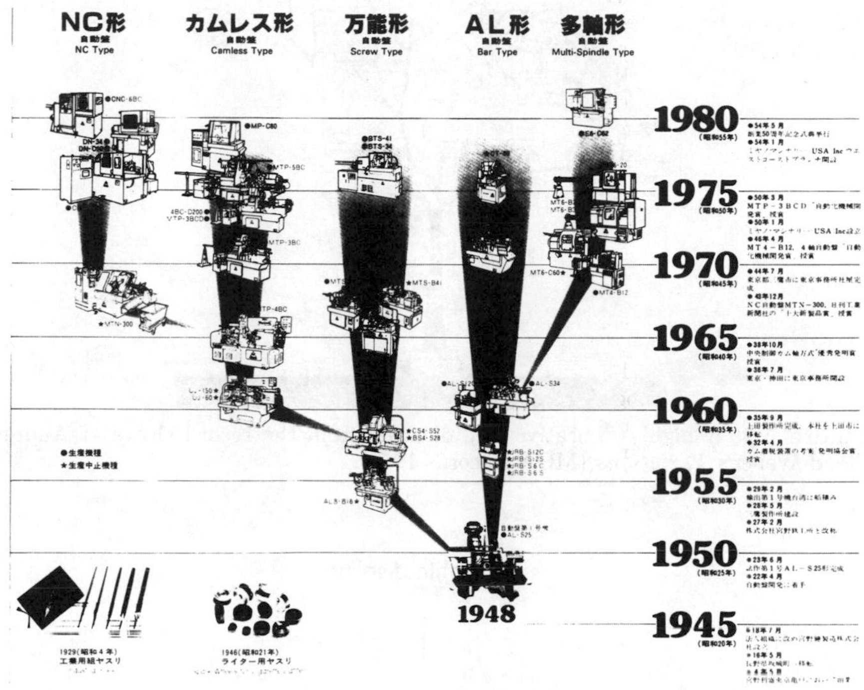

Figure 5.1. A hybrid time chart and flow chart with pictorial elements. From: *An outline of Miyano* (p. 6), a catalogue by Miyano Machinery, Japan.

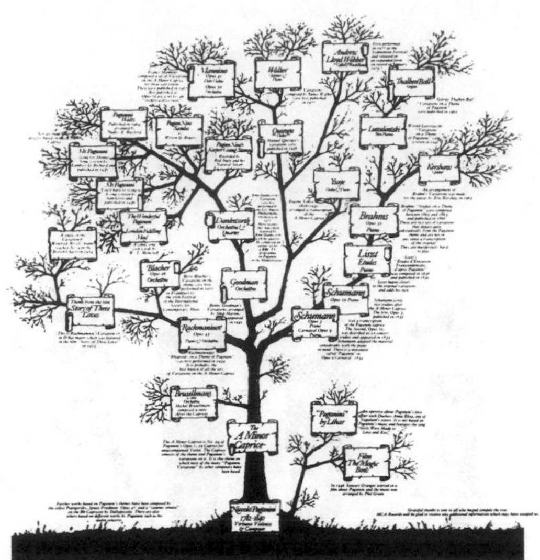

Figure 5.2. A highly figurative tree diagram from the record sleeve of Andrew Lloyd Weber's *Variations* (MCA Records 1978).

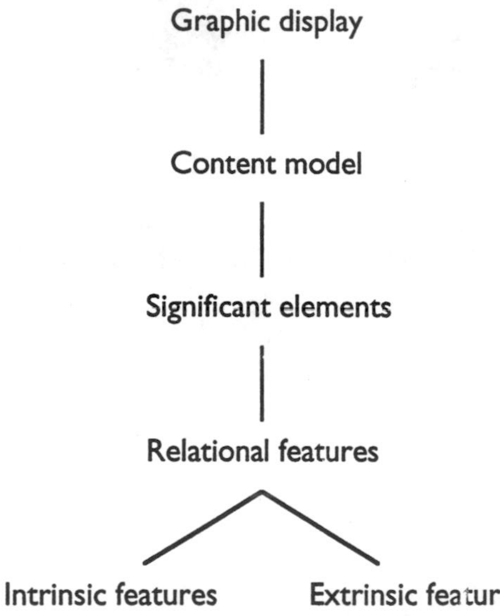

Figure 5.3. Diagrams can be analysed into various elements and features.

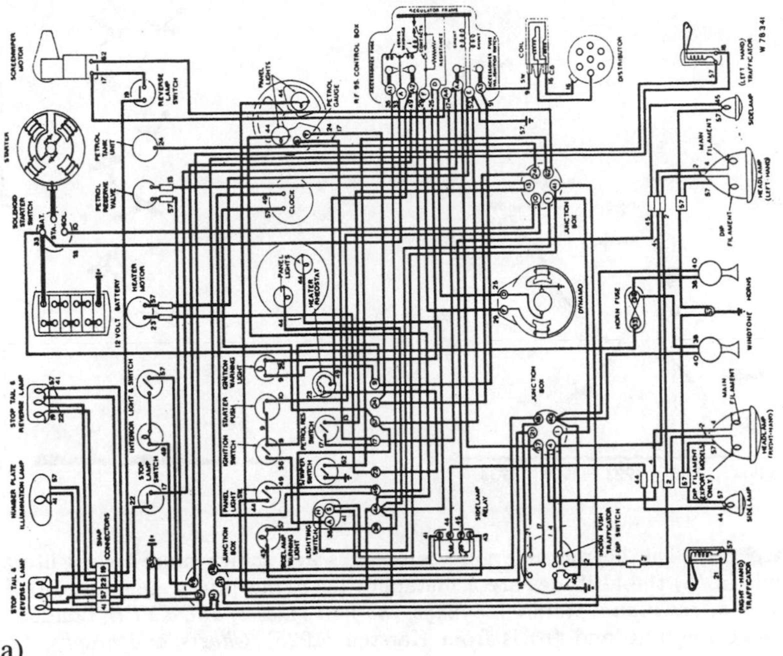

a)

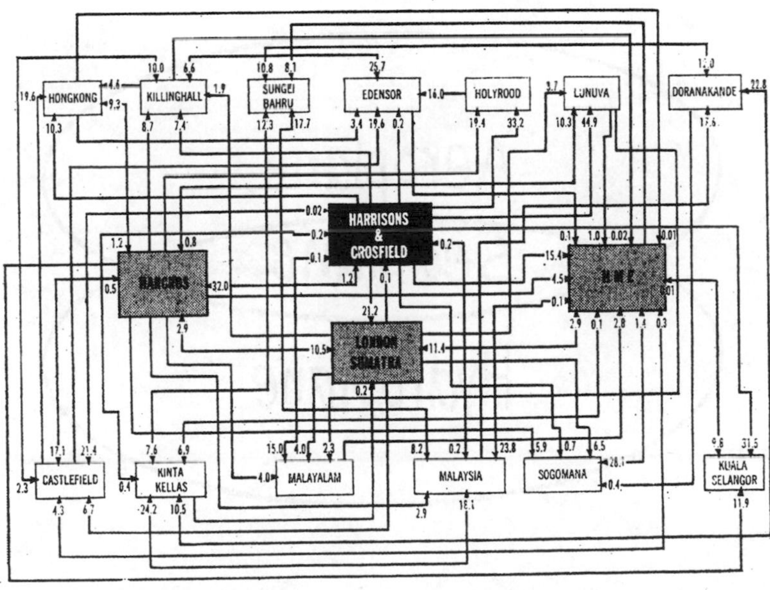

b)

Figure 5.4. The diagram (a) literally shows connecting wires, while (b) uses graphic metaphor to show ownership. Diagram (a) is from an Armstrong Siddley Motors service manual (1950). Diagram (b) is reprinted with permission from *The Sunday Times*, London © Times Newspapers Ltd., 29th January 1978.

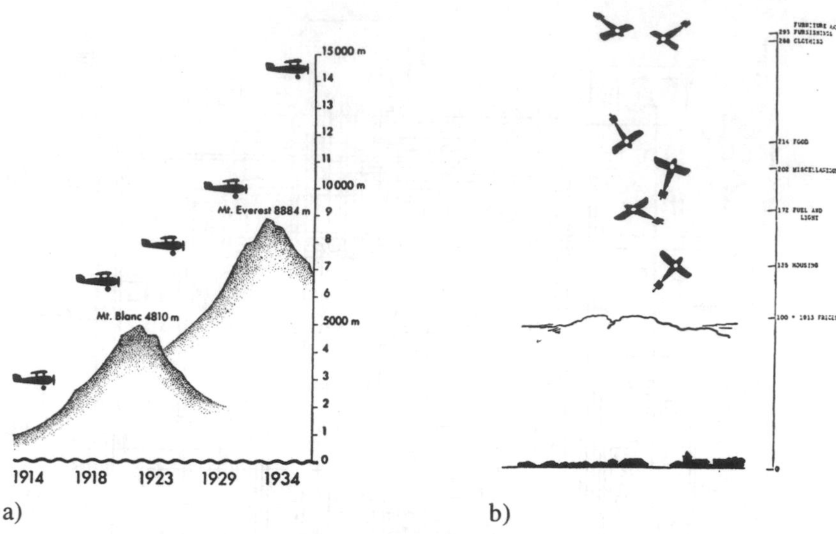

Figure 5.5. The heights flown by aeroplanes are shown more-or-less literally in (a), but in (b) the altitudes are a metaphor for rising prices. Diagram (a) is from Neurath *International picture language, the first rules of ISOTYPE*, London: Basic English Publishing and (b) is from Karsten (1925) *Charts and graphs*, London: Pitman.

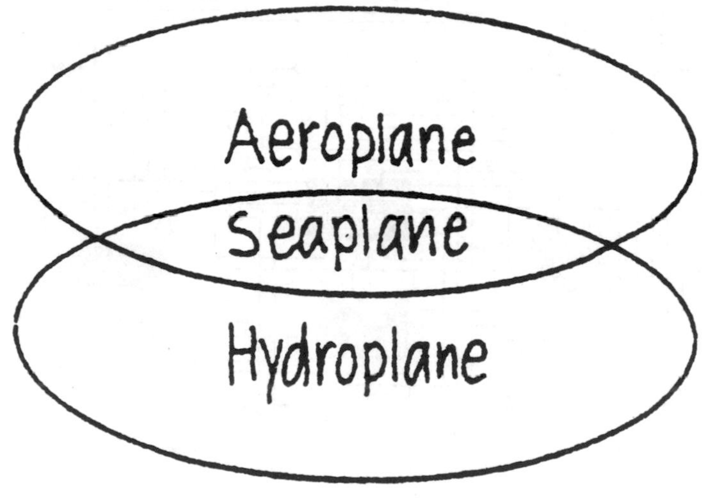

Figure 5.6. From [13] pp. 10–20.

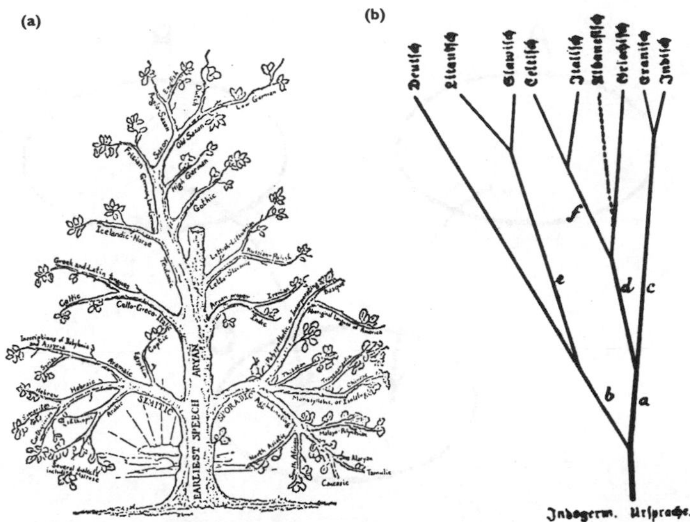

Figure 5.7. The schematised tree (b) and the more figurative one (a) both use the metaphor of branching to show relationships in linguistics. From, Stewart (1976), *Graphic representations of models in linguistic theory*, Bloomington: Indiana University Press.

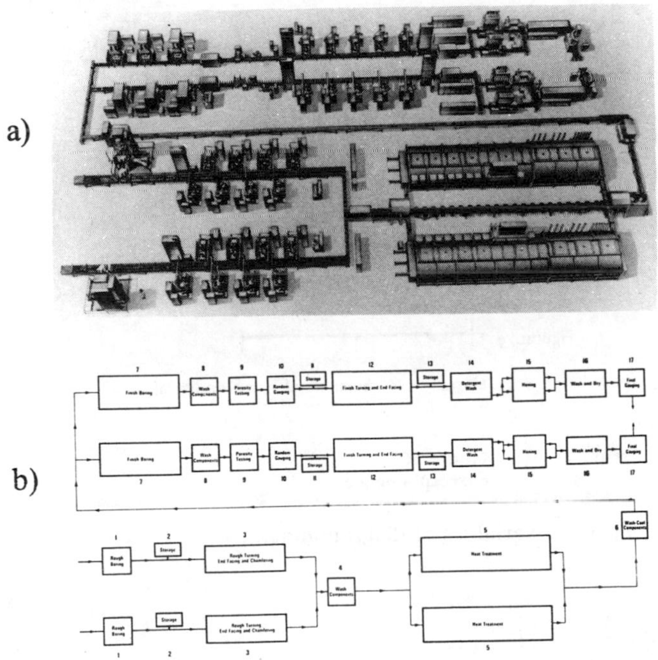

Figure 5.8. While both show the spatial layout of a factory machine line more-or-less literally, (a) is depicted figuratively, while (b) is not. From a Staveley Machine Tools catalogue.

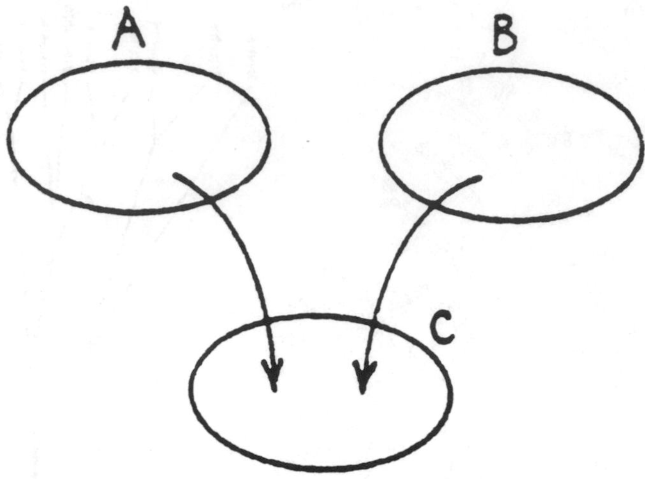

Figure 5.9. From [13] pp. 10–14.

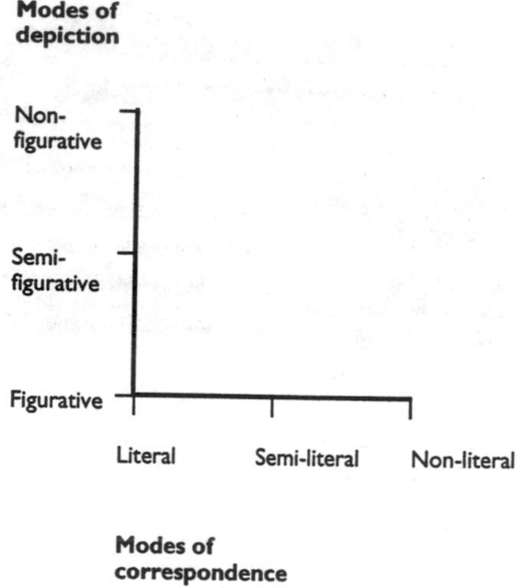

Figure 5.10. Two key variables of diagramming.

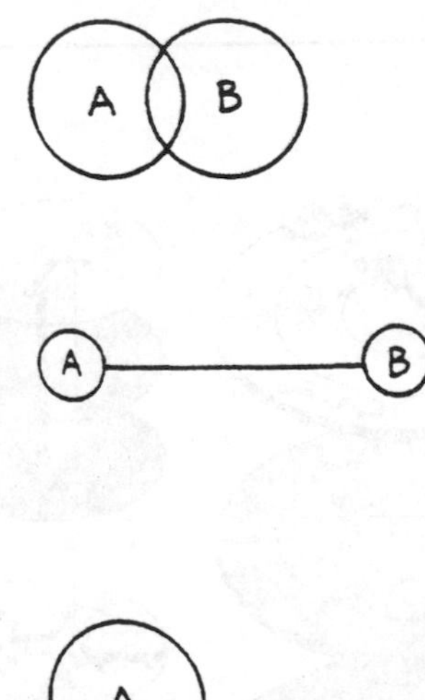

Figure 5.11. From [13] p. 8.

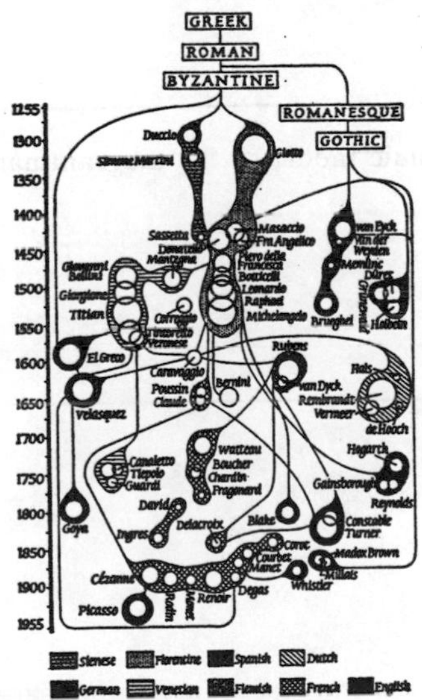

Figure 5.12. A hybrid time chart, flow chart and Venn diagram from *European Painting and Sculpture*, Eric Newton (Penguin Books 1941, Fourth edition, 1956) © The Estate of Eric Newton, 1941. Reproduced by permission of Penguin Books Ltd.

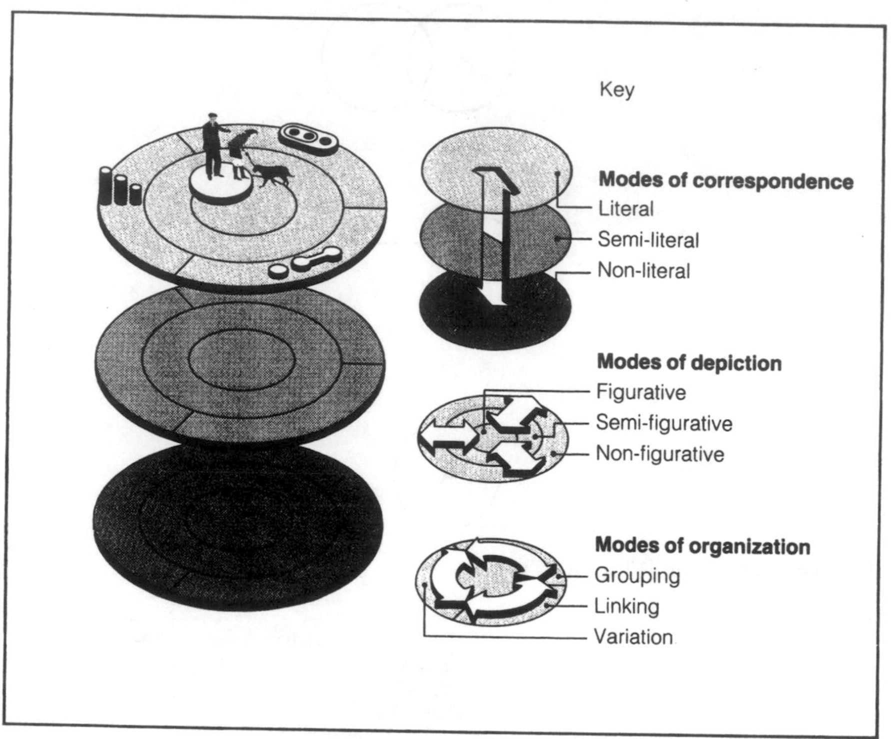

Figure 5.13. A taxonomic model of the diagrammatic modes of interpretation. From [13] p. 9.

Part II

Cognitive Aspects of Diagrams

Introduction

While diagrams have been around for hundreds of years, perhaps even thousands of years (depending on your definition), they have only been subjected to intensive scrutiny by psychologists for at most the past thirty years. The reasons for this are unclear. On one hand, there have been pressures from without, the rise of computing technology, and in particular recent advances of graphical interfaces, have raised questions concerning the utility and design principles of visual metaphors for interacting with computers. On the other hand the imagery debate of the 1970s and 1980s spurred debate as to precisely what it means to "think visually". Although the debate focused on the nature of our internal representation ("in the head") when apparently reasoning about visual phenomena, in its latter stages a weak consensus arose as to the reality or even the utility of formulating theories of mental visualisation in terms of one unique representation.

This is hardly the whole picture, and strong proponents of the linguistic and imagistic camps still preside in the cognitive psychology community today. However, one of the most exciting outcomes of the endeavour to resolve the imagery debate was a growth of interest in the relationship between internal and external representations, and the birth of the diagrammatic reasoning community itself. To talk of a diagram being an external representation and to explain its utility in terms of the reduced load imposed on internal cognition, both in terms of reduced memory load and the facility to make inferences perceptually, is now mainstream cognitive psychology. Furthermore, to design graphical interfaces and computer visualisations by reference to such cognitive phenomena is widely viewed as the first-principle approach of choice.

In the opening chapter of this part (Chapter 6), Blackwell asks "How is using diagrams related to things that happen inside our heads?" Establishing the nature of this relationship is surely the goal for all cognitive studies of diagrams, and Blackwell's chapter presents a survey of both the experimental and theoretical work in psychology that has addressed this question. His account is non-technical, in the sense that it is deliberately pitched at the diagrams researcher without a formal training in psychology, but its historical perspective and Socratic style is a useful summary for experts and non-experts alike.

Diagrams as a non-visual device is still a little-explored topic, despite their immense significance for people with visual impairments. Furthermore, as with other topics in cognitive psychology, in particular the imagery debate, experimentation on subjects with cognitive dysfunctions can often give rise to compelling observational data and predictive tests. In Chapter 8, Ungar explores the use of tactile maps by blind and visually impaired subjects in an attempt to verify the conjoint retention hypothesis that spatial/perceptual information and linguistic/verbal information are coded by distinct processes, separated in memory. The hypothesis has important theoretical implications, as well as obvious educational applications. The results diverge from those of the theory's originators, and raise a number of interesting questions as to the nature of the appropriate experimental set-up.

Chapters 11, 10 and 12 address the design and comprehension of traditional two-dimensional graphs. In Chapter 11, Zacks, Levy, Tversky and Schiano, report a comprehensive survey of the properties of graphs produced for academic and popular journals, as well as major US newspapers. Codifying over 8000 graphs they establish the importance of graphs in the print media. Furthermore, even with the advent of freely available (and highly configurable) software for information design, and contrary to many popular critiques of modern graphic design, Zacks and his co-authors found that the percentage of graphs that employ the features often criticised by the "experts" is very low. It seems clear therefore that were a significant degree of expert editorial control to be exercised (as in the case of the sample publication surveyed) classic design mistakes would rarely be made; however, whether the same can be said of publications where the editorial control is looser remains to be established.

The theme of traditional graphs for representing quanititative information is continued in Shah's contribution (Chapter 10), in which the author considers the role of format, content and individual difference in graph comprehension. The model of graph comprehension presented by Shah is something akin to traditional views of text comprehension. This involves a translation between the elements of graphical primitives and the conceptual representation of the quantitative relations, through repeated cycles of association. Through a number of empirical studies, Shah identifies the important role in the process of a user's prior knowledge of the domain, expectations and graph-reading expertise, as well as pointing to practical implications of the framework.

In Chapter 1 Sloman identified that an immediate goal of diagrams research should be to identify the nature of the interaction between visual and linguistic reasoning, and in Chapter 12 Toth and Lewis address this issue through an investigation of the role of working memory in diagrammatic reasoning and decision making. Their emipirical framework tests Baddeley's model of working memory (i.e. phonological loop, visuospatial sketchpad and executive control system) but extends a traditional methodology to examine the role of the external representation of the task and its interaction with the internal representation. Through an array of verbal, visual and mental suppression conditions they conclude that, regardless of the nature of the presentation of the gain-loss problem used as the task, solving such problems utilises both visual and verbal resources of working memory.

The final three chapters of this part address the issue of using external representations of what are essentially mechanical reasoning problems. Chapter 13 concentrates on the traditional gear, belt and pulley system, i.e. on kinematic reasoning about static displays. Through the use of eye-tracking technology, and the resulting trace of the locus of visual attention (or rather, the centre of the visual field), it is found that independent raters can predict both the primary axis of force transmission of the system and whether the subject solved the problem correctly. The authors of the study do not attempt to place their results in the context of any particular theory of external cognition, since its principal significance concerns the use of eye tracking in studying cognition of external representations – and it is hard to deny that this is a technology that is likely to have an increasingly important impact on empirical diagrams research.

Continuing thc topic of reasoning about physical systems, Kozhevnikov, Hegarty and Mayer (Chapter 9) present some important groundwork research on the relationship between subjects' spatial abilities and their success at solving different classes of kinematic reasoning problems. They identify that different classes of kinematics problems require distinct cognitive skills; for example, their results distinguish between extrapolation and frame of reference problems (which are correlated with spatial imagery abilities), and speed prediction and more graphical problems which pertain to mathematical reasoning abilities and an explicit knowledge of physical laws. To conclude this theme, and "cognitive issues" section, Suwa and Tversky (Chapter 14) explore how both expert and novice designers shift their focus of attention in their own sketches. Although the sample used is small compared with the usual empirical samples, the work makes convincing arguments as to a number of aspects of sketching. These include the freeing up of cognitive resources by externalising basic elements of a design and the promotion of the ability of a designer to perform a number of calculations, and inferences as to the physical and functional implications of early design commitments.

Although it is widely accepted that research in diagrammatic reasoning is all the better for being a multidisciplinary affair, the role of the psychologist is arguably the most important one. Cognitive psychology provides the methodologies and theories through which all empirical studies of diagram use proceed, yet often the relevance of the results of other fields to psychology itself is not clear. One of the explicit intentions of this volume is to demonstrate that such research can only be enriched by closer scrutiny of, on the one hand, the often tacit knowledge of the practitioners, and on the other hand formalisation of their theories (and thereby the ability to test the predictive powers of psychological theories) by computer scientists and mathematicians.

6. Psychological Perspectives on Diagrams and their Users

Alan F. Blackwell

How is using diagrams related to things that happen inside our heads? The study of diagrammatic reasoning often focuses on computational models of diagram use rather than on studies of human performance. This chapter presents a survey of the empirical and theoretical research that has investigated these processes. It considers the origins, interpretation and manipulation of diagrams, treating diagrams as a notational system that can be used and studied in an experimental context. This is a review of the experimental psychology literature rather than a complete theoretical framework, and is particularly intended as an introduction to the field for those who are commencing research projects into diagrammatic reasoning, having come from a discipline other than cognitive psychology.

6.1 Introduction

Any cognitive study of information other than text or speech encounters the lingering remnants of the imagery debate, as examined in the collection edited by Block [10]. Much early research on diagram use was motivated by one of the entrenched positions in this debate. The main views can be simplified to two alternatives. Either:

1. if seeing a diagram causes an image-like mental representation, it is the nature of this internal representation that ought to be studied (possibly without even using any external diagram); alternatively,
2. if a diagram causes a mental representation just like any other representation formed in response to text or speech, research should instead concentrate on the ways that we interact with the external diagram.

The structure of this chapter reflects these alternative perspectives of previous research. Some characteristics of diagrams are independent of their location, while others are best studied in terms of mental representations (using the techniques of cognitive neuroscience) in terms of external manipulations (using task-based experiments), or in terms of perceptual processing (using psychophysical or decision-based experiments).

6.2 Diagrams as Notation

How do diagrams differ from text or pictures? This section describes their formal properties (whether internal or external), but also where those properties come from and how they support reasoning.

6.2.1 What are Diagrams?

We can describe diagrams according to their status in the world, in thought or in communication. All three are informed by Peirce's [60] analysis of signs.

Ittelson [39] has described how perception of *markings* differs from other perceptual tasks. Markings do not occur naturally – they are human artefacts whose function lies in the intention behind their creation. Ittelson distinguishes these types: *designs* are decorative, *writings* carry meaning by agreed conventions, and *depictions* evoke objects or experiences. *Diagrams* are separate. They provide non-visual information in a visual form. Diagrams are like writing – in depending on agreed conventions – but unlike writing in that the overall form affects the interpretation.

Bertin [6] describes graphics in terms of the ways in which ink can be distributed on a surface. These *variables of the plane* include location on X and Y axes plus a number of "Z axis" variations. These include the size of a mark, density and colour of ink, or orientation and shape. His analysis is based on information-carrying potential; information to be communicated must be mapped to these variables. Where information is topographical, X and Y are devoted to location. Ordered values must be mapped to an ordered scale – size or density, if X and Y have been allocated – while unordered values can be mapped to differential variables such as colour or shape. Bertin's characterisation of graphical variables and correspondences is only one of a number of such taxonomies – a wide range of these is discussed in the separate chapter by Blackwell and Engelhardt.

Goodman [31] offers a general account of representational *symbol systems.* In order to say whether two messages are different, the characters of a notation must be both disjoint and unambiguous. Syntactical disjointedness is distinguished from syntactically dense (or analogue) notations, which have infinitely many characters – between any two there can be a third. In many symbol systems, further information is added when the message is interpreted, and this can result in a message that is semantically dense, even though syntactically differentiated. Goodman places diagrams in the class of *models*, which are analogue (semantically dense) in some dimensions, but digital (or syntactically disjoint) in others.

6.2.2 Where do Diagrams Come From?

Resemblance. The most naive account (more likely to be made of diagrams with a strongly pictorial element) is simply that they resemble the things they refer to. Goodman [31] was particularly concerned to address this ***fallacy of depiction***. Denotation, the core of representation, does not rely on resemblance. If it did, it would not be possible to depict things that do not exist, so cannot be resembled. Ittelson [39] challenges the ***pictorial assumption*** in psychology: that "the processes involved in the visual perception of the real world and the processes involved in the visual perception of pictures are identical", and Scaife and Rogers [66] warn of the ***resemblance fallacy*** – the intuition that diagrams resemble the visual world.

Metaphor. Gibbs [28] has found experimental evidence for mental images that underlie common idioms, and uses this to support Lakoff's [44] ***conceptual metaphor*** theory – that abstract concepts are derived from embodied experience in the physical world. This applies especially to spatial metaphors for abstract concepts, as Lakoff has observed in the case of diagrams [45]. The use of metaphor to structure representations is familiar in computer systems [11]. The now ubiquitous computer icon was intended as a metaphor, in the context of the Pygmalion system – "a visual metaphor for computing. Instead of symbols and abstract concepts, the programmer uses concrete display images, called 'icons' " [75]. Recent research by the author does, however, cast some doubt on the universal utility of metaphor to computational diagram users [8,9].

The Frame Problem. How do icons make a computer more usable? Shneiderman [71] defined ***direct manipulation***: objects are continuously represented on the screen, the user acts on them, and actions have immediately visible impact. By comparison to symbolic references in verbal commands, diagrams do not suffer from the ***frame problem*** – the consequences of any action are apparent in the representation [50]. Lewis [48] argues that humans are "attuned" to constraints in our physical environment, and this helps us recognise how to use diagrams.

6.2.3 What do Diagrams Provide?

Locality and Labels. Larkin and Simon [46] described a computational model that directly uses diagram features. Diagrams group related information in the same area, so searches can be constrained to the vicinity of a goal. Correspondences can also be established from topological relationships – unlike symbolic systems, where they are found by searching for related labels.

Expressive Power and Specificity. Stenning and Oberlander [77] argue that diagrams aid cognitive processing because of their *specificity* – the way in which they limit abstraction. Diagrams have fewer interpretations, so are more tractable than unconstrained textual notations. Goodman [31] similarly noted that interpretation of language involves a potentially infinite search for meaning because it is syntactically differentiated but semantically dense.

Pragmatics. Theoretical work on notations seldom considers the question of usability. Green's *cognitive dimensions* [33] provide a vocabulary for discussing the way that notations are used. They are based on the observation that every notation highlights some kinds of information at the expense of obscuring others, and that recognition of these trade-offs can guide notation designers.

6.3 Diagrams as Thoughts

There have been many comparisons of verbal and visual tasks. They include the way that differing capabilities are distributed through the brain, the way that people choose strategies for different tasks, and psychometric measures to investigate variation between individuals.

6.3.1 Variation Within the Brain

The distinction between visual and verbal representations in the brain is often described as a simple dichotomy. This may be easier to implement as a computational model, but even simple experiments show that pictures and propositions cannot clearly be divided.

Hemispheric Specialisation. One of the most widely known dichotomies in cognitive neuroscience is that of hemispheric specialisation in the brain. Advocates of diagrams have even suggested that the right hemisphere is "needlessly at rest and underutilised" when using text [72]. Evidence of hemispheric specialisation comes from neurological studies, functional imaging, and presentation of stimuli in one half of the visual field. Some simple visual tasks can require more time when carried out by the left hemisphere. Ratcliff [63] found that patients with right hemisphere lesions are slower at a simple image inversion task. Kosslyn et al. [43] found that spatial judgements are faster when a stimulus is presented in the left visual field (right hemisphere) and categorical ones faster when presented to the left hemisphere. Baker et al. [5] used functional imaging to find that spatial planning tasks result in more right hemisphere blood flow.

Interaction of Verbal and Visual, What and Where. Despite hemispheric specialisation, there are many tasks in which verbal and visual information is combined. Paivio's *dual coding* theory [58] elaborated the observation that memory improves when a concrete image can be associated with a verbal task. Spatial coding of words was observed by Santa [65] – words are encoded verbally if they are arranged linearly, but a non-linear spatial arrangement impairs judgement of words far more than images.

In addition to visual and verbal, we can distinguish categorical and coordinate, or "what" and "where" information. Categorical information might not be verbal – object identity is often encoded in terms of an image of the object. Farah et al. [24] describe this dissociation in a case study of a brain-damaged patient who has severely impaired memory for images, but normal spatial memory. Tresch et al. [82] found a similar dissociation using a dual task experiment: object memory was impaired by a colour judgement task, and spatial memory by a motion detection task. Mecklinger and Müller [56] have measured differences in neural activity when subjects memorise either identity or location of objects in a grid. This is supported by Mishkin et al. with neuroanatomical evidence for separate cortical pathways that process object vision and spatial vision [57].

Further evidence for complex interactions is found in apparently spatial tasks. Anooshian and Seibert [2] demonstrated that navigation is not a "pure" spatial system – route memory depends on visual landmarks. Neither is memory for node and link diagrams purely visual [15]. Even if subjects are asked to make purely visual comparisons, topological complexity of a diagram affects performance.

Axes and Orientation. Spatial axes seem to play a role that is complementary to visual and verbal codes in human cognition. They introduce a categorical element by dividing space into distinct regions. Hayward and Tarr [35] showed that memory for spatial locations seems to be associated with horizontal and vertical axes. When subjects were asked to make judgements about relative positions of points on a map, accuracy was improved when near an axis. McNamara [54] created axes by placing strings across a room in which the location of objects was to be remembered. These strings distorted position judgements toward the axis, while also influencing mnemonic coding.

6.3.2 Variation in Individual Capability

Are there innate differences between individuals in image manipulation skills, or in the ability to transfer information between visual and verbal modes? Do such differences have any effect on real-world tasks? The ability to rotate a mental image is controversial, as it shows a larger gender difference than any other psychometric test [34], but is this ability useful?

Cognitive Styles: Visualiser and Verbaliser. As with hemispheric specialisation, there was a popular dichotomy between *cognitive styles* in early research on individual differences. A classic experiment compares judgement speed using pictures and sentences. MacLeod et al. [52] were able to predict performance by dividing SAT scores into spatial and verbal tasks. Shah and Miyake [70] have devised a *spatial span* measure – memory for a sequence of locations. This too is correlated with visualisation tasks, but not with verbalisation.

It is possible to compensate for these differences by task-specific training, however. Frandsen and Holder [27] identified subjects with lower scores on a spatial manipulation test, and trained them to use Venn diagrams. The previously observed difference between visualisers and verbalisers disappeared. Even without training, people often optimise their performance by choosing either a verbal or visual strategy appropriate to their own abilities.

Gender Correlations. Paivio and Clark [59] found a systematic element in strategy choice: more males used imagery for dynamic problem-solving tasks, while more females used it for static memory tasks. Delgado and Prieto [20] carried out a large study looking for possible strategic differences that might account for gender differences in mental rotation, but could only confirm Halpern's [34] observation that this is the cognitive task exhibiting the most general gender-related performance difference. Silverman et al. [73] carried out a cross-cultural study, aimed at finding different patterns of gender difference as a result of differences in gender associations between cultures. Once again, the same gender effect was found in all groups studied.

6.3.3 Development of Strategies

Although there appear to be innate differences between the strategies that individuals choose in visual reasoning tasks, strategy is also affected by education, expertise and culture.

Developmental Studies. The ability to interpret a denotative relationship between a representation and the real world is not innate. DeLoache and her collaborators have found that children learn to transfer relationships from a scale model or picture to the world between the ages of two and three years [18], while very young infants may attempt to interact with some two-dimensional representations as though they were three-dimensional objects [19]. Presson [62] records the development of spatial reasoning skills for use in *secondary spatial reasoning* tasks such as map reading.

Studies of drawing in children observe the development of depictive conventions. Many features of childhood productions, such as arms and legs connected to the head, can be attributed to undeveloped planning skills [81] or motor skills [85]. Nevertheless, some special cases show the development of representational strategies analogous to linguistic skills [40]. Scott and Baron-Cohen [69] found that autistic children are unable to draw imaginary objects. This is interesting in the light of Goodman's theory of depiction. Does it mean that, for autistics, drawing can only be depictive? Would this also prevent them from using some diagrammatic conventions? More work here could provide valuable insights, especially in light of the suggestion that poor adult drawing performance results from perceptual encoding rather than from deficiencies in motor performance [16].

The Nature of Expertise. Lowe's [51] study of weather maps found that novices read surface notational features, while experts are primarily aware of underlying structure, including information not visible in the notation. Expert circuit designers also use configural information independently of spatial organisation [22], while being well aware of which graphical properties can express relatedness (for example) independently of connectedness [61].

How is expert knowledge encoded and related to the visual aspects of a diagrammatic notation? Computational models of diagram usage access previous solutions via visual elements of the diagram that look the same. McDougal and Hammond [53] indexed diagrams based on overall configuration, Koedinger and Anderson [42] indexed using local detail, and Thagard, Gochfeld and Hardy [80] simply matched the structure of visually adjacent elements. Each of these models is supported by empirical studies. If solutions are accessed by spatial configuration, decreased performance is observed when it is not available [12]. Chambers [13] describes the importance of local detail in determining the interpretation of a figure, and Beveridge and Parkins [7] demonstrated that a diagram emphasising structure of a problem maximises discovery of an analogical solution.

More recent computational models emphasise the skill of experts in integrating information from multiple representations [78], and some studies have found specific evidence for sophisticated mental imagery strategies derived from visual representation tools such as an abacus [37] or chessboard [64].

Cultural Differences. Many diagrams rely on expert knowledge, but are there any universal diagram properties? Tversky, Kugelmass and Winter [83] asked English, Hebrew and Arabic-speaking children and adults to illustrate various quantities and points in time. They found several universal conventions. Adults from all cultures show quantity increasing from bottom to top of a page. Children sometimes show quantity horizontally, but still not top to bottom. More unexpectedly, a left-to-right time direction was preferred by all children too young to read, regardless of whether they came from a culture with left-to-right or right-to-left writing systems.

Evidence for cultural differences in imagery skills comes from Kearins' [41] studies of Australian Aboriginal children, who score poorly on supposedly "culture fair" IQ tests. She devised a test in which children had to memorise, then reconstruct, the positions of objects in a grid. The performance of Aborigines not only equalled, but greatly exceeded that of Europeans. Kearins observed strategic differences that may explain the results; European children appeared to construct linguistic encodings (speaking under their breath). Aborigines, in contrast, appeared to form visual images of the scene, which were available for mental scanning during reconstruction of the grid.

6.4 Application of Diagrams

The first two parts of this chapter have discussed general properties of diagrammatic notations, and specific properties of mental representations. The main claims for diagrammatic reasoning depend, however, on the way that these two are combined in diagram applications.

6.4.1 Information Input: Diagrams as Models

Denis [21] defines the situations in which it is useful to build an analogue model of a problem. When modelling a physical situation, there may be physical dimensions which can be directly represented in an image, allowing comparative judgements. A similar strategy can be used for abstract problems, if there are abstract dimensions that can be treated spatially. In this case there is a trade-off between the benefit of accessing information from an image, and the cost of transforming the abstract problem. Huttenlocher's [38] experiments are an early demonstration of the way that a verbal description of a problem can result in visualisation of the problem rather than a grammatical representation. These "three term series" problems showed that people often reason about relative height from images rather than propositional terms.

When an illustration of a problem situation is given, it can form the core of a spatial representation – so the main contribution of a diagram may be that it reduces the cognitive load of assigning abstract data to appropriate spatial dimensions. For example, Glenberg and Langston [29] found that where information about temporal ordering is only implicit in text, a flow diagram will reduce errors in answering questions about that ordering.

6.4.2 Information Processing: Diagram Manipulation

Once a problem is represented in diagrammatic form, how is the diagram used? The experiments described above simply involve "reading off" an answer by inspecting a model. Inferences can also be made by directly transforming an image, without converting information from the image into propositional form. Ullman [84] proposed a number of *visual routines* that could be used to derive information directly from an image, although it is difficult to demonstrate that these are involved in human reasoning.

An algorithm that has been observed in the use of mechanical diagrams is *mental animation* of the depicted machine. Hegarty [36] used a gaze-tracking procedure to find that inferences were made about a diagram of ropes and pulleys by imagining the motion of the rope along a *causal chain*. Schwartz [68] has been able to influence when subjects choose an animation strategy. If a device is represented with a realistic illustration, his subjects made judgements at a speed proportional to the amount of motion, indicating that they were mentally animating the device. If the problem is presented in simple geometric terms, judgements were made in constant time, suggesting an alternative strategy. Sims and Hegarty [74] have presented evidence that mental animation does employ visuospatial working memory resources.

One of the most extensively investigated image algorithms is seen in Finke's research on the use of images for creative synthesis [25]. In these experiments, subjects are shown a set of geometric shapes, and asked to suggest a creative combination of them. Finke's model of creativity claims that new configurations can be generated and explored by manipulating and combining images.

6.4.3 Information Output: Verbalisation from Diagrams

Once an image-like model has been constructed, and a problem solution found by manipulating it, how is the solution reported? The image could be copied out as an external diagram, but it is more common to report problem solutions verbally. This is described by Levelt [47] as the *speaker's linearisation problem*. He proposed that, given a diagram to describe verbally, people make a *gaze tour*, guided by connectivity, with diagram elements mentioned in the order they are visited.

The gaze tour is based on the observation by Linde and Labov [49] that New Yorkers, asked to describe their apartments, list the rooms in order of walking through them. They conclude from this that topological structure is represented in terms of events, but Taylor and Tversky [79] provide a far richer model of how spatial descriptions are structured. Depending on the configuration of the space to be described, either a *route* structure, a *survey* structure, or a mixture of the two can be used. Linde and Labov's observation of route structure, they claim, resulted simply from the fact that most New York apartments have a linear arrangement of rooms.

Evidence of verbalisation strategies also comes from working memory experiments. Baddeley's [4] model of working memory defines a *phonological loop* and a *visuospatial sketchpad*. The phonological loop depends on verbal articulation speed; the VSSP should be independent of articulation. However, Smyth and Scholey [76] also measured a correlation between articulation speed and locations remembered in a diagrammatic array. This suggests the kind of verbalisation strategy that Kearins [41] thought might be culture-specific.

Interaction between verbalisation and images can compromise diagrammatic reasoning. Many investigations of problem solving have asked experimental subjects to "think aloud" [23]. Schooler et al. [67] found that solutions to insight problems fall by 25% when thinking aloud. If such problems are normally solved using imagistic processes, verbalisation may impose inappropriate coding.

6.4.4 The Need for Externalisation

This chapter started by asking whether diagrams are representations "inside the head", or simply markings in the world. There is ample empirical evidence for image-based mental representations, both visual and spatial, that can be used to carry diagrammatic information. Are external representations even necessary, given these internal strategies? Research on mental models indicates that an illustration can help form an appropriate image when working with abstract information, but does that mean that an external diagram is only of transitory use when encountering a new type of problem?

When We Need an External Image. Evidence for the importance of external representations comes from an experiment by Chambers and Reisberg [14]. Subjects who memorised an ambiguous picture were only able to report one of two possible interpretations on the basis of their memorised image. If they then copied it onto paper, they could immediately see the alternative interpretation. This result has been controversial, as it raises the question of how much the mental image is like a visual image. Chambers [13] has noted more recently that the ambiguous picture they used must be reoriented for the alternative interpretation. She suggested that orientation information is associated with the image, so the two cannot be separated until the image is externally perceived.

In an expert domain, Davies [17] has shown that computer programmers rely on the ability to inspect their own previous productions as they create a program. This supports the parsing/gnisrap theory of how external representations are used as a perceptual extension of working memory [32]. Visual sketches are also considered to be critical aids to creative design by professional groups such as architects [26,30].

When Internal Images Suffice. The Chambers and Reisberg [14] result prompted a vigorous response from researchers who believe that mental images must be reinterpretable. Anderson and Helstrup [1] investigated whether an external representation would improve performance in the Finke creative synthesis task. They allowed half of their subjects to doodle on paper when generating creative combinations. They found that the availability of an external image did not result in improved creativity. This suggests an opposite conclusion from that of Chambers and Reisberg.

Diagrams in Context. There is still much need for investigation and debate on the relationship between diagrams as external representations and internal representations. We know something about the general properties of diagrammatic representations. There is some evidence of how expert diagram users employ external representations. And we are still designing experiments to test the nature and capacity of mental representations. It is already clear that they are far more complex than would be suggested by the imagery debate. Research continues into more sophisticated models of diagram use in context. Zhang [86], for example, has proposed a computational model in which the external representation can be used to infer novel information through perceptual biases while internal representations are used for low-cost simulation of future changes to the external representation.

The greatest danger is that we produce cognitive models that account only for the limited evidence from one of these richly interacting streams of investigation. Schwartz [68] and DeLoache and Marzolf [18] both provide examples of minor variations that make a critical difference in the ability to use a diagram as an internal representation. Such issues cannot easily be explained by current computational models. In order to address them we need to broaden our scope of enquiry to include cultural conventions, theories of metaphor and pragmatics, and the working practices of the technical specialists who are the largest population of diagram users.

References

1. Anderson, R.E. and Helstrup, T. (1993). Visual discovery in mind and on paper. Memory and Cognition 21:283–293.
2. Anooshian, L.J. and Seibert, P.S. (1996). Diversity within spatial cognition: Memory processes underlying place recognition. Applied Cognitive Psychology 10:281–300.
3. Arnheim, R. (1970). Visual thinking. London: Faber & Faber.
4. Baddeley, A.D. (1986). Working memory. Oxford: Oxford University Press.
5. Baker, S.C., Rogers, R.D., Owen, A.M., Frith, C.D., Dolan, R.J., Frackowiak, R.S.J. and Robbins, T.W. (1996). Neural systems engaged by planning: A PET study of the Tower of London task. Neuropsychologia 34:515–526.
6. Bertin, J. (1981). Graphics and graphic information processing. Berlin: Walter de Gruyter.
7. Beveridge, M. and Parkins, E. (1987). Visual representation in analogical problem solving. Memory and Cognition, 15:230–237.
8. Blackwell, A.F. (1998). Metaphor in diagrams. Unpublished PhD thesis, Cambridge University.
9. Blackwell, A.F. and Green, T.R.G. (1999). Does metaphor increase visual language usability? In Proceedings IEEE symposium on visual languages VL'99, Tokyo, September 1999, pp. 246–253.
10. Block, N. (1981). Imagery. Cambridge, MA: MIT Press.
11. Carroll, J.M. and Thomas, J.C. (1982). Metaphor and the cognitive representation of computing systems. IEEE Transactions on Systems, Man and Cybernetics 12:107–116.
12. Carroll, J.M., Thomas, J.C, Miller, L.A. and Friedman, H.P. (1980). Aspects of solution structure in design problem solving. American Journal of Psychology 93:269–284.
13. Chambers, D. (1993). Images are both depictive and descriptive. In B. Roskos-Ewoldsen, M. Intons-Peterson and R.E. Anderson (Eds), Imagery, creativity and discovery. Amsterdam: North Holland, pp. 77–97.

14. Chambers, D. and Reisberg, D. (1985). Can mental images be ambiguous? Journal of Experimental Psychology: Human Perception and Performance 11:317–328.
15. Chechile, R.A., Anderson, J.E., Krafczek, S.A. and Coley, S.L. (1996). A syntactic complexity effect with visual patterns: Evidence for the syntactic nature of the memory representation. Journal of Experimental Psychology: Learning, Memory and Cognition 22:654–669.
16. Cohen, D.J. and Bennett, S. (1997). Why can't most people draw what they see? Journal of Experimental Psychology: Human Perception and Performance 23:609–621.
17. Davies, S.P. (1996). Display-based problem solving strategies in computer programming. In W.D. Gray and D.A. Boehm-Davis (Eds), Empirical studies of programmers: Sixth workshop. Norwood, NJ: Ablex, pp. 59–76.
18. DeLoache, J.S. and Marzolf, D.P. (1992). When a picture is not worth a thousand words: Young children's understanding of pictures and models. Cognitive Development 7:317–329.
19. DeLoache, J.S., Pierroutsakos, S.L., Uttal, D.H., Rosengren, K.S. and Gottlieb, A. (1998). Grasping the nature of pictures. Psychological Science 9:205–210.
20. Delgado, A.R. and Prieto, G. (1996). Sex differences in visuospatial ability: Do performance factors play such an important role? Memory and Cognition 24:504–510.
21. Denis, M. (1991). Imagery and thinking. In C. Cornoldi and M.A. McDaniel (Eds), Imagery and cognition. New York: Springer-Verlag.
22. Egan, D.E. and Schwartz, B.J. (1979). Chunking in recall of symbolic drawings. Memory and Cognition 7:149–158.
23. Ericsson, K.A. and Simon, H.A. (1985). Protocol analysis: Verbal reports as data. Cambridge, MA: MIT Press.
24. Farah, M.J., Hammond, K.M., Levine, D.N. and Calvanio, R. (1988). Visual and spatial mental imagery: Dissociable systems of representation. Cognitive Psychology 20:439–462.
25. Finke, R.A., Pinker, S. and Farah, M.J. (1989). Reinterpreting visual patterns in mental imagery. Cognitive Science 13:51–78.
26. Fish, J. and Scrivener, S. (1990). Amplifying the mind's eye: Sketching and visual cognition. Leonardo 23:117–126.
27. Frandsen, A.N. and Holder, J.R. (1969). Spatial visualization in solving complex verbal problems. Journal of Psychology 73:229–233.
28. Gibbs, R.W. Jr (1996). Why many concepts are metaphorical. Cognition 61:195–324.
29. Glenberg, A.M. and Langston, W.E. (1992). Comprehension of illustrated text: Pictures help to build mental models. Journal of Memory and Language 31:129–151.
30. Goldschmidt, G. (1991). The dialectics of sketching. Creativity Research Journal 4:123–143.
31. Goodman, N. (1969). Languages of art: An approach to a theory of symbols. London: Oxford University Press.
32. Green, T.R.G., Bellamy R.K.E. and Parker, J.M. (1987). Parsing and Gnisrap: A model of device use. In Empirical studies of programmers: Second workshop. Norwood, NJ: Ablex, pp. 132–146.
33. Green T.R.G. and Petre M. (1996). Usability analysis of visual programming environments: A "Cognitive Dimensions" approach. Journal of Visual Languages and Computing 7:131–174.

34. Halpern, D.F. (1992). Sex differences in cognitive abilities. Hillsdale, NJ: Erlbaum.
35. Hayward, W.G. and Tarr, M.J. (1995). Spatial language and spatial representation. Cognition 55:39–84.
36. Hegarty, M. (1992). Mental animation: Inferring motion from static displays of mechanical systems. Journal of Experimental Psychology: Learning, Memory and Cognition 18:1084–1102.
37. Hishitani, S. (1990). Imagery experts: How do expert abacus operators process imagery? Applied Cognitive Psychology 4:33–46.
38. Huttenlocher, J. (1968). Constructing spatial images: A strategy in reasoning. Psychological Review 75:550–560.
39. Ittelson, W.H. (1996). Visual perception of markings. Psychonomic Bulletin and Review 3:171–187.
40. Karmiloff-Smith, A. (1990). Constraints on representational change: Evidence from children's drawing. Cognition 34:57–83.
41. Kearins, J.M. (1981). Visual spatial memory in Australian Aboriginal children of desert regions. Cognitive Psychology 13:434–460.
42. Koedinger, K.R and Anderson, J.R. (1990). Abstract planning and perceptual chunks: Elements of expertise in geometry. Cognitive Science 14:511–550.
43. Kosslyn, S.M., Koenig, O., Barrett, A., Cave, C.B., Tang, J. and Gabrieli, J.D.E. (1989). Evidence for two types of spatial representations: hemispheric specialization for categorical and coordinate relations. Journal of Experimental Psychology: Human Perception and Performance 15:723–735.
44. Lakoff, G. and Johnson, M. (1980). Metaphors we live by. Chicago: University of Chicago Press.
45. Lakoff, G. (1993). The contemporary theory of metaphor. In A. Ortony (Ed.), Metaphor and thought (2nd ed). Cambridge, UK: Cambridge University Press, pp. 202–251.
46. Larkin, J.H. and Simon, H.A. (1987). Why a diagram is (sometimes) worth ten thousand words. Cognitive Science 11:65–99.
47. Levelt, W.J.M. (1981). The speaker's linearisation problem. Philosophical Transactions of the Royal Society B 295:305–315.
48. Lewis, C.M. (1991). Visualization and situations. In J. Barwise, J.M. Gawron, G. Plotkin and S. Tutiya (Eds), Situation theory and its applications. Stanford University: CSLI, pp. 553–580.
49. Linde, C. and Labov, W. (1975). Spatial structures as a site for the study of language and thought. Language 51:924–939.
50. Lindsay, R.K. (1988). Images and inference. Cognition 29:229–250.
51. Lowe, R.K. (1993). Diagrammatic information: Techniques for exploring its mental representation and processing. Information Design Journal 7:3–18.
52. MacLeod, C.M., Hunt, E.B. and Mathews, N.N. (1978). Individual differences in the verification of sentence–picture relationships. Journal of Verbal Learning and Verbal Behavior 17:493–507.
53. McDougal, T.F. and Hammond, K.J. (1995). Using diagrammatic features to index plans for geometry theorem-proving. In J. Glasgow, N.H. Narayanan and B. Chandrasekaran (Eds), Diagrammatic reasoning. Menlo Park, CA: AAAI Press, pp. 691–709.
54. McNamara, T.P. (1986). Mental representations of spatial relations. Cognitive Psychology 18:87–121.
55. McNamara, T.P., Halpin, J.A. and Hardy, J.K. (1992). The representation and integration in memory of spatial and nonspatial information. Memory and Cognition 20:519–532.

56. Mecklinger, A. and Müller, N. (1996). Dissociations in the processing of "what" and "where" information in working memory: An event-related potential analysis. Journal of Cognitive Neuroscience 8:453–473.
57. Mishkin, M., Ungerleider, L.G. and Macko, K.A. (1983). Object vision and spatial vision: Two cortical pathways. Trends in Neurosciences 6:414–417.
58. Paivio, A. (1971). Imagery and verbal processes. New York: Holt, Rinehart & Winston.
59. Paivio, A. and Clark, J.M. (1991). Static versus dynamic imagery. In C. Cornoldi and M.A. McDaniel (Eds), Imagery and cognition. New York: Springer-Verlag, pp. 221–245.
60. Peirce, C.S. (1903/1932). Collected papers, Vol. II: Elements of logic. C. Hartshorne and P. Weiss (Eds). Cambridge, MA: Harvard University Press.
61. Petre, M. and Green, T.R.G. (1990). Where to draw the line with text: Some claims by logic designers about graphics in notation. In D. Diaper, D. Gilmore, G. Cockton and B. Shackel (Eds), Proceedings of Interact '90. Amsterdam: Elsevier, pp. 463–468.
62. Presson, C.C. (1987). The development of spatial cognition: Secondary uses of spatial information. In N. Eisenberg (Ed.) Contemporary topics in developmental psychology. New York: Wiley, pp. 77–112.
63. Ratcliff, G. (1979). Spatial thought, mental rotation and the right cerebral hemisphere. Neuropsychologia 17:49–54.
64. Saariluoma, P. and Kalakoski, V. (1997). Skilled imagery and long-term working memory. American Journal of Psychology 110:177–202.
65. Santa, J.L. (1977). Spatial transformations of words and pictures. Journal of Experimental Psychology: Human Learning and Memory 3:418–427.
66. Scaife, M. and Rogers, Y. (1996). External cognition: How do graphical representations work? International Journal of Human Computer Studies 45:185–214.
67. Schooler, J.W., Ohlsson, S. and Brooks, K. (1993). Thoughts beyond words: When language overshadows insight. Journal of Experimental Psychology 122:166–183.
68. Schwartz, D.L. (1995). Reasoning about the referent of a picture versus reasoning about the picture as the referent: An effect of visual realism. Memory and Cognition 23:709–722.
69. Scott, F.J. and Baron-Cohen, S. (1996). Imagining real and unreal things: Evidence of a dissociation in autism. Journal of Cognitive Neuroscience 8:371–382.
70. Shah, P. and Miyake, A. (1996). The separability of working memory resources for spatial thinking and language processing: An individual differences approach. Journal of Experimental Psychology: General 125:4–27.
71. Shneiderman, B. (1983). Direct manipulation: A step beyond programming languages. IEEE Computer (August):57–60.
72. Shu, N.C. (1988). Visual programming. New York: Van Nostrand Reinhold.
73. Silverman, I., Phillips, K. and Silverman, L.K. (1996). Homogeneity of effect sizes for sex across spatial tests and cultures: Implications for hormonal theories. Brain and Cognition 31:90–94.
74. Sims, V.K. and Hegarty, M. (1997). Mental animation in the visuospatial sketchpad: Evidence from dual-task studies. Memory and Cognition 25:321–332.
75. Smith, D.C. (1977). Pygmalion: A computer program to model and simulate creative thought. Boston, MA: Birkhauser.
76. Smyth, M.M. and Scholey, K.A. (1996). The relationship between articulation time and memory performance in verbal and visuospatial tasks. British Journal of Psychology 87:179–191.

77. Stenning, K. and Oberlander, J. (1995). A cognitive theory of graphical and linguistic reasoning: logic and implementation. Cognitive Science 19:97–140.
78. Tabachnek-Schijf, H.J.M., Leonardo, A.M. and Simon, H.A. (1997). CaMeRa: A computational model of multiple representations. Cognitive Science 21:305–350.
79. Taylor, H.A. and Tversky, B. (1996). Perspective in spatial descriptions. Journal of Memory and Language 35:371–391.
80. Thagard, P., Gochfeld, D. and Hardy, S. (1992). Visual analogical mapping. In Proceedings of the 14th Annual Meeting of the Cognitive Science Society, pp. 130–135.
81. Thomas, G.V. and Silk, A.M.J. (1990). An introduction to the psychology of children's drawings. Hemel Hempstead: Harvester Wheatsheaf.
82. Tresch, M.C., Sinnamon, H.M. and Seamon, J.G. (1993). Double dissociation of spatial and object visual memory: Evidence from selective interference in intact human subjects. Neuropsychologia 31:211–219.
83. Tversky, B., Kugelmass, S. and Winter, A. (1991). Cross-cultural and developmental trends in graphic productions. Cognitive Psychology 23:515–557.
84. Ullman, S. (1984). Visual Routines. Cognition 18:97–159.
85. vanSommers, P. (1984). Drawing and cognition. Cambridge, UK: Cambridge University Press.
86. Zhang, J. (1997). The nature of external representations in problem solving. Cognitive Science 21:179–217.

7. Combining Semantic and Cognitive Accounts of Diagrams

Corin A. Gurr

Theories of diagrammatic reasoning typically seek to account for either the formal semantics of diagrams, or for the cognitive advantages which diagrams hold over other forms of representation. Regrettably, almost no theory exists which accounts for these issues jointly, nor how they affect one another. This chapter sets out the basis for such a combined theory, the main parts being: a principled exploration of the fundamental components of diagrammatic languages; semantic and cognitive perspectives on reasoning in diagrams; and the relation between these three parts. This chapter thus lays out a larger context than is generally used for examining the use of diagrams in reasoning or communication. A context in which detailed studies of sub-problems – here, what it is that makes diagrams *effective* – may be embedded.

7.1 Requirements for a Combined Account

Theories of diagrammatic reasoning have become more prevalent in recent years, as this volume gladly illustrates. A collection of both seminal and contemporary papers on diagrammatic reasoning, from cognitive and computational perspectives, is found in [1]. An excellent overview of the different approaches, historical and recent, to this field of research is found in [2].

The majority of theories of diagrams fall into two broad categories. The first are motivated by the desire to provide a justification for diagrammatic reasoning in formal proofs. Such theories are primarily concerned with providing an account of the correspondence between diagrams and some formal semantics for them. The second category of theories is concerned with explaining the impact of diagrams upon human cognition; seeking to explain what advantages diagrammatic representations hold for the reasoner over other forms of representation.

Theories of the first kind (for example, [3–5] and the collection in [6]), in common with the formal semantic studies of natural language from which they derive their methods, generally ignore issues concerning the cognitive complexity of inference, and indeed leave out any consideration of the inferential mechanisms that operate over the sentences or diagrams of the languages whose semantics is being specified. Theories of the second kind (for example, [7–12]) generally lack a fully specified formalism and semantics on which they could base a computational account of how the system of representations is embedded in a user's performance of some task.

This chapter suggests the basis for a combined semantic and cognitive theory of diagrammatic representation and reasoning. A particular aim is the provision of a combined account of the relative *effectiveness* of diagrammatic languages. We commence in the next section with a principled consideration of the semantics of diagrams, through exploration of the similarities and differences between the fundamental components of diagrammatic and traditional textual languages. Section 7.3 examines issues relating to the effectiveness of reasoning in diagrams, particularly from a semantic perspective. Section 7.4 introduces the primary issues of cognitive effectiveness, and relates these to those issues raised in the previous two sections. While a complete exploration of cognitive effectiveness is beyond the scope of this paper, Section 7.5 highlights the major remaining issues. Finally, Section 7.6 concludes with a summary of the major issues for a combined theory of diagrams.

7.2 Components of Diagrammatic Languages

An attempt, not to claim any complete theory of diagrams, but rather to sketch a larger structure which accounts for the processes of using diagrams in reasoning or communication, and to clarify the relations between its component problems, is given in [13] based upon an examination of the analogies (and dis-analogies) between diagrammatic languages and, more classical, textual languages (ranging in complexity from simple algebras, classical logics, more esoteric logics and ultimately to full-blown natural languages). We review that sketch here, emphasising how it demonstrates that aspects of 'semantics' are more broadly distributed over the components of diagrammatic languages than in the traditional textual case.

According to Morris [14], the study of systems of communication can be divided into three parts: syntax, semantics and pragmatics. Syntax is devoted to "the formal relation of signs to one another"; semantics to "the relation of signs to the objects to which the signs are applicable" and pragmatics to "the relation of signs to (human) interpreters". While the notion of "signs" can be read very widely, most work in pragmatics has followed in the footsteps of syntax and semantics, and focused on language use. This section examines syntactic, semantic and pragmatic issues in a theory of diagrams. Firstly, however, it is necessary to consider something even more fundamental than the syntax of diagrams: their vocabularies.

7.2.1 Diagrammatic Vocabularies

In textual languages the notions of vocabulary, syntax and semantics are readily separable. The syntactic rules which permit construction of sentences may be completely independent of the chosen vocabulary, and may be clearly delineated from a definition of semantics. For example, let P be a propositional logic whose vocabulary consists of the propositions p and q, and the symbols $\wedge$ and $\neg$ representing "and" and "not" respectively. The syntactic and semantic rules for P tell us, respectively, how to construct and interpret formulas using this vocabulary. However, we may substitute the symbols $\{X, Y, \&, \tilde{}\}$ for $\{p, q, \wedge, \neg\}$ throughout P to produce a logic which is effectively equivalent. Alternatively, we could retain the vocabulary and syntax of P, while altering the semantics to produce a vastly different logic.

In diagrammatic languages the concepts of vocabulary, syntax and semantics do not separate so clearly. For example, a diagrammatic vocabulary may include shapes such as circles, ellipses, squares, arcs and arrows, all of differing sizes and colours. These objects often fall naturally into a hierarchical typing which will almost certainly constrain the syntax and, furthermore, inform the semantics of the system – similarly in the case of spatial representing relations, such as transitivity, which are part of the vocabulary but clearly constrain the construction of potential diagrams and will likely be mapped to semantic relations with similar logical properties.

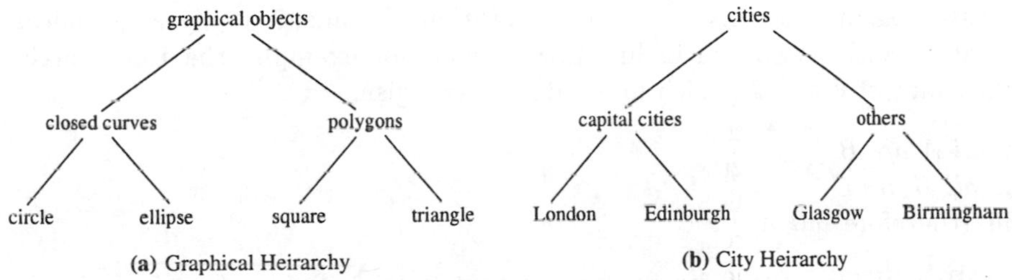

Figure 7.1. Matching syntactic and semantic hierarchies.

A fundamental distinction between textual and diagrammatic vocabularies is in the categorial nature of the latter. Essentially, a naturally ordered type hierarchy holds over many graphical symbols – something that cannot, to any significant extent, be said of textual symbols. This hierarchy may be 'directly' exploited by the semantics of symbols so as to reflect the depicted domain.[1] For example, Fig. 7.1(a) shows a hierarchy of graphical symbols in

[1] Note that this partitioning of symbols into direct and indirect categories should not be interpreted as suggesting that all visual representations, whether texts or diagrams, are either exclusively direct or indirect. Clearly "pure" languages, containing only direct or indirect symbols, are possible. Such examples, however, are merely extremes in a spectrum of representations which employ both direct

which types are ordered from top to bottom, though not from right to left. A situation which mapped these symbols to cities, such as those of Fig. 7.1(b), could exploit this type structure, preserving the ordering of types in each domain across the mapping. The advantage of this constraint is obviously that anything which holds for cities should hold for capital cities, and anything which holds for those should hold for London, say; and the analogous generalisations apply on the graphical side.

This example demonstrates that, purely as a consequence of choice of symbols for a diagrammatic vocabulary, certain inferences from any given representation will arise, as it were, "for free". Exploiting the categorial nature of diagrammatic vocabularies in this way requires that the structure inherent in the diagrammatic vocabulary is systematically matched to a relevant structure in the semantics. This systematic mapping of structure in the representing domain (diagrammatic language, in this case) to semantic structures is typically referred to as "systematicity".

7.2.2 Syntax and Semantics

A second fundamental aspect of diagrammatic languages, which distinguishes them from textual ones, is that generally the representing relations between diagrammatic tokens are "directly" semantically interpreted. That is to say that, unlike textual languages, there is typically a direct mapping from some representing relation in a diagram to the relevant semantic relation. The classic example of this is the representation of transitivity by a graphical relation such as spatial inclusion. Consider, for example, the Euler circle diagrams of Fig. 7.2, which illustrate the syllogism:

i *all A are B*
ii *all B are C*
iii (therefore) *all A are C*

Here the transitive, irreflexive and asymmetric relation of set inclusion is represented by the similarly transitive, irreflexive and asymmetric graphical relation of proper spatial inclusion in the plane. In textual languages the relationships between tokens are necessarily expressed by the concatenation relation, which must then be interpreted by some intermediary abstract syntax that captures the desired relationship. As a consequence of this interpretation of abstract syntax, concatenation has no uniform semantic interpretation.

A striking feature of many *effective* diagrams, and a consequence of direct interpretation, is that the spatial relations between their tokens share logical properties with the (not necessarily spatial) relations between denoted objects in the target domain. The classic example of representational efficacy arising from this preservation of constraints is the representation of set inclusion

and indirect symbols. It is between the extremes of this spectrum, rather than at them, that most visual representations exist.

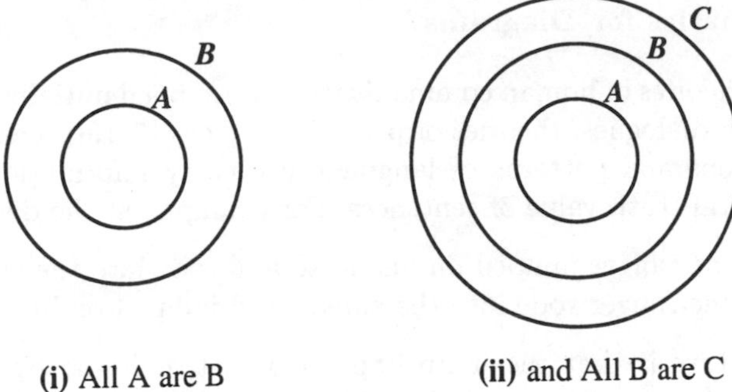

(i) All A are B (ii) and All B are C

Figure 7.2. Transitivity in Euler circle diagrams.

(transitive, irreflexive and asymmetric) by spatial inclusion in the plane, as in the Euler circle diagrams of Fig. 7.2. Since proper inclusion is transitive, irreflexive and asymmetric, its efficacy in representing set inclusion is obvious.

This matching of properties between represented and representing (diagrammatic) relations is, again, an example of systematicity. Notice that what we have so far is common to typical discussions about the semantics of graphics. Indeed, account for the systematicity principle may be seen as fundamental to approaches such as [15,16], and also to discussions of analogical reasoning, as in [17]. However, Wang and Lee [16] stress the importance of attention to the properties of graphical relations, such as transitivity: it may be a source of trouble if a transitive relation is used to depict one which is intransitive, or vice versa. It is in general not possible to map all relations exactly, and in different cases different properties are the important ones, so hard-and-fast rules are difficult to devise.

The systematicity principle, as we have shown, may be interpreted as applying at many different levels to the semantics of a representation system. For example, at one level it may refer to the matching of logical properties between representing and represented relations (such as representing set inclusion by spatial inclusion). At a more abstract level it may refer to the preservation of some type hierarchy between represented and representing worlds, as with the domain and graphical types of Figs. 7.1(b) and 7.1(a), respectively. Systematicity, in the sense of a structuring of semantics, is certainly not exclusively the preserve of diagrammatic languages. The distinction is that, as diagrammatic vocabularies may possess a hierarchical type structure and diagrammatic relations may be directly semantically interpreted, diagrams, thus, have a head start over textual representations in the systematicity stakes. However, a significant consequence of this is that the systematicity of diagrammatic languages will vary from language to language, while in the textual case it is more-or-less universal.

7.2.3 Pragmatics for Diagrams

In linguistic theories of human communication, developed initially for written text or spoken dialogues, theories of pragmatics (see [18, 19]) seek to explain how conventions and patterns of language use carry information over and above the literal truth value of sentences. For example, in the discourse:

1. (a) The lone ranger jumped on his horse and rode into the sunset.
   (b) The lone ranger rode into the sunset and jumped on his horse.

(1a)'s implicature is that the jump happened first, followed by the riding. By contrast, (1b)'s implicature is that riding preceded jumping. In both (1a) and (1b), implicatures go beyond the literal truth conditional meaning. For instance, all that matters for the truth of a complex sentence of the form P *and* Q is that both P and Q be true; the order of mention of the components is irrelevant. Pragmatics, thus, helps to bridge the gap between truth conditions and 'real' meaning. This concept applies equally well to diagrams. Indeed, there is a recent history of work which draws parallels between pragmatic phenomena that occur in natural language, and for which there are established theories, and phenomena occurring in visual languages [20–22].

Studies of digital electronics engineers using CAD systems for designing the layout of computer circuits demonstrated that the most significant difference between novices and experts is in the use of layout to capture domain information [23]. In such circuit diagrams the layout of components is not specified as being semantically significant. Nevertheless, experienced designers exploit layout to carry important information by grouping together components which are functionally related. By contrast, certain diagrams produced by novices were considered poor because they either failed to use layout or, in particularly "awful" examples, were especially confusing through their misuse of the common layout conventions adopted by the experienced engineers. The correct use of such conventions is thus seen as a significant characteristic distinguishing expert from novice users. These conventions, termed "secondary notations" by Petre and Green [23], are shown by Oberlander [22] to correspond directly with the graphical pragmatics of [21].

More recent studies of the users of various other visual languages, notably visual programming languages, have highlighted similar usage of graphical pragmatics [11]. A major conclusion of this collection of studies is that the correct use of pragmatic features, such as layout in graph-based notations, is a significant contributory factor in the comprehensibility, and hence usability, of these representations.

In summary, as this section has demonstrated, aspects of semantics are more broadly distributed throughout the components of diagrammatic languages than in their traditional textual counterparts. The key for effective diagrammatic languages is that they should exploit this difference.

7.3 Relating Representations to Reasoning Tasks

A primary argument put forward to justify the claim of diagrammatic representation systems being more effective than textual ones is that certain inferences are somehow more immediate, or even are automatic, in diagrams. In such representational systems conclusions appear "for free", as compared with textual systems where a logical inference must be made to produce the conclusion. For example, inferring from the information *all A are B; all B are C* the conclusion *all A are C* is arguably a more straightforward inference in the direct representation system of Fig. 7.2 (Euler's circles) than in the textual case. It can be argued that this is due to the fact that construction of diagram 7.2(ii) automatically includes the representation of the conclusion *all A are C*, and thus the information appears for free.

This argument is given a formal account by Shimojima [24], where apparent inferences such as that of Fig. 7.2(ii) are termed inferential "free rides". A free ride is defined as a form of side effect of the manipulation of a diagram. When a sequence of valid operations cause some consequence to become manifest in a diagram, where that consequence was not explicitly insisted upon by the operations, a free ride occurs. For the syllogism illustrated by the Euler circle diagram of Fig. 7.2 for example, the free ride manifests itself when we construct Fig. 7.2(ii). Fig. 7.2(i) is drawn according to the premise *all A are B*, with the circle *A* inside the circle *B*. From the premise *all B are C* we next draw the circle *C* so as to completely surround the circle *B* (Fig. 7.2(ii)). We observe that the circle *C* completely surrounds *A*, and thus conclude that *all A are C*. Note, however, that none of the instructions for constructing Euler circle diagrams which caused us to draw Fig. 7.2(ii) explicitly insists that we should draw circle *C* so as to surround circle *A*. This is a semantically meaningful (and correct) fact which was entailed for "free" by virtue of the construction rules applied to the premises. This then, is a free ride. However, this issue is more complex than it may appear at first glance.

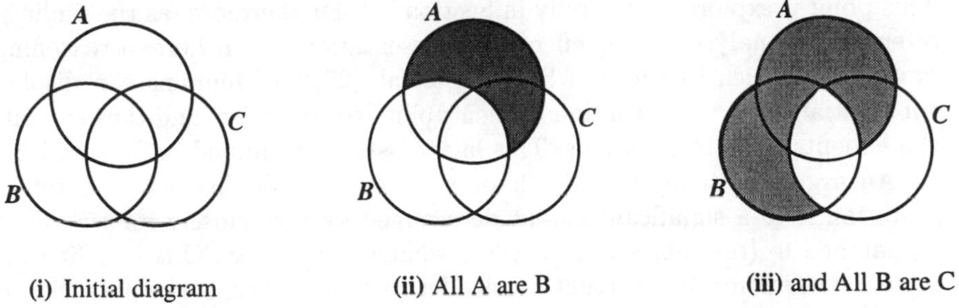

Figure 7.3. Transitivity in a Venn diagram (shaded areas indicate empty sets).

Consider the Venn diagram of Fig. 7.3, which represents the earlier syllogism of Fig. 7.2 (*all A are B; all B are C* therefore *all A are C*). In this diagram precisely the same graphical relation (proper spatial inclusion) represents the logical relation of set inclusion. In the same manner as for Fig. 7.2(ii), construction of the final Venn diagram of Fig. 7.3(iii) automatically includes the representation of the conclusion *all A are C*. However, it would be disingenuous to claim that, for a human reasoner, there would be no difference whatsoever between recognising this conclusion in Fig. 7.2(ii) and recognising it in Fig. 7.3(iii). In the light of this example, it would seem more accurate to term such occurrences "cheap rides" rather than free rides; with the addendum that some rides are cheaper than others.

We take the view that reasoning is a two-stage process[2]: Firstly, the process of constructing a new diagram which composes premises typically results in a number of inferences being made, all of which are visible in the resulting diagram. Secondly, the desired conclusion must be recognised in the resulting diagram. In diagrammatic languages the diagram resulting from an inference step typically contains numerous logical inferences. For example, both the diagrams of Fig. 7.2(ii) and Fig. 7.3(iii) contain the required conclusion (that *all A are C*), but they also contain a number of other potential conclusions. Recognising the desired conclusion is not automatic and ease of recognition can vary substantially between differing representational systems, as it does in this case.

A significant issue here is the task for which the diagrammatic notation is being used. In the case of solving syllogisms such as the above, the task is a relatively simple one and the Euler circle diagrams of Fig. 7.2 are significantly more effective than the Venn diagrams of Fig. 7.3. However, this is because the representation is well matched to the task. Venn diagrams are more subtle than Euler circle diagrams, being more sophisticated than necessary for this particular task, but capable of benefiting reasoning in other, more complex, tasks for which Euler circle diagrams are insufficient. A major point to note here is that, for a human reasoner, what significantly assists the reasoning process is *recognising* the inherent properties of a particular system of representation. Properties must be recognised for their benefits to be available. This point is explored more fully in Section 7.4. Furthermore, as the findings of empirical analyses of the effect of representations upon human reasoning have shown (such as those of Stenning et al. [25, 26]), humans can display substantial individual variation in their ability to recognise, and thus exploit, the benefits of representations. This latter issue is examined in Section 7.5.

An argument taken by this chapter is that the effectiveness of a representation is to a significant extent determined by how closely its semantics is matched to (resembles in structure) what it represents. One benefit that certain diagrammatic representations offer to support this is the potential to directly capture pertinent aspects of the represented artifact (whether this

[2] "Three-stage process" might be a more accurate statement, but we do not consider here the initial stage of selecting a particular combination of inference rules and premises. Our argument here is primarily concerned with the performance of a reasoning *step*, rather than issues concerning reasoning *strategy*.

be a concrete artifact or some abstract concept). To clarify this argument we must explain what is meant here by "pertinent".

By "pertinent" aspects of the represented artifact, we refer to those which are relevant to particular reasoning tasks. We argue that reasoning is strongly influenced by the structure of the representation within which one reasons. Where the structure of a representation matches the primary concepts over which one must reason, reasoning is made easier. Conversely, having the "wrong" structure in a representation will interfere with reasoning, making it more difficult. This argument is supported by a number of empirical studies of users employing different representations for similar tasks. Notable among these are Stenning and Yule's study [26] of the representations, including numerous diagrammatic representations, used in solving logical syllogisms. The study examined students' errors in syllogistic reasoning, and demonstrated that a syllogism's representation was a significant predictor of error. Similarly, a study by Zhang and Norman [12] examined the effect of alternative representations employed in solving the Tower of Hanoi problem. The purpose of this study was to assess the interplay between internal (cognitive) and external (diagrammatic, in these cases) representations. An examination of the results reveals that students were most successful when using those external representations which most accurately reflected the problem task, and were less successful when the representations were either incomplete (being too abstract) or unsupportive of basic reasoning tasks.

7.4 Effectiveness as Constraints over Interpretation

Arguing that effective diagrams are those which are closely matched to what they represent leads us to consider how we may measure the closeness of a particular match. In this section we briefly review a framework, originally introduced in [15], which provides such a means of measurement. However, to satisfactorily combine such a semantic account of effectiveness in diagrams with a cognitive account requires considering the relation between this semantic account and issues concerning the human recognition of semantic effectiveness; that is, issues related to the process by which a user translates to, or from, the information and inferences as represented and (the user's understanding of) the world which this actually represents.

A major difference between diagrammatic and textual systems, explored more fully in [20], concerns the means by which constraints over interpretation determine the effectiveness of a representation system for the human reader/reasoner. The suitability, learnability and usability of diagrammatic representations (and hence effectiveness) are profoundly influenced by issues of human reaction to representations. It is the constraints over interpretation which determine the suitability of a particular representation for a task of reasoning. For diagrams these constraints form three dimensions. The first of these concerns their *origins*: whether the constraints are intrinsic to the interpretation of the medium, or are of conventional, external origin. The second dimension of constraints concerns their *point of operation*; that is, whether constraints apply to, and aid reasoning, over individual representations within a system of representation; or whether they apply to the representational system as a whole. The final dimension of constraints concerns their *availability* to the human who reasons with the system. We review next the framework of [15] for measuring the closeness of match between representation and represented. We then relate this account to Section 7.2's account of diagram semantics, and to the dimensions of constraints over interpretation.

7.4.1 The Relation Between Representation and Represented

Different forms of representation may vary in how closely they correspond to that which they represent. Diagrams are typically noted as corresponding quite closely: being "homomorphisms" or even "isomorphisms". A discussion in [15] of whether or not diagrams are homomorphisms (or isomorphisms) considers both diagrams and the artifacts they represent as "α-worlds": collections of objects and sets of relations between them. Morphisms are examined in the context of the mappings (in either direction) between such abstractions of the structure in diagrams and the structure of the semantic domain. For a mapping from represented to representation to be homomorphic every relation over semantic objects must be accurately represented in the representation. "Accurate" in this context means that a relation is represented between objects in the representation if it holds for those objects in the semantics, and is not represented if it does not hold. Stenning and Inder [27] also recognise this property as being of importance, referring to representational systems for which every possible representation has this property as systems possessing *exhaustiveness*. For a mapping from representation to semantics to be homomorphic every relation over objects in the representation must accurately correspond to some relation over semantic objects.

Differing representations and representational systems may clearly vary in the closeness of the correlation between represented and representing. It is easily seen that non-homomorphic representational systems are likely to be unreasonably intractable, as there would be no guarantee of any correlation between the representing and represented objects and relations. Consequently, the framework of [15] particularly explores the spectrum ranging from homomorphic (in effect, structurally *similar*) representations to isomorphic (structurally *equivalent*) ones, highlighting significant points on that spectrum. Thus effective (closely matched) representations are those which are closer to being isomorphisms. For example, let us assume that the correlation between a diagram and that which it represents is an isomorphism. The implication of this for the human agent who interprets the diagram is that their interpretation correlates precisely and uniquely with the artifact (concrete or abstract) being represented. By contrast, where the correlation is not an isomorphism then there may potentially be a number of different target artifacts which would match the interpretation.

7.4.2 Directness, (Iso)morphism and the Origin of Constraints

An important issue when considering an isomorphic representation is whether the isomorphism is intrinsic, or must be enforced by external constraints.

A *B* *C* *D*

Figure 7.4. Diagram with intrinsic representation of integer ordering.

Consider a representation system for integer ordering, illustrated by the diagram in Fig. 7.4. In this diagram particular integers (say, the integers 1–4) are represented by labelled squares (*A* to *D* respectively), and the less-than relation is represented by left-of. The relation left-of is isomorphic in properties to the less-than relation, being irreflexive, anti-symmetric and transitive. Such isomorphisms were first referred to as *intrinsic* representations by Palmer [28]. This illustrates our first dimension of constraints over interpretation: their origin, being either intrinsic – as here – or externally imposed.

7.4.3 The Point of Operation of Constraints

Choosing to represent a relation with an isomorphic (graphical) relation does not by itself guarantee that every ensuing representation is an isomorphism.

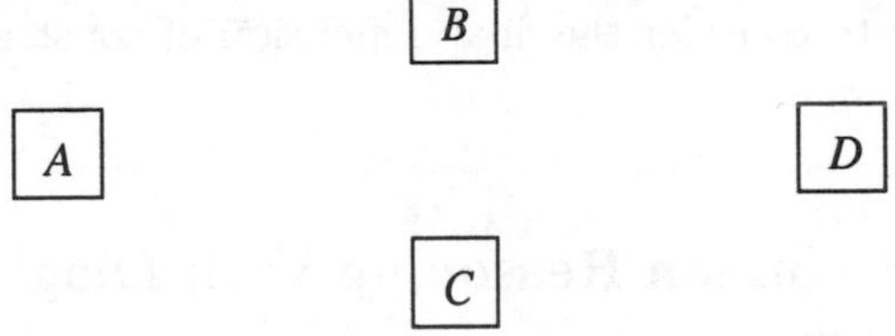

B

A *D*

C

Figure 7.5. Intrinsic, yet non-isomorphic, diagram representing integer ordering.

As an example, consider the diagram of Fig. 7.5, which is constructed by the same representational system as that of Fig. 7.4. In this diagram neither of the squares labelled *B* and *C* is left of the other. Thus neither integer represented by these two squares is less than the other. Several interpretations are possible: that the diagram is incomplete (the relation between *B* and *C* is perhaps unknown); that distinct squares do not necessarily denote distinct integers (*B* and *C* denote the same integer); or perhaps that the diagram is simply invalid (non-homomorphic). Whatever the explanation, the diagram of Fig. 7.5 is clearly not an isomorphic representation of any set of integers. Thus while certain diagrams constructed using this representational system may be isomorphisms, the system is not isomorphic in all cases. This illustrates our second dimension of constraints over interpretation: their point of operation. The constraints in the above representational system operate at the level of (certain) individual diagrams, but not at the level of the entire system.

7.4.4 Relating Directness and Morphism to Constraints

Section 7.3 argued that a significant determiner of effectiveness in a representation (or representational) system is that it should be closely matched to the represented artifact for desired reasoning tasks. The framework of [15], which explores homo- and iso-morphic representations, provides a means of measuring the closeness of this matching. Shimojima's free rides may be seen as a consequence of a particularly closely matched representation. However, as also argued in Section 7.3, such rides are more accurately termed "cheap" rather than free, and it is the constraints over interpretation (origin, point of operation and availability) which determine the cost of a particular ride.

Typically, effective diagrams may exploit the potential benefits of directness, as discussed in Section 7.2, to intrinsically constrain their interpretations. The classical example of a free ride in Fig. 7.2 illustrates this, demonstrating that constraints of intrinsic origin are cheaper (more obvious to the reader) than those of external origin. Similarly, constraints that operate over an entire system of representation imply cheaper rides than those which only operate over certain diagrams within a system – as in the former case the reader may be confident that the constraint applies to *every* diagram in the representational system. However, neither of these dimensions of constraints will have much impact if they are not readily apparent to the reader. This leads us naturally to consider the final dimension of constraints: their availability.

7.5 Studies of Human Reasoning with Diagrams

Understanding the availability of constraints for the human reader – recognition of intrinsic or systemic semantic constraints – is a broad field of study. Indeed, identification of the most "appropriate" means of representing particular information could be said to be the fundamental goal of Graphic Design. Clearly a complete exploration of this issue is beyond the scope of this chapter. Instead, this section focuses on a further, significant, remaining cognitive issue: individual differences in human response to representations.

The previous sections have argued that the difficulty of the different stages of the reasoning process vary substantially between the textual and diagrammatic cases. Furthermore, in the diagrammatic case (where recognition of a conclusion is the significant factor) the effectiveness of different representational systems may be at least in part predicted by the complexity of this stage. These arguments are supported by empirical studies of students solving syllogisms with a variety of representations (including textual logic, Venn diagrams and Euler circle diagrams). The results, reported in [26], do indeed show that – at a gross level – diagrammatic representations are more effective than textual ones, and that, once the system is learned, Euler circle diagrams are more effective than Venn diagrams. However, this is by no means the whole story.

While at a gross level diagrammatic representations of syllogisms appear to win out over textual ones, examining the data in detail throws up some highly significant variations between individual students. Continuations of the studies reported by [26] have revealed that different students can make significantly different numbers of errors in the separate processes of translating logical sentences to diagrams, manipulating the diagrams, and reading off conclusions. Furthermore, one can predict, based upon individual cognitive differences between the students (such as holist/serial and spatial/non-spatial aptitude) in which process a particular student is most likely to make errors.

This impact of individual cognitive differences upon reasoning ability, in particular its relevance to diagrammatic reasoning, is even more vividly demonstrated by a study of student users of Hyperproof [29]. Hyperproof is a computer program, created for teaching first-order logic, which uses multimodal graphical and textual methods. Logical sentences about objects are represented in a "Block's World" of polyhedra on a checkerboard. A distinctive feature of Hyperproof is that it has "graphical" rules which permit users to transfer information to and fro, between the graphical and textual modes.

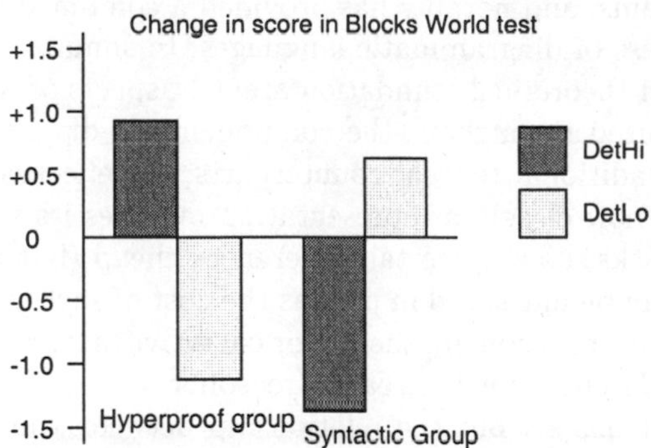

Figure 7.6. Change in reasoning ability of different teaching groups.

Stenning et al. [25] report a study of Stanford undergraduates learning first-order logic from Hyperproof, as compared with the "Syntactic Group", a control class learning from the textual-only part of the Hyperproof program (with the graphical window turned off). Figure 7.6 presents one notable result.

While the study was set up to examine the utility of the graphical representation in teaching first-order logic, a surprising result was just how strong were the interactions between students' prior styles of reasoning and the outcomes of different methods of teaching. Those students ("DetHi" students) who were adept at analytical reasoning (as indicated by prior tests) found the graphics-enabled version of Hyperproof to be an extremely effective way of teaching logic. Conversely, the (in general, equally able) students who scored lower on the pretests ("DetLo" students) found the textual-only Hyperproof far more effective. Most surprisingly, the results, seen in Fig. 7.6 (indicating relative change in pre- and post-test scores) indicate that for those students exposed to the system which did not suit their preferences, their reasoning performance had actually deteriorated by the end of the course.

The moral to be drawn from these results is that even the most effective representations are not equally effective for everybody. This suggests that a key factor in ensuring the effectiveness of a representational system (visual or otherwise) is in making the advantages of that system equally available to all users. A second moral to draw from this study is that individual differences cannot be discounted when assessing effectiveness. It should be noted that, in the data presented in Fig. 7.6, the DetHi and DetLo students pretty much balance one another out. Thus, if individual differences were ignored, we would, erroneously, be unable to note any significant difference between the effectiveness of the Hyperproof versus the Syntactic teaching courses.

7.6 Summary and Conclusions

This paper has presented the basis for a combined semantic and cognitive theory of diagrams, and notably has provided a combined account of the relative effectiveness of diagrammatic languages. In summary, the major issues in our combined theoretical foundation are: (i) aspects of semantics are more broadly distributed throughout the components of diagrammatic languages than in their traditional textual counterparts; (ii) effectiveness of reasoning is tied up with how closely a representation matches its semantics and supports desired tasks, making certain inferences cheap (but seldom free); thus effectiveness may be measured in part as the cost of such inferences; (iii) constraints over interpretation are the major cognitive factors further influencing the "cost" of inferences for the reader/reasoner.

This chapter has set out the relationship between the above issues: how the origin and point of operation of constraints relates to our semantic account of effectiveness, and the role played in this by fundamental diagrammatic aspects such as semantic directness. Finally we have set out the remaining crucial factors which determine the cognitive effectiveness of representations. The key point here is that, for the human reasoner, the advantageous properties must be easily recognisable for their benefits to be available. As the studies of the previous section indicate, human reasoners vary significantly in their ability to recognise and exploit both the meta-systemic properties and the more fundamental systematicity of diagrammatic languages.

The conclusion we draw here is that while the above properties give visual and diagrammatic representations the potential to be highly effective in reasoning, this effectiveness will only be realised if these properties are available to the human reasoner. Directness, systematicity, free rides and intrinsically isomorphic representations will all facilitate object-level reasoning with visual representations; but it is the meta-logical fact that these properties can be seen to be inevitable which has the more important impact on the effectiveness of the representational system.

Acknowledgements

The development of the ideas presented here has been greatly assisted by numerous discussions with colleagues in the Human Communication Research Centre; in particular the contributions of John Lee and Keith Stenning have proved invaluable. This work was supported by EPSRC Grant # GR/L37953 "Understanding Software Architectures".

References

1. Glasgow, J., Narayan, N.H. and Chandrasekaran, B. (1995). Diagrammatic reasoning: Cognitive and computational perspectives. Cambridge, MA: MIT Press.
2. Narayan, N.H. and Hübscher, R. (1998). Visual language theory: Towards a human-computer interaction perspective. In K. Marriot and B. Meyer (Eds), Visual language theory. Berlin: Springer, Ch. 3.
3. Myers, M. and Konolige, K. (1995). Reasoning with analogical representations. In J. Glasgow, N.H. Narayan, and B. Chandrasekaran (Eds), Diagrammatic reasoning: Cognitive and computational perspectives. Cambridge, MA: MIT Press, pp. 273–302.
4. Shin, S-J. (1995). The logical status of diagrams. Cambridge, UK: Cambridge University Press.
5. Sowa, J.F. (1993). Relating diagrams to logic. In Conceptual graphs for knowledge representation: Proceedings of 1st international conference on conceptual structures. LNAI 699. Berlin: Springer.
6. Allwein, G. and Barwise, J. (1996). Logical reasoning with diagrams. New York: Oxford University Press.
7. Blackwell, A.F. and Green, T.R.G. (1999). Does metaphor increase visual language usability? In 15th IEEE symposium on visual languages (VL'99). Los Alamitos CA: IEEE Computer Society Press, pp. 246–253.
8. Blackwell, A.F., Whitley, K.N., Good, J. and Petre, M. (in press). Cognitive factors in programming with diagrams. Artificial Intelligence Review (special issue on thinking with diagrams).
9. Campbell, K.J, Collis, K.F. and Watson, J.M. (1995). Visual processing during mathematical problem solving. Educational Studies in Mathematics 28:177–194.

10. Hegarty, M. (1992). Mental animation: Inferring motion from static displays of mechanical systems. Journal of Experimental Psychology: Learning, Memory and Cognition 18(5):1084–1102.
11. Petre, M. (1995). Why looking isn't always seeing: Readership skills and graphical programming. Communications of the ACM 38(6):33–45.
12. Zhang, J. and Norman, D. (1994). Representations in distributed cognitive tasks. Cognitive Science 18:87–122.
13. Gurr, C.A. (1999). Effective diagrammatic communication: Syntactic, semantic and pragmatic issues. Journal of Visual Languages and Computing 10(4):317–342.
14. Morris, C.W. (1938). Foundations of a theory of signs. In O. Neurath, R. Carnap and C. Morris (Eds), International encyclopedia of unified science. Chicago: Chicago University Press, pp. 77–138.
15. Gurr, C. (1998). On the isomorphism, or lack of it, of representations. In K. Marriot and B. Meyer (Eds), Visual language theory. Berlin: Springer, Ch. 10.
16. Wang, D. and Lee, J. (1993). Visual reasoning: Its formal semantics and applications. Journal of Visual Languages and Computing 4:327–356.
17. Gentner, D. (1983). Structure-mapping: A theoretical framework for analogy. Cognitive Science 7:155–170.
18. Gazdar, G. (1979). Pragmatics: Implicature, presupposition and logical form. New York: Academic Press.
19. Levinson, S.C. (1983). Pragmatics. Cambridge, UK: Cambridge University Press.
20. Gurr, C., Lee, J. and Stenning, K. (1998). Theories of diagrammatic reasoning: Distinguishing component problems. Mind and Machines 8(4):533–557.
21. Marks, J. and Reiter, E. (1990). Avoiding unwanted conversational implicature in text and graphics. In Proceedings of the eighth national conference on artificial intelligence (AAAI-90). Menlo Park, CA: AAAI Press, pp. 450–456.
22. Oberlander, J. (1996). Grice for graphics: Pragmatic implicature in network diagrams. Information Design Journal 8(2):163–179.
23. Petre, M. and Green, T.R.G. (1992). Requirements of graphical notations for professional users: Electronics CAD systems as a case study. Le Travail Humain 55:47–70.
24. Shimojima, A. (1996). Operational constraints in diagrammatic reasoning. In J. Barwise and G. Allwein (Eds), Logical reasoning with diagrams. New York: Oxford University Press, pp. 27–48.
25. Stenning, K., Cox, R. and Oberlander, J. (1995). Contrasting the cognitive effects of graphical and sentential logic teaching: Reasoning, representation and individual differences. Language and Cognitive Processes 10(3/4):333–354.
26. Stenning, K. and Yule, P. (1997). Image and language in human reasoning: A syllogistic illustration. Cognitive Psychology 34(2):109–159.
27. Stenning, K. and Inder, R. (1995). Applying semantic concepts to analysing media and modalities. In J. Glasgow, N.H. Narayan and B. Chandrasekaran (Eds), Diagrammatic reasoning: Cognitive and computational perspectives. Cambridge, MA: MIT Press, pp. 303–338.
28. Palmer, S.E. (1978). Fundamental aspects of cognitive representation. In E. Rosch and B.B. Lloyd (Eds), Cognition and categorisation. Hillsdale, NJ: Lawrence Erlbaum Associates, pp. 259–303.
29. Barwise, J. and Etchemendy, J. (1994). Hyperproof. Stanford: CSLI Publications.

8. Tactile Maps and a Test of the Conjoint Retention Hypothesis

Simon Ungar
Mark Blades
Christopher Spencer

Kulhavy and his colleagues found that when a map and related factual information were learned together, the probability of recalling the factual information was greater than when information was learned without a map, or with a list of place names. They account for this finding with their "conjoint retention" hypothesis – a corollary of Paivio's "dual coding" theory. The present study extended this research by including a group of blind and visually impaired participants who learned a tactile map. Twelve blind and visually impaired participants and forty-eight sighted participants learned either a map (map condition) or a list of place names (list condition) for either 10 minutes or 2 minutes and then heard a text describing places on the map/list. After a filled pause, participants were asked to recall information from the text and, in the map condition, to make a reconstruction of the map. Kulhavy's original finding was replicated for sighted participants who studied the map/list for 10 minutes. However, sighted participants exposed to the map/list for 2 minutes and blind participants performed at the same level with both the map and with the list. In all cases, differences between conditions were small. Further analyses revealed that encoding of the map's structure, a crucial variable in Kulhavy's model, may not have been a major factor in determining recall of factual information.

8.1 Introduction

Tactile maps and diagrams have a long history of practical use by blind and visually impaired people. It is likely that informal diagrams have been scratched, carved or constructed from the early days of human civilisation. It is only comparatively recently, however, that attempts have been made to understand the psychological processes underlying the use of tactile maps, and the implications that this might have for the education of blind and visually impaired people or for the design of tactile diagrams. As recently as 1932, it was still argued that people who had been blind from birth were unable to comprehend space, and that therefore any form of spatial description (i.e., maps and diagrams) would be pointless [26].

More recently, a number of researchers have shown that even people with no visual experience whatsoever can acquire spatial abilities which are equivalent to those of sighted people, albeit qualitatively different (e.g., [13, 22]). This has given rise to a renewed interest in tactile communication of spatial concepts in maps and diagrams [2, 3, 7, 9, 11, 20, 29, 33, 34].

Our previous research has focused on tactile maps, and in particular the ways in which blind and visually impaired people acquire information from tactile maps and the strategies they employ while using maps to make wayfinding decisions during navigational tasks [4, 29–31, 33, 34]. The present study focused on the way in which tactile maps are represented mentally and asked whether such representations can for the basis for conjoint retention of map and text information.

8.2 The Conjoint Retention Hypothesis

In a series of studies, Kulhavy and his colleagues examined what they refer to as the "conjoint retention" hypothesis, a corollary of Paivio's "dual coding" hypothesis according to which spatial/perceptual information and linguistic/verbal information are coded by qualitatively distinct processes, and are ultimately stored separately in memory [23]. When related spatial/perceptual and linguistic/verbal information is encoded, a dual representation is established in memory containing the 2 types of information, which are coded conjointly. Overlap between spatial and linguistic information will allow mutual cueing of one type of information by the other. The main prediction is that recall for either type of information will be improved when related spatial and linguistic information are encoded conjointly, e.g., when students learn a map together with textual information about the mapped place (see reviews of this research in [16, 17]).

In a typical study, participants viewed a cartographic map with instructions to learn it as an intact unit, and then either read or heard a text containing facts related to features of the map. After a filled pause, participants were asked to free-recall as many text facts as they could. It was generally found that participants who saw a map could recall more text facts than participants in other conditions who saw a list of map features, map fragments or no map at all (e.g., [16, 17]). According to Kulhavy, "overall performance increases because the dual perceptual/linguistic representation provides a richer cueing and retrieval base for the learner to draw from during recall" [15, (p. 30)]. This proposition, which we might term the weak version of conjoint retention, states that any stored visuospatial information which is related to any verbal–linguistic information in memory should give rise to mutually facilitated recall.

However, the claims of the conjoint retention hypothesis are stronger than this. The phenomena are expected to apply only to information learned from a physically present map or picture, and not to mental images elaborated on the basis of, for instance, a rich verbal spatial description of an environment. Neuropsychological work on the generation of mental images (e.g., [10, 28]) suggests that an essentially visual component persists in a stored mental image, which distinguishes it from verbally derived information in memory which presumably retains at least some of the temporal characteristics of verbal information. To test this part of the hypothesis, learning from an actual cartographic map was compared with learning from a good verbal description of spatial locations to determine their relative facilitation of recall for related text [15, 25]. It was found that facilitation was considerably greater in the case of the physical map.

8.3 Educational Implications

If diagrams and maps function for sighted people as the conjoint retention hypothesis suggests, they may also provide a useful pedagogical aid for blind and visually impaired people. However, it is known that tactile information (as opposed to visual information) is acquired sequentially over time and must be integrated to provide coherent spatial information [12,35]. For this reason, tactile information may have more in common with verbal information, because it is initially received temporally/sequentially and requires further processing to combine the information into an integrated spatial representation. According to the strong version of conjoint retention, it is only information which is acquired simultaneously, as a whole (i.e., visual information) that can facilitate text recall. If this is the case, learning from a tactile map would not be expected to result in facilitation for recall of related text facts.

There has been considerable debate in the mental imagery literature about whether people who are congenitally totally blind can form mental images which are at least functionally analogous to those of sighted people [8, 32]. Several researchers have shown that congenitally blind and sighted people perform similarly on a range of tasks, such as mental rotation [5,6,21,24,31], and scanning [14], which require the manipulation of mental images of a spatial character. It has also been shown that blind people can use tactile cartographic maps to make useful judgements about the spatial structure of their environments [9, 30]. Such findings suggest that mental imagery may have a general spatial basis which is used to code information from tactile and visual modalities, rather than separate dedicated coding systems for each modality, each of which retains the essential characteristics of its associated modality.

If the conjoint retention hypothesis is robust, as it appears to be from the many successful replications of the phenomenon (e.g., [16, 17]), the method can provide a useful means of investigating mental imagery in people who are blind or visually impaired. If blind people are capable of forming mental images of spatial or pictorial displays (such as tactile maps) which are functionally equivalent to the images derived from visual input, then we should expect to find the conjoint retention phenomena in blind people who learn a tactile map in conjunction with related text.

Another aspect of what we have termed the strong version of conjoint retention is the prediction that facilitation will be dependent upon accurate encoding of the absolute position of a feature within the frame of the map [16, 17]. For instance Kulhavy et al. [19, (p. 166)] claimed that recall of absolute location of features "is an index of how accurately structure is represented in the student's image of the map".

A number of studies have tested the relationship between the criterion of structure and memory for text facts by calculating the conditional probability of correctly recalling a particular text fact given that the related map feature has been correctly placed in a map reconstruction task – expressed as $P[fact/feature]$ [1,18,19,25,27]. According to the strong version of conjoint retention, this relationship should be close.

However, the interpretation of this conditional probability is somewhat problematic. The probability of obtaining a high mean value of $P[fact/feature]$ purely by chance becomes greater if few locations are placed correctly in the map reconstruction task. More significance can be attributed to a high value of $P[fact/feature]$ when a low value is obtained for another probability expression; namely that of recalling a text fact when the map feature was not placed correctly, or $P[fact/not(feature)]$.

In a review of a large number of studies of conjoint retention, Kulhavy et al. [16] cite values for $P[fact/feature]$ ranging between 0.53 and 0.85. However, $P[fact/not(feature)]$ was only calculated in one study where a value of around 0.37 was obtained. That this is considerably lower than its related $P[fact/feature]$ value of 0.80 provides some support for the dependence of facilitation of fact recall upon a structurally intact representation of the map.

The present study is a further exploration of the conjoint retention hypothesis. In our experiment we replicated Kulhavy's [16,17] study of conjoint encoding of a map and related factual text. Two groups of sighted people and a group of blind and visually impaired people took part in the experiment. One group of sighted people (Print-2 group) performed the experiment exactly as Kulhavy's participants had done (cf. [1,18,19,25]) and were allowed 2 minutes to look at the map or the list. The blind and visually impaired participants (Tactile-10 group) and the second sighted group (Print-10 group) followed the same procedure, except they were allowed 10 minutes to explore the map or list.[1]

The following hypotheses were made: following the results of the previous studies, we expected both groups of sighted participants to perform better when they associated the text with a map rather than with a list. As there

[1] Pilot studies revealed that 10 minutes was approximately the minimum exposure time needed for blind and visually impaired participants to explore the entire map. With shorter exposure times any difference between blind and sighted groups might have been due to the blind participants having failed to find all the map locations.

was no previous research on conjoint retention in blind and visually impaired people, we made no specific predictions about their performance. If the imagery of blind people is functionally equivalent to that of sighted people one would expect to find a facilitatory effect of map learning on text recall in both the sighted and the blind groups. If, on the other hand, specifically visual imagery is crucial for conjoint retention, we would expect the map to facilitate text recall in the sighted participants but not in the blind participants.

8.4 Experiment

8.4.1 Participants

There were 60 participants in the experiment. Twelve were blind or visually impaired (Tactile-10 group), and their individual characteristics are given in Table 8.1. Forty-eight sighted participants responded to advertisements placed in a university department of psychology, and were paid for participation. The sighted participants had a mean age of 25 : 0 and ranged from 16 : 2 to 48 : 11 and were randomly assigned to two groups of 24 participants (Print-10 group and Print-2 group).

Table 8.1. The characteristics of the blind and visually impaired participants involved in Experiment 1

Participant	Age(y : m)	Degree of impairment	Onset of impairment
LC	22 : 4	Totally blind	Birth
KH	42 : 2	Totally blind	Birth
KT	40 : 0	Totally blind	Birth
OY	30 : 3	Residual vision	Birth
EG	44 : 11	Residual vision	32 years
CS	34 : 0	Totally blind	Birth
CD	48 : 5	Totally blind	Birth
OS	56 : 3	Totally blind	20 years
JT	25 : 5	Residual vision	8 years
QK	13 : 6	Totally blind	Birth
MS	15 : 0	Totally blind	Birth
EY	15 : 1	Residual vision	Birth

8.4.2 Materials

The maps used in the experiment were of two fictional towns (referred to as River Town and Railway Town), and are shown in Fig. 8.1. Each map consisted of 18 labelled point symbols, a main road represented by a broad line, four minor roads represented by thinner lines and a labelled line feature (river or railway) represented by a thick broken line. Tactile maps, constructed using the microcapsule process, were identical to the print maps, except that the lines were raised and the labels were in Braille. All maps were 29.6cm × 21cm (A4).

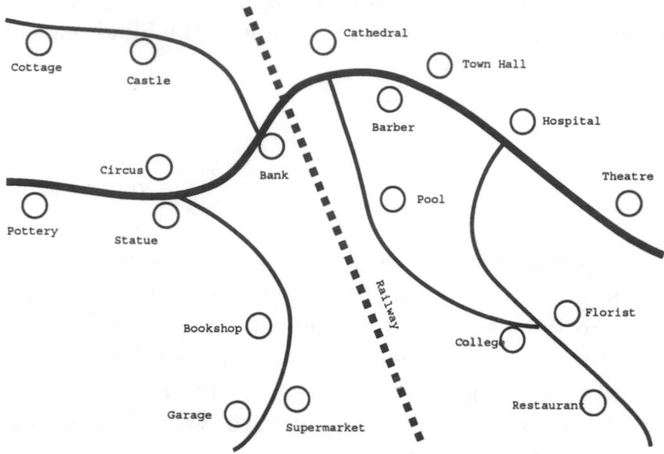

Figure 8.1. Map of Railway Town used in the experiment.

The lists consisted of all the labelled items on each map printed down the left-hand side of an A5 sheet of paper. Identical lists in Braille were also constructed.

A descriptive text was generated for each map – 36 factual statements were randomly assigned to each of the 36 map locations (18 on each map). The texts thus consisted of a series of statements like the following examples: "The cathedral was built in the nineteenth century"; "The book shop is a brick building"; "The farm is said to be haunted". The texts were recorded onto audio cassettes by a male reader using professional recording equipment. The texts were read at the rate of approximately one statement per 5 seconds. Thus the recorded version of each description lasted 90 seconds. The tapes were played to the participants on a portable stereo cassette player, which was placed opposite the participant on the same table-top on which the map or list was placed. The order in which places names appeared was separately randomised for the map, the list and the description.

In the map condition, participants estimated the positions of map locations using 1 cm diameter counters which they stuck to an A4 size plastic-covered board. Labels (print or Braille) were also provided so that the participants could label the counters once they had been positioned on the board.

8.4.3 Procedure

All the blind and visually impaired participants used tactile materials (Tactile-10 group) throughout the experiment and all the sighted participants used print materials (Print-10 group and Print-2 group). A repeated-measures design was used for the Map and List conditions. Order of presentation of the two conditions and of the two towns were counterbalanced across participants. A period of one week separated the two conditions.

Participants performed all tasks while seated at a desk or table. During the initial instruction phase the task was explained to them. They were told that they would have to learn a map/list for 10 minutes (for the Tactile-10 and Print 10 groups) or for 2 minutes (for the Print-2 group), after which they would hear a tape giving information about the places on the map/list. They were told that they would then have to solve mental arithmetic problems, after which they would be required to recall as much information from the tape as possible. Two additional instructions were given in the map condition: first participants were asked to try to form a complete mental image of the map, as if they were drawing a picture of the whole map in their head; second they were informed that they would have to make a copy of the map. These instructions were similar to those given in previous studies (e.g., [27]).

Once the instructions had been given, the map/list was attached to the table in front of the participant. He or she was given either 10 minutes (for the Tactile-10 and Print-10 groups) or 2 minutes (for the Print-2 group) to learn the map/list, before the tape was played just once. The map/list remained on the table while the tape was played and for 30 seconds after. The map/list was then removed, and the participant was asked to solve mental arithmetic problems for 60 seconds. These problems involved multiplying two single-digit numbers together and then adding a third single digit number.

The participant was then asked to recall as much about the tape as possible; i.e., as many of the places and their associated facts as she could remember. The recall was recorded into tape. Then, in the map condition only, participants were asked to use the board and counters to estimate the positions of places on the map. Finally, participants were asked to describe first how they learned the map/list, and then to describe how they learned the facts from the tape.

8.4.4 Analysis

Because the data did not fulfil parametric assumptions, non-parametric models were used for all analyses. Between-group effects were examined using the Mann–Whitney U test and within-group effects were examined using the Wilcoxon test. For between-group effects, comparisons were made between groups sharing one characteristic – i.e., between Tactile-10 and Print-10, and between Print-10 and Print-2.

8.4.5 Results

Free Recall: Places. Participants received a score of one for each place name correctly recalled. Table 8.2 gives the mean number of places remembered by participants across experimental groups and conditions. Overall, there was no difference in number of place names correctly recalled between the map and text conditions (mean score: *map* = 15.6, *text* = 15.7; $Z = -0.31$; N.S.). This was also true for the Print-10 group ($Z = -0.57$; N.S.) and for the Print-2 group ($Z = -0.78$; N.S.). However, the Tactile-10 group recalled significantly fewer place names in the map condition than in the list condition ($Z = -2.85$; $p < 0.05$). In the map condition, the Print-10 group recalled significantly more places than both the Tactile-10 group ($Z = -2.10$; $p < 0.05$) and the Print-2 group ($Z = -4.65$; $p < 0.001$). In the list condition, the Print-10 group recalled significantly more places than the Print-2 group ($Z = -4.53$; $p < 0.001$). The Print-10 group and the Tactile-10 group did not differ ($Z = -1.02$; N.S.).

Table 8.2. Mean number of places correctly recalled by participants in the three experimental groups and across the two learning conditions (map and list)

	Learning Conditions			
	Map		List	
Groups	Mean	SD	Mean	SD
Tactile-10	15.9	2.0	17.4	1.0
Print-10	17.2	0.9	16.9	1.5
Print-2	14.0	2.77	13.6	2.47

Free Recall: Facts. Participants received a score of one point for each of the text facts that they correctly associated with its corresponding map/list location. A lenient scoring criterion was used, whereby participants were considered to be correct if they remembered the substance or gist of the original fact (cf. [19]). Table 8.3 gives the mean number of text facts remembered by participants across experimental groups and conditions.

There was no overall difference between the map and text conditions (mean score: *map* = 10.3, *text* = 10.2; $Z = -0.31$; N.S.), however, the Print-10 group recalled significantly more facts in the map condition than in the list condition ($Z = -2.06$; $p < 0.05$). There was no significant difference between the two conditions in number of facts recalled for either the Tactile-10 group ($Z = -1.83$; N.S.) or for the Print-2 group ($Z = -0.09$; N.S.). There were no significant differences between groups in either condition.

Accuracy of Map Reconstruction. Participants' map reconstructions were scored for number of features correctly located. Correct location was defined as any recalled feature that was within 2 cm from the location of

Table 8.3. Mean number of text facts correctly recalled by participants in the three experimental groups and across the two learning conditions (map and list)

	Learning conditions			
	Map		List	
Groups	Mean	SD	Mean	SD
Tactile-10	10.0	3.5	11.8	4.1
Print-10	11.4	3.6	10.1	3.8
Print-2	9.5	3.75	9.5	2.99

the same feature on the original map. This criterion was chosen as it lies between those used in studies by Kulhavy and his colleagues which have used criteria of 1.27cm [18,19] and 2.54cm [1].[2] The mean accuracy scores for the three groups are shown in the first column of Table 8.4. The Print-10 group made significantly more correct location estimates than the Tactile-10 group ($Z = -2.72$; $p < 0.001$) and than the Print-2 group ($Z = -3.64$; $p < 0.001$).

Conditional Probabilities (Map Condition). For each participant, we calculated the probability that a text fact would be recalled given that the related map feature was correctly located in the map reconstruction task – $p[fact_c|feature_c]$. We also calculated the probability that a text fact would be recalled correctly given that the related feature was not placed correctly in the reconstruction task – $p[fact_c|feature_i]$. The relative magnitude of these two probabilities gives an indication of strength of the relation between accuracy of representation and text recall. If the value of $p[fact_c|feature_c]$ is high and the value of $p[fact_c|feature_i]$ is low, this gives support to the conjoint retention hypothesis. Any other pattern of results would not support this hypothesis.

The values of both conditional probabilities for each of the experimental groups are given in the right-hand columns of Table 8.4. The values obtained here highlight the difficulty of interpreting such conditional probabilities out of context. Taken alone, the relatively high values of $p[fact_c|feature_c]$ suggest that fact recall is contingent on correct placement of features on the map reconstruction. However, this finding is qualified by the even higher values of $p[fact_c|feature_i]$ suggesting that participants were at least as likely to recall a fact correctly whether or not they correctly placed the corresponding feature on the map reconstruction. This is most marked in the Print-2 group, and it was this group that precisely replicated the procedure of previous studies.

---

[2] Kulhavy gives no systematic rationale for his choice of criterion in the various studies. Most of his maps were approximately US letter size (21.6cm × 27.9cm) and therefore the variation in the criteria used is hard to account for. As we were attempting to replicate Kulhavy's procedure as closely as possible in a single study with a similar sized map, it was decided to choose a conservative, intermediate criterion.

Table 8.4. Mean accuracy in the reconstruction task and mean conditional probabilities for the three experimental groups in the map condition

	Features		$p[fact_c\|feature_c]$		$p[fact_c\|feature_i]$	
Group	Mean	SD	Mean	SD	Mean	SD
Tactile-10	4.1	3.34	0.52	0.36	0.62	0.22
Print-10	7.0	3.15	0.63	0.26	0.66	0.21
Print-2	2.9	3.35	0.27	0.35	0.53	0.23

8.5 Discussion

The results provide only limited support for the weak version of the conjoint retention hypothesis. Learning a map did selectively facilitate recall of text facts by sighted participants who studied the map for 10 minutes (Print-10 group). However, although significant, the advantage of the map condition was small in practical or educational terms; sighted participants recalled on average only 10% more facts in the map condition than in the list condition. This contrasts with comparable studies by Kulhavy and his colleagues in which participants remembered about 50% more text information after learning a map than after learning either a list of features [1,25] or a series of individual features [18,19]. The possible explanation that the advantage of a map is greater when exposure times are shorter was discounted by the performance of the Print-2 group (a direct replication of Kulhavy's procedure) who showed no advantage in the map condition.

The fact that the Tactile-10 group performed no better in the map condition than in the text condition might provide support for the claim of the strong version of conjoint retention that the process exploits specifically visual properties of spatial memory. In other words, facilitation only occurs when a map has been learned visually. However, one would have no reason for choosing this explanation over the more conservative one that the tactile group simply did not form as good a mental image of the map as did the sighted, regardless of the specific form of imagery used by the different groups.

The latter possibility is supported by the fact that map reconstruction scores were lower for the tactile group than for the print group. It is likely that the blind and visually impaired group were generally less familiar with maps, and so may not have acquired suitable strategies for learning information from maps. This latter possibility is consistent with previous findings (e.g., [34]).

The conditional probabilities that were calculated provided no support for the strong version of conjoint retention. All participants were about as likely to recall a text fact whether they had or had not correctly placed the corresponding map feature on their reconstruction. Examination of the participants' individual scores and conditional probabilities suggests that this effect is not simply due to the generally poor performance on the reconstruction task; even those participants who estimated the locations of places very accurately did not show any dissociation between the two conditional probability scores.

If we accept the conjoint retention hypothesis, and that spatial structure is important for facilitating factual recall, it may be that absolute position of features on a reconstructed map is not an ideal measure of the structure of a person's representation [36]. We might examine other more sophisticated measures such as relative location. Future research might exploit the recent developments in spatial analysis software to examine such factors.

Alternatively, there may be other aspects of a map than spatial structure per se which facilitate learning of associated information. For instance, the map might more readily evoke images of streets and buildings of a familiar town. This framework of meaning, which is not specifically spatial but is more subjective and personal, may have formed a more effective basis than the list for learning the text information. Conversely, maps may not be so immediately evocative for blind and visually impaired people who are likely to be less familiar with their conventions.

8.6 Conclusions

In summary, blind, visually impaired and sighted participants were given a standard test of conjoint retention [16, 17] in which informational text was learned conjointly with either spatially structured place information (a map) or unstructured place information (a list of place names). Sighted participants who studied the map and the list for 10 minutes (Print-10 group) recalled more after learning the map. Although significant, the differences between conditions were small. No facilitatory effect was found for blind and visually impaired participants (Tactile-10 group) or for sighted participants who studied the map and list for only 2 minutes (Print-2 group). For all groups in the map condition, participants were no more likely to recall a given text fact when the corresponding feature had been correctly located in the map reconstruction task than when the feature had not been correctly located.

The present study was the first to look at conjoint retention by blind and visually impaired people. Although the dissociation between the performance of these participants and the Print-10 group on text recall might imply that actual visual imagery is necessary for conjoint retention, such a conclusion is not warranted. The tactile map users showed no advantage in the map condition and were less accurate in the map reconstruction task, though it is possible that, given enough time to learn the map more completely, the resulting image of the map would have facilitated text recall in the same way as it did for the Print-10 group. Blind and visually impaired people receive considerably less exposure to maps; for instance, they are unlikely to have access to maps as used in advertising, on public transport or in books. They therefore tend not to have acquired or invented strategies for organising and learning information from maps [31]. Future research might consider conjoint retention in a group of highly experienced tactile map readers.

In the case of the sighted participants, the results provide only limited support for the weak version of conjoint retention. Although a small advantage for the map was found for the Print-10 group, this did not seem to be contingent upon a highly accurate mental image of the map. Furthermore, the results of the Print-2 group provided no support for the conjoint retention hypothesis. These results indicate that further studies of conjoint retention should be carried out before the hypothesis is considered to be robust.

Acknowledgements

This research was funded by an award from the Economic and Social Research Council grant R000234891. We are indebted to all the people who gave up time to take part in this research, as well as for the valuable and insightful comments made by many of our participants. We are also grateful to Raymond Kulhavy and William Stock for their invaluable comments during the preparation of this paper.

References

1. Abel, R.R. and Kulhavy, R.W. (1986). Maps, mode of text presentation, and children's prose learning. American Educational Research Journal 23:263–274.
2. Andrews, S.K. (1983). Spatial cognition through tactual maps. In Proceedings of the 1st international symposium on maps and graphics for the visually handicapped. Washington, DC: Association of American Geographers.
3. Bentzen, B.L. (1982). Tangible graphic displays in the education of the blind persons. In W. Schiff and E. Foulke (Eds), Tactual perception: A sourcebook. Cambridge, UK: Cambridge University Press.
4. Blades, M., Ungar, S. and Spencer, C. (in press). Map using by adults with visual impairments. Professional Geographer.
5. Carpenter, P.A. and Eisenberg, P. (1978). Mental rotation and frame of reference in blind and sighted individuals. Perception and Psychophysics 23:117–124.
6. Dodds, A. (1983). Mental rotation and visual imagery. Journal of Visual Impairment and Blindness 77:16–20.
7. Dodds, A. (1988). Tactile maps and the blind user: Perceptual, cognitive and behavioural factors. In Proceedings of the 2nd international symposium on tactile maps and graphics for visually impaired people. Nottingham: Nottingham University Press.
8. Ernest, C.H. (1987). Imagery and memory in the blind: A review. In McDaniel and Pressley (Eds), Imagery and related processes. New York: Springer.
9. Espinosa, M-A., Ungar, S., Ochata, E., Blades, M. and Spencer, C. (1998). Comparing methods for introducing blind and visually impaired people to unfamiliar urban environments. Journal of Environmental Psychology 18:277–287.
10. Farah, M. (1984). The neurological basis of mental imagery: A componential analysis. Cognition 18:245–272.

11. Hampson, P.J. and Daly, C.M. (1989). Individual variation in tactile map reading skills: some guidelines for research. Journal of Visual Impairment and Blindness 83:505–509.
12. Heller, M.A. (1991). Haptic perception in blind people. In M. Heller and W. Schiff (Eds), The psychology of touch. Hillsdale, NJ: Lawrence Erlbaum Associates.
13. Juurmaa, J. (1973). Transposition in mental spatial manipulation: A theoretical analysis. American Foundation for the Blind Research Bulletin 26:87–134.
14. Kerr, N.H. (1983). The role of vision in "visual imagery" experiments: Evidence from the congenitally blind. Journal of Experimental Psychology: General 112:265–277.
15. Kulhavy, R.W., Lee, J.B. and Caterino, L.C. (1985). Conjoint retention of maps and related discourse. Contemporary Educational Psychology 10:28–37.
16. Kulhavy, R.W. and Stock, W.A. (1996). How cognitive maps are learned and remembered. Annals of the Association of American Geographers 86:123–145.
17. Kulhavy, R.W., Stock, W.A. and Kealy, W.A. (1993). How geographic maps increase recall of instructional text. Educational Technology Research and Development 41:47–62.
18. Kulhavy, R.W., Stock, W.A., Verdi, M.P., Rittschof, K.A. and Savenye, W. (1993). Why maps improve memory for text: The influence of structural information on working memory operations. European Journal of Cognitive Psychology 5:375–392.
19. Kulhavy, R.W., Woodward, K.A., Haygood, R.C. and Webb, J.M. (1993c). Using maps to remember text: An instructional analysis. British Journal of Educational Psychology 63:161–169.
20. Landau, B. (1986). Early map use as an unlearned ability. Cognition 22:201–223.
21. Marmor, G.S. and Zaback, L.A. (1976). Mental rotation by the blind: Does mental rotation depend on visual imagery? Journal of Experimental Psychology: Human Perception and Performance 2:515–521.
22. Millar, S. (1994). Understanding and representing space: Theory and evidence from studies with blind and sighted children. Oxford: Oxford University Press.
23. Paivio (1986). Mental representations: A dual coding approach. Oxford: Oxford University Press.
24. Röder, B., Rösler, F., Heilund, M. and Hennighausen, E. (1993). Haptic mental rotation tasks performed by blind and sighted individuals. Zeitschrift Für Experimentelle und Angewandte Psychologie 40:154–177.
25. Schwartz, N.H. and Kulhavy, R.W. (1981). Map features and the recall of discourse. Contemporary Educational Psychology 6:151–158.
26. Senden, S.M.v. (1932). Space and sight: The perception of space and shape in the congenitally blind before and after operation. Glencoe, IL: Free Press.
27. Stock, W.A., Kulhavy, R.W., Peterson, S.E., Hancock, T.E. and Verdi, M.P. (1995). Mental representations of maps and verbal descriptions: Evidence they may affect text memory differently. Contemporary Educational Psychology 20:237–256.
28. Tippett, L.J. (1992). The generation of visual images: A review of neuropsychological research and theory. Psychological Bulletin 112:415–432.
29. Ungar, S. (in press). Cognitive mapping without visual experience. In R. Kitchin and S. Freundschuh (Eds), Cognitive mapping: Past, present and future. London: Routledge.
30. Ungar, S., Blades, M., Spencer, C. and Morsley, K. (1994). Can visually impaired children use tactile maps to estimate directions? Journal of Visual Impairment and Blindness 88:221–233.

31. Ungar, S.J., Blades, M. and Spencer, C. (1995). Mental rotation of a tactile layout by young visually impaired children. Perception 24:891–900.
32. Ungar, S., Blades, M. and Spencer, C. (1996). The construction of cognitive maps by children with visual impairments. In J. Portugali (Ed.), The construction of cognitive maps. Dordrecht: Kluwer.
33. Ungar, S., Espinosa, A., Blades, M., Ochata, E. and Spencer, C. (1997). Use of tactile maps by blind and visually impaired people. Cartographic Perspectives 28:4–12.
34. Ungar, S., Blades, M. and Spencer, C. (1997). Strategies for knowledge acquisition from cartographic maps by blind and visually impaired adults. Cartographic Journal 34:93–110.
35. Warren, D.H. (1984). Blindness and early childhood development. New York: American Foundation for the Blind.
36. Winn, W. (1991). Learning from maps and diagrams. Educational Psychology Review 3:211–247.

9. Spatial Abilities in Problem Solving in Kinematics

Maria Kozhevnikov

Mary Hegarty

Richard Mayer

This study investigates the relationship between students' spatial abilities and their ability to solve problems in physics, specifically in kinematics. The approach taken is to consider spatial ability not as single and undifferentiated, but composed of different components. The hypothesis is that different types of kinematics problems require different spatial abilities. Sixty undergraduate psychology students, who had not taken any physics courses at college level, took a battery of cognitive tests measuring different spatial skills, verbal ability and mechanical reasoning. In addition, students were presented with a series of kinematics problems by means of a written problem solving questionnaire. Analyses of students' responses indicated that different types of kinematics problems require different cognitive skills. It was found, for instance, that extrapolating complex two-dimensional motion correlates significantly with spatial visualisation ability, whereas inferring direction of motion from a graph correlates with spatial orientation ability. However, performance on other types of kinematics problems (e.g., evaluating an object's speed and some types of graph problems) do not correlate with spatial abilities, indicating that they may require mostly semantic knowledge of physics laws or mathematical reasoning.

9.1 Introduction

Historically, there is much evidence that spatial imagery plays a central role in physics conceptualisation processes and in scientific discoveries. Research that analyses the process of physics discoveries such as Galileo's laws of motion, Maxwell's laws, Faraday's electromagnetic field theory, and Einstein's theory of relativity emphasises the extensive use of spatial imagery by physicists and its crucial function in these discoveries [25, 26]. The majority of physics problem solving strategies involve spatial representations in the form of graphs, diagrams, or physical models. Free-body diagrams, field lines, and energy levels are examples of spatial constructs that do not exist in the real world, but help physicists to understand and predict its phenomena.

This gives rise to the idea that visual–spatial abilities may play an important role in students' physics learning and problem solving. Supporting this idea, research on expert–novice differences in problem solving noted that "experts draw more diagrams" while solving physics problems [33, (p. 101)] and that the diagrams "can support extremely useful and efficient computational processes" [21, (p. 99)]. Furthermore, the United States Employment Service [8] includes physics in the list of occupations requiring a high level of spatial ability, i.e., the ability to perform spatial transformations with mental images or their parts. There is also considerable evidence that spatial ability is an important criterion in the prediction of students' achievement in mathematics and a wide range of technical subjects [24]. However, surprisingly little attention has been devoted to understanding the role of spatial imagery in physics problem solving. Even research on expert–novice problem solving, which mentions the importance of visual–spatial representations in physics (e.g., [6,11]), focuses mostly on verbal aspects of problem representation and semantic knowledge of physics laws.

The goals of this research are (1) to provide evidence that spatial imagery is used as a strategy in problem solving in physics by examining the relationship between students' spatial abilities and problem solving in kinematics, and (2) to examine which spatial ability factors contribute to solving different types of kinematics problems. Kinematics was chosen as the topic of this study because it depends on a diversity of visual–spatial representations, such as graphical schematic representations (vectors of force or velocity; and graphs of motion) as well as concrete physical representations (blocks, pulleys, or springs).

9.2 Theoretical Background

9.2.1 Problem Solving in Kinematics

Although there has been considerable research on students' difficulties in solving kinematics problems, this research has focused primarily on the interaction between students' prior knowledge or "common-sense" beliefs about the physical world, and formal physics instruction [7,23]. For example, one debate in this literature has concerned the extent to which common-sense beliefs reflect consistent and deeply ingrained theories about the physical world (e.g., [13]) or whether they reflect fragmented and situational understanding of physical phenomena (e.g., [9,29]). Despite the large number of studies on students' difficulties in solving kinematics problems, almost no attempt has been made in the educational literature to relate students' errors in kinematics problem solving to their spatial abilities. However, there is evidence from research in cognitive psychology that the evaluation of complex kinematics events is related to individual differences in spatial ability. Isaak and Just proposed that as kinematics events become more complex, imagining these events should be more dependent on visual–spatial processing demands [17]. Consistent with this view, they found that students' susceptibility to incorrect judgements about complex kinematics events such as rolling motion is related to their spatial ability level.

9.2.2 Spatial Ability Factors

Individual differences in spatial ability have been studied since the 1920s, when spatial ability was first differentiated from general intelligence, and from verbal and numerical abilities [34]. Isolating a spatial factor led to attempts to analyse and break it down further into component factors of spatial ability. The broadest reviews of psychometric studies of spatial ability (e.g., [4,22,24]) converge in the conclusion that there is evidence for two or three major spatial abilities factors.

The first factor, *spatial relations* (sometimes called *speeded rotation*), involves the ability to engage rapidly and accurately in mental rotation processes that are necessary to judge the identity of a pair of stimuli. Tests loading on the spatial relations factor require one to rotate the stimulus, rather than imagining the oriented self [22]. The second factor, *spatial visualisation*, reflects processes of apprehending, encoding, and mentally manipulating spatial forms [22]. Spatial visualisation ability is also defined as the ability to transform spatial images accurately [18, 27]. The tests loaded on the spatial visualisation factor are administered under relatively unspeeded conditions and are much more complex than tests loading on the spatial relations factor. The third proposed factor, *spatial orientation*, involves the ability to imagine how a stimulus array will appear from another perspective [22, 24]. In a true spatial orientation test, the subject must imagine that he/she is reoriented in space, and then make some judgment about the situation [4].

Beginning in the 1970s, researchers have applied cognitive psychology methods and theories to the study of spatial ability. One focus of this research has been to interpret individual differences in spatial abilities in terms of theories of working memory. Baddeley and Lieberman proposed a theory that working memory consists of a central executive and at least two "slave systems": the phonological loop which is specialised for storing and processing verbal information, and the visuospatial sketchpad, which is specialised for storing and processing visual–spatial information [2]. According to this model, visual–spatial imagery processes take place in the visuospatial sketchpad and are limited by its capacity [1]. It has been proposed that spatial ability tests can provide a measure of visual–spatial working memory capacity, and that people who differ in spatial abilities also differ in performance on tasks that involve spatial imagery. For example, Salthouse et al. [30] assumed that individual differences in spatial visualisation arise due to the simultaneous processing and storage demands of carrying out a sequence of spatial transformations on an imagined object, which tax the supply of visual–spatial working memory resources. Similarly, Shah and Miyake proposed that spatial working memory has limited resources, and that storage and processing demands imposed by spatial ability tests compete for those resources [32].

9.2.3 Mechanical Ability Factors

In 1928, Cox (in [34]) proposed a mechanical ability factor, which underlies the mental processes involved in comprehending mechanical relations, mechanical models, technical tasks, and even spatial visualisation tests. Subsequently, the fact that there is a close relationship between spatial abilities and mechanical ability has been repeatedly noted by numerous researchers (see [24] for a review).

Mechanical reasoning tests include many items about kinematics (e.g., predicting the direction of motion and speed of mechanical system components). Theories of how people reason about mechanical systems have suggested that this reasoning involves simulating the behaviour of mechanical systems using spatial imagery processes. First, selective interference studies found that performance on mechanical reasoning tasks showed a significant interference with maintenance of spatial but not verbal information in working memory [31]. Second, there is a high correlation between spatial visualisation and mechanical reasoning tests [14]. On the basis of this evidence, it was suggested that mechanical reasoning processes take place within the visual–spatial component of working memory and are limited by its computational capacity [31].

The above research suggests that mechanical reasoning involves the extensive use of spatial processes and is closely related to spatial ability. Mechanical reasoning tests are also strong predictors of college students' success in final physics exams [3]. For these reasons, the relation of different types of mechanical reasoning problems to physics problem solving and spatial ability was also examined in this research.

9.3 Research Hypotheses

First we hypothesise that there is a strong relationship between students' performance on spatial ability tests and solving problems in kinematics. We assume that spatial imagery is a capacity-constrained process taking place in visual–spatial working memory [1, 32]. Since problem solving in kinematics often involves visualising complex motion and mentally manipulating graphs and diagrams, it should depend on an individual's visual–spatial resources, which are related to his/her spatial abilities.

Based on psychometric studies of spatial ability, we consider spatial ability not as single and undifferentiated, but as composed of different spatial abilities [4,22]. Therefore we do not merely predict a simple relation between spatial ability and kinematics problem solving. Rather, we suggest that solving different types of kinematics problems requires different spatial abilities. For instance, the ability to visualise complex two-dimensional motion may differ from the ability to solve problems that require translation from one system of reference to another. We also suggest that there are other types of kinematics problems whose solutions are based on semantic knowledge of physics laws or verbal–logical reasoning, rather than on spatial imagery.

9.4 Method

The participants were 60 undergraduate psychology students, who had not taken any physics courses at the college level. They were recruited from the Psychology Subject Pool at the University of California, Santa Barbara.

9.4.1 Instrumentation

The materials consisted of a pre-test questionnaire and eight paper-and-pencil tests measuring a number of different spatial ability factors, verbal ability, and mechanical ability. In addition, students were presented with a series of kinematics problems by means of a written problem solving questionnaire.

Pre-test Questionnaire. The pre-test questionnaire included questions about students' high school physics background, age, and gender.

Kinematics Questionnaire. To classify kinematics problems into categories that may require different visual–spatial abilities, we reviewed qualitative kinematics problems presented in physics textbooks and in mechanics diagnostic tests aimed to assess students' understanding of the laws of mechanics (e.g., the Mechanics Diagnostic Test of Halloun and Hestenes [13] and the Force Concept Inventory Test of Hestenes et al. [16]). Five different classes of kinematics problems were distinguished (see examples in Appendix I):

1. *Extrapolation problems* involved predicting the motion of an object from an observed to an expected path. All extrapolation problems in our questionnaire involved complex two-dimensional motion, that is, processing motion in horizontal and vertical directions, as well as the integration of these motions to represent the system's overall motion.
2. *Speed problems* required the student to make a qualitative judgement about an object's absolute velocity, for instance, to determine if the object's velocity is constant, increasing, or decreasing.
3. *Graph problems* involved relating one type of graph to another, relating a graph to the real-world situation it represents, and matching verbal information with relevant features of a graph.

4. *Frame of reference problems* involved the translation from one system of reference to another.
5. *Comparison problems* required students to compare one object's motion (e.g., the object's trajectory, velocity, or acceleration) to that of another.

The questionnaire included 15 kinematics problems: 3 problems of each of the above types. The internal reliability of the test is 0.81 (Cronbach's alpha).

Spatial Ability Tests. Students' levels of spatial relations were assessed using the *Card Rotation Test* and the *Cube Comparison Test* [10]. The Card Rotation Test consists of 10 items. Each item requires participants to view a two-dimensional target figure and judge which of the five alternative test figures are planar rotations of the target figure (as opposed to its mirror image) as quickly and as accurately as possible. The Cube Comparison Test consists of 21 items. Each item presents two drawings of cubes, with letters and numbers printed on their sides. Participants must judge whether the two drawings could show the same cube from different orientations.

Spatial visualisation abilities were assessed by the *Paper Folding Test* and the *Form Board Test* [10]. The Paper Folding Test consists of 10 items. Each item shows successive drawings of two or three folds made in a square sheet of paper. The final drawing shows a hole being punched in the folded paper. The task is to select one of five drawings to show how the punched sheet would appear when fully opened. The Form Board Test consists of 24 items. Each item of the Form Board Test presents five shaded drawings of pieces, some or all of which can be put together to form a figure presented in outline form. The task is to indicate which of the pieces, when fitted together, would form the outline figure.

Students' level of spatial orientation was assessed by means of the *Perspective Taking Test* [20].

A configuration of seven visually distinct objects was drawn on an 8.5X11 inch sheet of paper. On each item, the subject was asked to take a particular perspective in the display and indicate the direction to a target object. The test consisted of 10 items. The internal reliability of the test (Chronbach's alpha) is 0.83. The deviation of the actual direction of the target from the direction pointed by subjects (absolute directional error) was calculated for each trial. A participant's total score was his/her average directional error across all trials subtracted from 180 (the maximal directional error).

Verbal Ability Test. Participants' verbal ability was measured by means of the *Advanced Vocabulary Test*, which measures "availability and flexibility in the use of multiple meanings of words" [10, (p. 163)]. The Advanced Vocabulary Test consists of 18 items. On each item the task is to indicate which of five numbered words is nearest in meaning to a given word.

Mechanical Reasoning Test. The mechanical reasoning test was made up of 22 items of the type used in tests of mechanical comprehension, such as the *Bennet Test of Mechanical Comprehension* [3]. Each item showed a diagram of a mechanical system with several components and arrows indicating the motion of two of the system components. The task was to determine whether the motion of the two components was consistent, i.e., whether one component would move in the direction shown, given that the other was moving in the direction shown. The items in the test were classified into three groups according to Hegarty and Kozhevnikov's [15] classification of three different types of mechanical systems (see examples in Appendix II):

1. *Linear mental animation tasks.* The items in this category show mechanical systems with at least five mechanical components, and all components move in the same two-dimensional plane. They can be solved by a piecemeal strategy involving a chain of interferences in which the motion of only one component is inferred at a time. Therefore, the need to simultaneously store and process a large amount of spatial information is not a limiting factor in linear animation and it is not predicted by spatial ability [15].
2. *Three-dimensional animation tasks.* The items in this category show mechanical systems in which the components move in different non-parallel planes. These items require the ability to perceive the position and configuration of mechanical components and to imagine their motion in three-dimensional space. They can also be solved by breaking the system down into a series of components that are animated sequentially. Hegarty and Kozhevnikov [15] found that ability to solve these problems depended on spatial orientation ability, i.e. ability to reorient oneself in space to imagine viewing the machine from different perspectives.
3. *Complex animation tasks.* These items require the ability to keep several mechanical components in mind and animate them in parallel. Therefore we would expect solution of these problems to be highly demanding of visual–spatial processing resources, and they have been found to be correlated with spatial visualisation ability [15].

The test included six linear animation tasks, eight three-dimensional animation tasks, and eight complex animation tasks. The internal reliability (Cronbach's alpha) for linear mental animation items was 0.64; for three-dimensional animation tasks it was 0.75, and for complex animation tasks it was 0.65.

9.4.2 Procedure

The participants were tested in small groups of up to six students per session. First, they completed the pre-test questionnaire. Then, they were administered the kinematics questionnaire, which took about 40 minutes to complete, although the subjects were not placed under any time restriction. Then, participants completed the Card Rotation Test, the Cube Comparison Test, and the Paper Folding Test, in that order. Each of these tests was preceded by the standard instructions for that test and took 3 minutes to complete. Then, participants took the Form Board Test, which took 8 minutes, the Advanced Vocabulary Test, which took 4 minutes, and the Perspective Taking Test, which took 5 minutes. Finally, the participants completed the Mechanical Reasoning Test, which took 15 minutes.

9.5 Results

9.5.1 Correlation Analysis

Correlations between the cognitive factors (spatial ability, verbal ability, and mechanical reasoning tests) and kinematics problems are presented in Table 9.5.1.

Table 9.1. Correlation coefficients between problems and cognitive ability tests

	E	S	G	F	C	K
Paper folding	**0.35	0.25	0.18	−0.11	*0.29	**0.33
Form board	*0.34	0.22	0.20	0.16	0.20	*0.30
Card rotation	−0.02	0.04	0.12	−0.07	0.16	0.07
Cube comparison	0.20	0.06	0.19	0.11	0.19	0.19
Perspective taking	0.18	0.16	*0.26	0.25	0.18	*0.29
Linear animation	0.08	0.01	0.05	0.04	*0.39	0.16
3D animation	0.19	0.12	0.21	**0.36	0.28	*0.34
Complex animation	**0.53	**0.47	**0.32	0.25	**0.36	**0.54
Verbal ability	0.02	0.15	**0.38	0.05	0.25	0.24

*Correlation significant at the 0.05 level (2-tailed).
**Correlation significant at the 0.01 level (2-tailed).
E: Extrapolation; *S*: Speed; *G*: Graphs; *F*: Frame of reference; *C*: Comparison; *K*: Kinematics.

Overall Kinematics Questionnaire. In general, performance on the whole kinematics questionnaire correlated with the paper folding test ($p = 0.01$), form board test ($p < 0.05$), spatial orientation test ($p < 0.05$), three-dimensional animation tasks ($p < 0.01$), and complex animation tasks from the mechanical reasoning test ($p = 0.001$). These results strongly suggest that spatial visualisation and spatial orientation are important abilities in physics problem solving. In contrast, speeded rotation tests (card rotation and cube comparison) did not correlate significantly with any types of kinematics problems. It seems that the ability to solve mental rotation problems quickly is not as crucial in problem solving in kinematics as spatial visualisation and spatial orientation ability.

Extrapolation Problems. As Table 9.1 shows, solving extrapolation problems correlates highly with both spatial visualisation tests: the paper folding test and the form board test ($p < 0.01$). The high correlation between performance on extrapolation problems and spatial visualisation might be due to a need for high spatial-processing resources to solve extrapolation problems. All extrapolation problems in the questionnaire involved complex two-dimensional motion, i.e., processing motion in horizontal and vertical directions, as well as the integration of these motions to represent the system's overall motion. Limited capacity of visual–spatial working memory may cause the subjects either to neglect one of the motions or to integrate these motions incorrectly.

Speed Problems. No significant correlation was found between performance on speed problems and any of the spatial abilities. These results suggest that the evaluation of an object's speed is a type of inference that cannot be carried out within visual–spatial working memory alone. It is possible that speed problem solutions depend more on semantic knowledge of physics laws than on spatial imagery processes. This interpretation is consistent with evidence that visualisation allows inference about the direction of motion, but not about its rate. For instance, Fallside [12] found that when subjects viewed animations of a mechanical system, they were able to detect motion inconsistencies when components were moving in the wrong direction, but were seldom able to detect motion inconsistencies of components that were moving in the right direction but at the wrong speed.

Graph Problems. No significant correlation was found between performance on graph problems and either the paper folding test ($p = 0.16$) or the form board test ($p = 0.15$). It is plausible that for most subjects, who have not studied physics, but have had a lot of mathematics courses, it was easier to solve graph problems by applying purely verbal–analytical strategies. This was supported by further interviews with students who reported the use of mathematical and verbal–logical strategies to solve graph problems [19]. And, indeed, as Table 9.1 shows, performance on graph problems correlated significantly with verbal ability ($p < 0.01$).

Further analysis of students' responses showed that many students had difficulty in determining the correct direction of an object's motion from a graph, especially when a negative velocity was involved. About 50% of all students confused forward and backward motion, as well as upward and downward motion. The fact that a significant correlation ($p < 0.05$) was found between the students' score on graph problems and the perspective taking test supports the view that the ability to determine the direction of motion from the graph is somewhat tied to spatial orientation ability.

Frame of Reference Problems. A marginal correlation between performance on frame of reference problems and the perspective taking test ($p = 0.05$) suggested that spatial orientation ability is important in solving frame of reference problems. A limitation of the perspective taking test is that it showed only a two-dimensional situation, whereas frame of reference problems in the physics questionnaire were not limited to two-dimensional cases. This interpretation is supported by the result that there was a high correlation between frame of reference problems and three-dimensional mechanical reasoning tasks ($p < 0.01$). This correlation might be explained by the fact that both types of problems depend on the ability to reorient oneself in space.

Comparison Problems. Solving the comparison problems requires the student to break down the objects' trajectories into small units, corresponding to different time intervals, and then compare the motion of the two objects in each of the intervals successively. This strategy is similar to the piecemeal mental animation strategy used to solve linear mental animation problems [15], according to which the student decomposes the system, and animates system components in a sequence corresponding to the causal sequence of events. There is a high correlation between comparison problems and linear mental animation tasks ($p < 0.01$), supporting the view that both tasks involve problem decomposition and piecemeal comparison.

9.5.2 Factor Analysis

In order to explore more thoroughly the relationships between kinematics problems and cognitive factors, a factor analysis was performed on the data set. Although the sample size was small for factor analysis, it did meet the four-cases-to-one variable rule of thumb proposed by Cattell [5]. The principal diagonal method of factorisation was used to obtain the initial factor matrix. As a result, four factors were extracted. The final correlation matrix, presented in Table 9.2, was obtained using Varimax rotation of axes. Loadings under 0.32 (10% overlapping variance between a variable and a factor) are omitted.

Interpretation of Factors. Perspective taking, three-dimensional animation tasks, form board, and paper folding tests are all highly loaded on the first factor. Therefore, it is reasonable to identify Factor I as a *spatial factor.* As can be seen in Table 9.2, extrapolation problems, frame of reference problems, and complex animation tasks are also loaded on the spatial factor. The high loading of frame of reference and extrapolation problems on the spatial factor supports the idea that individual differences in solving these kinds of problems are at least somewhat due to differences in students' spatial ability level.

The second factor might tentatively be labelled as a *physics factor.* All types of kinematics problems are loaded to a greater or lesser extent on this factor. Complex mental animation is also highly weighted on this physics factor. Solution of complex mental animation tasks, although requiring high spatial visualisation ability, seems also to be substantially facilitated by knowledge of physics laws.

Table 9.2. Factor loadings for principal factors extraction and Varimax rotation of four factors

Variable	I	II	III	IV	Communality
Spatial layout	0.858				0.858
Form board	0.708				0.523
Paper folding	0.706				0.635
Linear mental animation			0.766		0.759
3D animation	0.810				0.774
Complex animation	0.556	0.572			0.723
Verbal ability				0.876	0.836
Extrapolation problems	0.505	0.644			0.745
Speed problems		0.878			0.790
Graphs problems		0.511	0.358	0.599	0.750
Comparison problems		0.359	0.780		0.768
Frame of reference	0.520	0.346			0.618

Factor III, on which linear mental animation tasks loaded substantially, could be described as a ***linear visual–spatial processing factor***. The high loading of spatial tasks that involve decomposition and piecemeal processing (comparison problems and linear mental animation tasks) has verified the importance of linear visual–spatial thinking in physics problem solving. Factor IV might be considered to be a ***verbal ability factor***. The only type of physics problems loaded on this factor is graph problems.

9.6 Discussion

The results of this study indicate that different types of kinematics problems require different cognitive skills. Solution of extrapolation and frame of reference problems seems to be highly related to spatial imagery. However, other types of kinematics problems (e.g., evaluating an object's speed and some types of graph problems) may require mostly semantic knowledge of physics laws or mathematical reasoning.

The results of this study give support to the idea that spatial imagery is an important component of problem solving in kinematics. The aspects of spatial imagery that promote problem solving in kinematics are image accuracy and the capacity of visual–spatial imagery resources (as measured by spatial visualisation tests), but not the speed of processing spatial information (as measured by spatial relations tests).

Another finding of this research is that a piecemeal visual–spatial processing factor has been isolated by factor analysis from other spatial imagery factors. The principal distinction between this factor and other spatial imagery components is that it meaures the ability to decompose a problem and process the information piecemal, a process that does not require high spatial processing resources, but is similar to verbal–logical thinking. This factor is important in solving comparison problems that require linear visual–spatial thinking.

There is a possibility that the relationships found in this study between students' levels of spatial ability and their ability to solve physics problems are not causal relationships. That is, increasing students' spatial ability levels may not be sufficient to increase their ability to learn and understand physics. Indeed, it is possible that learning physics may increase students' spatial ability. For example, experience associated with analysing graphs and resolving diagrams may significantly contribute to the development of students' spatial imagery skills. Further research is needed to address these issues. This research might use converging measures to the ones used here, for example, protocol analysis of how students solve physics problems and their reports of the use of imagery during problem solving.

The focus of this study was on the relationship between spatial ability and problem solving in kinematics. Further research is needed to explore the relations between spatial abilities and performance in other types of physics problems. The identification of general cognitive factors as well as specific spatial imagery skills that affect problem solving in different areas of physics can enable science educators and curriculum developers to combine appropriate visual–spatial and mathematical representations in textbooks, computer simulations, and in the classroom.

Acknowledgements

This research was partially supported by the Office of Naval Research under contract N00014-96-10525 to the University of California, Santa Barbara.

References

1. Baddeley, A. (1992). Is working memory working? The fifteenth Barlett lecture. Quarterly Journal of Experimental Psychology 44A(1):1–31.
2. Baddeley, A.D. and Lieberman, K. (1980). Spatial working memory. In R. Nickerson (Ed.), Attention and performance, Vol. VIII. Hillsdale, NJ: Lawrence Erlbaum Associates.
3. Bennet, C.K. (1969). Bennet mechanical comprehension test. New York: Psychological Corporation.
4. Carroll, J.B. (1993). Human cognitive abilities: A survey of factor-analytical studies. Cambridge University Press, UK: Cambridge.
5. Cattell, R.B. (1952). Factor analysis: An introduction and manual for the psychologists and social scientists. New York: Harper.
6. Chi, M.T.H. and Glaser, R. (1988). The nature of expertise. Hillsdale, NJ: Lawrence Erlbaum Associates.
7. Clement, J. (1983). A conceptual model discussed by Galileo and used intuitively by physics students. In D. Gentner, and A. Stevens (Eds), Mental models. Hillsdale, NJ: Lawrence Erlbaum Associates, pp. 325–339.

8. Dictionary of Occupation Titles (1991). U.S. Department of labor, employment and training administration. U.S. Employment Service, Career Press, Washington, DC.
9. diSessa, A.A. (1988). Knowledge in pieces. In G. Forman and P. Pufall (Eds), Constructivism in the computer age. Hillsdale, NJ: Lawrence Erlbaum Associates, pp. 49–70.
10. Ekstrom, R.B., French, J.W. and Harman, H.H. (1976). Manual for kit of factor referenced cognitive tests. Princeton, NJ: Educational Testing Service.
11. Ericsson, K.A. and Smith, J. (1991). Toward a general theory of expertise. Cambridge University Press, UK: Cambridge.
12. Fallside, D.C. (1988). Understanding machines in motion. Unpublished doctoral dissertation, Carnegie Mellon University, Pittsburgh, PA.
13. Halloun, I.A. and Hestenes, D. (1985). The initial knowledge state of college physics students. American Journal of Physics 53:1043–1055.
14. Hegarty, M. and Sims, V.K. (1994). Individual differences in mental animation during mechanical reasoning. Memory and Cognition 22:411–430.
15. Hegarty, M. and Kozhevnikov, M. (1999). Spatial abilities, working memory and mechanical reasoning. In J. Gero and B. Tversky (Eds), Visual and spatial reasoning in design. Preprints of the International Conference in Design, MIT, Cambridge, MA, 15–17 June, pp. 221–241.
16. Hestenes, D., Wells, M. and Swackhamer, G. (1992). Force concept inventory. American Journal of Physics 30:141–154.
17. Isaak, M.I. and Just, M.A. (1995). Constrains on the processing of rolling motion: The curtate cycloid illusion. Journal of Experimental Psychology: Human Perception and Performance 21:1391–1408.
18. Kosslyn, S.M., Care, K.R. and Wallach, R.W. (1984). Individual differences in mental imagery ability: A computational analysis. Cognition 18:195–243.
19. Kozhevnikov, M. (1999). Students' use if imagery in solving qualitative problems in kinematics. Unpublished doctoral dissertation, Technion, Haifa, Israel.
20. Kozhevnikov, M. and Hegarty, M. (1999). Perspective taking ability is distinct from mental rotation ability. In 40th annual meeting of the Psychonomic Society, Los Angeles, CA, 18–21 November.
21. Larkin, J.H. and Simon, H.A. (1987). Why a diagram is (sometimes) worth the thousand words. Cognitive Science 11:65–100.
22. Lohman, D.F. (1988). Spatial abilities as traits, processes, and knowledge. In R. J. Stenberg (Ed.), Advances in the psychology of human intelligence. Hillsdale, NJ: Lawrence Erlbaum Associates, pp. 181–232.
23. McCloskey, M. (1983). Nave theories of motion In D. Gentner and A. Stevens (Eds), Mental models. Hillsdale, NJ: Lawrence Erlbaum Associates, pp. 229–324.
24. McGee, M.G. (1979). Human spatial abilities: Psychometric studies and environmental, genetic, hormonal, and neurological influences. Psychological Bulletin 86:889–918.
25. Miller, A.I. (1986.) Imagery in scientific thought. Cambridge, MA: MIT Press.
26. Nerssesian, N.J. (1995). Should physicists preach what they practice? Science & Education 4:203–226.
27. Poltrock, S.E. and Agnoli, F. (1986). Are spatial visualization and visual imagery ability equivalent? In R.J. Sternberg (Ed.) Advances in psychology of human intelligence. Hillsdale, NJ: Lawrence Erlbaum Associates, pp. 255–296.
28. Psychological Corporation (1990). The mechanical reasoning test of the differential aptitude test. New York.

29. Ranney, M. (1994). Relative consistency and subjects' "theories" in domains such as nave physics: Common research difficulties illustrated by Cooke and Breedin. Memory and Cognition 22:494–502.
30. Salthouse, T.A., Babcock, R.L., Mitchell, D.R.D., Palmon, R. and Skovronek, E. (1990). Sources of individual differences in spatial visualization ability. Intelligence 14:187–230.
31. Sims, V.K. and Hegarty, M. (1997). Mental animation in the visuospatial sketchpad: Evidence from dual-task studies. Memory & Cognition 25:321–332.
32. Shah, P. and Miyake, A. (1996). The separability of working memory resources for spatial thinking and language processing: An individual differences approach. Journal of Experimental Psychology: General 125:4–27.
33. Schultz, K. and Lochhhead, J. (1991). A view from physics. In M.U. Smith (Ed.), Toward a unified theory of problem solving: Views from the content domains. Hillsdale, NJ: Lawrence Erlbaum Associates, pp. 99–114.
34. Smith, M. (1964). Spatial ability: Its educational and social significance. London: University of London Press.

Appendix I

Example of Extrapolation Problem (I) and Speed Problem (II)

I. The accompanying figure shows a rocket coasting in space from a to b in the direction of the dotted line. Between a and b, no outside forces act on the rocket. When it reaches point b, the rocket fire its engines at a right angle to line ab and it keeps firing with a constant force for a certain amount of time T. At this time the rocket will reach point c in space. Which of the paths below will the rocket follow from b to c?

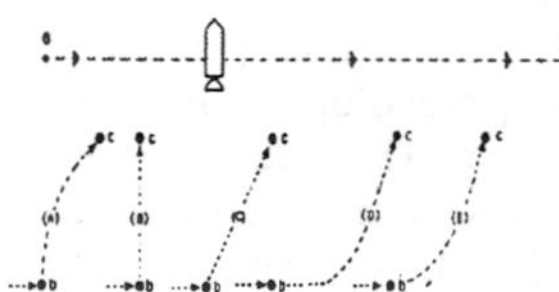

Figure 9.1.

II. As the rocket moves from b to c, its speed is:
(a) constant;
(b) continuously increasing;
(c) continuously decreasing;
(d) increasing for a while, and constant thereafter;
(e) constant for a while, and decreasing thereafter.

Example of Graph Problem

Here is a graph of an object's motion. Which sentence is a correct interpretation?

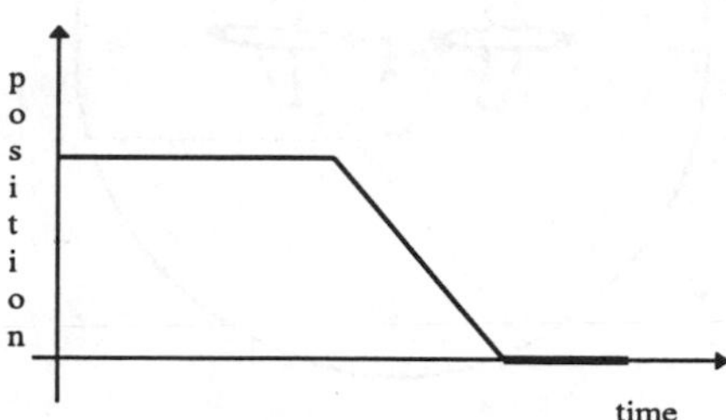

Figure 9.2.

(a) The object rolls along a flat surface. Then it rolls forward down a hill, and then finally stops.
(b) The object doesn't move at first. Then it rolls forward down a hill and finally stops.
(c) The object is moving at a constant velocity. Then it slows down and stops.
(d) The object doesn't move at first. Then it moves at constant speed and then finally stops.
(e) The object moves along a flat area, moves backward down a hill, and then it keeps moving.

Example of Frame of Reference Problem

A flight trainer stands at point C as shown in Fig. 9.3. He sends two identical aeroplanes A and B to do a flight around the earth. The arrow represents the direction of earth's rotation.
Plane A moves in a direction that is opposite to the earth's rotation and plane B moves in the direction of earth's rotation. The planes start at point C and move at the same speed. Assume that there is no wind and ignore air resistance. Which plane will the trainer see first?

(a) Plane A.
(b) Plane B.
(c) Both at the same time.

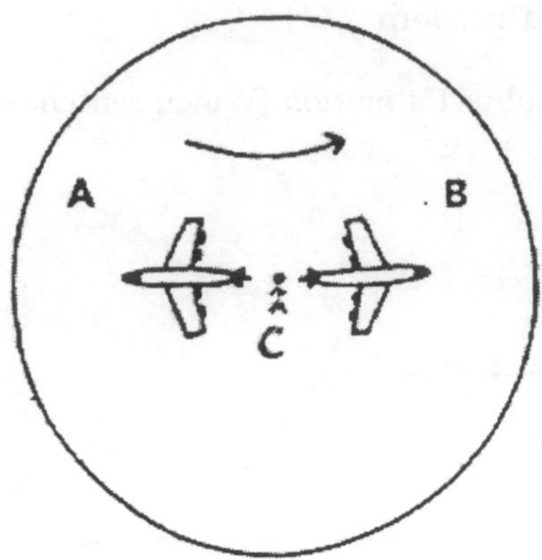

Figure 9.3.

Example of Comparison Problem

The position of two cars at successive 0.20-second time intervals are represented in the drawing below. Is there a moment at which the cars have the same speed?

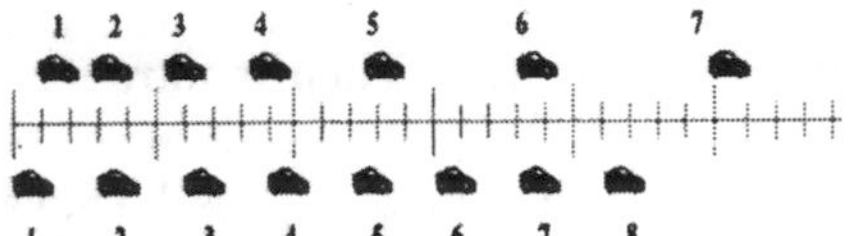

Figure 9.4.

(a) No.
(b) Yes, at instant 2.
(c) Yes, at instant 5.
(d) Yes, at instants 2 and 5.
(e) Yes, at some time during interval 3 and 4.

Appendix II

Task Loaded on the Linear Mental Animation Factor.

When the handle is turned in the direction shown, in which direction will the final gear turn? All gears are fixed in their centres.

Figure 9.5.

Task Loaded on the Three-Dimensional Factor.

As pulley X turns, in which direction will pulley Y turn?

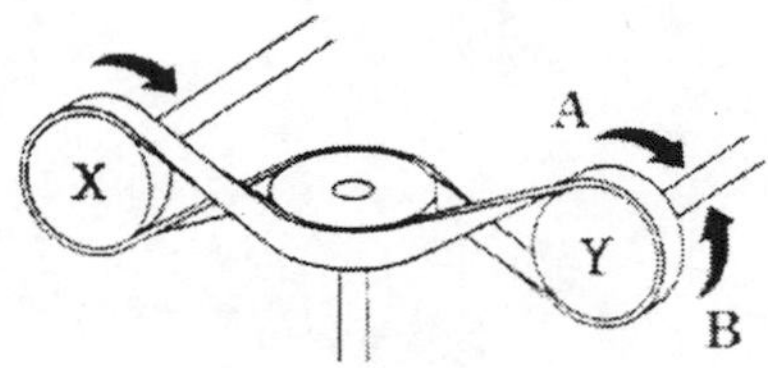

Figure 9.6.

Task Loaded on Spatial Visualisation Factor.

When the little wheel turns around, the big wheel will:

(a) turn in direction A;
(b) turn in direction B;
(c) move back and forth.

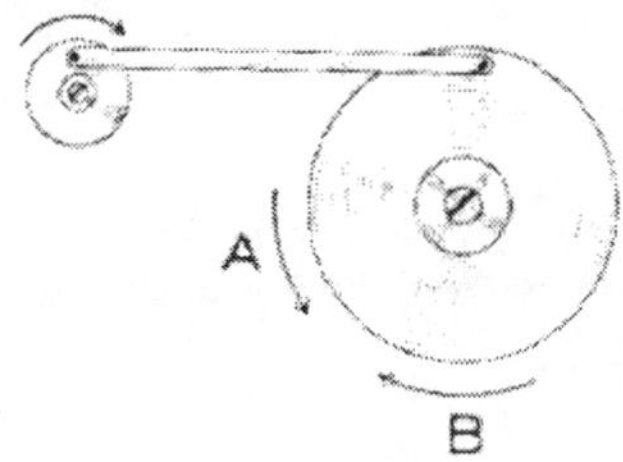

Figure 9.7.

10. Graph Comprehension: The Role of Format, Content and Individual Differences

Priti Shah

Graphs are used extensively to facilitate the communication and comprehension of quantitative information, perhaps because they seem to exploit natural properties of our visual system such as the ability to process large amounts of information in parallel. Rather than a holistic pattern recognition process, however, research has found that graph comprehension is a complex, interactive process akin to text comprehension. Viewers form a mental model of the quantitative information displayed in the graph through serial, iterative cycles of identifying and relating the graphic patterns to associated variables. Furthermore, graph comprehension is not only constrained by bottom-up perceptual features of the graphical display, but is also influenced by top-down factors such as the viewer's expectations about, or familiarity with, the graph's content. Finally, individual differences in graph comprehension skill and domain knowledge interact with the bottom-up influences such that highly skilled graph viewers are less influenced by both the bottom-up and top-down visual characteristics.

10.1 Introduction

Graphical displays are one of the primary means for the representation, communication, and dissemination of quantitative information. Graphs are used to depict mathematical functions, display data from social and natural sciences, and specify scientific theories. As a result, graphical displays are used extensively in textbooks, scientific journals, and the popular media.

What are the processes by which viewers interpret graphs? What makes some graphs easy to understand for some people, and what makes other graphs difficult to understand? The goal of this paper is to outline a cognitive model of graph interpretation, based on a series of empirical studies. An understanding of how viewers interpret graphs, and the factors that make them easy and difficult for different populations, may help us to solve many practical, as well as theoretical problems. The practical problems include how graphs might be created by graphic designers, scientists, and automatic data display systems to more effectively communicate quantitative information [25].

10.2 The Model

A model of graph comprehension will share some characteristics of more general models of diagram interpretation. However, graphs are unique compared to other diagrams and visual displays. They are based on *rational imagery*, meaning that information that is presented is systematically related to the graphic representation [1]. The relation is neither arbitrary, as is the relation between words and concepts, nor a first-order isomorphism, as is the relation between pictures and their referents [1]. Graphs can be distinguished from partially abstract diagrams that are meant to depict visuospatial information (for example, mechanical diagrams of pulley systems or biological diagrams of the functioning of the circulatory system) because graphs represent some *quantitative* property of either concrete objects or abstract concepts. The relation between a represented concept and the graph is based on an analogy between quantitative scales and visual dimensions such as length, colour, or area in which the visual dimensions are usually analogue representations of this quantitative information [1, 12, 20]. Thus, in the continuum of different forms of written information, graphs are more abstract than pictorial diagrams, but still represent information in an analogue, non-arbitrary fashion.

Researchers have suggested that it is this "rational" , or "natural" link between quantitative and spatial information that makes graphs particularly well-suited for representing quantitative information [18,20,30]. Indeed, graph comprehension can seem easy, particularly when a trend or quantitative relationship is explicitly represented in the visual features of the graph [13] or when a viewer has already learned the association between a graphic feature and a quantitative relationship, such as "an upwardly curved line indicates an accelerating relationship" [20]. In the graph in Fig. 10.1, for example, the reader will easily detect the decelerating relationship between batting average and baseball player's income. In order to understand the same relationship in the table below, by contrast, a viewer must effortfully compute the relative differences in the numbers in the cells.

However, both task analyses of graph comprehension [1, 20], as well as errors viewers make in graph comprehension [9,24,25], suggest that a complex set of cognitive processes underlies the apparently effortless comprehension of graphs [11]. In addition, viewers' internal representations of data are often inaccurate or incomplete. This is true of not only college students, but also of populations with significant experience in data analysis and interpretation [24]. Consider, for example, the graph in Fig. 10.1 once again. Although the graph explicitly depicts the increasing, decelerating, relationship between batting average and income that most experienced graph viewers readily

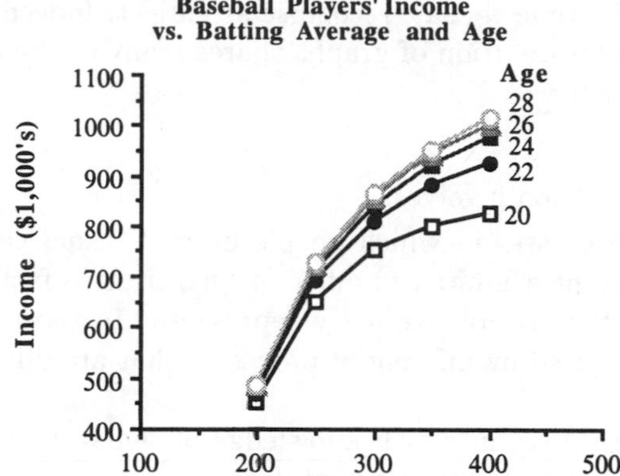

Batting Average

		200	250	300	350	400
Age	20	450	650	750	800	825
	22	470	690	810	880	925
	24	480	710	840	920	975
	26	485	720	855	940	1000
	28	487.5	725	862.5	950	1012.5

Figure 10.1. A graph and table depicting the same data: Baseball players' income (in $1,000s) vs. Batting average and Age. In the graph above the decelerating, increasing relationship between batting average and income is explicitly plotted in the x–y lines. In the table, the deceleration is inferred by the decreasing differences between the numbers in the table.

comprehend, the graph also depicts a similar relationship between age and income that is much more difficult to comprehend.

A fundamental issue in graph research, then, is the characterisation of the interpretation processes and the principled identification of the characteristics of graphic formats, data sets, and interpretation tasks that influence the interpretation processes. In this chapter, I review a number of studies in which my colleagues and I examine how task and graph characteristics lead to different internal representations of data – representations that support relatively effortless and automatic retrieval of some quantitative concepts, and the effortful and complex induction of other quantitative concepts.

These studies show that the interpretation of many commonly used graphs is the result of a complex sequence of cognitive processes – processes that are highly limited and systematically biased. The time it takes to form even a simple interpretation of some graphs in our studies can be about 30 seconds, or the same amount of time as reading and understanding a paragraph of text, rather than the time to, say, recognise an object. Indeed, this research suggests that the interpretation of graphs shares many of the characteristics of text comprehension.

Specifically:

1. Graph comprehension involves:
   - Bottom-up processes in which people extract visual chunks that explicitly represent a limited number of quantitative facts or relations. Information that is not explicitly represented in those visual chunks must be computed by inferential processes that are difficult and error prone.
   - Top-down processes in which knowledge of semantic content also influences viewers' interpretations of data.
2. The interaction of top-down and bottom-up processes is individually applied to different chunks, so that the interpretation process is serial and incremental, rather than automatic and holistic.
3. Individual differences in graph comprehension skill and domain knowledge interact with bottom-up influence of graphic format such that highly skilled graph viewers are less influenced by both the visual characteristics of the graphs.

In the next section, I describe some of the empirical evidence supporting each of these conclusions.

10.3 Empirical Support for the Model

10.3.1 Bottom-Up Processes

As discussed above, one of the oft-touted advantages of graphical displays is the fact that they "take advantage of the visual perceptual processing system". This general statement has been made in reference to a wide varicty of graph comprehension tasks and graphical displays, from scientific discovery [10, 16] to exploratory data analysis [29]. The assumption is that graphical displays, because they depict quantitative information visually, make explicit certain quantitative facts and relations that may not have been previously apparent or obvious from other media such as text or tables. For example, chaotic patterns of weather were discovered when multiple time-series plots were placed on top of one another [10]. The similarities and slight differences between the visual patterns suddenly "popped out", demonstrating the potential power of the perceptual aspect of graph comprehension

Thus, previous theoretical and computational approaches to graphical display comprehension (e.g. [5,6,13,17,20]) imply that graphical displays are most useful when they make quantitative information perceptually obvious. Therefore, information can be retrieved "automatically"; so that cognitive effort is minimised, and the perceptual advantage of graphs over other ways of presenting quantitative information is maximised [13,20]. By implication, the visual characteristics of the graphical display, then, should have an influence on what kinds of information are easy and difficult to comprehend from a graph.

Research that has compared the speed and accuracy of identifying different quantitative facts and relations from different graphic formats has supported this view (e.g. [7,8,14]). For example, Carswell and Wickens [4] found that bar graphs and other "separable" displays were better suited for identification of individual facts. By contrast, line graphs, and other displays that integrated two or more variables, were better suited for tasks that required synthesis.

In recent research, my colleagues and I have examined how the characteristics of graphical displays influence not just the speed and accuracy of making perceptual judgments, but viewers' ability to describe, interpret, and explain quantitative relations. In these studies, we have begun to specify, for different graphic formats, exactly what visual features are "mapped" more or less automatically to quantitative conclusions, and what information is more difficult to retrieve and must be inferred by complex processes. We examined expert and novice viewers' interpretations of a number of commonly used graphic formats: line graphs, bar graphs, divided bar charts, and "three-dimensional" wireframes. For each of these formats, we have found that viewers are able to retrieve a limited number of quantitative facts or relations that are highly constrained by the kinds of visual chunks made explicit by the display. When information is not explicitly represented in a particular graphic format, viewers have tremendous difficulty understanding that information and are often unable to do so. Below, I describe a few of the studies in which we specify the way in which the graphic format influences viewers' interpretations of data.

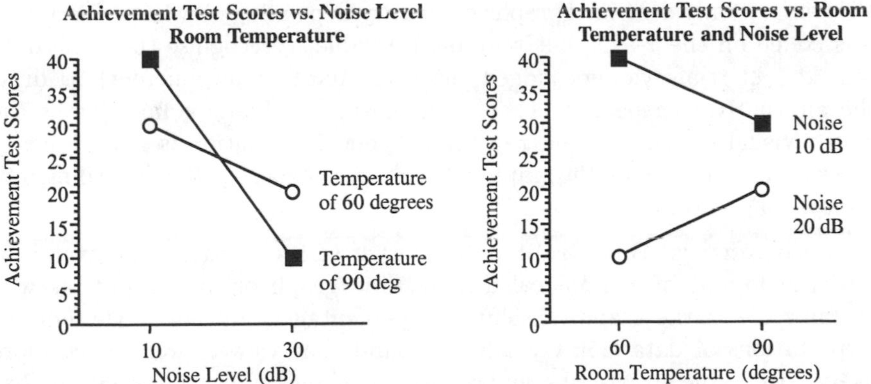

Figure 10.2. Two perspectives the same data set in which the main visual features of the graphs differ greatly, leading to qualitatively different interpretations of data.

Line Graphs. In one series of studies, we examined the interpretation of line graphs, specifically, the limitations that influence a viewer's comprehension of these line graphs [24]. In these studies, viewers were asked to briefly describe or explain a series of individually presented three-variable line graphs, such as the graphs in Figs 10.1 and 10.2. The viewers' descriptions and/or explanations were coded according to the type and amount of information they included about the relations presented in the graphs.

Overall, the results suggested that the comprehension of line graphs involves abstracting a limited set of propositions that describe the functional relations depicted by the x–y lines of the graphs; viewers rarely described more than nominal or ordinal information about the parameter on the curve (in Fig. 10.2(a), the relationship between room temperature and achievement test scores). A typical description of the graph in Fig. 10.2(a), for example, is that

- achievement test scores decrease as noise level increases when it is 60 degrees;
- achievement test scores decrease as noise level increases when it is 80 degrees;
- achievement test scores decrease more when it is 60 degrees than when it is 80 degrees.

But the graph in Fig. 10.2(b) elicits qualitatively different verbal description focusing on its x–y lines:

- achievement test scores decrease with room temperature for low (10 dB) noise levels;
- achievement test scores increase with room temperature for high (30 dB) noise levels;
- achievement test scores for low (10 dB) noise levels are higher than for high (30 dB) noise levels.

Viewers not only described graphs differently depending on what information was coded on the x-axis, but were often unable to recognise the same data (on 32% of trials viewers judged the same data to be different) or draw the alternative perspective. These studies suggest that for line graphs the major visual chunks are the x–y lines. When information is not explicitly presented in the lines on the graphs viewers, even experts, often have difficulty interpreting data.

Line Graphs vs. Bar Graphs. In another series of studies, we examined how the format of a graphical display (line graph or bar graph), as well as the scale of the graph (absolute vs. percentage), influenced viewers' interpretations of data [25]. Overall, we found that viewers were much more likely to be able to describe and answer questions about information that is explicitly represented in graphs. When graphs require any kind of mental computation, such as integrating information across a display or translating from an absolute to percentage scale, viewers have tremendous difficulty and were often unable to do so. More specifically, these studies characterised the kinds of visual chunks that are retrieved for line graphs and bar graphs. As in the previous studies, we found that line graphs emphasise x–y trends. By contrast, bar graphs emphasise comparisons that are closer together on the display. Finally, we found that line graphs are more biasing (emphasising the x–y relations), while bar graphs are more neutral.

Line Graphs vs. Wireframes. A third series of studies examined the kinds of interpretations viewers gave to line graphs and three-dimensional wireframe graphs such as the graph in Fig. 10.3. These studies demonstrate that the internal representation of line graphs and wireframe graphs emphasise different properties of the data. A typical description of a line graph includes descriptions of the x–y lines, including differences between lines and changes along the x-axis. Viewers provide more varied descriptions of wireframe graphs, often describing maxima, minima, and/or shape of the data space in addition to some information about the quantitative relations. Thus, wireframe graphs, particularly the landscape-like complex graphs in this study, lead to qualitatively different interpretations that emphasise configural properties rather than quantitative relations.

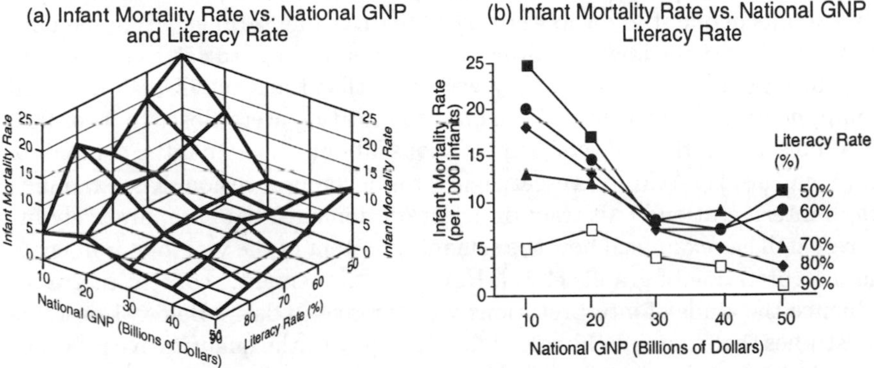

Figure 10.3. The wireframe graph in (a) supports a spatial representation of data in which the important graphical features, such as maxima and minima, are represented. By contrast, the line graph in (b) supports a propositional representation that emphasises a decreasing relationship between year and mortality rate.

Summary. A number of studies demonstrate that the perceptual characteristics of the visual display, in particular the kinds of visual chunks that are retrieved in comprehending different graphic formats, influence viewers' interpretations of data. Thus, just as characteristics of text, such as coherence [28], influence what kinds of inferences readers can easily make, the characteristics of the graphical display influence what kinds of quantitative inferences graph viewers can make. In designing graphical displays, as in designing text, it is not merely enough that information is presented technically correct, but also that it is designed to effectively communicate the relevant quantitative information.

10.3.2 Top-Down Processes

A second major characteristic of our model of graph comprehension is that, in addition to a bottom-up influence of the characteristics of the graphical display, there is a top-down influence on the semantic content of the quantitative information that is depicted in the graph. Again, this aspect of a model of graph interpretation has parallels to models of text comprehension. Specifically, models of text processing incorporate the notion that readers' prior knowledge, expectations, and goals influence the process by which they read and the kinds of information that they comprehend and remember about a passage (e.g., [28]).

In the general case, it appears that viewers encode and remember pictures and diagrams differently depending on their knowledge and expectations. For example, verbal labels have long been known to distort viewers' memory for ambiguous pictures. More recent evidence suggests that viewers have schemas for graphs and maps that distort their representations of them. For example, participants in one study who were asked to draw line "graphs" from memory tended to distort the lines and draw them as being closer to 45° than the lines originally were [22, 27]. When they were told that the same display depicted a map, however, they distorted the lines so that they were closer to 0° or 90°.

What about the viewers' expectations about the semantic content of graphical displays? Much research in graph comprehension has examined graph interpretation in abstract or arbitrary domains, but comparatively little research has examined how the semantic content of the variables influences the interpretation of graphs [4, 15]. How does the semantic content of graphs influence the kinds of interpretations viewers give to data? Below, I describe two studies that examine the role of familiarity with the quantitative relations presented in graphs as well a viewer's causal expectations.

Familiarity. The premise of the first study was that viewers would be more likely to interpret relationships between variables for which viewers had expectations about general trends, such as number of car accidents, number of drunk drivers, and traffic density, compared to variables for which viewers did not have any expectations, such as ice cream sales, fat content, and sugar content [23]. The results from this study suggest that when viewers had particular expectations, they were likely to describe those relationships (for example, as drunk driving increases, car accidents increase), ignoring "idiosyncratic" data points such as local maxima and minima that were inconsistent with the general expected trends. By contrast, when viewers did not have expectations, they were less likely to describe general trends, and more likely to describe local maxima and minima. These results suggest that viewers' familiarity with quantitative trends influenced whether or not they would describe those trends.

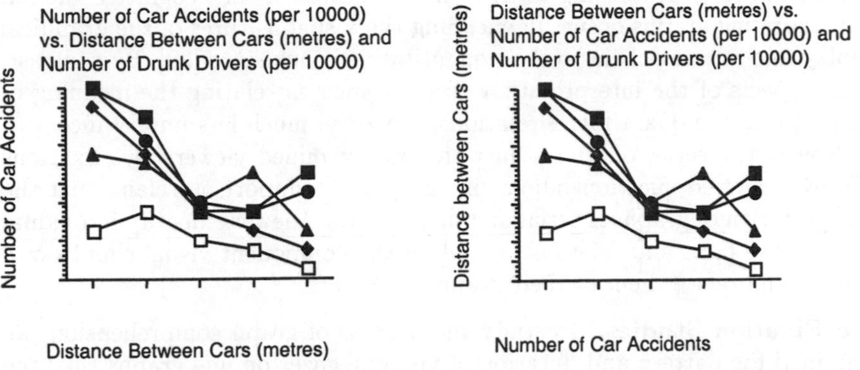

Figure 10.4. (a) shows a conventionally plotted graph that is consistent with most viewers' expectations of the causal relations between these variables. In (b), the x- and y-axes are reversed.

Expectations about Causal Relations. In a second experiment, we examined viewers' expectations of causal relations that are depicted in graphs [23]. The assumption was that, given a set of familiar variables, viewers are likely to have some expectations about the directionality of causal relations. For example, one expectation is that increased rates of drunk driving or decreased distance between cars cause car accidents and not vice versa. In this study, viewers were presented with graphs that depicted data about common topics, in which the likely dependent variable (the number of car accidents) was plotted on the y-axis (conventional) as shown in Fig. 10.4(a), or on the x-axis (reversed), as shown in Fig. 10.4(b).

When the position of graphic variables is conventional, viewers were able to use their knowledge about the graphic format and the variables to make accurate inferences about the quantitative relations. However, when graphs are plotted so that the data are inconsistent with viewers' expectations about causal relations, viewers frequently described relations that were not actually depicted in a graph but are consistent with their expectations or models about the causal relations. For example, even though the graph in Fig. 10.4(b) does not depict a relationship between the number of drunk drivers and the number of car accidents, most viewers would expect a relationship between those two variables, and novice graph viewers inaccurately described this relationship on 93% of the trials.

10.3.3 The Serial Nature of Graph Interpretation

Models of graph interpretation that are based on task analyses tend to emphasise the holistic, pattern recognition aspects of graph interpretation (as reviewed in [11]). According to these models, most of the "cognitive" action in interpreting graphs occurs in encoding the visual features of the graphical display and relating them to their quantitative conclusions [20]. By contrast, other aspects of the interpretation process, such as relating the meaning of the graph to the quantitative relations, are given much less importance.

However, a series of studies in which we examined viewers' eye fixations as they answered questions about line graphs [2] support our claim that the interpretation of graphs is serial and incremental. Viewers identify individual quantitative facts and relations, based on the component visual chunks of a display, and relate them to their graphic referents.

Eye Fixation Studies. To study the process of graph comprehension, we examined the pattern and duration of viewers' gazes on line graphs (like the ones shown in Figs 10.1 and 10.2) as they described and answered questions about graphs. The results demonstrated that the comprehension of graphs is complex, with viewers spending the majority of the time interpreting a graph relating information from the lines on the graph to their referents, rather than viewing the patterns of lines themselves. Furthermore, the results supported a model of graph interpretation in which viewers serially identify each individual visual chunk and relate it to its graphic referents (the variable names). This iterative model can predict the distribution of viewers' gazes across different parts of a graph as well as the total number of gazes required to interpret graphs that vary in complexity.

10.3.4 Individual Differences

The final characteristic of our model of graph interpretation is that individual differences in graph comprehension skill and domain knowledge interact with bottom-up influence of graphic format such that highly skilled graph viewers are less influenced by both the visual characteristics of the graphs [26]. This aspect of our model is supported by a recent study in which we compared interpretations of line graphs and bar graphs provided by skilled and unskilled graph viewers. Graph skill was measured by a standardised test of graphing skill, the Test of Graphing in Science [19]. In order to compare domain knowledge differences, half the participants saw line graphs and bar graphs that included familiar semantic content, such as "Age of owner vs. value of car (in dollars) and the rate of car thefts". The assumption was that these participants would have some knowledge of these familiar variables. The other half of the participants saw graphs that had no semantic content, only letters such as "A vs. B and C", so they could not have any domain knowledge.

Overall, graph-reading skill and domain knowledge both interacted with graph format. Unskilled graph viewers, as well as viewers with no domain knowledge, typically provided surface-level descriptions of the data (for example, describing each line individually in a line graph). By contrast, skilled graph viewers and those who had some domain knowledge made some inferences about the data. In particular, they were more likely to provide some general trend (or main effect) information, such as the fact that as people get older they have more expensive cars. Although such a description is less detailed than a surface description, it requires the viewers to make an inference and to be less influenced by the way in which information is plotted.

10.4 Conclusions

In summary, cognitive research supports a model of graph comprehension in which interpreting a graph involves translating the visual features of a graph to a conceptual representation of the quantitative information via multiple, integrated cycles of identifying quantitative relations and the variables associated with them. The model assumes that graph viewers have knowledge about different graphic formats that influences and supports the interpretation process. The relative ease or difficulty in interpreting a graph occurs because two kinds of processes are involved in comprehending different kinds of information from a graph. If the graph supports visual chunks that the viewer can map to relevant quantitative information, then it may be automatically retrieved. If not, quantitative information must be computed by inferential processes that consist of a number of retrieval and comparison substeps. When graph viewers have less experience interpreting graphs, they may be forced to rely on semantic knowledge even if this knowledge is not consistent with the data. Thus, the model proposes that viewers' knowledge about graphic formats, and expectations about the relationships between particular variables, have a top-down influence on the kinds of interpretations that viewers give to graphs.

This model had a number of theoretical and practical implications. Theoretically, this research suggests that a computational model of graph interpretation will need to incorporate the following features:

1. The ability to support the automatic retrieval of some quantitative facts (if straight line, linear relationship).
2. Individual identification of each quantitative function or fact.
3. Individual differences in knowledge influence about what visual features imply.
4. The ability to identify and compare values to infer relations.
5. Limited capacity in the amount of information viewers can maintain.
6. Interactivity: perceptual properties influence interpretation; knowledge about graphs and quantitative relations influence what viewers encode.

There are also a number of practical implications of graphic design relevant for both people [25] and automatic data display systems [21]. Different ways of presenting data can influence what is easy to retrieve and therefore what viewers are likely to encode. In designing graphical displays one should use knowledge of the bottom-up influence of graphic format to design displays that maximize retrieval and minimize difficult inferential processes. In addition, because viewers are able to comprehend a limited amount of information, an individual graph should contain at most a couple of relevant quantitative concepts. Finally, graphic design should be tailored to the audience; it is particularly important to pay attention to the characteristics of graphical displays for novice graph viewers.

Acknowledgements

This research was supported by a James S. McDonnell Postdoctoral Fellowship for Cognitive Studies in Educational Practice, a fellowship from the National Academy of Education, and Office of Naval Research grant # N00014-98-1-0350. I would like to acknowledge the contributions of Patricia Carpenter, Mary Hegarty, Richard E. Mayer, and Destiny Shellhammer, who collaborated on aspects of this research. I would also like to thank Mary Hegarty and James Hoeffner for comments on an earlier version of this paper.

References

1. Bertin, J. (1983). Semiology of graphics: Diagrams networks maps (W. Berg, trans.). Madison, WI: University of Wisconsin Press.
2. Carpenter, P.A. and Shah, P. (1997). A model of the perceptual and conceptual processes in graph comprehension. Under review.
3. Carswell, C.M., Emery, C. and Lonon, A.M. (1993). Stimulus complexity and information integration in the spontaneous interpretation of line graphs. Applied Cognitive Psychology 7:341–357.
4. Carswell, C.M. and Wickens, C.D. (1987). Information integration and the object display: An interaction of task demands and display superiority. Ergonomics 30:511–527.
5. Casner, S.M. (1990). Task-analytic design of graphic presentations. Unpublished doctoral dissertation, University of Pittsburgh, Pittsburgh, PA.
6. Casner, S.M. and Larkin, J.H. (1989). Cognitive efficiency considerations for good graphic design. In Proceedings of the Cognitive Science Society. Hillsdale, NJ: Lawrence Erlbaum Associates.
7. Cleveland W.S. and McGill, R. (1984). Graphical perception: Theory, experimentation, and application to the development of graphical methods. Journal of the American Statistical Association 77:541–547.
8. Cleveland, W.S. and McGill, R. (1985). Graphical perception and graphical methods for analyzing scientific data. Science 229:828–833.
9. Culbertson, H.M. and Powers, R.D. (1959). A study of graph comprehension difficulties. Audio Visual Communication Review 7:97–100.

10. Gleick, J. (1987). Chaos: Making a new science. New York: Penguin Books.
11. Guthrie, J.T., Weber, S. and Kimmerly, N. (1993). Searching documents: Cognitive processes and deficits in understanding graphs, tables, and illustrations. Contemporary Educational Psychology 18:186–221.
12. Hegarty, M., Carpenter, P.A. and Just, M.A. (1991). Diagrams in the comprehension of scientific texts. In R. Barr, M.L. Kamil, P. Mosenthal, and P.D. Pearson (Eds), Handbook of reading research, Vol. 2. New York: Longman.
13. Larkin, J. and Simon, H. (1987). Why a diagram is (sometimes) worth ten thousand words. Cognitive Science 11:65–99.
14. Legge, G.E., Gu, Y., and Luebker, A. (1989). Efficiency of graphical perception. Perception and Psychophysics 46:365–374.
15. Leinhardt, G., Zaslavsky, O. and Stein, M.K. (1990). Functions, graphs, and graphing: Tasks, learning, and teaching. Review of Educational Research 60:1–64.
16. Lewandowsky, S. and Spence, I. (1989). The perception of statistical graphs. Sociological Methods and Research 18:200–242.
17. Lohse, G.L. (1993). A cognitive model of understanding graphical perception. Human–Computer Interaction 8:353–388.
18. MacDonald-Ross, M. (1977). Graphics in texts. Review of Research in Education 5:49–85.
19. McKenzie, D.L. and Padilla, M.J. (1986). The construction and validation of the Test of Graphing in Science (TOGS). Journal of Research in Science Teaching 23:571–579.
20. Pinker, S. (1990). A theory of graph comprehension. In R. Freedle (Ed.), Artificial intelligence and the future of testing. Hillsdale, NJ: Lawrence Erlbaum Associates, pp. 73–126.
21. Roth, S.F. and Hefley, W.E. (1993). Intelligent multimedia presentation systems: Research and principles. In M. Maybury (Ed.), Intelligent multi-media interfaces, pp. 13–58.
22. Schiano, J.D. and Tversky, B. (1992). Structure and strategy in encoding simplified graphs. Memory and Cognition 20:12–20.
23. Shah, P. (1995). Cognitive processes in graph comprehension. Unpublished doctoral dissertation.
24. Shah, P. and Carpenter, P.A. (1995). Conceptual limitations in comprehending line graphs. Journal of Experimental Psychology: General 124:43–62.
25. Shah, P., Mayer, R. and Hegarty, M. (in press). Which graphs are better? Textbook graphs as aids to knowledge construction. Journal of Educational Psychology.
26. Shah, P. and Shellhammer, D. (1999). The role of domain knowledge and graph reading skills in graph comprehension. Presented at the 1999 meeting of the Society for Applied Research in Memory and Cognition, Boulder, CO.
27. Tversky, B. and Schiano, D.J. (1989). Perceptual and conceptual factors in distortions in memory for graphs and maps. Journal of Experimental Psychology: General 118:387–398.
28. van Dijk, T.A. and Kintsch, W. (1983). Strategies of discourse comprehension. New York: Academic Press.
29. Wainer, H. and Thissen, D. (1981). Graphical data analysis. Annual Review of Psychology 32:191–241.
30. Winn, B. (1987). Charts, graphs, and diagrams in educational materials. In D. Willows and H.A. Houghton (Eds), The psychology of illustration. New York: Springer.

11. Graphs in Print

Jeff Zacks
Ellen Levy
Barbara Tversky
Diane Schiano

Diagrams for presenting quantitative data are an important component of print communication. Their rate of use is high and rising. This reflects in part the recent development of software tools for generating data graphics. These programs allow a wide range of choices for data visualisation – some of which may be ugly or ineffective. How has graph usage evolved during this period? A survey of graph usage in academic journals, magazines, and newspapers during the years 1985–1994 revealed several dynamic trends in the characteristics of data graphics, as well as robust differences between media. However, graph features that have been singled out by experts as poor choices, such as "3-D" rendering, do not seem to be on the rise.

11.1 Introduction

Graphical presentations of quantitative information are a striking and ubiquitous component of print media. A well-chosen graphic can effectively communicate a large amount of information efficiently [18], and can make that information perceptually salient and memorable. Many popular news publications, and most scientific journals, make extensive use of data graphics. Recently, three developments in the study and technique of graphing have impacted on our understanding of how graphics function.

In the past two decades a cottage industry has developed around critiques of poor design in statistical graphics. Jibes at inefficient, misleading, or inelegant graphs come from designers [26], statisticians [8, 31], and perceptual psychologists [16]. In all these cases, criticism is accompanied by constructive insight into how to produce graphics that are efficient, perspicuous, and elegant.

At the same time, the behavioural study of how people extract quantitative information from visual displays has begun to blossom. This tradition has its roots as far back as at least the 1930s (see [20] for an excellent review), but since the early 1980s there has been a flowering of both empirical research (e.g. [5, 6, 10, 12, 23–25, 30]) and theoretical analysis [2, 9, 19, 22].

As critical analysis and behavioural research have expanded, the technologies employed by the producers of graphs have metamorphosed. At the time Bertin's[1] [4], Cleveland's [9], and Tufte's [26] major works appeared, computer technologies for creating quantitative graphics were the exclusive province of major media design shops and well-funded scientists, many in private industry. Since then, the explosion of computing power and the development of the graphical user interface have put sophisticated tools for producing statistical graphics in the hands of everyday producers and consumers of data.

These three developments have contributed substantially to what we know about how data graphics work, and have provided a wealth of new abilities to produce ever more sophisticated figures. They also raise a question: How are graphs actually being used in contemporary communications? This question is relevant to the critical enterprise, because there is little interest in critiquing that which does not occur; also, there is reason to worry that the new technologies have made it not simply easier to produce graphs, but particularly easier to produce *bad* ones. It is relevant to the behavioural study of graphs, because empirical and theoretical understanding is most valuable for those graphical formats that are frequently used. Finally, it is important for producers of tools for creating visualisations, because it is of interest to know what kinds of graphs are of interest to different communities, and it is worthwhile to identify areas that are poorly served by current technologies.

The research described here focuses on the use of graphs in print media. This is a question that has received limited attention in previous work. Cleveland [8] surveyed the use of graphs in scientific publications. He classified graphics as depicting one, two, or three variables. Most graphs (83%) showed two variables. There was marked variability in the amount of space per page devoted to graphs, ranging from 31% to 0%, depending on the publication. Cleveland also examined graphs from a sample of *Science* magazine in detail, and was dismayed to find that 30% had errors of construction, reproduction quality, explanation, or discriminability of items. Tufte [23] surveyed 15 news magazines and newspapers from around the world, and coded their graphs as relational (linking two or more variables) or not. Relational graphs were taken to be more graphically sophisticated than non-relational ones. No publication in Tufte's sample contained more than 10% relational graphs, and no publication from the United States even reached 1%.

The purposes of the current project were threefold: first, to characterise the types of graphs typically used in current print publications; second, to compare different media in their use of data graphics; third, to examine dynamic trends in the features of graphs – especially those that have been facilitated by new computing technologies and/or those that have been targeted by critics of graphical practices as bad form. These issues were addressed by

[1] Bertin's book was published in English in 1983. It originally appeared in French, in 1967 [2].

sampling a range of print media during a 10-year period, from 1985 to 1994, and coding a large number of graphs on a large number of features.

11.2 Method

We began by selecting 17 publications as the database to be studied. To allow comparisons between different publication media, we selected seven academic journals, six magazines, and four daily newspapers. The journals were the *Journal of the American Statistical Association*, *Educational Review*, the *Journal of Experimental Psychology: Learning, Memory and Cognition*, the *Journal of Political Economy*, *Nature*, and *Science*. The magazines were *Byte*, *Fortune*, *Newsweek*, *Popular Mechanics*, the *Economist*, *Time*, and *US News and World Report*. The newspapers were the *Los Angeles Times*, the *New York Times*, the *Wall Street Journal*, and the *Washington Post*. In each category, publications were selected to cover a range of interests and to be representative of the medium. Area experts were consulted in the selection of the scientific journals. For each publication, the first issue with a January date and the first issue with a July date from each of the years 1985–1994 were selected for coding, for a total of 340 issues coded (17 publications × 2 issues/year × 10 years).

Each issue to be coded was examined by one of two trained coders (a small number of graphs that were missed by the coders were coded by the first author). Coders examined each publication for graphs. A graph was defined to be a figure on the page that used spatial configuration to represent numerical information. Each graph in the publication was identified by page number and recorded. The coder then rated each figure on 39 features, described in Table 11.1. Questions about coding were resolved by discussion between the two coders and the first two authors.

As can be seen in Table 11.1, the features coded sorted into eight groups. The first group (Type) captured the general type of graph. Often a given graph would be of more than one type (e.g., a figure that is both a bar and a line graph, as in Fig. 11.1). The second group (Dimensionality) captured the apparent dimensionality of the figure. The third group (General) captured some general features of the figure. The fourth group (Orientation) captured whether the figure contained vertically oriented or horizontally oriented elements (or both, or neither). The fifth, sixth, and seventh groups captured some specific features of the X, Y, and Z variables, respectively (if each existed). The final group (Legend) captured properties of the legend. Figures 11.1 and 11.2 give examples of graphs coded according to the coding system.

In addition to the experimental data, each of the two coders also coded one issue of *Scientific American* from January 1995. Inter-coder reliability was good. For the 11 graphs in this issue, the coders agreed on 417 (97%)

Table 11.1. Features of graphs coded in this study. Groups of features are indicated in the left column, individual features in the middle column, and their definitions in the right column

	Feature	Description
Type	Point	True if the figure contained data elements that were points whose height and width were not informative.
	Line	True if the figure contained data elements that were lines whose width was uninformative.
	Bar	True if the figure contained data elements that were rectangular areas of consistent height or width.
	Pie	True if the figure contained a data element that was a circle or ellipse whose area was subdivided from the centre.
	Pictograph	True if the figure used the scale of a picture, or the number of a set of repeated pictures, to convey a numerical value.
	Cartograph	True if the figure contained a map onto which numerical data were projected (by any technique). If the cartograph was a weather map, this was noted.
	Stem and leaf	True if the graph was a stem-and-leaf plot, as described by Tukey [28] or any variant.
	Box and whiskers	True if the graph was a box-and-whiskers plot, as described by Tukey [28] or any variant.
	Other	True if the graph was of a type other than those listed above.
Dimensionality	Simple	True if the figure contained data elements with negligible enclosed area (e.g. points, lines, curves).
	Area	True if the figure contained data elements that formed enclosed areas (shaded or not) and did not give the appearance of depth (e.g. rectangles, irregular polyhedra).
	Volume	True if the figure contained data elements that gave the appearance of solid volumes.
	Surface	True if the figure contained data elements that gave the appearance to surfaces with extent in depth, but negligible volume.
General	Colour vs b/w	Coded as "colour" if the figure contained multiple hues, else "black and white".
	Repeated	If the figure was a member of a group that used a similar frame that was repeated separated by space, coded as true. If the figure contained multiple data sets drawn on the same connected axes, the number of data sets depicted was drawn.

	Feature	Description
	Background picture	True if the figure contained a background picture.
	Error bars	True if the figure contained error bars.
Orientation	Horizontal	True if the figure contained elements that depicted numerical values using horizontal extent.
	Vertical	True if the figure contained elements that depicted numerical values using vertical extent.
	Stacked	True if the figure contained elements that depicted numerical values using the distance between adjacent elements.
X Variable	Qualitative	True if the dimension represented a nominal variable.
	Quantitative	True if the dimension represented an ordinal, interval, or ratio variable. For discrete quantitative variables, the number of levels was noted.
	Time series	True if the quantity described was a time or date.
	Labelled data points	True if any the data points were accompanied by text labels.
	Grid	True if the figure contained a horizontal and/or vertical grid.
Y Variable		*(Same features as "X Variable" above.)*
Z Variable		*(Same features as "X Variable" and "Y Variable" above.)*
Legend	Qualitative	True if the dimension represented a nominal variable. For legends, the number of levels in the legend was recorded, and the dimension(s) to which the legend referred (x, y or z) was also noted.
	Quantitative	True if the dimension represented an ordinal, interval, or ratio variable. For discrete quantitative variables, the number of levels was noted. For legends, the dimension(s) to which the legend referred (x, y, or z) was also noted.
	Time series	See above.
	Colour map	True if the legend was a colour map.

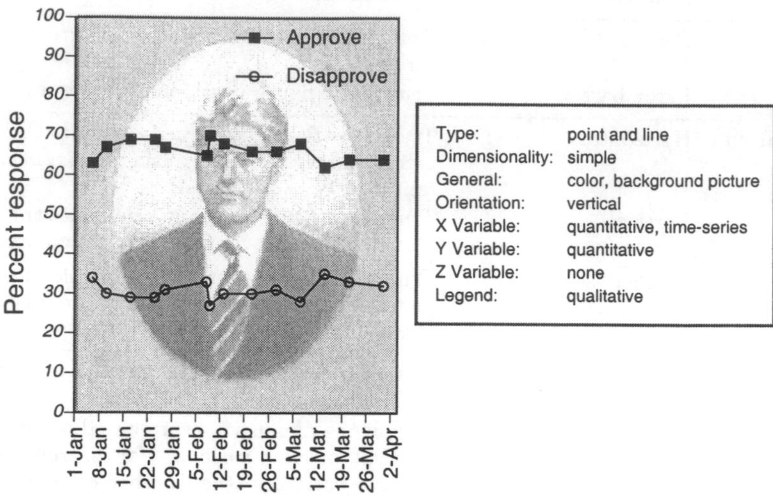

Figure 11.1. Example graph to demonstrate the coding scheme. (Data are presidential approval ratings for January–March 1999.)

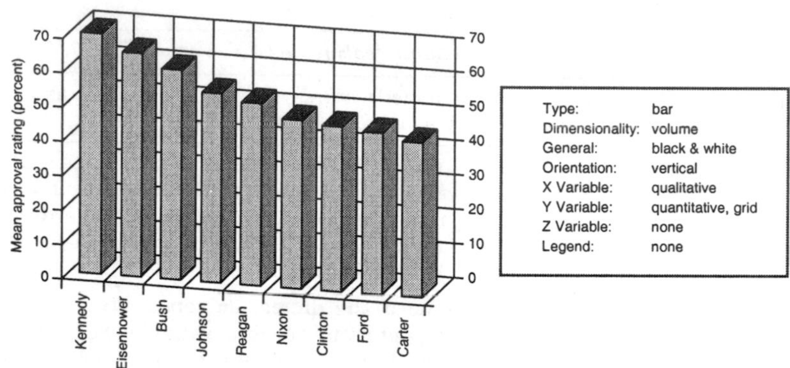

Figure 11.2. Second example graph to demonstrate the coding scheme. (Data are mean presidential approval ratings for the last nine presidents.)

of the 429 features rated. The 12 differences in coding can be summarised as: one disagreement on graph type (giving rise to two differences in coding), one disagreement on colour vs. black-and-white, one disagreement as to whether two graphs were repeated, one disagreement on whether a graph was a time series, three disagreements on whether data points were labelled, one disagreement on whether there was a y-axis grid, one disagreement about whether a figure contained a legend, and one disagreement on the number of levels in a legend.

Data were entered into a database and tabulated by year and issue. Means were calculated by year, publication, and medium.

11.3 Results

11.3.1 Number of Graphs per Issue

The sample consisted of 8159 graphs (5250 from academic journals, 1365 from magazines, and 1544 from newspapers). The mean number of graphs per issue was 24.3. The number of graphs per issue during the period sampled rose sharply for academic journals (from 34.7 in 1985 to 61.2 in 1994) and in newspapers (from 10.1 in 1985 to 24.5 in 1994). For magazines, the number of graphs per issue held roughly to its mean of 9.75 graphs per issue. (See Fig. 11.3.) These data support the notion that graphs are an important means of communication across the range of print media, and their importance has increased in recent years.

11.3.2 Types of Graphs

As can be seen in the prior discussion of the coding scheme employed here, there are an impressive variety of types of data graphics available to authors and designers. The different media here appear to utilise these different graph types in systematically different ways. In the sample, journals employed line graphs (72.5%) and point graphs (46.6%) almost exclusively, while magazines and newspapers used mainly line graphs (49.3% for magazines, 50.1% for newspapers) and bar graphs (44.1% for magazines, 27.9% for newspapers). The latter two media also occasionally used pie graphs, and newspapers also occasionally employed cartographs (primarily weather maps) and other graph types (see Fig. 11.4). Note that for each medium the graph types sum to more than 100%, as a given graph can be an instance of more than one type (e.g., a scatterplot with a curve fit would be both a point and a line graph).

11.3.3 Apparent Dimensionality and "3-D" Effects

Graphs can be rendered in a number of different styles. One way to vary the rendering style of a figure is to manipulate the apparent dimensionality of

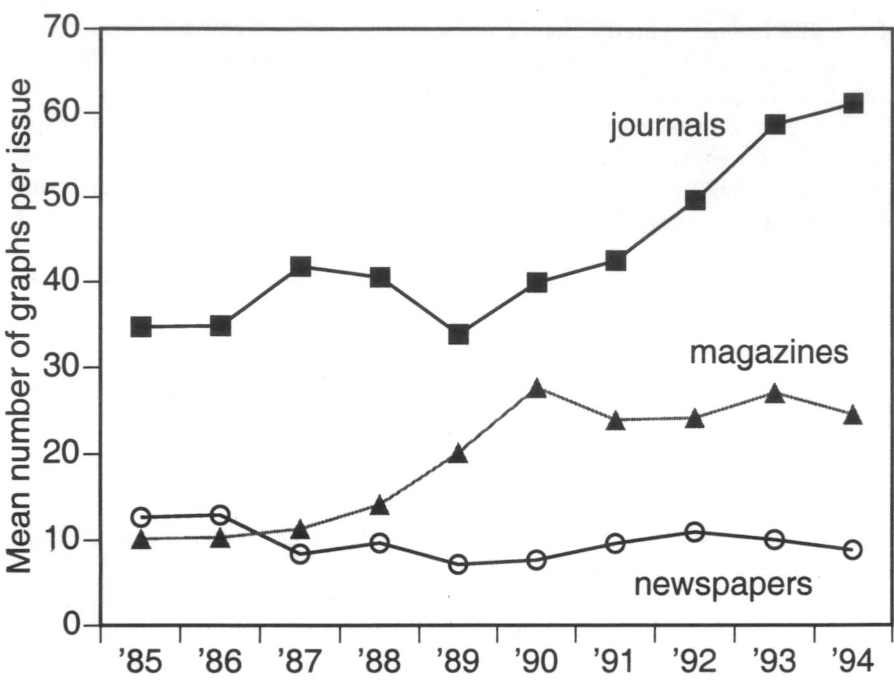

Figure 11.3. Number of graphs per issue by medium and year.

the data elements (see Section 11.2 above). Graphic designers [26], perceptual theorists [16], and experimentalists have all spilled ink on the relative merits of different rendering styles. Tufte's argument for rendering data using minimalist data elements ("simple" dimensionality, in our terminology) has been influential. The experimental data suggest that in fact viewers of data graphics are more accurate in making magnitude judgements from simple or area graphs rather than volume graphs, but that this effect is relatively small and not particularly robust [32]. Given the range and strength of opinion regarding this issue, it is of interest to see how graphs of varying dimensionality are used across different media.

In the publications sampled, "3-D" graphs were quite rare. Volume graphs made up only 1.6% of our sample, and surface graphs only 1.5%. Academic journals chose primarily simple graphs (85.3%), while magazines and newspapers employed a mix of simple and area graphs (for magazines 47.7% simple and 50.0% area; for newspapers, 49.3% simple and 36.1% area). Furthermore, as Fig. 11.5 shows, the proportion of volume and surface graphs did not increase markedly over the period surveyed. It does, however, seem that for magazines the relative proportion of simple graphs increased, at the expense of area graphs (see Fig. 11.6).

There were systematic relationships between graph type and apparent dimensionality. Figure 11.7 indicates that point and line graphs tended to be

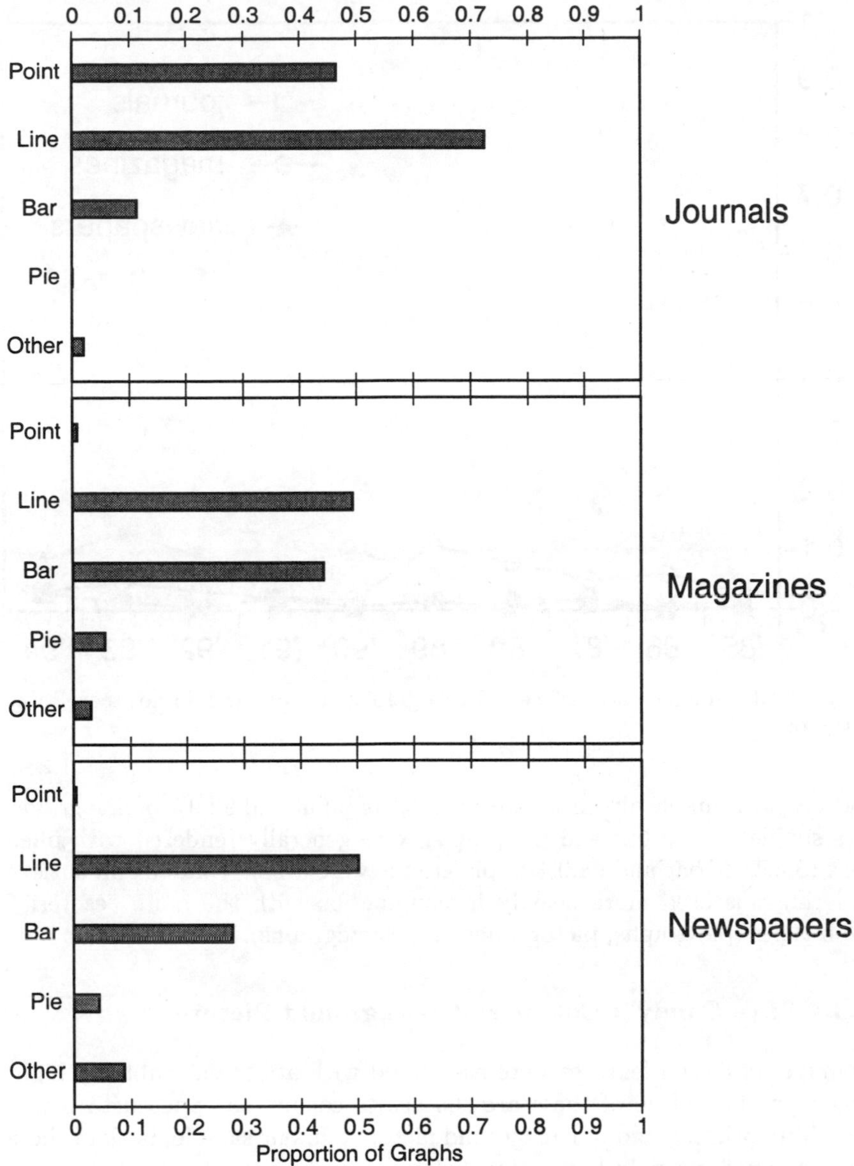

Figure 11.4. Distribution of graph types by medium. Each pane shows the proportion of graphs of each type for one medium. (Pictographs, cartographs, stem-and-leaf plots, and box plots are collapsed into the "Other" category because they occurred infrequently.)

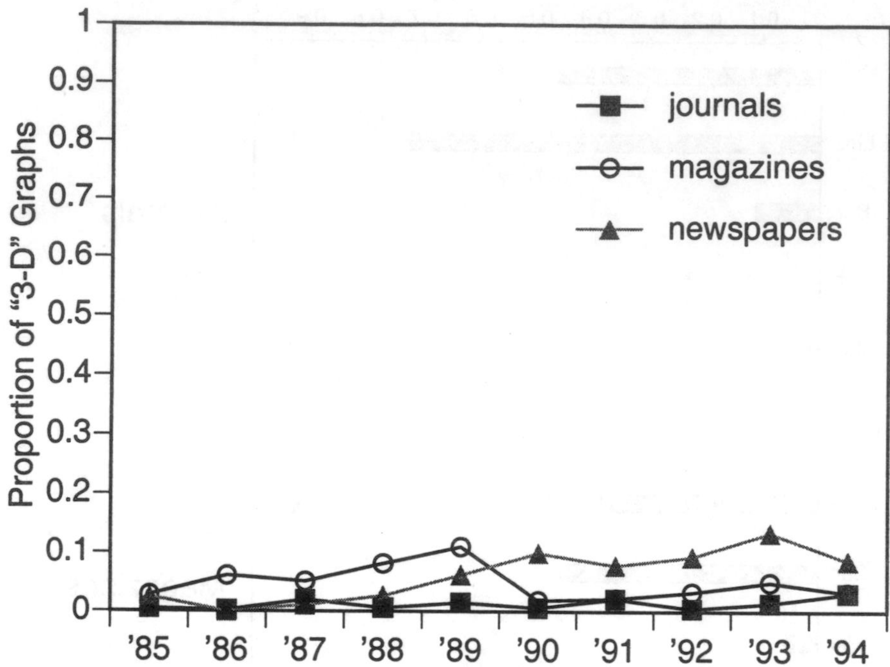

Figure 11.5. Volume and surface ("3-D") graphs are rare and do not seem to be on the rise.

rendered with simple dimensionality (97.1% of point and 92.9% of line graphs were simple), while bar and pie graphs were generally rendered with filled areas (85.5% of bar and 64.0% of pie graphs were area). Volume and surface rendering was used more heavily in conjunction with the more "esoteric" graph types (pie graphs, pictographs, and cartographs).

11.3.4 "Eye Candy": Colour and Background Picture

A number of design features were associated with attention-grabbing, high-impact graphics. Two of these are the use of colour (as opposed to black-and-white) printing and of background pictures. In our sample, both of these features varied strongly by medium. Colour graphs were dominant in magazines, constituting 69.9% of the graphs appearing there, while they almost never appeared in academic journals (1.0%) or newspapers (0.0%). There was some indication that colour graphs increased in prevalence in magazines (see Fig. 11.8). Background pictures were rare over all: only 1.4% of the figures sampled contained them. Magazines were more likely to contain background pictures (7.7%) than journals (0.1%) or newspapers (0.6%).

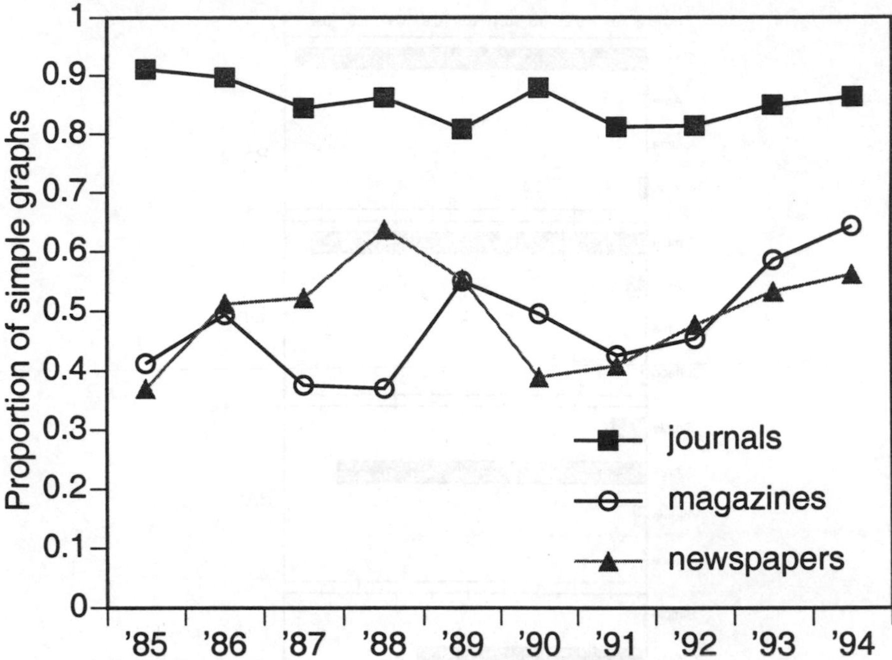

Figure 11.6. The prevalence of simple graphs in magazines and newspapers showed a modest increase.

11.3.5 Features that Support Quantitative Precision: Error Bars, Grids, Labelled Data Points and Colour Maps

Graphs can be used to convey a rough general message or detailed data patterns. A number of features facilitate the latter usage. Error bars permit the viewer to make inferences about the stability of reported data values. Such information is especially valued in the sciences, so it is not surprising that they were used more in journals (10.7%) than in magazines or newspapers, where they never occurred in our sample. In journals, the use of error bars showed no increasing or decreasing trends.

The addition of a grid helps viewers make fine judgements about data values [16]. On the other hand, a grid adds clutter, disdained by Tufte [26] and other minimalists. Based on these considerations one might expect more frequent use of grids in scientific publications, in which accuracy is valued and design experience is rarer. Thus, it is somewhat surprising that grids (associated with either the X, Y, or Z axis) were least popular in academic journals (2.0%), and most popular in newspapers (71.6%), with magazines showing an intermediate disposition to use grids (38.8%); no medium showed any trends over time.

Similarly, labelling data points permits exact judgements of numerical values, but labelled data points (either X, Y, or Z values) were also rare

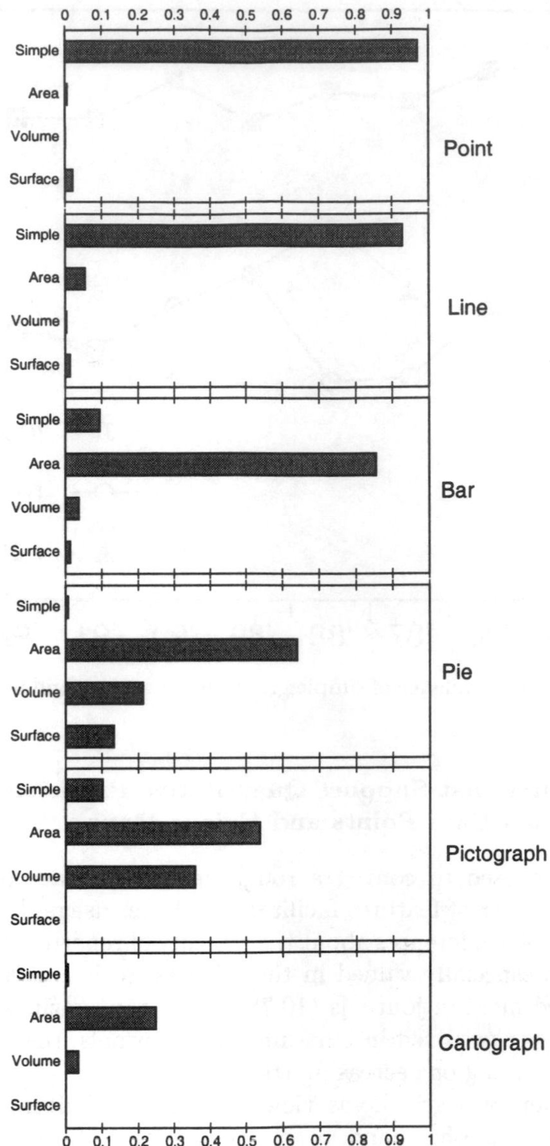

Figure 11.7. Distribution of apparent dimensionality by graph type. Each pane shows the proportion of graphs of one type that appeared in each apparent dimensionality. (Stem-and-leaf plots, box plots, and those classified as "other" are omitted because they occurred infrequently, as are graphs whose dimensionality could not be classified.)

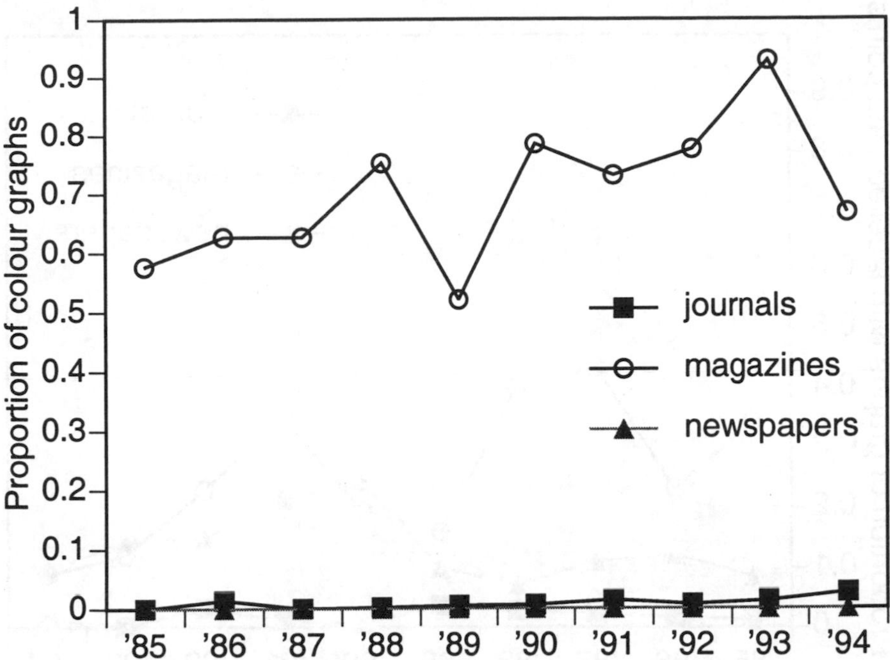

Figure 11.8. Colour graphs appeared predominantly in magazines, where they may be on the rise.

in academic journals (1.7%), and more prevalent in magazines (26.0%) and newspapers (14.6%). There was some indication that labelled data points are becoming less common in magazines (see Fig. 11.9).

Colour maps, which permit the display of quantitative information about an additional variable, were rare in all three media (0.7% for journals, 0.1% for magazines, and 0.0% for newspapers).

11.3.6 Number and Type of Variables

As Descartes originally demonstrated, a figure on a page can be an exquisite medium for depicting the relationship between two variables. The habit of assigning one variable to the horizontal axis and another to the vertical axis is deeply ingrained in Western visual literacy. Indeed, horizontal and vertical lines have a privileged status in perception [13]. Not surprisingly, the vast majority (94.9%) of graphs in this sample depicted relationships between two variables. This figure agrees reasonably well with Cleveland's [8] figure of 83% for scientific publications. However, 11.6% of graphs in magazines depicted only one variable. These were typically figures that picked out a set of values along a single continuum, as in a time-line. Such figures were very rare ($< 1\%$) in journals and newspapers.

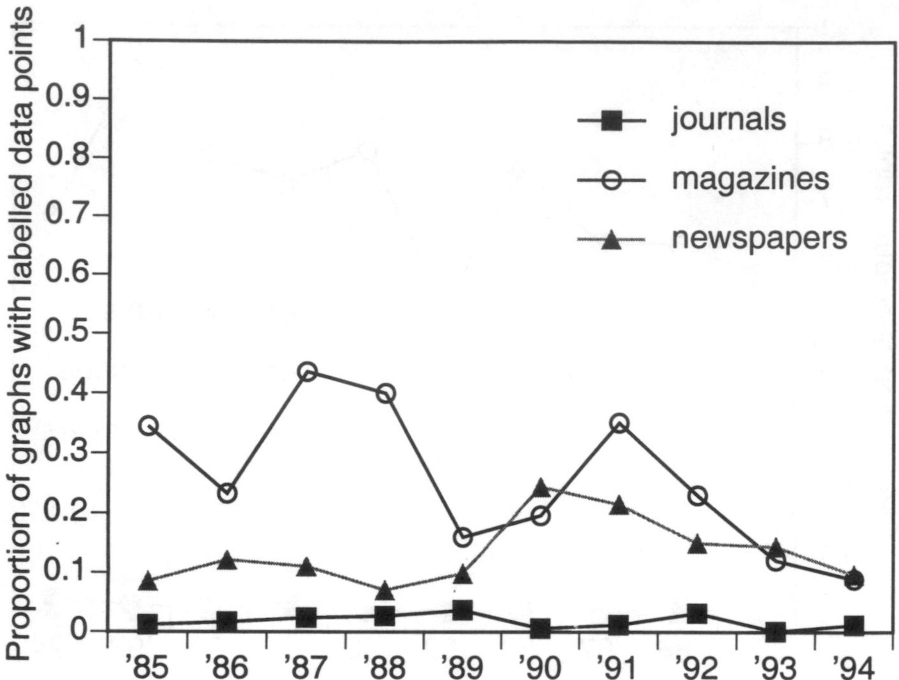

Figure 11.9. Data points with either X or Y values labelled were more prevalent in magazines and newspapers than academic journals, and seem to be on the decrease in magazines.

One basic function of a data graphic is the use of space to represent one or more quantitative variables [3]. It is therefore not surprising that most graphs in our sample (97.6%) included at least one quantitative variable. Quantitative variables can be plotted against other quantitative variables or against qualitative (categorical) variables. In our sample, figures in magazines were most likely to contain at least one qualitative variable in the X, Y or Z axes or in the legend (38.9%). Newspapers and journals also made some use of figures with qualitative variables (16.3% and 11.9%, respectively).

11.3.7 Repetition

Plotting several data sets on the same or similar sets of axes can be a powerful technique for making visual comparisons [9]. All the media studied made substantial use of this technique (35.9% of graphs in journals, 30.3% in magazines, and 46.6% in newspapers). During the period studied, the prevalence of repeated graphs in newspapers increased, and their prevalence in magazines decreased (see Fig. 11.10).

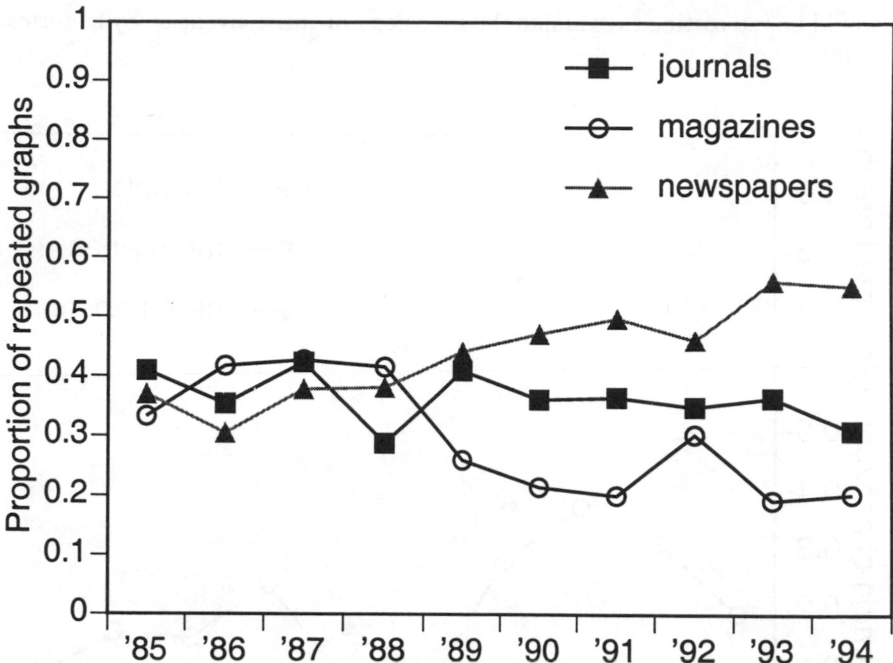

Figure 11.10. Repeated graphs became more prevalent in newspapers and less so in magazines during the time period sampled.

11.3.8 Orientation

Graphical elements vary in their orientation. As described in Section 11.2, a graphical element can depict a data value by its horizontal or vertical extent or location. Most of the figures in the sample (88.6%) used vertically oriented elements; a few (6.4%) contained horizontally oriented elements. As can be seen in Fig. 11.11, horizontally oriented elements appeared mostly in magazines, and seem to be on the decline. (A given figure could contain vertical elements, horizontal elements, neither, or both.)

What might go into the choice of vertical or horizontal orientation? Kosslyn [16, (p. 38)] suggests three principles. First, let reality decide between vertical and horizontal elements. Second, use horizontally oriented elements if the labels are too long to fit under a vertical display. Third, when in doubt use vertically oriented elements because that is the convention. This last principle may be related to the very general association of "up" with "more" across cognitive domains [7, 17, 29]. Given that some of the data being depicted surely have a natural horizontal orientation, and given that long labels indicate horizontal orientation, the dominance of vertical orientation suggests that designers' choice of orientation is being driven by social convention. It would be interesting to examine the genealogy of this convention by historical study.

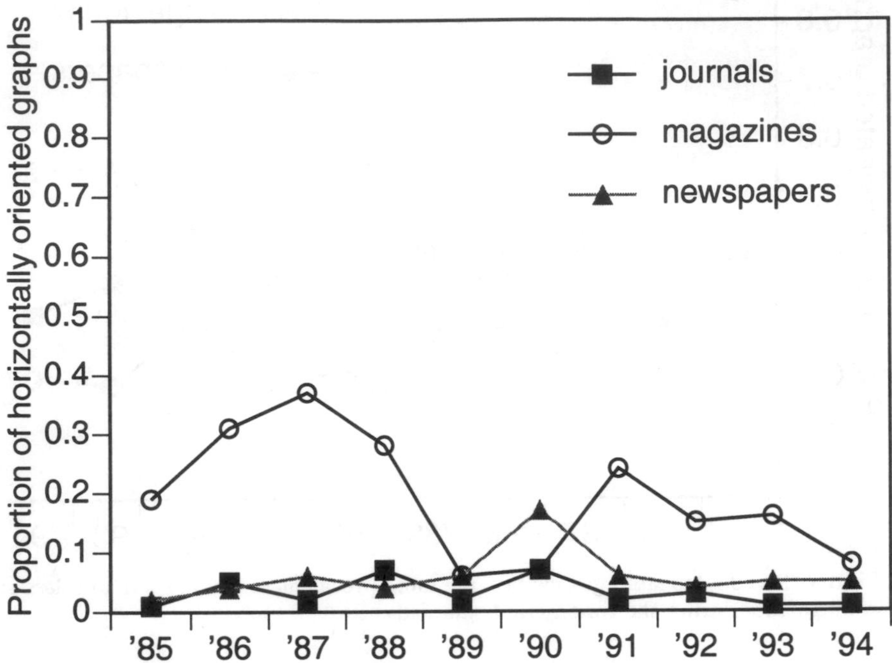

Figure 11.11. Graphs with horizontally oriented elements appeared mostly in magazines, and showed a decreasing trend.

11.3.9 Discussion

The most straightforward result of this study is its confirmation that data graphics are an important component of communication in print media. The average publication in our sample had more than 24 graphs per issue. Clearly, as a culture we rely heavily on data graphics to communicate quantitative information. Data graphics took on a wide variety of forms and styles, indicating a fertile ground for the study of their use.

Moreover, we stand at a particularly interesting time in the history of data graphics. Recent increases in the speed and power of personal computers have made powerful design tools available to the "manufacturers" of data, eroding the boundaries between author and designer. For example, in 2001 it takes about 10 minutes to download from the Internet and install a program that allows one to create 3-D rendered charts using texture-mapping, modelled objects and photorealistic background images [1]. Has this increase in the availability and sophistication of graphing technologies had profound effects on common usage?

One way graphing software may affect graphical communication is by simply making it easier for producers of print articles to generate data figures to accompany them. One would expect such effects to be most pronounced in scientific publications, in which the author(s) of an article also typically generate any figures that accompany it. During the period surveyed, the number of graphs per scientific journal issue nearly doubled, indicating that researchers are indeed taking advantage of the new tools.

But what kinds of graphs are being produced? In fact, the features identified with high-powered computing systems for data graphics seem to be the same ones that are abhorred by critics of graphical practice. Tufte's [26] well-known imperative to minimise the "data–ink ratio" provides a concise formulation of a minimalist ethic in design. Features such as added 3-D appearance, background pictures, and pictographs all violate this maxim. These are just the features one is likely to see on the packaging for an up-and-coming graphics software package.

Given that more powerful systems make it easier to produce graphs with precisely the features the experts decry, one might suspect (as we did) that graphs with these features are becoming more common. This intuition was *not* supported by these data. On the contrary, features such as 3-dimensional appearance and background pictures were rare and do not seem to be increasing in popularity. Pictographs were seldom used. The modal graph in our sample was a simple line graph. Perhaps the impression of increasing use of graphics that are more glitzy than communicative has come from a few blatant examples.

Just as graphics tended to be conservative in style, the data they represented tended to be plain in structure. Rather than depicting complex high-dimensional interactions, most figures in the sample showed relationships between two variables.

The choices by authors and designers as to how to render data are probably governed both by features in the data and by conventions particular to their community. Some design choices are constrained by the data to be depicted. Categorical variables are typically not used with point graphs, probably because interpreting them would require difficult visual discriminations. A large number of levels of a categorical variable makes drawing a pie graph unwieldy, and long category names encourage the use of horizontally oriented figures, to allow room for the labels. Other design choices may reflect the influence of one's cultural community or of the genre of communication. For example, in our sample scientific journals used almost exclusively simple rendering styles, while magazines and newspapers tended to use a mix of simple and area rendering. (Such effects may also reflect differences between the media in the kinds of data being graphed.) For a discussion of how conventions can emerge in graphic communication see [33].

In sum, worries about negative recent developments in the design of data graphics were not supported by this survey. The graphics coded were by and large simple and elegant in exactly the ways advocated by designers [26,27] and psychologists [16]. To the extent that developments in computing hardware and software have influenced data graphics, these influences don't seem to be obviously negative.

The placement of sophisticated tools in the hands of producers of graphs offers a unique opportunity for experts in graphical design and perceptual psychology to influence the production of data graphics. Systems for data graphics differ in the features they afford. By embedding principles of graphical perception in software packages, one might hope to guide authors' and designers' choices in felicitous directions. One example of such a tight integration between graphical perception and software design is the Trellis graphics library for S-Plus [21]. Designed for exploratory data analysis, this system reflects the influence of a large body of research on the part of its designers [9]. The application of a similar approach to packages for publication graphics for scientific authors and media designers would no doubt be of great value.

In advocating that designers and perceptual psychologists build their understanding of data graphics into computer systems, one major caveat is in order. Our experience has been that most variables of interest in graphical design interact with several other variables. This means that simply designing a system with a good default value for each feature is likely to be insufficient. Rather, good solutions are likely to resemble case-based reasoning systems [15] (see also [11,14]), in which exemplars of situations similar to the one at hand are identified and elegant solutions to those examplar situations (as identified by research or design) are recommended.

Acknowledgements

Thanks to Dianne Beck for assistance in data analysis. This research was partially supported by the Office of Naval Research, Grant no. NOOO14-PP-1-O649 to Stanford University and by Interval Research Corporation.

References

1. Adrenaline Software, Inc. (1993). [Computer software]. Charts Pro 1.04. Québec, Canada.
2. Bertin, J. (1967). Sémiologie graphique: Les diagrammes, les réseaux, les cartes. Paris: La Haye Mouton.
3. Bertin, J. (1980). The basic test of the graph: A matrix theory of graph construction and cartography. In P.A. Kolers, M.E. Wrolstad and H. Bouma (Eds), Processing of visible language 2. New York:Plenum Press, pp. 585–604.
4. Bertin, J. (1983). Semiology of graphics (W.J. Berg, trans.). Madison, WI: University of Wisconsin Press.

5. Carswell, C.M. (1992). Choosing specifiers: An evaluation of the basic tasks model of graphical perception. Special issue: Visual displays. Human Factors 34:535–554.
6. Casali, J.G. and Gaylin, K.B. (1988). Selected graph design variables in four interpretation tasks: A microcomputer-based pilot study. Behaviour & Information Technology 7:31–49.
7. Clark, H.H. and Clark, E.V. (1977). Psychology and language: An introduction to psycholinguistics. New York: Harcourt Brace Jovanovich.
8. Cleveland, W.S. (1984). Graphs in scientific publications. American Statistician 38(4):261–269
9. Cleveland, W.S. (1985). The elements of graphing data. Monterey, CA: Wadsworth Advanced Books and Software.
10. Cleveland, W.S., Harris, C.S. and McGill, R. (1983). Experiments on quantitative judgments of graphs and maps. Bell System Technical Journal 62:1659–1674.
11. Fish, D. and McCartney, R., this volume.
12. Gattis, M. and Holyoak, K.J. (1996). Mapping conceptual to spatial relations in visual reasoning. Journal of Experimental Psychology: Learning, Memory, and Cognition 22:231–239.
13. Howard, I.P. (1982). Human visual orientation. New York:Wiley.
14. Kerpedjiev, S., this volume.
15. Kerpedjiev, S., Carenini, G., Green, N., Moore, J. and Roth, S. (1998, October). Saying it in graphics: From intentions to visualisations. Paper presented at the EEE symposium on information visualisation (InfoVis '98), Research Triangle Park, NC.
16. Kosslyn, S. (1993). Elements of graph design. New York:Freeman.
17. Lakoff, G. and Johnson, M. (1980). Metaphors we live by. Chicago: University of Chicago Press.
18. Larkin, J.H. and Simon, H.A. (1987). Why a diagram is (sometimes) worth ten thousand words. Cognitive Science 11:65–100.
19. Lohse, G.L., Walker, N. and Rueler, H.H. (1994). A classification of visual representations. Communications of the ACM 37:36–49.
20. Macdonald-Ross, M. (1977). How numbers are shown. AV Communications Review 25:359–409.
21. MathSoft (1995). [Computer software]. S-Plus 3.3. Cambridge, MA.
22. Pinker, S. (1990). A Theory of graph comprehension. In R. Freedle (Ed.), Artificial intelligence and the future of testing. Hillsdale, NJ: Lawrence Erlbaum Associates, pp. 73–126.
23. Shah, P. and Carpenter, P.A. (1995). Conceptual limitations in comprehending line graphs. Journal of Experimental Psychology: General 124:43–61.
24. Simkin, D. and Hastie, R. (1987). An information-processing analysis of graph perception. Journal of the American Statistical Association 82:454–465.
25. Spence, I. (1990). Visual psychophysics of simple graphical elements. Journal of Experimental Psychology: Human Perception and Performance 16:683–692.
26. Tufte, E.R. (1983). The visual display of quantitative information. Cheshire, CT: Graphics Press.
27. Tufte, E.R. (1990). Envisioning information. Cheshire, CT: Graphics Press.
28. Tukey, J.W. (1977). Exploratory data analysis. Reading, MA:Addison-Wesley.
29. Tversky, B. (1995). Cognitive origins of graphic conventions. In F.T. Marchese (Ed.), Understanding images. New York: Springer, pp. 29–53.
30. Tversky, B. and Schiano, D.J. (1989). Perceptual and conceptual factors in distortions in memory for graphs and maps. Journal of Experimental Psychology: General 118:387–398.

31. Wainer, H. (1984). How to display data badly. American Statistician 38:137–147.
32. Zacks, J., Levy, E., Tversky, B. and Schiano, D.J. (1998). Reading bar graphs: Effects of extraneous depth cues and graphical context. Journal of Experimental Psychology: Applied 4:119–138.
33. Zacks, J. and Tversky, B. (1999). Bars and lines: A study of graphic communication. Memory and Cognition 27(6):1073-1079.

12. The Role of Representation and Working Memory in Diagrammatic Reasoning and Decision Making

Jozsef A. Toth

C. Michael Lewis

In this chapter we present research into the role of visual and verbal working memory in visual reasoning. Eighty subjects participated in an experiment where 34 different gain–loss problems were represented in either text or graphical form. In order to test the role of each component of Baddeley's model of working memory, subjects performed secondary verbal, visual, and mental suppression tasks while reasoning about the problems represented in text or graphics, yielding six different conditions. In two control conditions, no suppression tasks were performed. Interference and preference reversals occurred in all six conditions involving suppression tasks, even though no interference was predicted in conditions involving the graphical representation coupled with the verbal suppression task and the text representation coupled with the visual suppression task. In the control conditions subjects made responses consistent with prospect theory with little interference. Response times were consistently slower in the four text conditions compared to the four graphics conditions. The visual and mental secondary tasks resulted in increased response times, respectively, but the verbal secondary task did not. The data suggests that certain graphical and text representations require both visual and verbal resources and the taxing of these resources results in perceptual–cognitive biases that sometimes favour a minimal inference strategy.

12.1 Introduction

Much has been written about working memory in cognitive science. In the realm of visual and diagrammatic reasoning, however, very little empirical evidence exists describing the role of this aspect of human information processing. The goal of this work is to understand the relationship between verbal and visual isomorphic representations and the different components of working memory that may manage relevant information as a decision maker reasons about a two-choice decision-making problem.

Baddeley's influential model of working memory consists of three components: (1) the articulatory rehearsal, or phonological loop, (2) the visuospatial sketchpad, and (3) the executive control system [1, 2]. Rather than serving as an intermediate repository for encoded data, working memory is part of a closed system and is instead governed by the central executive. By performing dual-task experiments whereby a load is placed on one of the three components by a secondary task, it has been demonstrated that, at some level of abstraction, these visual (sketchpad), verbal (articulatory loop), and controller-like (central executive) components indeed exist.

The work described herein will test this model of memory against behavioural data from a well-known theory in the decision-making literature known as prospect theory. According to the basic tenets of this theory, decision-making individuals tend to be risk-averse in situations involving gain and risk-seeking in situations involving loss, when deciding on alternatives such as the following:

- *Gain problem*: A: 4000 with probability 0.8; B: 3000 with probability 1.0
- *Loss problem*: C: −3000 with probability 0.9; D: −7000 with probability 0.45

Thus, in a situation involving gain, choices A and B above, the decision maker will usually prefer the more certain outcome (choice B) over the riskier or less certain outcome (choice A). Likewise, in a situation involving loss, choices C and D above, the decision-maker will typically choose the riskier loss (choice D) over the more certain loss (choice C). In the second pair above, although the certain loss of −3000 in choice C has a higher expected utility (−2700), decision makers typically go for the less certain option, choice D, with the lower expected utility (−3150). The prevailing theory suggests that the decision maker's perception of the problem alternatives – rather than adherence to normative principles as in Expected Utility Theory – contributes to these counter-intuitive results. Decision makers are thought to view the problems in terms of gains and losses. These perceptions are framed according to a neutral reference point, corresponding to the current asset position, which is assumed to be a null wealth (i.e., $0). One principle that appears to contribute to this phenomenon is the overweighting of certainty. In their words, "it appears that certainty increases the aversiveness of losses [by eschewing the certain loss and favouring the less probable loss] as well as the desirability of gains [by favouring the certain gain and eschewing the less probable gain]" [4, (p. 169)]. More recent experiments by Tversky and Kahneman [11] have revealed that the effect is reversed in cases involving low probabilities; risk-seeking for gains, and risk-averse for losses.

Three aspects of this problem that have received little or no attention, however, pertain to (a) the external representation of the gain–loss problem, (b) the relationship between the external representation and various cognitive and perceptual resources, including working memory, and (c) how the external representation bears on the corresponding internal representation, whether it be visual or verbal. To date, gain–loss problems have been represented in a quasi-tabular text form as seen in the two gain–loss examples above.

Regarding prospect theory and the gain–loss problem, at issue is which components of working memory, visual, verbal, or executive, are involved in the process of (1) viewing the external representation of the problem, (2) internal activation of the external problem constituents in either visual memory (sketchpad), verbal memory (articulatory loop), or both, (3) kinds of inferences drawn based on the information present in visual or verbal working memory, then making a decision (central executive), and (4) providing a response. These four stages of processing closely follow the experimental paradigm of Sternberg [10]. In the case of working memory, however, Baddeley assumes that a systemic interaction exists among these four components, whereas in Sternberg's model processing has been assumed to transpire serially, from (1) to (4). It has been determined in recent work in syllogistic reasoning that taxing the central executive component of working memory has effects on the decision-making process but taxing the sketchpad and articulatory loop does not [3].

In the case of the gain–loss problem it is expected that noticeable changes in decision-making behaviour will result if any of the three memory components are similarly taxed. Since the standard representation of the problem is text-based, it is assumed that verbal working memory will be most utilised during the solving of the problem. It can, however, also be assumed that certain individuals might consider such problems in a visual sense as well. Since magnitudes of wealth are considered, a decision maker may also associate a gain or a loss with a personalised mental depiction of the problem constituents (see also [5, 7]), but we expect the most salient effects of the external representation to predominate.

Also to be considered is the external representation of the gain–loss problem. The original representation, shown above, is in quasi-tabular text-based form. Re-representing the problem in a graphical form should have certain effects on the activation of visual versus verbal working memory. A graphical representation of the same problem should result in a higher activation of visual memory (sketchpad), but again, aspects of the problem may also activate verbal memory (articulatory loop) to a lesser extent, depending on the individual's predisposition to considering visual elements in a verbal fashion.

12.2 Method

12.2.1 Hypotheses

Text-Based External Representation. A problem externally represented in text should result in a primary activation of verbal working memory. Since the quantities ordinarily reasoned about, probability and pay-off, are postulated to be active in verbal working memory, and to a much lesser extent in visual working memory, a taxing of verbal resources via an articulatory suppression task should inhibit the reasoning process. This should result in interference and decisions that are inconsistent with prospect theory. The same should hold true for the loading of the central executive with a secondary task since this phase of processing (i.e., Sternberg's phase (3)) is presumed to be involved. If an attempt is made to load visual working memory with an articulatory suppression task, however, decisions should remain more consistent with prospect theory since it is presumed that with a text-based external representation the main interaction lies between the verbal and central executive components of the working memory system.

Diagrammatic External Representation. A problem externally represented in graphical form should result in a primary activation of visuospatial working memory. Since the quantities reasoned about, probability and pay-off, are postulated to be active in visual working memory, and to a lesser extent in verbal working memory, a taxing of visuospatial resources should inhibit the reasoning process and result in decisions that are inconsistent with prospect theory. The same should hold true for the loading of the central executive with a secondary task. If an attempt is made to load verbal working memory with a visuospatial suppression task, decisions should remain consistent with prospect theory since it is presumed that with a visual external representation the main interaction lies between the visual and central executive components of the working memory system.

12.2.2 Design

There were two independent variables in the 2 × 4 factorial design: (A) external representation of the gain–loss problem, and (B) working memory load. External representation of the gain–loss problem comprised two levels: (1) the default text representation as illustrated in Fig. 12.1, and (2) a diagrammatic representation in which probability and pay-off were presented in a graphical form (Fig. 12.2). Working memory load comprised four levels: (1) control, no load on the working memory system, (2) load on the articulatory loop with a secondary verbal suppression task, (3) load on the visuospatial sketchpad with a spatial suppression task, and (4) load on the central executive with a secondary mental suppression task. Thus, there were eight total experimental conditions. For the purposes of clarity, the eight conditions will be referenced as follows:

- *t*: Subject views a text-based representation of the gain–loss pair and chooses A or B.
- *t-verbal*: Same as *t* but subject also recites secondary verbal suppression task repeating the sequence "1, 2, 3, 4, 5" out loud while reasoning about the problem and making a choice.
- *t-visual*: Same as *t* but subject also performs secondary visual suppression task typing the sequence "4, 8, 6, 2" on the keypad in a clockwise fashion.

- *t-mental*: Same as *t* but subject utters a random sequence of digits from the set "1, 2, 3, 4, 5".
- *g*: Subject views a graphic representation of the gain–loss pair and chooses A or B.
- *g-verbal*: Same as *g* but subject also recites verbal suppression task.
- *g-visual*: Same as *g* but subject also performs visual suppression task.
- *g-mental*: Same as *g* but subject also performs mental suppression task.

12.2.3 Materials

Thirty-four two-choice gain–loss problems were randomly selected from a pool of 70 from the literature [4, 11]. Three- and four-choice problems also exist, but were excluded from the pool because of the exploratory nature of the work. Each of the 34 problems was represented in a text-based condition and in a graphics-based condition. The subject pool was divided into two groups. The first group performed conditions *t*, *g-verbal*, *t-visual*, and *g-mental*. The second group performed conditions *g*, *t-verbal*, *g-visual*, and *t-mental*. The four conditions in the experiment were presented as randomly ordered blocks and the 17 trials within each block were also randomised.

12.2.4 Subjects

Eighty subjects participated in the experiment, comprising graduate students, undergraduate students, and staff at the University of Pittsburgh. The subjects were assigned to each experimental session based solely on the criteria of availability. Subjects were paid five dollars or received extra credit for a course, which had been prearranged with the course instructor. In some cases, the subjects volunteered and did not receive any compensation.

12.2.5 Materials

The subjects were seated at a DEC 5000 Workstation with a 13-inch colour monitor. The software for the experiment was implemented in Harlequin Common Lisp using the Harlequin Common Application Programming Interface in the X Windows display environment. In both text (Fig. 12.1) and graphical (Fig. 12.2) conditions, the gain–loss constituents were depicted in black, alphanumeric characters and displayed in 10, 12 and 14-point courier roman bold font with a white background. In the graphical condition, probability was represented on the x axis and pay-off was represented on the y axis (Fig. 12.2). The two choices A and B along with the corresponding x, y dot were displayed in red and the rest of the display constituents (axes, tick marks, etc.) were in black on a white background. The design was intended to maintain as many isomorphisms between the text and graphic representations as possible [6,8]. Subdivisions of pay-off and probability were abstracted to minor hash marks rather than an explicit value (e.g., see Fig. 12.2). However, in the case of the probabilities 0.01 and 0.99, they were made explicit as a label in the 10-point font below the minor hash mark on the probability scale.

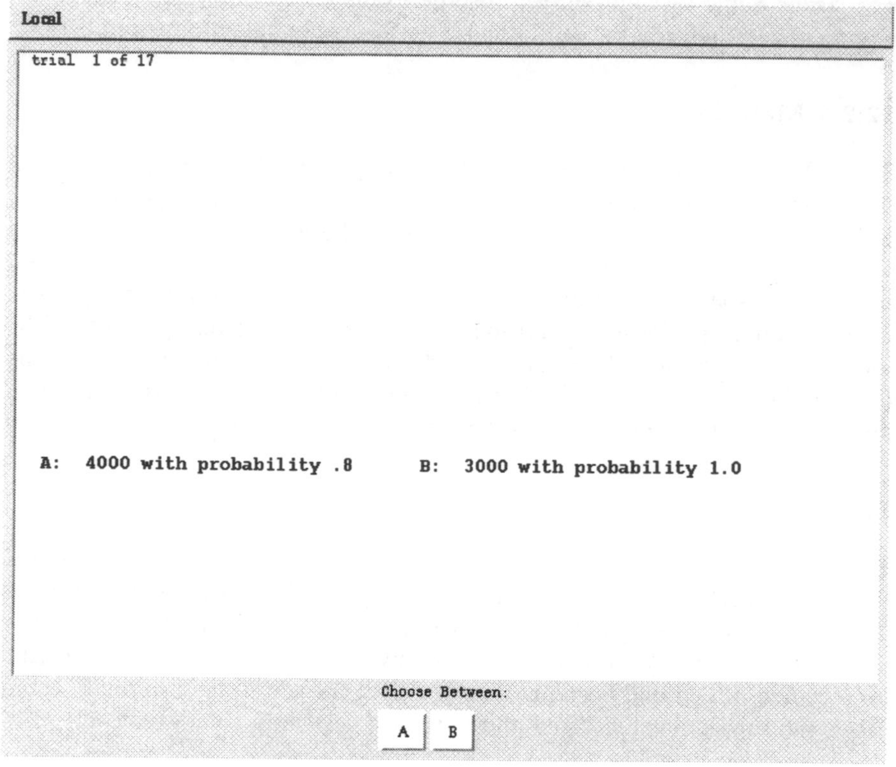

Figure 12.1. Layout of workstation screen for the experiment illustrating text-based representation of gain–loss problem.

12.2.6 Procedure

In the verbal suppression task, subjects uttered out loud the same sequence of digits (" 1, 2, 3, 4, 5") while presented with a text (i.e., *t-verbal*) or graphical (i.e., *g-verbal*) gain–loss problem on the computer display. The subject was instructed to keep pace with an activated metronome at the rate of one utterance per second. When the subjects were ready to choose one of the alternatives, they clicked either the "A" or "B" button in the display using the mouse. The problem disappeared from the display, and a new problem was presented. For the visual suppression task, subjects typed the sequence of four characters on the computer keypad portion of a keyboard in a clockwise circular fashion ("4, 8, 6, 2") with their non-dominant hand at the rate of one per second; again to the beat of the metronome. While still typing the sequence, the subject was presented with a gain–loss problem either in text (*t-visual*) or graphical form (*g-visual*). The subject then chose either "A" or "B" , the problem disappeared from the display, and a new problem was presented. For the executive controller suppression task, subjects were instructed to utter a random sequences of digits from the set of numbers 1 through 5 (e.g., "2, 1, 4, 5, 3 . . . ") in which each new sequence was unique, while presented with a gain–loss problem either in text (*t-mental*) or graphical (*g-mental*) form. Selection and problem display proceeded in the same manner as the other two tasks just described. In the two control conditions, subjects simply selected "A" or "B" from the text (*t*) or graphical (*g*) representation without performing a secondary task. The metronome remained active through the entire experiment as a task invariant. A block of practice trials was presented before the experiment began, which allowed the subject to learn each suppression task. All responses were written out to disk for subsequent analysis.

12.3 Results

Tables 12.1, 12.2, and 12.3 and Fig. 12.3 summarise the data from the experimental trials. $N = 20$ samples were recorded from the 80 subjects for each of the eight conditions. Table 12.1 displays the results of the likelihood ratio chi-square analysis using the SAS FREQ procedure. Each cell reflects the proportion of responses for the expected value to the left and right of the colon ":", and the p value below the pair, which is the result of the pairwise comparison of two conditions. Ten pairs of conditions were analysed. Only those gain–loss problems in which the observed responses varied significantly from the expected responses appear in the table along with the ID number from the original pool of 70. The expected responses were derived from responses to the control condition *t*. The observed responses were organised into two different categories: interference and reversal.

Interference responses are shown in Table 12.1 are in plain typeface. They are expressed as the ratio of observed responses to total responses (20). These responses varied significantly from the expected responses, but more than 50% of the 20 possible observed responses were still congruent with the expected response. These kinds of responses suggest error in which subjects were not able to provide the expected response owing to interference from the secondary task. The second kind of response, called a *reversal*, is indicated in boldface type. For the *reversals*, more than 50% of the observed responses were incongruent with the expected response.

Table 12.1 shows significant results from the likelihood ratio chi-square analysis using the SAS FREQ procedure($N = 20$, $d.f. = 1$). The first column

Table 12.1. Results of the likelihood ratio chi-square analysis using the SAS FREQ procedure

Pair ID:A; B	E	Expected choice response frequencies						
		t: g	*t: t-verbal*	*t: t-visual*	*t: t-mental*	*g: g-verbal*	*g: g-visual*	*g: g-mental*
0:.1,0; .9,50	B				.95:.75 .066	.9:1.0 .09		
3:0.5,0; .5,-50	A		1.0:.9 .09					
4:0.9,0; .1,50	B						.85:.6 .073	.85:.35 .001
11:0.5,0; .5,-100	A			1.0:.9 .09				
13:.75,0; .25,-100	A	.85:1.0 .036				1.0:.7 .002	1.0:.8 .014	
15:.95,0; .05,-100	A					.9:1.0 .09		
16:.01,0; .99,200	B			1.0:.9 .09	1.0:.9 .09			
19:.1,0; .9,-200	A			1.0:.85 .036	1.0:.9 .09		1.0:.9 .09	1.0:.85 .036
21:.5,0; .5,-200	A				1.0:.9 .09	1.0:.8 .014	1.0:.9 .09	1.0:.85 .036
23:.9,0; .1,-200	A	1.0:.85 .036	1.0:.85 .036	1.0:.9 .09	1.0:.9 .09			
24:.99,0; .01,200	B	.7:**.45** .108				**.45**:.75 .051	**.45**:.75 .051	
25:.99,0; .01,-200	A	1.0:.8 .014	1.0:.9 .09	1.0:.85 .036	1.0:.9 .09			
26:.01,0; .99,400	B		1.0:.9 .09		1.0:.75 .006			.95:.75 .066
27:.01,0; .99,-400	A					.9:1.0 .09		
28:.99,0; .01,400	B					.7:**.45** .108	.7:**.4** .055	.7:**.45** .108
31:.1,-50; .9,-100	A						.85:.6 .073	
35:.9,-50; .1,-100	A					.85:.5 .016		
41:.5,-50; .5,-150	A						.9:1.0 .090	
43:.75,-50; .25,-150	A		.8:.55 .088	.8:.55 .088				.85:.6 .073
47:.05,-100; .95, -200	A						.95:.7 .030	
50:.5,100; .5,200	B				.95:.7 .03			
56:.8,4000; 1.0,3000	B			.7:**.4** .055				
66:.5,1000; 1.0,500	B			.7:**.4** .055	.7:**.45** .108			.65:**.35** .056

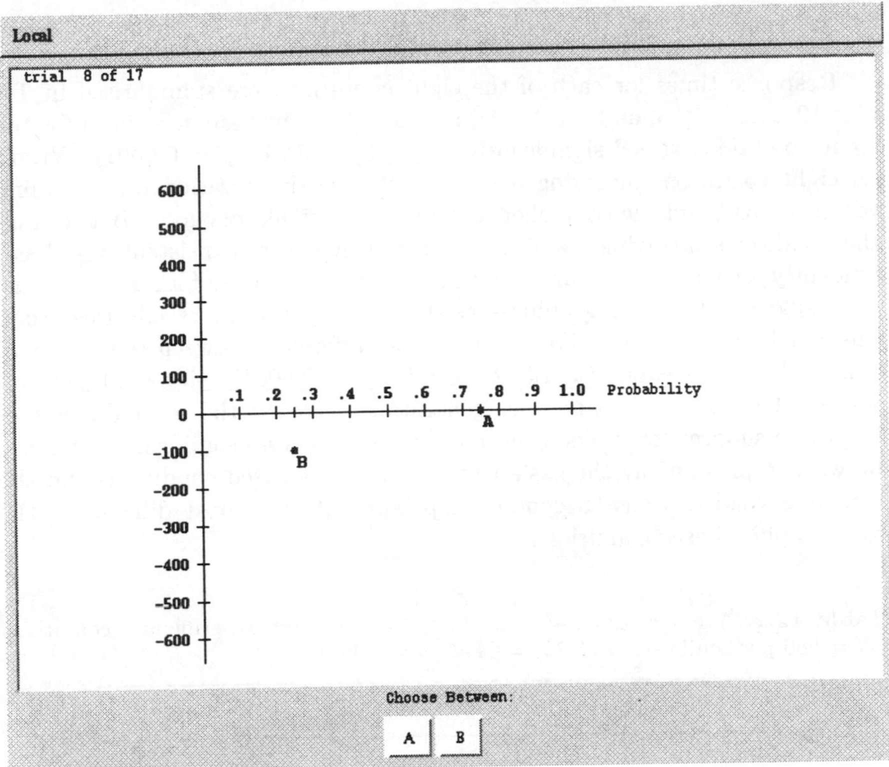

Figure 12.2. Layout of workstation screen for the experiment illustrating graphical representation of gain–loss problem. Pay-off and probability of choices A and B are represented spatially in a two-dimensional coordinate system. Choice A is 0 with probability 0.75 and choice B is -100 with probability 0.25.

indicates the gain–loss pair. The second column **E** shows the expected choice, A or B. The cells in the remaining seven columns show the results of the pairwise comparison expressed as a percentage of responses to the expected value (in column **E**) regarding the first condition to the left of the colon ":", the other condition to the right of the colon, and the *p* value directly below. The percentages are the ratio of observed responses to total responses. Normal typeface indicates interference, and boldface shows a preference reversal, e.g., for gain–loss pair 24 the expected response is B. In the *t:g* column the control condition *t* is compared to the control condition *g*. Fourteen of 20 subjects, or 0.7, chose B in *t* (the other 6 chose A), and 9 of 20 subjects, or 0.45, chose B in *g* (the other 11 chose A). The **0.45** is in boldface since this *g* appears to have caused a preference reversal.

For example, in the *g* v. *g-mental* pairwise comparison of gain problem 4 – where the expected response was B – in *g* 3 subjects chose A (0.9,0), and 17 subjects chose B (0.1,50) shown as 0.85 in the cell, but in *g-mental* 13 subjects chose A, and 7, or **0.35**, chose B. This data is highly suggestive of a reversal owing to the controller suppression task ($p = 0.001$).

Response times for each of the eight conditions are summarised in Tables 12.2 and 12.3, and illustrated in Fig. 12.3. The response times for the eight conditions varied significantly ($F(7, 79) = 64.46$, $p = 0.0001$). Within all eight conditions involving t and g, response times varied when loading the sketchpad and the controller, but not the articulatory loop. Between all the conditions involving t and g, response times were consistently and significantly longer for the four t conditions compared to the four g conditions. This suggests that the graphics-based reasoning was more effortless than the text-based reasoning. For instance, the difference between the means of *t-mental* and *g-mental* was 1.569 seconds ($p = 0.0001$). Many subjects reported that the random number generation task was the most difficult of the three suppression tasks, even though there was no significant difference between t, presumably the easiest of the four text-based conditions and the controller-loaded *g-mental* condition, presumably the most difficult of the four graphics-based conditions.

Table 12.2. Average response times (sec) for the eight experimental conditions ($N = 680$ per condition, $F(7, 79) = 64.46$, $p = 0.0001$)

		Secondary task			
		None	*Verbal*	*Visual*	*Mental*
Representation	*Text*	4.79	4.67	5.25	6.08
	Graphic	3.64	3.85	4.29	4.51

12.4 Discussion

The data described in the last section clearly indicates that the tasks designed for this experiment caused interference where none was expected and lacked interference in cases where it was predicted. For example, surprisingly little interference occurred in the *t-verbal* condition, but even proportionately more interference occurred in *g-verbal*, where hardly none was predicted. Likewise, the visuospatial suppression task caused interference in *t-visual* where little was thought to occur but interference did occur in *g-visual*. As predicted, interference occurred in *t-mental* and *g-mental*.

Regarding the articulatory suppression tasks, two explanations readily come to mind. In *t-verbal*, it may simply be the case that the repetition of the number sequence was not sufficient to cause the expected interference. In earlier experiments by Baddeley, interference increased only as the verbal task became more difficult. Regarding *g-verbal*, there are components of the graphical representation that are clearly verbal in nature, such as labels on

Table 12.3. Pairwise significance tests of mean difference absolute values using REGWQ method ($N = 680$ observations per pair, $d.f. = 5353$)

	t-verbal	t-visual	t-mental	g-none	g-verbal	g-visual	g-mental
t-none	.12 *n.s.*	.45 *.005*	1.28 *.0001*	1.16 *.0001*	.95 *.0001*	.50 *.04*	.29 *n.s.*
t-verbal		.57 *.005*	1.40 *.0001*	1.04 *.0001*	.83 *.0001*	.38 *.0001*	.17 *n.s.*
t-visual			.83 *.0001*	1.61 *.0001*	1.40 *.0001*	.96 *.0001*	.74 *n.s.*
t-mental				.24 *.0001*	2.23 *.0001*	1.79 *.0001*	1.57 *.0001*
g-none					.21 *n.s.*	.65 *.0001*	.87 *.0001*
g-verbal						.45 *.005*	.66 *.0001*
g-visual							.22 *n.s.*

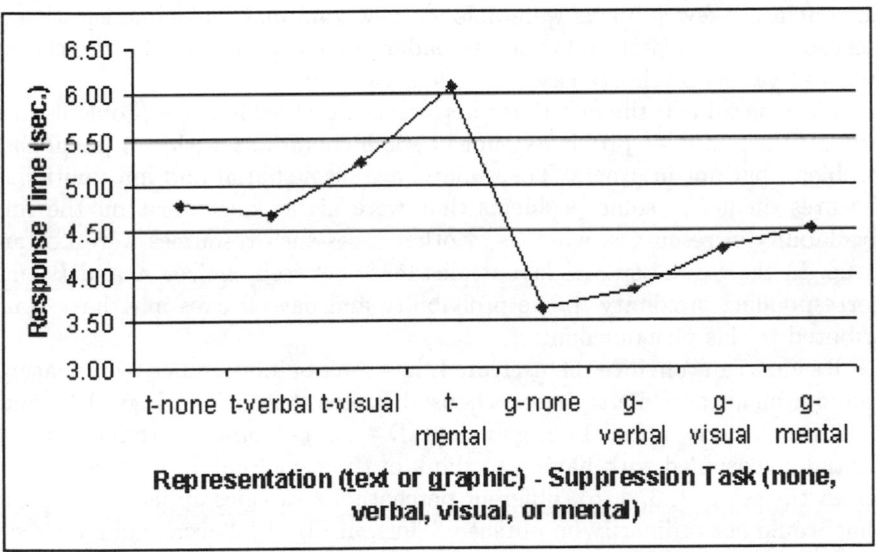

Figure 12.3. Mean response times for the eight experimental conditions. The secondary tasks, verbal, visual, and mental, respectively, resulted in longer response times for each modality.

the axes, and not as automatic as the verbal information presented in *t-verbal*. Thus, it might be the case that in the course of extracting available information from the graph, probability and pay-off, this extra processing is not as automatic as it is in the *t-verbal* case. It can be easily assumed that non-automatic information becomes ensnared in verbal working memory whereas automatic information does not [9]. More importantly, however, is the idea that *both* visual and verbal working memory appears to be active during the visual reasoning process.

Puzzling, however, is the interference, even reversals, generated in the *t-visual* condition (problems 56 and 66, $p = 0.055$). There appears to be a spatial component to the reasoning about the problem when presented in its text-based form; perhaps more than previously expected. Given the template:

A:w with probability x; B:y with probability z

the data suggests that the decision makers utilised the spatial relationships among the choice constituents to arrive at a decision. The secondary typing task and the random number generation task are activities that a typical subject performs on a less regular basis than talking and acting at the same time. In fact, many subjects exhibited great difficulties with these two tasks as opposed to the latter task. Thus, the interference appears to be genuine. In contrast, a few subjects were able to chew gum *and* shake one leg with a nervous tick while performing the secondary task and deciding on the choice alternatives, with relative ease – i.e., *four* tasks at once.

More puzzling is the fact that interference occurred in some problems but not in other, similar problems, and in some conditions within a particular problem, but not in others. There may have been spatial and informational features unique to some problems that were highly dependent on the full availability of resources, whereas in other cases such resources were not an issue. In the case of spatial layout, distances between choices A and B and corresponding proximity to the probability and pay-off axes may have contributed to this phenomenon.

Readers can convince themselves of the power of interference by repeating random numbers while trying to choose between A and B in Figs. 12.1 and 12.2. Particularly in Fig. 12.2 (gain–loss ID 4), in *g-mental* where the mind is partially consumed with having to think of the next random number to say (from the set 1, 2, 3, 4, 5), different perceptual strategies appear to pop up that would not ordinarily be pursued. Thus, in Fig. 12.2 even though choice B makes the most irrefutable sense (choice A lacks any pay-off), the heuristic that seems to arise is the selection of the choice that is farther to the right, a pay-off that is not associated with a minor hash mark (as it is in B, which requires additional visual inference), and that is anchored on the x axis. This is substantiated by the reversal in *g-mental* ($p = 0.001$) and the interference in *g-visual* ($p = 0.073$). It is presumed that such unique strategies account for much of the interference and reversals present in the data and is referred to as the "minimal inference" strategy.

Subjects reported developing other visual strategies for some of the problems. For instance, a pay-off further to the right and to the top of the display was better than any other pay-off. Indeed, interference with such heuristics does not appear in the data. Space does not permit an individual analysis of each gain–loss problem, but it is hoped that some of the thoughts provided in this section convince the reader that general theories about verbal and visual representations and their memory requirements are not yet within our reach. Only further research and the tuning of task and task constituents will further reveal the relationships among the entities described in this report. For instance, the addition of a third probability/pay-off choice, C, would further tax working memory.

12.5 Conclusions

In summary, it appears that in this study the visual, verbal, and reasoning components of memory were active when reasoning about text-based and diagrammatic representations. In tasks that are more visual in nature, such as chess, the components of working memory are more clearly defined such that no interference occurs while performing an articulatory suppression task, but prohibitive interference occurs when performing a visual suppression task [1]. In the domain described in this work, however, it appears that the gain–loss problem, whether represented in text, or graphically, utilises both visual and verbal components in working memory. More research in this domain will help to further elucidate many of the matters brought to light in this work.

Acknowledgements

Many thanks to Walter Schneider, Charles Perfetti, Hermina Tabachneck, Herbert Simon, Alan Lesgold, Terri Lenox, Dan Suthers, and Jeff Shrager. Special thanks to Michael Anderson for his guidance and patience during the preparation of this manuscript. This work was conducted while funded by the Defense Advanced Research Projects Agency Computer Aided Education and Training Initiative, under the title "Collaboration, Apprenticeship, and Critical Discussion: Groupware for Learning", Contract N66001-95-C-8621.

References

1. Baddeley, A. (1992). Is working memory working? Quarterly Journal of Experimental Psychology: Human Experimental Psychology 44A(1):1–31.

2. Baddeley, A. (1986). Working memory. New York: Oxford University Press.
3. Gilhooly, K.J., Logie, R.H. and Wynn, V. (1993). Working memory and strategies in syllogistic reasoning tasks. Memory and Cognition 21(1):115–124.
4. Kahneman, D. and Tversky, A. (1979). Prospect theory: An analysis of decisions under risk. Econometrica 47:262–271.
5. Koedinger, K.R. and Anderson, J.R. (1990). Abstract planning and perceptual chunks: Elements of expertise in geometry. Cognitive Science 14:511–550.
6. Kotovsky, K., Hayes, J.R. and Simon, H.A. 1985. Why are some problems hard? Evidence from tower of hanoi. Cognitive Science 7:248–94.
7. Larkin, J. and Simon, H.A. (1987). Why a diagram is (sometimes) worth ten thousand words. Cognitive Science 11:65–99.
8. Lewis, C.M. and Toth, J.A. (1992). Situated cognition in diagrammatic reasoning: Reasoning with diagrammatic representations. In Papers from the 1992 AAAI spring symposium. Technical report SS-92-02 47-52. Menlo Park, CA:AAAI Press.
9. Shiffrin, R. and Schneider, W.A. (1977). Controlled and automatic human information processing: II. Perceptual learning, automatic attending, and a general theory. Psychological Review 84(2):127–190.
10. Sternberg, S. (1969). The discovery of processing stages: Extensions of Donder's method. Attention and human performance II. Acta Psychologica 30:276–315.
11. Tversky, A. and Kahneman, D. (1992). Advances in prospect theory: Cumulative representation of uncertainty. Journal of Risk and Uncertainty 5:297–323.

Appendix: Gain–Loss Problems Utilised in Experiment

	Choice A		Choice B	
ID	Prob.	Payoff	Prob.	Payoff
0:	0.10	0	0.90	50
3:	0.50	0	0.50	-50
4:	0.90	0	0.10	50
8:	0.25	0	0.75	100
9:	0.25	0	0.75	-100
11:	0.50	0	0.50	-100
13:	0.75	0	0.25	-100
15:	0.95	0	0.05	-100
16:	0.01	0	0.99	200
17:	0.01	0	0.99	-200
19:	0.10	0	0.90	-200
21:	0.50	0	0.50	-200
23:	0.90	0	0.10	-200
24:	0.99	0	0.01	200
25:	0.99	0	0.01	-200
26:	0.01	0	0.99	400
27:	0.01	0	0.99	-400
28:	0.99	0	0.01	400
31:	0.10	-50	0.90	-100
32:	0.50	50	0.50	100
35:	0.90	-50	0.10	-100
37:	0.05	-50	0.95	-150
38:	0.25	50	0.75	150
39:	0.25	-50	0.75	-150
41:	0.50	-50	0.50	-150
43:	0.75	-50	0.25	-150
46:	0.05	100	0.95	200
47:	0.05	-100	0.95	-200
50:	0.50	100	0.50	200
53:	0.75	-100	0.25	-200
54:	0.95	100	0.05	200
56:	0.80	4000	1.00	3000
61:	0.90	-3000	0.45	-6000
66:	0.50	1000	1.00	500

13. Mechanical Reasoning about Gear-and-belt Diagrams: Do Eye-movements Predict Performance?

Leon Rozenblit
Michael Spivey
Julie Wojslawowicz

Building on work by Hegarty and colleagues, a series of studies examined participants' eye movements while solving mechanical reasoning problems presented as diagrams on a 2-D static display. In the first study, participants examined gear-and-belt systems. The eye movement trace carried sufficient information for an independent rater to predict the primary axis of orientation for the gear-and-belt system, and the direction in which force was transmitted through the system. In the second study, the record of a participant's eye movements was sufficient for independent raters to predict well above chance whether the participant solved each problem correctly. These exploratory studies demonstrate that eye movement patterns contain crucial information about the moment-by-moment cognitive processes that underlie mechanical reasoning and mental animation, and suggest fruitful directions for future research.

13.1 Introduction

Diagrams help us understand mechanical systems. Anyone who has ever stared at the assembly instructions for a mail-order product, or a line drawing of an automobile drive-train, knows that adding even the most meagre drawings to the text can greatly improve comprehension. Precisely why diagrams are so useful, however, remains an open question.

One tentative answer might be that diagrams represent spatial relationships in a format analogous to the mental models that people frequently develop when comprehending text. Compared to the ambiguities and memory demands involved in translating linguistic information into a mental model, diagrams may provide a more direct mapping onto the very format of internal representation that the human perceptual/cognitive system prefers to use.

This leads to the question of how the human perceptual/cognitive system uses mental models and diagrams. The way people mentally search elements of a mental mdel may correspond to the way they visually search elements of a diagram, and this correspondence may allow insight into how the mapping from diagrams to mental models is achieved. For example, recent eye-tracking work has shown that when subjects are imagining a dynamic scene or recalling a particular visual array, their eye movements (while staring at a blank screen) reflect the content of the imagined visual scene [3, 14]. With regard to the mechanical relationships between elements of such mental (or visual) models, previous work by Hegarty and Just [5], Hegarty and Sims [7], and Fallside and Just [4] suggests that eye movements are important indicators of the mechanical reasoning process for subjects attempting to reason about rope-and-pulley systems. Furthermore, much of the work on mechanical reasoning using reaction time and error rates as the primary dependent variables clearly implies a variety of interesting questions that can be answered with eye movement data [9–13].

Thus, several lines of evidence suggest that eye movements may contain a considerable amount of information about the mental models people use to understand the world around them. Specifically, tracking eye movements may give us some insight about how human beings understand complex mechanical systems, and the role diagrams play in the mechanical reasoning process.

We know that people look at diagrams that accompany text when they wish to understand a mechanical system. However, *what* people are looking at when they examine diagrams remains a mystery. Do they look at individual components, at the spatial relationships between the components, or at the contact points between the components? Do they examine the components in some logical sequence, or do they "randomly sample" components in no particular order?

Hegarty and Sims' (1994) work on rope-and-pulley systems [7] seems to provide a few tentative answers to the above questions: subjects performing a mechanical reasoning task with static diagrams of rope-and-pulley systems, accompanied by text descriptions, seem to examine the individual elements of the system in the order of the causal chain of events. However, whether Hegarty and Sims' results can be generalised to other types of mechanical systems, and to other types of mechanical reasoning tasks, remains an open question.

Fallside and Just [4], for instance, came to a somewhat different conclusion. They asked subjects to determine whether an animated pulley system was functioning in a realistic manner. Fallside and Just concluded that the subjects, at least initially, "randomly sampled individual elements, with replacement". After the subjects located the potential inconsistency, they shifted to a more systematic "confirmation mode". Of course, Fallside and Just used a very different task from Hegarty and Sims – an animated display with one unrealistically moving element, rather than a static diagram. Nevertheless, the discrepancy illustrates the potential problem with assuming that Hegarty and Sims' results will automatically generalise to other mechanical systems.

It is noteworthy that all the eye-tracking work on mechanical reasoning cited above was performed with diagrams of rope-and-pulley systems. While rope-and-pulley systems may be perfectly representative of a broad type of mechanical device, establishing principles that hold across different types of mechanical systems would be highly desirable.

In a series of exploratory studies, we showed subjects diagrams of gear-and-belt systems, and tracked their eye movements while they attempted to answer questions about the systems. Our basic assumption was that eye movements are used to encode and store information about the mechanical system.

The oculomotor system takes an active, and essential, role in the processing of visual information. It can determine the order, the kind, and the size of information chunks sent "upstream". We can improve our understanding of the oculomotor system's role in visual information processing by examining the quantity and the type of information present in the scan path.

We suggest that the oculomotor system plays the role of "foreman" in building the mental model from the elements in the diagram. The present studies were designed to test this assumption, and to demonstrate that the structure and sequence of the eye movement patterns do indeed reflect the structure and sequence of the mechanical reasoning process.

13.2 Eye-Tracking

Recently, a resurgence of interest in eye movements is providing a unique window onto a number of perceptual/cognitive skills, including visual working memory [1], visual imagery [3], spoken language comprehension [15], and sensorimotor integration [2]. Importantly, as noted by Viviani [16], what is necessary for interpreting a particular scan path is an independently motivated account of *why* the eyes have moved to a particular location at a particular time. Understanding the *sequence* of eye movements is at least as crucial as knowing which locations were fixated more than others.

There are two very important reasons to examine eye movement patterns in addition to the more traditional dependent variables in the cognitive sciences (e.g., accuracy, reaction time). First, eye movements are especially sensitive to probabilistic information. Thus, elements of a visual array that are briefly considered relevant for an initial cognitive strategy that eventually gets replaced by a different strategy (when the first strategy actually has no effect on final performance on the task, and in fact is rarely present in subjects' self-report) will nonetheless draw eye movements in the interim [15]. Second, eye movements are relatively automatic and not subject to voluntary control, thereby providing a relatively "honest" measure of moment-by-moment cognitive processes during problem solving in a visual context. It is only natural that this measure of performance be applied to mechanical reasoning.

In the present experiments, eye movements were monitored by an ISCAN eyetracker mounted on top of a lightweight headband. The camera provided an infrared image of the left eye sampled at 60 Hz. The centre of the pupil and the corneal reflection were tracked to determine the direction of the eye relative to the head. A scene camera, yoked with the view of the tracked eye, provided an image of the subject's field of view. Gaze position (indicated by cross-hairs) was superimposed over the scene camera image and recorded onto a Hi8 VCR with 30 Hz frame-by-frame playback. Accuracy of the gaze position record was about one degree of visual angle over a range of ± 25 degrees. In addition to the video playback, the eye camera's data stream was also used to reconstruct scan paths independent of the scene camera's view. So long as head movement during individual trials was not excessive, such reconstructions are relatively informative.

13.3 Study 1a – Guess the Axis

As a first step in looking at subjects' eye movements during mechanical reasoning with gear-and-belt systems, we wanted to test whether subjects' scan paths allowed one to determine some basic properties of the gear-and-belt system being observed. Do subjects tend to look at the majority of the mechanical system, or only at a few elements? Do their eye movements follow the predicted causal chain of events in the system, from gear to gear to gear? Does the scan path alone contain sufficient information to reconstruct the principal axis and the direction of the causal chain of a mechanical system?

13.3.1 Methods

Subjects. Five Cornell University undergraduates participated. The subjects were recruited through posters placed on campus and were offered five dollars or course credit for participation. Each subject participated in one experiment consisting of 24 trials.

Procedures. We showed subjects gear-and-belt diagrams on a computer screen. Each gear-and-belt diagram was arranged as a linear series of elements along a principal axis (horizontal or vertical). Each diagram also had an identifiable principal direction – i.e., the direction in which force is transmitted through the system (left to right, right to left, top to bottom, or bottom to top). The diagrams were counterbalanced so that one half had a vertical principal axis (PA) and one half had a horizontal PA. Further, half of all the vertical PA diagrams had a top-to-bottom principal direction, and the other half had a bottom-to-top principal direction; half the horizontal PA diagrams had a left-to-right principal direction (PD), and half had a right-to-left PD.

The diagrams consisted of gears, pulleys, belts and levers. Each diagram had an input lever and an end-box at opposite ends. After about 3 seconds from the beginning of each trial, subjects were given a pre-recorded voice command indicating the direction of force applied to the input lever. It was

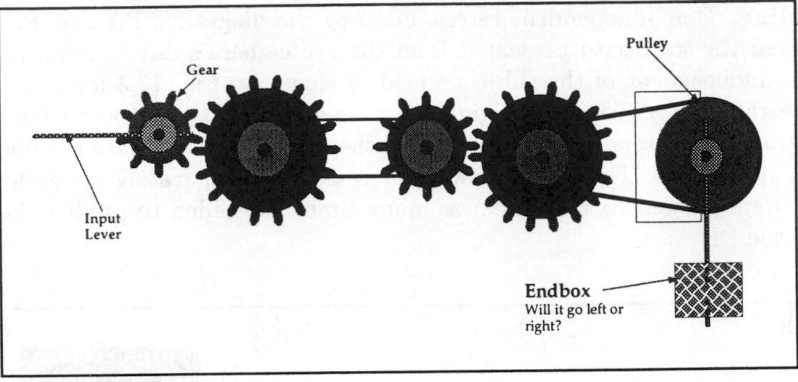

Figure 13.1. A practice example of a gear-and-belt diagram from the instructions and practice section of Study 1. In this display, the principal axis is horizontal, and the principal direction is left-to-right. If the pre-recorded voice command instructs the subject that the input lever (far left) moves downward, in what direction will the end-box move (left or right)?

the subjects' task to determine which way (right or left) the end-box would move when force was applied to the input lever in the direction indicated by the voice. See Figs 13.1 and 13.2. There were two practice trials and 24 experimental trials.

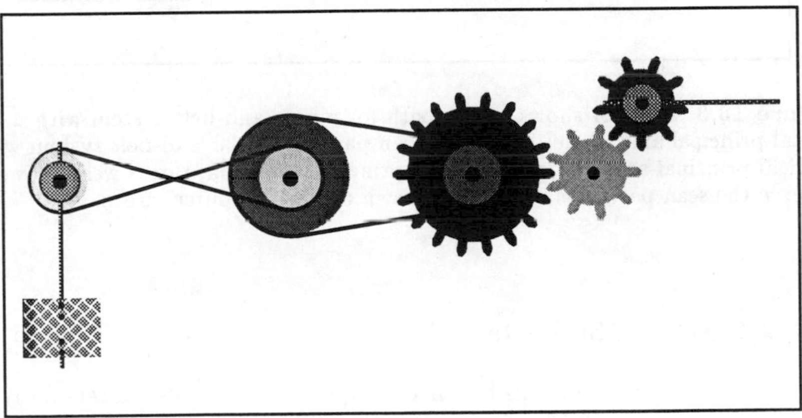

Figure 13.2. An example gear-and-belt diagram, in which the principal axis was horizontal, and the principal direction was right-to-left. In this display, the pre-recorded voice command instructed the subject that the input lever (far right) rotates downward. The participant's task is to determine whether the end-box moves left or right.

Coding. Four independent raters, blind to the diagram's PA and PD, examined the scan path presented from the eye camera's data stream in real time, independent of the subject's field of view. See Fig. 13.3 for examples. (The raters never saw the original diagram shown to the subject.) For each diagram, the raters had to guess, from the scan path information alone the PA and the PD. The raters viewed the scan paths separately for each trial and were allowed to view them as many times as needed to predict the PA and the PD.

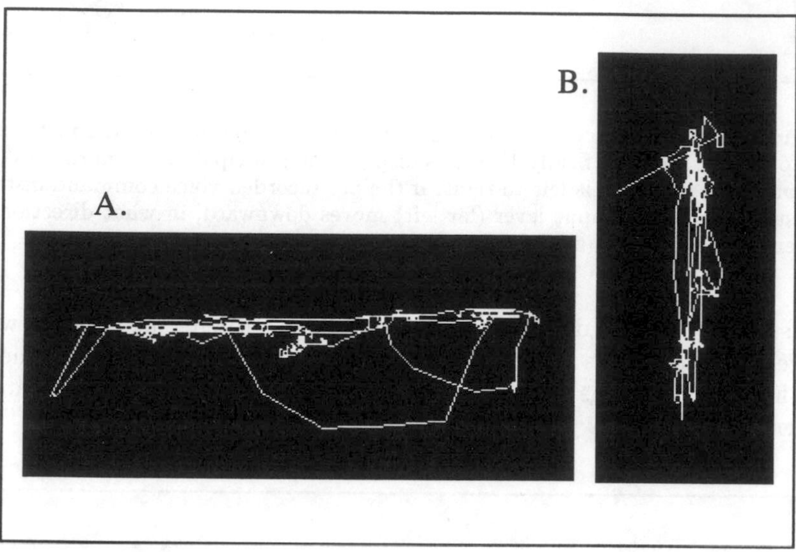

Figure 13.3. Panel A shows a scan path for a gear-and-belt system with a horizontal principal axis. Panel B shows a scan path for a gear-and-belt system with a vertical principal axis. For resolving the principal direction, raters were allowed to observe the scan paths being slowly redrawn on the computer screen.

13.3.2 Results of Study 1a

Raters viewed the scan paths for all 120 experimental trials. Raters correctly predicted the PA in 99.8% of all trials. The expected number of trials scored correct by chance is 50%. The worst-performing rater was correct on 99.2% of the trials; that rater was mistaken on only one of the 120 trials ($\chi^2 > 114.06$, $p < 0.0001$). The remaining three raters were correct on all of the trials. Needless to say, inter-rater reliability was extremely high.

Raters correctly predicted the PD in 97.4% of all trials, where the expected number of trials scored correct by chance would be 25% if we assume the rater does not know the principal axis, and 50% if we assume that he does. The latter assumption is highly plausible for trials where the raters correctly predicted the PA. Raters, in fact, correctly predicted PD on 97.4% of the trials on which they correctly predicted PA ($t = 64.999$, $p < 0.0001$) – far more than we would expect by chance. The lowest-scoring rater was correct on 95.8% of all trials; that rater was mistaken on 5 of the 120 trials. ($\chi^2 > 318$, $p < 0.0001$). Two raters were accurate on 97.4% of all trials, and one rater was accurate on 99.2% of all trials. Inter-rater reliability was, again, extremely high ($\chi^2 > 199$, $p < 0.0001$).

13.3.3 Discussion of Study 1a

The clear result of Study 1 is that the information in the scan path is sufficient to predict both the principal axis and the principal direction of force in a gear-and-belt system. Raters were extremely accurate when estimating the principal axis. They found the task almost trivially easy. In fact, examining the scan path makes the prediction fairly obvious to any untrained observer. Quite simply, the scan path tends to lie along a single, reasonably well-defined axis.

While the finding is far from surprising – we expect the subjects to look at the elements of the diagram, rather than at the blank portions of the screen, or off the screen – it gives us some useful information about how the subjects are performing the task. Note that the scan paths would have looked far less linear, and raters' job of guessing the PA might have been considerably more difficult, if the subjects spent most of their time looking at only a few of the diagram elements, if they glanced away from the diagram after examining each individual element, or if they focused entirely on the input or the output portion of the system. But because the information about the general axis of orientation of the gear-and-belt system is clearly present in the scan path, we can conclude that subjects spent most of their time looking at the diagrams, and examined most of the elements in the diagrams. Thus, while the findings agree with our expectations, they are an important first step towards establishing that *some* information about the system's layout was contained in the scan path.

The second result of Study 1 is much more intriguing. Raters were extremely accurate at predicting the principal direction of force through the system from the scan path information alone, although slightly less accurate than they were at predicting the principal axis. While our data indicate that the scan path contains information about the direction of force in the gear-and-belt system, we have no quantitative evidence of *how* the raters are able to extract the force information from the scan path.

Fortunately, debriefing the raters gave us plenty of qualitative insight about how they managed such accurate predictions. The raters reported attending to the order of fixations in the dynamic scan path display; they reasonably assumed that subjects looked at the input elements first, then moved on towards the output. That is, the raters wisely expected the eye movements to follow the direction of the force, from input to output. The raters' strategy proved extremely successful. Their success, in fact, allows us to conclude that the subjects examined the gear-and-belt systems in a methodical and consistent way: they started with the input, and tracked the direction of force, sequentially, through each element. This interpretation is consistent with Hegarty and Sims' findings with rope-and-pulley systems [7].

The raters were not, however, 100% accurate. Recall that they were somewhat less accurate at predicting the direction of force than at predicting the principal axis. What was the source of the inaccuracy? We suspect the raters' task was made more difficult by two properties of the scan path. First, the scan path rarely makes it clear when a subject has finished examining one element and moved on to another. Thus, within-element eye movements could easily be mistaken for between-element eye movements. Second, subjects often back-track: they proceed through the elements in a sequential manner, then pause, and restart their examination from some preceding element in the sequence. The back-tracks are obvious when you can see eye position over time superimposed on the diagram that the subject is examining. However, back-tracks are somewhat difficult to distinguish from other eye movements when viewing the scan path information alone. Ambiguity in element switches and back-tracking introduced noise into the scan path that made recovering the direction of force more difficult than recovering the primary axis. Nevertheless, the raters performed surprisingly well given the amount of noise from the two known sources.

Anecdotally, the raters emphasised some features of the scan path as "giveaways". Specifically, subjects spent a lot of time looking at contact points between the gears. The detailed examinations produced easily identifiable "clusters" in the scan path. As the subject moved from one gear-pair to the causally subsequent gear-pair, they left tell-tale marks in the scan path. The order in which the "clusters" appeared was, according to the raters, an extremely reliable indicator of the direction of force in the system. Unfortunately, the results of Study 1a provide no evidence that bears on the raters' anecdotal reports. Since accuracy was so high for all trials, and since all trials included some gears, we could not analyse the data from Study 1a to isolate the effect of the presence of gears on rating accuracy. We plan to address this issue explicitly in a follow-up study.

13.4 Study 1b – Guess the Axis Control

While the results of Study 1a were intriguing in their own right, we felt that a comparison with eye movements during a non-mechanical reasoning task would be more enlightening.[1] Specifically, nothing in Study 1a guaranteed that the results were specific to mechanical reasoning. The results could be

[1] We are grateful to the first reviewer for raising this issue.

common to any visual examination of objects with primary horizontal and vertical orientations and a primary direction of causal force. To test the claim that mechanical reasoning produced measurably different scan paths from non-mechanical reasoning about the same system we presented a different set of subjects with very similar diagrams, but asked them to perform a rather different reasoning task while we tracked their eye movements.

In Study 1b, we asked the subjects to evaluate whether a statement they heard about the gear-and-belt system was true or false. The subjects heard one of two pre-recorded voice statements: (1) "There are more gears than pulleys" or (2) "There are more pulleys than gears". The task was designed to parallel the task in Study 1a in several ways. Both tasks required the subject to decide between two alternatives. Both tasks involved sets of two verbal instructions that produced two different answers for the same diagram. Finally, both required subjects to reason about the presented diagram (typically over several seconds). However, the scan paths for Study 1a were usually a few seconds longer than those for Study 1b.

13.4.1 Methods

Subjects. Six Cornell University undergraduates participated. The subjects were recruited through posters placed on campus and were offered five dollars or course credit for participation. Each subject participated in one experiment consisting of 24 trials.

Procedures. We showed subjects gear-and-belt diagrams on a computer screen. The diagrams were either identical or very similar to the diagrams in Study 1a.[2] Each gear-and-belt diagram was arranged as a linear series of elements along a principal axis (horizontal or vertical). Each diagram also had an identifiable principal direction – i.e., the direction in which force is transmitted through the system (left to right, right to left, top to bottom, or bottom to top).

The diagrams were counterbalanced so that one half had a vertical principal axis (PA) and one half had a horizontal PA. Further, half of all the vertical PA diagrams had a top-to-bottom principal direction, and the other

[2] The new diagrams included all the old diagrams as a subset. The original set of diagrams used in Study 1a consisted of only three unique configurations of gears and belts. Each configuration was presented in four orientations with only slight modifications. The nature of the mechanical reasoning task made the orientation highly relevant to the answer the subject had to provide. Thus, subjects did not notice the similarities between the systems they were examining – the similarities were not relevant to the task. During the pilot phase of Study 1b, it became obvious that the new reasoning task made the similarities between the diagrams more salient. The pilot subjects had spontaneously verbalised that they noticed some diagrams were identical except for orientation. Thus, we had to double the number of unique diagrams to keep the subjects from memorising the answers to the reasoning questions.

half had a bottom-to-top principal direction; half the horizontal PA diagrams had a left-to-right principal direction (PD), and half had a right-to-left PD.

The diagrams consisted of gears, pulleys, belts and levers. Each diagram had an input lever and an end-box at opposite ends. After about 3 seconds from the beginning of each trial, subjects heard one of two pre-recorded voice statements: (1) "There are more gears than pulleys" or (2) "There are more pulleys than gears". The instructions defined "a gear" as "any element with teeth", and "a pulley" as "any element supporting a belt". The subjects had to determine whether the statement was true or false, and press the appropriate button. There were two practice trials and 24 experimental trials.

Coding. Four independent raters, blind to the diagram's PA and PD, examined the scan path constructed from the eye camera's data stream, independent of the subjects' field of view. See Fig. 13.3 for examples from Study 1a. (The raters never saw the original diagram shown to the subject.) For each diagram, the raters had to guess, from the scan path information alone, the PA and the PD. The raters viewed the scan paths separately for each trial and were allowed to view the diagrams as many times as needed to predict the PA and the PD.

13.4.2 Results of Study 1b

Three of the raters viewed the scan paths for all 144 experimental trials. The fourth rater viewed the scan paths for 120 experimental trials. (The fourth rater was one of the five original subjects for Study 1b, and was not permitted to rate his own trials.) Raters correctly predicted the PA in 92.0% of all trials. The expected number of trials scored correct by chance is 50%. The worst-performing rater was correct on 89.3% of the trials ($\chi^2 > 82.01$, $p < 0.0001$).

Raters correctly predicted the PD in 46.1% of all trials, where the expected number of trials scored correct by chance would be 50% if we assume the rater knows the principal axis. This assumption is very plausible for trials where the raters correctly predict the PA. Raters, in fact, correctly predicted PD on 50.3% of the trials on which they accurately predicted PA ($t = 0.089$, $p = 0.9294$) – no more than we would expect by chance. Tables 13.1 and 13.2 summarise the PA and PD results, for Studies 1a and 1b.

Raters in Study 1b were accurate at guessing PA. However, their PA guesses were significantly less accurate than the PA guesses of raters in Study 1a. The raters in Study 1b were not accurate at guessing PD; their PD guesses were significantly less accurate than the PD guesses of raters in Study 1a.

The subject's task in Study 1b involved a different kind of reasoning from the task in Study 1a. Nevertheless, we were not surprised that the raters in Study 1b could still do a very good job of guessing the PA of the system from the scan paths. Subjects rarely look at empty space. Thus, most scan paths are bound to convey some layout information about the examined objects.

Table 13.1. PD accuracy for correct PA trials (only 50% expected by chance)

	PD accuracy mean	Hypothesised mean	t-value	p-value
Study 1a Correct PA trials (469 trials)	0.974	0.5	64.999	< 0.0001
Study 1b Correct PA trials (508 trials)	0.502	0.5	0.089	< 0.9294

Table 13.2. Comparison of PA and PD accuracy for Studies 1a and 1b

	Study 1a	Study 1b	Mean difference	Critical difference	p-value
Mean PA accuracy	0.998	0.920	0.078	0.024	0.0002
Mean PD accuracy	0.974	0.461	0.513	0.053	0.0001

If fact, it is unlikely that layout information in the scan paths would be unique to mechanical reasoning. Any protracted examination of a system is likely to produce a sufficiently dense scan path for a rater to reconstruct some features of the system's layout. Nevertheless, we think the statistically significant difference between PA accuracy in Studies 1a and 1b is intriguing (see Table 13.2). The difference suggests that reasoning mechanically about a system may produce scan paths that contain more reliable layout information than non-mechanical reasoning about the same system.

The difference between Studies 1a and 1b in PD accuracy is dramatic. In fact, not only were the raters in Study 1a substantially better at guessing PD than were the raters in Study 1b, but the latter were no more accurate than we would expect by chance. The raters' poor performance in Study 1b is, again, not surprising. We did not expect subjects in Study 1b to systematically begin their visual inspection at the input, and move toward the output, as they generally did in Study 1a. However, it is important to establish that the systematic scan paths that follow the causal chain are specific to the mechanical reasoning task in Study 1a, and are not some general feature common to any visual examination of gear-and-belt diagrams.

13.5 Study 2 – Guess the Accuracy

If eye movements reflect the subject's problem solving procedure and the construction of the mental model, it may be possible to predict whether the subject solved the problem accurately by observing the eye position overlaid on the diagram throughout the solving process. To test that possibility, we presented 21 subjects with a series of gear-and-belt problems, recorded their eye movements, and asked independent raters to evaluate the eye movement patterns and to predict whether or not the subjects solved the problem correctly.

13.5.1 Methods

Subjects. Twenty-one Cornell University undergraduate and graduate students participated. The subjects were recruited through posters placed on campus and were offered five dollars or course credit for participation. Each subject participated in one experiment consisting of 34 trials.

Procedures. The diagrams in this study were similar to those of Study 1. However, they were not arranged in a line, along a single axis, because we wanted the largest possible elements, given the limitations of a 20-inch screen. Diagrams consisted of pulleys, gears and belts, and the number of elements presented ranged from two to five. Some diagrams consisted of only gears, some only pulleys, and other diagrams presented a combination of both gears and pulleys. Diagrams were counterbalanced for direction and number of elements. Again, each diagram consisted of both an input lever and an end-box at opposite ends. However, in this study, the voice was replaced by an arrow. The arrow pointed at the input lever in either an upward or downward direction, and the subjects were instructed that the arrow indicated the direction of force applied to the lever. One half of the diagrams had the arrow pointing upward and the other half had the arrow pointing downward. The subjects' task was to determine which direction the end-box would move (left or right) when force was applied to the input lever in the direction indicated by the arrow (see Fig. 13.4).

Coding. Three independent raters, blind to the subject's actual response ("left" or "right"), observed the videotapes containing the scene camera's view, with the eye position superimposed (as a cross-hair). The raters viewed the videotapes at a reduced speed. This allowed the raters to clearly follow the eye movement patterns of the participants. The raters attempted to predict, based on the above information alone, whether the subject correctly solved the gear-and-belt problem presented at the trial.

13.5.2 Results of Study 2

Three raters viewed all 714 trials by the 21 subjects. Fifty-three trials were excluded from the analysis because one or more of the raters determined that the track was too poor to score the trial (34 of the excluded trials were from one subject). The raters' predictions about the subjects' accuracy were correct substantially more often than would be expected by chance. Rater one was correct on 76.7% of all trials ($\chi^2 = 18.280$, $p < 0.0001$). Rater two

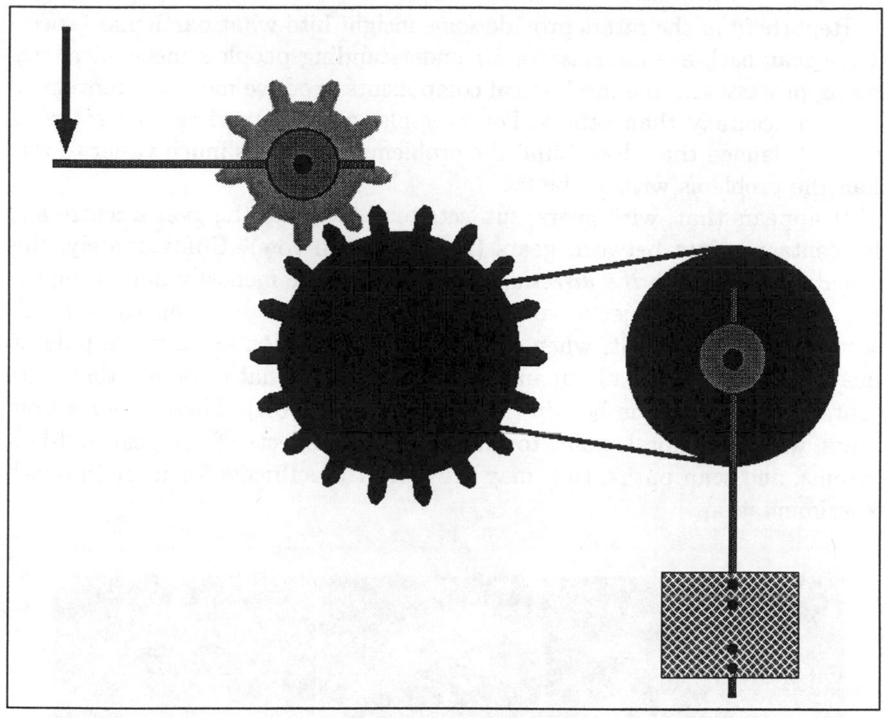

Figure 13.4. An example gear-and-belt diagram with three elements. The force arrow indicates that the input lever moves downward. The subject's task is to determine whether the end-box will move left or right.

was correct on 81.7% of all trials ($\chi^2 = 28.270, p < 0.0001$). Rater three was correct on 79.8% of all trials ($\chi^2 = 6.843, p < 0.0089$).

Inter-rater reliability was relatively good. Raters one and two agreed on 78.6% of all trials ($\chi^2 = 22.417, p < 0.0001$). Raters one and three agreed on 75.6% of all trials ($\chi^2 = 1.756, p < 0.1852$). Raters two and three agreed on 83.3% of all trials ($\chi^2 = 14.015, p < 0.0001$).

13.5.3 Discussion of Study 2

The subjects' eye movements during the solving of gear-and-belt problems provided sufficient information about the subjects' solution procedure for independent raters to estimate the subjects' accuracy. This result is encouraging for the goal of understanding how eye movements can provide a unique insight into mechanical reasoning. Nonetheless, so far these exploratory studies have merely shown that *some* information in the scan path can tell us about the moment-by-moment cognitive processes of mechanical reasoning; they have not pointed out *what that information is.*

Reports from the raters provide some insight into what particular aspects of the scan path are informative for understanding people's mechanical reasoning processes. Some mechanical components produce more eye movement clues to accuracy than others. For example, upon debriefing, the raters in Study 2 claimed that they found the problems with belts much easier to rate than the problems without belts.

It appears that, with gears, subjects tend to fixate the gear's centre and the contact points between gears (Fig. 13.5, top row). Unfortunately, this provides few clues to *the direction* in which they are mentally animating the gear. In contrast, subjects would often make a saccade along the forward moving section of a belt, when determining its rotation around two pulleys, thus providing a relatively unambiguous clue as to what direction they were mentally animating the belt in (Fig. 13.5, bottom row). These reports from raters, while anecdotal, point toward particular aspects of the gear-and-belt systems, and scan paths, that may prove fruitful subjects for more in-depth experimentation.

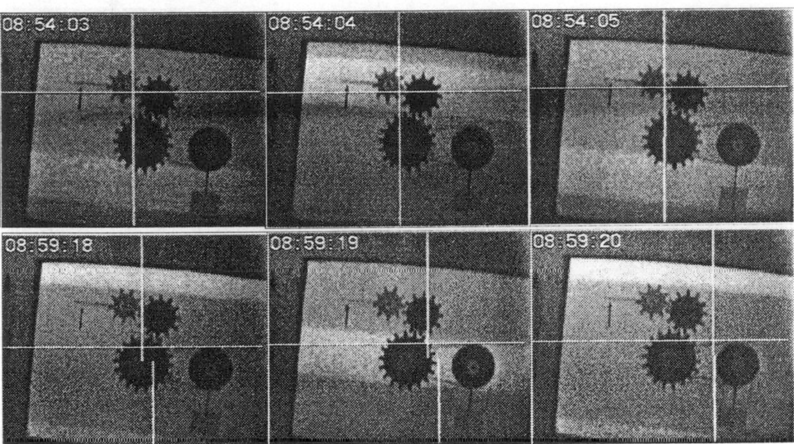

Figure 13.5. The top row of video frames shows the subject's field of view, with white cross-hairs indicating eye position, as he fixates the contact point between gears 1 and 2. The timecode displays minutes: seconds: frames (at 30 Hz). The bottom row of video frames shows, 5 seconds later, the subject making a saccade that traces (from left to right) the upper portion of the belt between the two pulleys.

Our stimulus set did not systematically vary the presence and absence of belts. Thus, the raters' intuition was difficult to test directly. Limited by the available data, we tried an indirect approach. If the raters' anecdotal accounts were correct, belts should be more or less helpful to the rater depending on their position in the causal chain. Belts closest to the output should be the most helpful and belts further from the output, less helpful.

Why should the position of the belt relative to the output matter to the raters' accuracy? If the raters primarily rely on saccade direction on belts to predict the direction of the mental animation, we would expect the raters to be more accurate with systems with belt components close to the output. Belts closer to the output should be more helpful to the rater than belts farther from the output because for near-the-output belts few sources of uncertainty remain between the mental animation of the belt and the subject's final output decision. Once the rater correctly identified the direction of the mental rotation of the last element, determining the subject's final decision should be relatively easy.

The analysis did provide some evidence that confirmed the rater's intuitions. For the 302 trials with one belt, the distance of the belt from the output was significantly (negatively) correlated to rater accuracy on that trial, once subject accuracy on the trial was controlled for. That is, raters were significantly more accurate at predicting subject performance on trials where the belt was close to the output than on trials where the belt was far from the output (model: trial accuracy = distance from output + correct score. Overall regression: $p < 0.0001$; belt's distance from output, partial $r = -0.367$, $P = 0.0018$).

13.6 General Discussion

Information contained in the scan path is especially interesting because of what it can tell us about the oculomotor system's crucial role in parsing and ordering the flow of visual information. Our two exploratory studies corroborate previous research (e.g., [4, 5, 7]) suggesting that eye movement patterns contain useful information about how people "think with diagrams". We also uncovered some hints about how eye movements differ when storing information about different elements in a gear-and-belt system.

Study 1 showed that information about the primary axis of a gear-and-belt system, and information about the primary direction of force in the system, can be recovered from the scan path. We have also noticed a prominent feature of the scan path: a tendency for fixations to cluster around gear–gear contact points. We suggest the tight clusters of fixations are tell-tales of subjects' mentally animating the gear pairs. If the suggestion is correct, the results of Study 1 closely parallel the previous findings of Hegarty and Sims [7], who reported that subjects performing a mechanical reasoning task about a rope-and-pulley system, using static diagrams and text, mentally animated the elements in the order of the causal chain.

In Study 2, we have demonstrated that a subject's eye movements contain sufficient information about his/her mechanical reasoning process to predict the subject's accuracy at better-than-chance level. We also made a preliminary effort to isolate the features of the eye movement information that were especially helpful. While our analysis is still in progress, a few preliminary observations are in order. A close look at the videotapes leads us to believe the raters' ability to predict subjects' answers was probably unrelated to the very end of each trial. That is, raters could not simply check the direction of the final glance. In fact, few subjects ever made eye movements to the end-box. Rather, the raters seemed to focus on eye movements around the belt components.

The subject's scan path around the belt component appears to be a relatively accurate indicator of the direction the subject thinks the belt will rotate. Raters report relying on belts to check the direction of the subject's mental rotation. Our post hoc analysis supports the raters' anecdotes – belts near the output seem to help rater accuracy more than belts further from the output.

Scan paths around gear elements, on the other hand, do not provide clear cues about the direction of mental animation. When examining the gears, subjects seem to fixate at the contact points between pairs of gears, and at the centres of gears. Why mentally animating different types of elements should involve very different types of eye movements is a fascinating question. We were so intrigued by the difference between gears and belts that we plan to look at the issue more closely in additional studies (see *Studies in Progress* below).

Thinking about the oculomotor system as an active and efficient encoder of visual information may help us understand the difference between eye movements associated with mentally animating gears versus belts. How could eye movements help efficiently encode information about mechanical systems? One way is by making saccades only when they capture a lot of information about the dynamics of the system. Note that the direction of a belt's rotation can be easily encoded as a single saccade along the flat section of the belt. In contrast, encoding the direction of a gear's rotation would require a complex series of short saccades along the circular perimeter. This observation leads us to speculate that studying eye movement patterns will be most informative for understanding mechanical and spatial reasoning in situations where one or two eye movements can efficiently help encode a particular causal relationship. Static diagrams of large linear displacements should produce more information-rich scan paths than diagrams of rotations. Studies to test the prediction, and several others, are in hand.

13.7 Studies in Progress and Future Directions

As expected, our exploratory studies raised more questions than they answered. We are now actively pursuing several of the hottest leads. First, we are investigating the difference between gears and belts suggested by our two exploratory studies. We suspect that the difference results from the differential efficiency with which eye movements can encode information about various types of mechanical components. Second, we are investigating whether giving subjects the direction of the input force via verbal instruction, rather than visually, changes the way they look at the diagram. Hegarty and Sims [7] suggest that it should, at least in the context of text–diagram integration. They propose that diagram comprehension and mental animation are two separate stages of thinking with diagrams. We suspect the differences, if they exist, are closely related to two somewhat divergent functions of mechanical reasoning: prognosis and diagnosis. In any case, eye movement information can provide valuable evidence for the reality of the proposed different stages of mechanical reasoning.

In the future, eye-tracking research may also provide new insight into the incorrect strategies people may use in mechanical reasoning tasks. Can individual differences in mechanical reasoning (such as those observed in [6]) be diagnosed from the eye movement patterns? Are poor mechanical reasoners ignoring important aspects of mechanical systems? If eye movements can reveal encoding failures, can the information be used to direct a poor mechanical reasoner's attention (and eye movements) so as to improve their performance? The questions are theoretically interesting, as well as practically significant. The potential for finding answers is an exciting reason to continue this line of research.

Acknowledgements

This work was supported by a Sloan Fellowship in Neuroscience to the second author. We are grateful to David Dunning, Diana Fitek, Rachael Joyce, Melinda Tyler, Beth Vinluan, and Chris Wheeler for assistance in experimental design and data collection.

References

1. Ballard, D.H., Hayhoe, M.M. and Pelz, J.B. (1995). Memory representations in natural tasks. Journal of Cognitive Neuroscience 7:66–80.
2. Ballard, D.H., Hayhoe, M.M., Pook, P.K. and Rao, R.P. N. (1997). Deictic codes for the embodiment of cognition. Behavioral and Brain Sciences 20:723–767.
3. Brandt, S.A. and Stark, L.W. (1997). Spontaneous eye movements during visual imagery reflect the content of the visual scene. Journal of Cognitive Neuroscience 9:27–38.
4. Fallside, D.C. and Just, M.A. (1994). Understanding the kinematics of a simple machine. Visual Cognition 1(4):401–432.
5. Hegarty, M. and Just, M.A. (1993). Constructing mental models of machines from text and diagrams. Journal of Memory and Language 32:717–742.
6. Hegarty, M., Just., M.A. and Morrison, I.R. (1988). Mental models of mechanical systems: Individual differences in qualitative and quantitative reasoning. Cognitive Psychology 20:191–236.
7. Hegarty, M. and Sims, V.K. (1994). Individual differences in mental animation during mechanical reasoning. Memory and Cognition 22:411–430.

8. Johnson-Laird, P.N. (1983) Mental models: Towards a cognitive science of language, inference, and consciousness. Cambridge, MA: Cambridge University Press.
9. Kaiser, M.K. and Calderone J.B. (1991). Factors influencing perceived angular velocity. Perception and Psychophysics 50:428–434.
10. Kaiser, M.K. and McCloskey, M. (1986). Development of intuitive theories of motion: Curvilinear motion in the absence of external forces. Developmental Psychology 22:67–71.
11. Kaiser, M.K., Proffitt, D.R., Whelan, S.M. and Hecht, H. (1992). Influence of animation on dynamical judgements. Journal of Experimental Psychology 18:669–690.
12. Schwartz, D.L. and Black, J.B. (1996a). Analog imagery in mental model reasoning: Depictive models. Cognitive Psychology 30:154–219.
13. Schwartz, D.L. and Black, J.B. (1996b). Shuttling between depictive models and abstract rules: Induction and fallback. Cognitive Science 20:457–497.
14. Richardson, D.C. and Spivey, M.J. (2000). Representation, space and Hollywood Squares: Looking at things that aren't there anymore. Cognition 76(3):269–295.
15. Tanenhaus, M., Spivey-Knowlton, M., Eberhard, K. and Sedivy, J. (1995). Integration of visual and linguistic information during spoken language comprehension. Science 268:1632–1634.
16. Viviani, P. (1990). Eye movements in visual search: Cognitive, perceptual, and motor control aspects. In E. Kowler (Ed.), Eye movements and their role in visual and cognitive processes. Amsterdam: Elsevier Science.

14. How do Designers Shift their Focus of Attention in their Own Sketches?

Masaki Suwa
Barbara Tversky

External representations serve as visual aids for problem solving and creative thinking. Past research has enumerated some of the features of external representations that enable this facilitation. We have questioned how and why architectural design sketches facilitate exploration of design ideas, by conducting protocol analyses of designers' reflections on their own sketching behaviour. Our previous analyses of their protocols revealed that skilled designers, once they shift attention to a new part of a sketch, are able to explore related thoughts more extensively than novice designers. How do they keep focused on related thoughts? What are the driving forces for successive exploration? We examined the types of information that expert and novice designers considered during and between chunks of related thoughts. We found that focus shifts driven by consideration of information about spatial relations led to successful exploration of related thoughts. We relate these results to some aspects of facilitation by the externality of sketches.

14.1 Introduction

14.1.1 Characteristics of External Representations

External representations such as diagrams, sketches, charts, graphs, and even hand-written memos not only serve as memory aids, but also facilitate and constrain inference, problem-solving and understanding. A geometry diagram in theorem-proving tasks is a typical example; it highlights plausible inference paths [6], visually cues necessary knowledge structure [11,15], and constrains assimilation of new knowledge [23]. Petre [16] showed that good use of graphical representations in programming environments prevents programmers from mis-cueing and mis-understanding. Architectural sketches also, the topic in this paper, serve as visual aids for design thinking in many ways [14]. Architects put ideas down on paper and inspect them. As they inspect their own sketches, they discover visual cues that suggest ways to refine and revise ideas. This cycle – sketch, inspect, revise – is like having a conversation with one's self [19].

How do the externality and visibility of representations facilitate problem-solving and creative endeavours? A number of scholars have provided insights on the facilitatory value of external representations (e.g., [3,4,7,8,10,13,16, 20,22,25–27]. Our own analysis and the research at hand lead us to highlight three functions of external representations. First, externalisations facilitate memory, both short-term working memory and long-term memory. They reduce working memory load by providing external tokens for the elements that must otherwise be kept in mind, freeing working memory to perform mental calculations on the elements rather than both keeping elements in mind and operating on them. A second memory function of external representations is to remind the user of conceptual knowledge necessary for problem solving and of other similar situations that may promote creativity [9]. Because diagrams are richer in information than descriptions, they may call to mind a wider range of associations. Moreover, as they are visual and spatial, they are especially appropriate for stimulating visual and spatial associations.

Next, external representations may promote both visuospatial and metaphoric calculation, inference, and insight (e.g., [4,13,25–27]). For example, inferences, either literal or metaphoric, about size, distance, and direction are easily made from diagrams. Calculations requiring counting or sorting or ordering are easily made by rearranging external spaces [10]. Insights, especially those based on proximity, grouping, and common fate, may be facilitated by inspection of diagrams. Externalisation of visual ideas allows them to be inspected, which promotes reorganisation, reconceptualisation, and reformulation of the same visual display (cf. [17]).

Finally, externalising a set of ideas forces some organisation, specificity, and coherence to a set of concepts [20], which, in turn, by inspection, may lead to new discoveries (cf. [7,8,19,22]). Thus, constructing externalisations of a set of concepts serves a function similar to modelling them.

There are a number of advantages to studying the use of external representations in real-world domains where the external representations are produced by problem solvers and inherent in the problem solving. The domain we have chosen is that of architectural design. There is some similarity even across cultures in the ways that architects use sketches in designing. Their early sketches – those that are useful in solving the most elemental and essential problems – differ greatly from the architectural plans and models presented to the public [5,12,14,18]. First, they are sketchy. That is, they are not at all specific. Rather, they tend to use blobs with indeterminate shape to indicate possible spatial arrangements. Next, they are two-dimensional, typically beginning with a plan or overview, a horizontal view. Only after determining the spatial arrangement do architects turn to the vertical plane. As Arnheim [1] noted, the horizontal plane of architectural drawings defines the plane of function; it shows how people will interact with the environment. The vertical plane is the plane of appearance; it shows how the environment will look to observers.

We have been studying the early stages of the design process by presenting a problem – the design of an art museum – to practising architects and to architectural students. After working on the design, participants viewed videos of their design sessions and reported what they were thinking as they drew. We have been analysing those protocols [25], and we will report further analyses here. Design sketches are a particularly interesting domain to study as they serve a dual purpose to the designer: they are used to express and demonstrate design ideas and they serve as a graphic display to be inspected to critique, to refine, and to generate further ideas [9].

14.1.2 Review of the Previous Findings on Design Processes

Our previous research had two main results. The first was to classify the types of information that participants perceive and report thinking about as they sketched [24, 25]. The information fell into three major categories – emergent properties, spatial relations and functional thoughts – each with subcategories to be discussed. Importantly, emergent properties and spatial relations are visual in nature, whereas functional thoughts are inherently non-visual. Designers, especially expert designers, read them off from the visual display.

The second result was to identify the basic units and structure of the design process. The design process proceeds in cycles. First, designers turn their attention to a new design topic from which they start to explore conceptually related topics. When that topic has been exhausted, designers turn their attention to a new topic. We call each fragment of design thought, which is the smallest unit of design processes, a "segment". A segment, whether consisting of one sentence or many, is defined as one coherent statement about a single item/space/topic. We call a set of contiguously occurring segments that are conceptually related to each other a "dependency chunk". See [25] for the precise definition of "dependency chunk". The basic idea is that if a segment "A" is not related to any segments in the conceptually related sequence that immediately precedes the segment A, then the set of previous segments are grouped into a dependency chunk, and the segment "A" is treated as a focus shift.

The entire design process consists of many dependency chunks (see Fig. 14.1). Some chunks consist of relatively many segments and others a few. In extreme cases, chunks consist of only one segment. We call it an "isolated" segment. In such cases, a designer shifted attention to a new topic, but failed to explore any related thoughts, and then shifted to another.

Segments are classified into two types, as shown in Fig. 14.1. One is the first segment in each dependency chunk. It represents a topic to which the designer's attention has been newly shifted. We call it a "focus shift" segment. The remaining segments that follow a focus shift segment are called "continuing" segments.

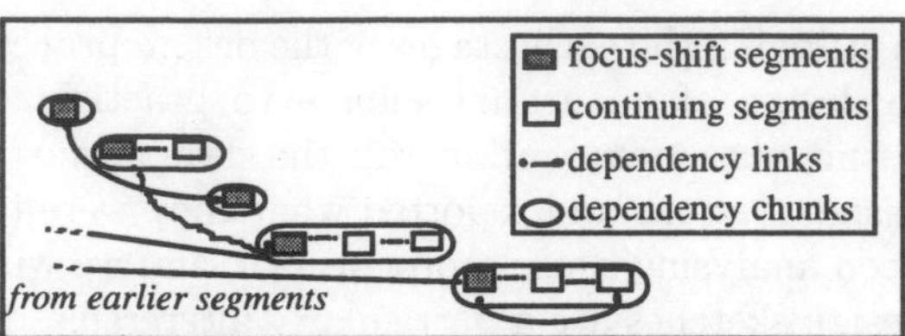

Figure 14.1. A schematic diagram of segments, conceptual dependency and dependency chunks.

We found evidence that skilled designers had more and longer dependency chunks than novices [24, 25]. Once experts shift their focus to a new topic, they are more capable of exploring related thoughts. Forming longer chunks leads to deeper and more substantial exploration, and thus contributes to the success of a design process.

14.1.3 The Goal of this Chapter

One mark of expertise in architectural drawings as well as other skilled domains is larger chunks of knowledge [2]. So, why and how are expert architects able to explore related thoughts more successfully? What are the driving forces that allow them to form longer dependency chunks? We address this question, reducing it to two precise questions. First, what kinds of focus shifts are likely to prolong subsequent related thoughts? Second, what types of within-chunk thoughts are likely to prolong chunks of related thoughts? To address these questions, we compared the types of information that occurred in longer and shorter chunks.

14.2 Experiment

14.2.1 The Architectural Design Task

The experiment consisted of two tasks: a design task and a report task. Five practising architects and six architectural students participated. In the design task, each participant worked on designing an art museum through successive sketches for 45 minutes. They were provided with a simple diagram representing an outline of the site, in which they were supposed to arrange not only a museum building but also a sculpture garden, a pond, a green area, and a parking lot. The building was required to have entrance(s), a ticket office(s), display rooms for about 100 paintings, a cafeteria, and a gift shop. Participants were told that the curator wants this to be a museum of "light, air and water". They were supposed to use freehand sketches as a tool for designing. Their sketching activity was videotaped.

Following the design task was the report task. While watching their own videotapes, participants were asked to remember and report what they were thinking as they drew each portion of each sketch. Participants were not interrupted with questions during the report. We recorded the participants' voices as well as videotaped the screen itself on which not only their sketching activity in the design task but also their pointing gestures in the report task were visible. The content of their verbal report is used for our analysis. Their pointing gestures helped clarify what they were reporting on.

14.2.2 Information Subclasses

The contents of participants' verbal protocols were encoded into subclasses of major categories. The determination of the major categories, i.e. emergent properties, spatial relations and functional thoughts, and their subclasses was based on empirical and theoretical considerations (see [25]).

Emergent properties denote depicted elements and their visual features. The subclasses are spaces in a sketch, physical entities, structural components, shapes, sizes, and materials/textures. Spatial relations denote spatial arrangements among two or more depicted elements. The subclasses are horizontal spatial relations, vertical spatial relations and horizontal global relations. The first two are local in contrast to the third; they are relationships among locally existing elements. Vertical relations primarily refer to structural components that are constructed on top of others. Global relations include locational relationships between the entire site and specific spaces on the site; directions depicted elements' face, and rough spatial organisation of elements in the site.

The subclasses of functional thoughts are practical roles, abstract features, reactions of people, circulation of people and cars, view issues, and light issues. Practical roles denote the roles that designed elements functionally play in a museum. Abstract features denote sensations and emotions that the designer himself feels from visual aspects of sketches. Reactions denote the influence and impact that visitors to the museum would have from the design. The last three subclasses are typical considerations in architectural designs.

14.2.3 Encoding of Verbal Protocols

For each participant, we divided the entire protocol into segments. Then, by analysing conceptual dependency among segments, we identified the dependency chunks and thereby the distinction between focus shift and continuing segments. Further, for each segment, we encoded its semantic content into information subclasses. Typically, the protocol for a participant consisted of a few hundred segments. The number of encoded subclasses varied from a few hundred to a thousand, depending on the participant.

14.3 Results

14.3.1 Length of Dependency Chunks: Architects vs. Students

We first wanted to reconfirm the difference in the length of chunks between skilful designers and novices because we added a number of practising architects' data to our previous corpus. First, we calculated the average number of segments in a chunk (ANS) for each participant, and compared it between architects and students. The ANS averaged over the five architects was 2.5 with a standard deviation of 0.48, whereas the ANS averaged over the six students was 1.9 with a standard deviation of 0.18. A statistically significant difference is recognised, $t(9) = 2.46$ ($p < 0.025$).

Then, for more precise confirmation, we examined to what extent longer streams of related thoughts are dominant in participants' design process: we calculated the ratio of the number of segments belonging to "long" chunks[1] to the total number of segments in the protocol for each participant. The advantage of using this ratio for more precise analysis over the ANS value is that unlike the ANS value the ratio is not affected by the configuration within "shorter" chunks, that is, the distribution of the number of "isolated" segments and the number of chunks whose length is 2. Table 14.1 shows that the average ratio of the five architects is statistically greater than that of the six students, $t(9) = 2.16$ ($p < 0.05$). Thus, one salient characteristic distinguishing expert from novice designers is longer streams of segments that are conceptually related.

At the same time, we found that the students happen to divide into two groups by chunk length, as shown in Table 14.1, $t(4) = 2.56$ ($p < 0.05$). The three students with higher ratios actually fit within the range of 1σ around the average of architects. So, we added the three students with significantly longer chunks to the sample of five architects to examine characteristics of long chunks. The remainder of the analyses are on these eight "experts".

Table 14.1. The ratio of the number of segments belonging to long chunks to the total number of segments

Architects' average (%)	S.D.	Students' average (%)	S.D.		
62.8	11.5	48.9	8.4		
		Students' 1–3 average (%)	S.D.	Students' 4–6 average (%)	S.D.
		55.0	5.3	42.8	6.3

[1] We defined "long" chunks as those whose length is equal to or greater than 3 segments. This is justified by the finding that the ANS value averaged over the architects, who produced longer chunks than students, was 2.5.

14.3.2 Information Categories: Focus Shift vs. Continuing Segments

Shifting attention to a new topic and exploring thoughts related to a current topic are different mental activities. We examined the difference in terms of the types of information that frequently occur in focus shift and continuing segments. For each participant, we counted the number of information categories per 10 segments, for both kinds of segments. Figure 14.2 shows the average frequency over the eight participants for both segments for each of the three major categories. Emergent properties are more frequent in focus shift segments than in continuing segments. Spatial relations are more frequent in continuing segments.

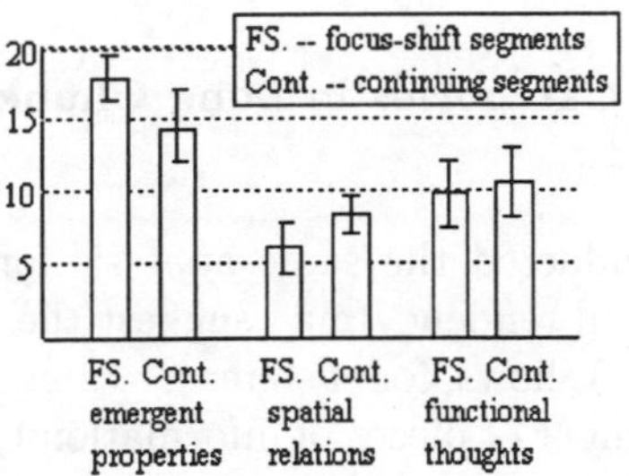

Figure 14.2. Average number of pieces of information belonging to the category per 10 segments, for each major category for both kinds of segments.

Then, we decided to analyse these phenomena in more detail for each information subclass of each major category within subjects. The reasons are as follows. First, the type(s) of information subclass that each participant frequently thought of in focus shift or continuing segments may differ from each other, and therefore investigations across subjects may suppress individual differences. Second, the number of subjects is small, so it is difficult to reach statistical significance for the entire group of subjects. However, the number of information subclasses per subject is large, so reaching statistical significance within subjects is feasible. Most of our analyses will be based on the numbers of participants for whom a particular pattern of data is significant.

We examined, for each participant for each information subclass of each major category, whether or not the information subclass occurs statistically more or less frequently in focus shift segments than in continuing segments. Table 14.2 shows, for each major category, the number of participants for whom at least one subclass of the major category was more frequent in either type of segment. For most participants, at least one subclass of emergent properties was more frequent in focus shift segments, and at least one subclass of spatial relations in continuing segments.

This result suggests an overall tendency that emergent properties were more characteristic of focus shift segments, and that spatial relations were more characteristic of continuing segments.

Table 14.2. The number of participants, for each major category, for whom at least one subclass of the category was more frequent in either type of segment

	Emergent properties	Spatial relations	Functional thoughts
Focus shift >>continuing	7	2	2
Continuing >> focus shift	1	6	3

14.3.3 Information Categories in Long Chunks: Focus Shift vs. Continuing Segments

Following this, we conducted the same analysis for long chunks only. Differences from the overall tendency may suggest the necessary conditions for long chunks. Figure 14.3 shows, for each major category for both kinds of segments, the average number of pieces of information belonging to the category per 10 segments.

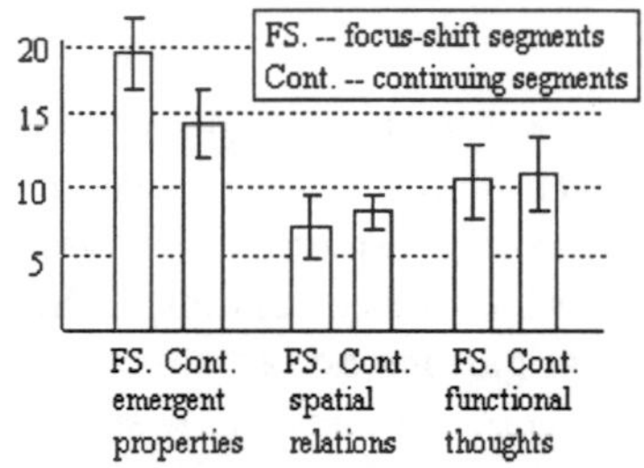

Figure 14.3. Average number of pieces of information belonging to the category per 10 segments, for each major category for both kinds of segments in long chunks.

For emergent properties, the same tendency as the overall one is observed for long chunks as well: emergent properties were more frequent in focus shift segments than in continuing segments. For spatial relations, however, the tendency is less salient for long chunks: spatial relations were as frequent in continuing segments as in focus shift segments.

Then we examined, for each participant for each information subclass of each major category, whether or not the information subclass occurs statistically more or less frequently in the focus shift segments of long chunks than in the continuing segments of long chunks. Table 14.3 is similar to Table 14.2, showing the number of participants, for each major category, for whom at least one subclass of the category was more frequent in either type of segment. For most participants, at least one subclass of emergent properties was important in the focus shift segments of long chunks as well. Comparing with Table 14.3, however, fewer participants (2 instead of 6) exhibited the tendency that at least one subclass of spatial relations was more frequent in the continuing segments of long chunks than in the focus shift segments of long chunks.

Table 14.3. The number of participants, for each major category, for whom at least one subclass of the category was more frequent in either type of segment of long chunks

	Emergent Properties	Spatial relations	Functional thoughts
Focus shift >>continuing	6	2	3
Continuing >> focus shift	1	2	1

The results may be summarised as follows. The overall tendency that spatial relations are more frequent in continuing segments than in focus shift segments derives not from long chunks but from shorter chunks. Further, both types of segment in long chunks possess similar characteristics in terms of the frequency of the subclasses of spatial relations.

14.3.4 Information Categories in Focus Shift Segments: Long Chunks vs. Shorter Chunks

These results motivated us to compare long chunks with shorter chunks in terms of the frequency of information subclasses. One comparison is between the focus shift segments of long chunks and those of shorter chunks. Another is between the continuing segments of long chunks and those of shorter chunks. As far as the latter comparison is concerned, we found no significant differences. This is consistent with the fact that, for all the major categories, the density of pieces of information belonging to each category in the continuing segments of long chunks (shown in Fig. 14.3) is almost the same as those in the continuing segments of the entire protocol (shown in Fig. 14.2). Here, the comparison between the focus shift segments of long chunks and those of shorter chunks is reported.

Table 14.4 shows, for each major category, the number of participants for whom at least one subclass of the category was significantly more frequent in the focus shift segments of long chunks or in those of shorter chunks. First, the number of participants for whom at least one subclass of emergent properties was more frequent in the focus shift segments of long chunks is five, whereas the corresponding number for those of shorter chunks is three. This means that, for some participants, thinking of certain subclasses of emergent properties in focus shift segments may prolong the subsequent related thoughts, but may rather shorten them for other participants. In contrast, for a greater number of participants, thinking of subclasses of spatial relations is more associated with long chunks than with shorter ones. For functional thoughts, the tendency is mixed; thinking of subclasses of functional thoughts may or may not be more associated with long chunks than with shorter chunks. To sum, it seems that thinking about subclasses of spatial relations in focus shift segments is likely to lengthen the subsequent chunks.

Table 14.4. The number of participants, for each major category, for whom at least one subclass of the category was more frequent in the focus shift segments of long chunks or in those of shorter chunks

	Emergent Properties	Spatial relations	Functional thoughts
Long >> shorter	5	5	5
Shorter >> long	3	1	2

Table 14.5 shows, for each participant, the subclasses that he or she more frequently thought of in the focus shift segments of long chunks than in those of shorter chunks. Importantly, the subclasses that each participant relied on as a driving force for successive exploration differed from each other, yet each participant relied on at least one subclass of spatial relations or functional thoughts. Further, more than one participant relied on horizontal local spatial relations, abstract features, or practical roles. This suggests the significance of these subclasses in focus shift segments as a driving force of the subsequent exploration, but significant subclasses may not be limited to these because of the small number of participants.

14.4 Discussion

14.4.1 Emergent Properties in Focus Shift Segments

With a sample of five pratising architects and three architectural students with especially long chunks, it is difficult to make statistical comparisons across subjects. Instead, we examined statistically significant patterns within subjects, and found some similarities suggestive of generalities across subjects. First, it seems that thinking about emergent properties was commonly associated with focus shift more than continuing segments. This tendency was salient also when we observed only long chunks.

Table 14.5. The information subclasses, for each participant, that were statistically more frequent in the focus shift segments of long chunks than in those of shorter chunks

Participants	Subclasses of emergent properties	Subclasses of spatial relations	Subclasses of functional thoughts
Architect 1		Vertical	
Architect 2	Structure		Practical roles, lights
Architect 3		Global	Abstract features
Architect 4			Practical roles, circulation
Architect 5	Structure	Horizontal	
Student 1	Spaces	Horizontal	
Student 2	Structure		Views
Student 3	Shape	Horizontal	Abstract features

14.4.2 Spatial Relations in Focus Shift Segments and Continuing Segments

Second, thinking about spatial relations was commonly associated with continuing more than focus shift segments, when we observe the overall tendency throughout the entire protocol. This tendency, however, drives from shorter chunks. Actually, in long chunks, spatial relations were not more associated with either type of segment. Investigating the comparison between the focus shift segments of long chunks and those of shorter chunks, we found a tendency that spatial relations were more associated with the focus shift segments of long chunks than with those of shorter chunks. Put differently, thinking of spatial relations in focus shift segments may prolong the subsequent related thoughts.

14.4.3 Locations and Relations in External Representations

How should these findings be interpreted? Sequences of related thoughts may be longer either because the information in the focus shift segment is especially successful in stimulating related thoughts or because the primary information used in the continuing segments is useful for continuing related thoughts, or both.

First, thinking about emergent properties seemed to stimulate shifting focus to a new topic. Sketches facilitate this type of action. Emergent properties are visible in or suggested by sketched elements. Simply because the tokens for designed elements are externally represented and thus visible, they become the target of focus shifts. And simply because those elements are organised in locations in a sketch in a specific way, the discovery of implicit empty spaces is facilitated.

Second, thinking about spatial relations, especially horizontal local spatial relations, both in focus shift segments and during the very exploration of related thoughts was found to be important. Generally speaking, there are three possible cases in which a participant thinks of spatial relations in focus shift segments. First, he or she revisits a spatial relation that was considered before. Second, he or she tries to create a new element in an empty space, regarding the spatial relation between the element and other existing elements. Third, he or she newly discovers an implicit, thus unintended, spatial relation between more than two existing elements. Sketches facilitate these types of actions as well. A specific organisation of elements in a sketch brings the participant's attention to previously attended relations, and even facilitates the discovery of empty spaces and new relations.

14.4.4 Function in External Representations

Whereas using external representations to facilitate thinking about entities or elements and their spatial array is straightforward, thinking about functional aspects of the situation from the external representation requires associations and inferences from the visuospatial display to things that are not purely visual or spatial. In the present analysis, functional thoughts are not more associated with any specific type of segment. However, our previous analyses of the protocols [25] showed that experts are more adept at perceiving functional thoughts from external representations than novices. This result is suggestive of an important role of external representations. Visual aspects of external representations facilitate associations, reminding and inferences that are necessary for calling to mind non-visual aspects of design thoughts.

14.5 Conclusion

We opened by discussing some of the benefits of external representations in thinking and problem solving. As we have shown in our detailed analyses of designers' retrospective reports of their drawing activity in design, sketches serve most of these roles for architects in the design process. Sketches facilitate memory by externalising the basic design elements, freeing the designer to think about the emergent properties of the elements, the spatial arrangements among them, and the functional implications of the elements and their spatial arrangement. Sketches promote calculations, inferences, and insights by serving as a spatial display on which those mental operations can be performed, and by promoting mental operations based on spatial factors such as proximity, grouping, distance, direction, common fate, and continuity. Finally, sketches facilitate design by inducing designers to be explicit and specific to a certain degree, and thereby to benefit from unintended discoveries.

Acknowledgements

We would like to express our thanks to J.P. Protzen and Y.E. Kalay at the University of California, Berkeley, his former student, Jinwon Choi, and P. Teicholz at CIFE, Stanford University, for putting us in contact with students of architecture and practising architects, and to the staff at the Department of Architecture of UC Berkeley for lending space and equipment to us. This research was supported by the Office of Naval Research, Grant no. NOOO14-PP-1-O649 to Stanford University.

References

1. Arnheim, R. (1977). The dynamics of architectural form. Berkeley, CA:University of California Press.
2. Chase, W.G. and Simon, H.A. (1973). Perception in chess. Cognitive Psychology 4:55–81.
3. Clement, J. (1994). Use of physical intuition and imagistic simulation in expert problem solving. In D. Tirosh (Ed.) Implicit and explicit knowledge. Norwood, NJ:Ablex.
4. Cox, R. and Brna, P. (1995). Supporting the use of external representations in problem-solving: The need for flexible learning environments. Journal of Artificial Intelligence in Education 6(2).
5. Fraser, I. and Henmi, R. (1994). Envisioning architecture: An analysis of drawingd. New York:Van Nostrand Reinhold.
6. Gelernter, H. (1963). Realization of a geometry-theorem proving machine. In E.A. Feigenbaum and J. Feldman (Eds) Computer and thought. New York:MacGraw-Hill.
7. Goel, V. (1995). Sketches of thought. Cambridge, MA:MIT Press.
8. Goldschmidt, G. (1991). The Dialectics of sketching. Creativity Research Journal 4(2):123–143.
9. Goldschmidt, G. (1994). On visual design thinking: The vis kids of architecture. Design Studies 15(2):158–174.
10. Kirsh, D. (1995). The intelligent use of space. Artificial Intelligence 73(1–2):31–68.
11. Koedinger, K.R. and Anderson, J.R. (1990). Abstract planning and perceptual chunks: Elements of expertise in geometry. Cognitive Science 14:511–550.
12. Landay, J.A. and Myers, B. (1995). Interactive sketching for the early stages of user interface design. In Human factors in computing systems: CHI'95 conference proceedings. New York:ACM Press, pp.43–50.
13. Larkin, J. and Simon, H.A. (1987). Why a diagram is (sometimes) worth ten thousand words. Cognitive Science 11:65–99.
14. Laseau, P. (1989). Graphic thinking for architects and designers (2nd edn). New York:Van Nostrand Reinhold.

15. McDougal, T. and Hammond, K. (1992). A recognition model of geometry theorem-proving. In Proceedings of the 14th annual conference of the cognitive science society. Hillsdale, NJ:Lawrence Erlbaum Associates, pp. 106–111.
16. Petre, M. (1995). Why looking isn't always seeing: Readership skills and graphical programming. Communications of the ACM 38(6):33–44.
17. Reisberg, D. (1987). External representations and the advantages of externalizing one's thoughts. In Proceedings of the 9th annual meeting of the Cognitive Science Society, pp. 281–293.
18. Robbins, E. (1994). Why architects draw. Cambridge, MA:MIT Press.
19. Schon, D.A. and Wiggins, G. (1992). Kinds of seeing and their functions in designing. Design Studies 13:135–156.
20. Stenning, K. and Oberlander, J. (1995). A cognitive theory of graphical and linguistic reasoning: Logic and implementation. Cognitive Science 19:97–140.
21. Suwa, M., Gero, J. and Purcell, T. (1998). The roles of sketches in early conceptual design processes. Proceedings of 20th annual conference of the Cognitive Science Society. Hillsdale, NJ:Lawrence Erlbaum Associates, pp. 1043–1048.
22. Suwa, M., Gero, J. and Purcell, T. (in press). Unexpected discoveries and s-invention of design requirements: Important vehicles for a design process. Design Studies.
23. Suwa, M. and Motoda, H. (1994). PCLEARN: A model for learning perceptual-chunks. In Proceedings of the 16th annual conference of the Cognitive Science Society. Hillsdale, NJ:Lawrence Erlbaum Associates, pp. 830–835.
24. Suwa, M. and Tversky, B. (1996). What architects see in their design sketches: Implications for design tools. In Human factors in computing systems: CHI'96 conference companion. New York:ACM Press, pp. 191-192.
25. Suwa, M. and Tversky, B. (1997). What do architects and students perceive in their design sketches? A protocol analysis. Design Studies 18(4):385–404.
26. Tversky, B. (1995a). Cognitive origins of graphic conventions. In F.T. Marchese (Ed.), Understanding images. New York:Springer, pp. 29–53.
27. Tversky, B. (1995b). Perception and cognition of 2D and 3D graphics. In Human factors in computing systems: CHI'95 conference companion. New York:ACM Press, p. 175.

Part III

Formal Aspects of Diagrammatic Reasoning

Introduction

The following part presents selected chapters on formal aspects of diagrammatic reasoning. There are many aspects of diagrammatic reasoning that can (and should) be formalised, including issues in syntax and semantics as well as cognitive and computational aspects of diagram processing. Since cognitive aspects are dealt with in a separate part of the present volume, the following part will focus on syntactic and semantic aspects and their implications for computational implementations.

Much of the work on formalising diagram notations is motivated by the application of diagrams in mathematical reasoning. Though the usefulness of diagrams for the mathematical thought process is rarely doubted and many outstanding mathematicians have devised their own diagrammatic notations for particular domains, pictures are traditionally only accepted as illustrations or thought aids by most working mathematicians. Despite the existence of systems like Euler circles, Venn diagrams, Peirce's existential graphs and Frege's Beweisschrift, the prevailing opinion is that visual methods cannot be accepted as valid devices of mathematical reasoning, since they are not sufficiently formalised. A major line of contemporary research in diagrammatic reasoning therefore attempts to establish the status of diagrams as reasoning devices that are acceptable in proofs and other formal mathematical arguments. Obviously this requires the construction of diagram systems with rigorously defined syntax and semantics as well as the construction of visual proof calculi on the basis of which an analysis of the soundness and completeness of these systems can be given. The prototypical example of a class of diagram systems for which this has successfully been performed are variants of Venn diagrams.

The work by Swoboda in Chapter 21 continues this important line of research. While sound and complete systems of (extended) Venn diagram have been presented previously by other authors elsewhere, this chapter describes for the first time an implementation of such a system on a formal basis, so that the system behaviour is provably correct (up to implementation errors).

Apart from concerns about their formal status, another hurdle for the acceptance of diagrammatic reasoning systems in mathematics is the fact that, like for any other notations, their usage is a skill that can require a considerable amount of learning. Simple systems may be interpreted rather intuitively, but complex systems require a more refined way of interpretation. Partially for this reason, almost all diagrammatic notations that have gained an established place in mathematical discourse are comparatively simple systems of weak expressiveness, such as Euler circles. Notations of higher expressiveness and consequentially of higher syntactic complexity have rarely been adopted by working mathematicians. Shin's work in Chapter 17 shows that the perceived complexity of a diagrammatic notation may sometimes be due to an inadequate way of reading it. She argues that even very expressive diagrammatic systems can be useful tools for deductive tasks, because many implications that have to be inferred explicitly in sentential representations can directly be read off the visual representation. However, the key to this is to find not only a correct, but also an appropriate reading or interpretation of the notation. Shin formalises an alternative reading for Peirce's existential graphs, a complete first-order language, and shows how this reading sheds a new light on the utility of existential graphs as a formal (and comprehensible!) reasoning device.

Given the aim to use diagrammatic notations as mathematical proof devices, diagrammatic theorem proving could be seen as the penultimate touchstone of diagrammatic reasoning in mathematics. Obviously, there are two possible places for diagrams in theorem proving: On one hand, we can try to completely formalise diagrammatic deductions themselves and to prove theorems exclusively by manipulating these diagrams. This would mean that the inference in a theorem prover would still work by elementary syntactic manipulation, but that the (standard) sentential system of representation is replaced by a diagrammatic system. The second possibility is to use diagrams as an addition to classical theorem proving. In such a model the proof proper would still be performed with a classical sentential notation, but a diagrammatic representation of the problem could be used to extract information about *how* the proof can proceed. In this model the use of diagrams in the proof system would be much closer to the traditionally excepted usage of diagrams in mathematical discourse.

The first approach is taken in Chapter 18, where Jamnik describes Diamond, an interactive diagrammatic proof checker. A basic proof in Diamond is an initial diagram and a finite, linear sequence of diagrams that follow from the initial diagram via a certain class of diagram manipulations. A schematic diagrammatic proof is a proof that involves an inductive sequence of statements for which a particular diagram provides one instance of that sequence. The system is able to synthesise some inductive schemata from diagrams of n-dimensional arrays. Proofs in Diamond operate exclusively on the level of syntactic manipulation of diagrams without mapping these to a secondary representation.

The second approach to diagrammatic theorem proving is taken by Barker-Plummer in Chapter 19. He describes a theorem prover, the &/Grover system, which consists of two components: a theorem prover, called &, and a diagram interpretation component, called Grover. While Grover uses diagrams to extract assumptions about which sub-proofs should be performed, the actual formal proof is entirely sentential and performed in a standard way. Essentially, the diagram is used as guidance for the proof search strategy, i.e. as meta-information on the proof.

A diagrammatic theorem prover, as well as any other system that employs diagrammatic input, must somehow interpret the diagrams it is given. In most existing systems, the implementation of such diagram interpretation components is done in an ad hoc fashion. This raises the question whether the formal specification of syntax and semantics of diagram notations and their computational implementation can be brought closer by using declarative, executable specifications for diagram notations. This is the aim of the work by Haarslev et al. presented in Chapter 22. Obviously, any formalisation of diagram notations must combine a way to model the basic spatial properties of the diagram objects with a way to model how diagrams of increasing complexity can be built from simpler constituents. The approach by Haarslev et al. uses an integration of description logic with reasoning on concrete domains to marry an abstract symbolic reasoning mechanism with a qualitative spatial reasoning method. In contrast to what the name seems to suggest, this does not mean that the modelling starts from concrete geometries which could be given, for example, in terms of real-valued arithmetics. Instead, the laws of geometry are modelled as constraints on a set of qualitative spatial conditions such as *inside, outside* etc. The combination of these two types of reasoning leads to a logic that is expressive enough to formalise syntax and semantics of many kinds of diagrams and that can, due to its decidability, directly be used as the basis of diagram interpretation in computational systems. The utility of this method is demonstrated by applying it to the interpretation of visual query languages for spatial information systems.

The choice for a particular modelling of spatial properties is crucial for any kind of diagrammatic reasoning system, because different kinds of notations exploit different kinds of spatial properties. Venn diagrams, for example, use only topological properties, whereas engineering drawings often use concrete metric information. Therefore the choice of spatial modelling deeply influences the applicability of any formalism.

In Chapter 16 Giavitto and Valencia investigate another possible modelling of spatial diagram properties. Their approach is based on the use of combinatorial algebraic topology for spatial modelling. They particularly analyse the use of such models for a task that could be another candidate for a touchstone of diagrammatic reasoning systems: solving visual reasoning problems as they appear in typical IQ tests.

An important part of formalising diagrammatic reasoning is the specification of the dynamic aspects of diagram notations. This is obvious when we define valid inferences in diagrammatic reasoning by admissible transformations of diagram configurations, such as individual inference steps in Venn notations or execution steps in automaton diagrams. Dynamic aspects are also important in several other contexts, for example when the interaction with a diagrammatic system has to be described.

One way to formalise the behavioural and/or dynamic aspects of diagram notations is by rule-based specifications that detail how a diagram can be modified. The work by Meyer in Chapter 15 describes such a rule-based model that attempts to bridge between formalisation and implementation of dynamic diagram systems. This approach introduces a declarative logic language that allows the definition of dynamic aspects of diagrammatic notations by embedding diagrammatic term structures into a Horn clause logic. The promise of such an approach is a highly declarative programming paradigm for the implementation of diagram systems in which the formal specification of a diagram notation can also be used as an executable specification.

Rule-based specifications of diagram dynamics as well as any other formalisation of diagram dynamics that defines "actions" of diagram modification must deal with the problem of embeddings. The question is: Given a part of a diagram that is to be replaced with a new sub-diagram or is to be modified in any other way, how does the new sub-diagram have to be embedded into the context? Traditionally most transformation systems have used some fixed embedding mechanism. It is then the responsibility of the specification designer to utilise this embedding mechanism in a sensible way, i.e. to write only such rules that do not produce undesired or unexpected effects. In Chapter 20 Foo proposes a new approach to this problem. He uses a first-order modelling to derive a notion of local extent in diagrams, which allows to automatically compute diagrammatic regions that can safely be manipulated in isolation. Such an approach could not only be used to improve the specification of diagrammatic systems, but it could also prove beneficial in the implementation of intelligent diagrammatic systems that have to support user interaction.

The chapters in this part can only highlight a few selected approaches that are representative of major current trends in the formalisation of diagrammatic reasoning. No book could cover every possible aspect of the formalisation of diagrammatic reasoning – after all almost any kind of reasoning could be performed diagrammatically. The issue is complicated further by the fact that we do not yet have an answer to the question of what are the proper building blocks for a unified theory of diagrammatic reasoning. Currently we are faced with a collection of several theories which handle different individual aspects. It will be a matter for future research to clarify their relations and interconnections.

15. Diagrammatic Evaluation of Visual Mathematical Notations

Bernd Meyer

This chapter discusses the specification of diagrams and diagram transformations with picture logic, a visual Horn clause language. Formalisation techniques for diagrammatic languages have previously mainly been investigated for the specification of *static* visual syntax. For reasoning about many types of diagrams, however, a formalisation of their *dynamic* aspects is indispensable. This is particularly true for many diagrammatic mathematical notations, because their evaluation rules or consequence relations correspond to visual or graphical transformations. The chapter presents constraint-based extensions of picture logic which render it suitable for the specification of such diagram notations and the required transformations.

15.1 Diagrammatic Reasoning and Formalisations

Despite the fact that visual reasoning and visual intuition is often one of the major factors in mathematical discovery, the prevailing conviction among mathematicians is that visual methods can only serve as inspiration for discovery and as illustrations for proofs but never as valid arguments or proofs themselves, because they are not based on a well-defined, closed set of reasoning methods. Nevertheless, the potential that visual expression holds for mathematical language and even for mathematical proofs has been realised and has been demonstrated several times. A superb collection of examples is [1], but even the introduction to this book states that "of course, 'proofs without words' are not really proofs".

However, in recent years there has been an increasing amount of discussion on whether the rigorous dismissal of pictures as formal mathematical devices is really justified. For a discussion of these points see [2–5]. It is certainly clear that visual notations which we intend to use as valid mathematical devices need a proper formalisation. Examples of such notations are Venn diagrams and Euler circles, Peirce's α-β-calculus, various visual notations for Church's λ-calculus [6, 7], but also more exotic notations, such as boundary logic [8]. As a general principle, the same holds for more technical notations, for example specification languages, such as state charts [9], and visual programming languages, such as Pictorial Janus [10]. Such notations can only be fully valid tools with the same rights as their linguistic counterparts, if we can formalise them adequately.

Some ground-breaking approaches have attempted to establish the status of diagrammatic notations as fully valid mathematical reasoning devices by rigorously formalising them [2,3,11,12]. Beyond doubt, this represents a big leap forward towards establishing diagrammatic notations as valid mathematical systems. However, each of these approaches was aimed at a particular diagrammatic system and the formalisation methods chosen were targeted towards this specific system. A general framework or meta-language for the definition of diagrammatic mathematical notations has not yet been established. From a computational perspective it is important to find such a framework, since it gives rise to general methods for the computational implementation of diagrammatic notations.

The present chapter explores the possibility of such a framework. It presents a visual logic programming language for diagram handling and discusses its application to the definition and evaluation of visual mathematical notations. A logic-programming based approach seems an excellent candidate for this framework, since it bridges between mathematical reasoning and computational implementation. Logic-based specification methods have already proven their suitability for linguistic representations and their realm can be extended to the domain of diagrams by making pictures a new domain for logic programming.

An important property of our specification language is that it is a diagrammatic language itself. In general, we can distinguish between reasoning *about* diagrams and reasoning *with* diagrams. An integration of both directions into reasoning about diagrams with visual methods seems promising, for if we aim at using diagrams as reasoning tools, one of their natural places should be where the domain of reasoning is diagrammatic itself. Such an integration of reasoning with diagrams about diagrams is given in the framework discussed here.

15.2 Formalisation of Static Diagram Notations

Extensions of logic grammars have been used before to specify visual languages (e.g. in [13–16]). Our approach extends this idea and provides a more general integration of logic programming with visual expressions by integrating diagrammatic structures as first-class data into a full constraint logic programming framework. This integration, called *picture logic*, has undergone several revisions in recent years. It was born as a visual set rewriting mechanism integrated into Prolog [17] based on picture matching instead of unification. Full picture unification [18,19] was added later to allow a better integration with logic programming. Experiences with this formalism have shown that standard logic programming is only sufficient as the basis of the framework, when its usage is restricted to syntactic specification of static diagrams. When we attempt to specify dynamic aspects of diagram languages, i.e. the way diagrams are changed or modified, it turns out that ***constraints*** play a vital role. In general, spatial or geometric constraints can be used to automatically propagate changes of one part of a diagram to other parts. As a simple example consider a Venn diagram in which a circle has to be moved. In order to maintain the semantics of the diagram, it is necessary to move other circles accordingly, so that the intersections are maintained. The best way to capture such a behaviour is by using constraint-based specifications.

However, in a straightforward naive extension of picture logic to a constraint-logic based framework, dynamic diagram notations can still cause problems. This is because in a rule-based approach the manipulation of a diagram can lead to temporary inconsistencies in intermediate transformation stages which can only be resolved in a later transformation phase. In a logic-based framework, such inconsistencies, even if only transitory, are prohibitive, since a derivation must not lead to inconsistent states. Technically, in a logic-programming framework, the derivation would fail as soon as an inconsistency occurs and it would not be possible to resolve it at a later derivation stage.

The extension of picture logic to a constraint-based framework therefore consists of two steps: (1) a transition from logic programming to constraint logic programming allows to handle spatial relations properly; (2) a meta-programming technique is used to establish a transaction concept which allows to avoid transient inconsistencies when rewriting a diagram.

We first review the basic framework of picture logic. The general idea behind it is to introduce a new kind of term structure for the description of diagrams into logic programming. These new terms, called ***picture terms***, have to be regarded as partially specified example pictures, much like normal terms in logic programming are partially specified terms. Picture terms are diagrams themselves and can be unified with other diagrams. By embedding picture terms into a logic programming language we obtain a diagrammatic logic language for the specification of diagrams.

Picture terms consist of two types of entities: visual constants (picture objects) and visual variables (for picture objects).[1] The entities in a picture term are in implicit spatial relationships that can be inferred from the depiction of the term.

The underlying formal model for picture terms are graphs with typed object nodes and typed relationship edges according to the definition of a picture vocabulary $V = (OT, RT)$ of graphical object types OT and spatial relation types RT. Assuming the vocabulary $(\{circle, line, label\}, \{touches : circle \times line,\ attached : label \times line\})$, Fig. 15.1 shows a picture term and the picture term graph to which it corresponds.

[1] We adopt the convention that variable names start with upper-case letters and constants with lower-case letters.

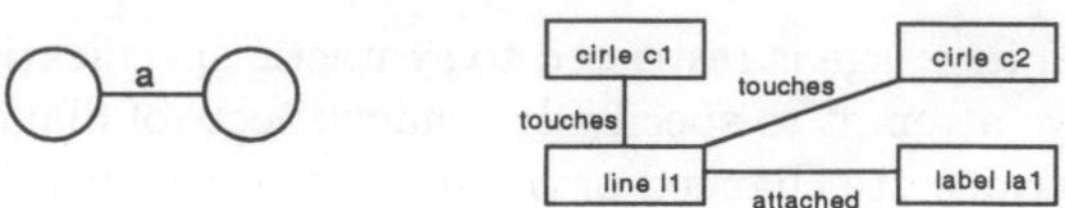

Figure 15.1. Picture term graph.

The entities explicitly denoted in the picture term are called the "foreground". Every foreground object can be equipped with an arbitrary number of attributes. A *circle c*, e.g., could have attributes *c.centre* → *point* and *c.radius* → *real*. Attributes can be any (possibly partial) data structure carrying additional information for either geometry handling or semantic interpretation purposes.

Two additionally types of special variables can be used in picture terms if we need to refer to the context of the foreground objects: a "background" variable (depicted as a solid frame around the term) can be used to denote an unknown object context. This is the analogue of, for example, a list rest in normal logic programming. A second context variable, called the "frame", is used to denote the set of spatial relations between objects in the background and foreground objects. It is depicted as a dashed frame around the term.

Picture terms are integrated with logic programming by defining a second kind of unification that is applied to picture terms. This unification essentially performs a subgraph unification of two picture term graphs by finding a variable substitution π (called projection) for picture object variables that makes two picture term graphs identical up to some context which is contained in the background variable. It essentially solves the following simplified equation for π:

$$\pi(P \oplus B \oplus F) = \pi(P' \oplus B' \oplus F')$$

where $< P, B, F >$ and $< P', B', F' >$ are the picture terms to be unified. P (P') is the graph corresponding to the explicitly given picture objects and their relations (the foreground), B (B') is the graph corresponding to the background, the frame F (F') is the set of relation edges connecting nodes in P (P') with nodes in B (B'), π is the variable projection, and $\oplus$ is a merge operation for graphs. It is important to note that both context variables, background and frame, are partial data structures; i.e., an existent background or frame can be extended by new objects or relations during unification.

With picture unification, picture terms can be used in a logic program anywhere a normal term can be used. The rule in Fig. 15.2, for example, is taken from a specification that defines the language accepted by a non-deterministic finite state automaton solely by applying visual transformations. Here, it can be observed how the transition to a diagrammatic language makes the semantics of the specification apparent.

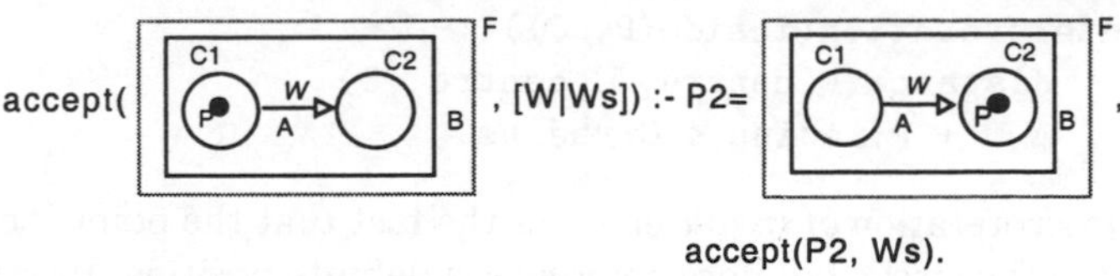

Figure 15.2. A picture logic rule.

One advantage of such a specification is that it can be used for animation of diagrams without extra costs. If, for example, the instantiation of the first argument of *accept* is visualised for each inference step, then an animated execution of an NFA results.

This is the point where geometry and therefore arithmetic constraints come into play: When the above rule is applied, which moves the marker from one state to the next state, the actual coordinates of the new marker position are unknown. Since formally the transformation of the picture is only given by a transformation of the corresponding picture term graph, we only know that the point P is no longer inside of $C1$ but now inside of $C2$. Picture term graphs define only abstract spatial relations; nothing is known about the absolute coordinates which are needed if the term has to be visualised. Of course, in this trivial case it would be easy to assign the new coordinates of P by simply computing the centre of $C2$, if its coordinates are known. However, in general this is not a good idea for a number of reasons: (1) Viewing the computation of the new coordinates as an assignment is not well aligned with the paradigm behind a logic language, since we should not assume a procedural model of computation and in particular no order of execution. (2) Even if we were willing to accept this, arithmetics is poorly integrated with logic programming and conflicts with back-tracking, unbound variables, etc. (3) In most cases it is not possible to calculate the new or changed geometric properties on the spot, because the problem is underspecified. In the above case, for example, we would only know that P has to be inside of $C2$ but not where exactly it is located within $C2$. Putting it simply into the centre might generate a layout that conflicts with other objects which could be put into $C2$ later.

Obviously a more general form of integration with geometry is required, and this comes quite naturally with the transition to a constraint logic framework. Using constraint logic programming (CLP), we can give the abstract spatial relations a geometric semantics by interpreting them as arithmetic constraints on the geometric attributes of the involved objects.

Every relation type in the vocabulary can either have an interpretation or be uninterpreted (in which case it is handled as a purely abstract relation as above). An interpretation is defined by a normal CLP-clause involving arithmetic constraints. The *inside* : *point* × *circle* relation is, e.g., interpreted as:

```
interpretation(inside(P, C)) :-
    distance(P.centre, C.centre, D)
    & D + P.radius < C.radius.
```

This interpretation of *inside* enforces the fact that the point is completely contained in the circle, but does not give it a definite position. Interpretations can also be symbolic, not directly involving any constraints:

```
interpretation(left_touching(X,Y)) :-
    touching(X,Y) & left_of(X,Y).
```

During the unification of pictures the interpretations of relations are enforced. For every relation in a picture that is the result of some unification the interpretation attached to this relation is evaluated. By this the unification is effectively constraining the values of the attribute variables of the involved objects. Our language is based on a logic programming language that directly handles constraints over real intervals, CLP(RI) [20], so that we can leave the actual constraint solving entirely to the underlying CLP language.

With interpretations we have introduced a new, additional source why picture term unification can fail: if the interpretations of some relations in the picture are not compatible, the underlying constraint solver will detect the inconsistency and reject the unification. For example, without interpretations, a picture term could well contain the relations $inside(X,Y)$ and $outside(X,Y)$ at the same time, because they are handled as "meaningless", unrelated predicates. With the appropriate interpretation the contradiction will be detected and the CLP derivation will simply fail (or back-track) at this point. In short, if every relation has an appropriate geometric interpretation no geometrically inconsistent pictures can occur.

A system of constraints can either be fully constrained, in which case it has exactly one solution, under-constrained, in which case it has more than one solution, or over-constrained so that it does not have any solutions. In general, it is very difficult to write diagram specifications that produce only fully constrained geometries. The main reason for this is conflicting objectives in the layout. For example, we want the tokens in a Petri net diagram to be in the centre of the place marker. However, if there is more than one token in the place, only one of them can be in the centre and the others have to be moved to different positions. For such reasons, the specification will usually either be over-constrained or under-constrained.

For an over-constrained system to produce a solution, some of the conflicting constraints have to be relaxed automatically during the derivation. This can, for example, be achieved by extending the CLP paradigm with hierarchical constraints such as in the HCLP framework [21,22]. As yet we have not investigated methods to support over-constrained specifications in our framework. A good overview of recent research into over-constrained systems can be found in [23].

Another way to avoid over-constrained systems is to intentionally remove the possible conflicts from the specification. This results in under-constrained solutions. Under-constrained systems are of no harm to the derivation, but they do not produce sufficiently instantiated geometric variables for a concrete layout of the picture. If an under-constrained picture has to be displayed, the geometric attributes have to be instantiated with concrete values first. In the simplest case the underlying constraint solver can be forced to instantiate the geometric attributes with concrete values. Since the CLP(RI) solver works on real intervals, such a function can readily be achieved by an interval-splitting algorithm and is indeed part of the CLP(RI) system. Of course, this will only produce a rather arbitrary layout that is consistent with the given constraints. There is no control over which layout is produced if several are possible. Therefore, it will in general be necessary to use specialised domain-specific layout modules to generate meaningful and aesthetically pleasing layouts for under-constrained pictures. For some application areas, in particular for graphs [24], such layout techniques are well investigated. However, in general automatic layout is a hard problem and computationally expensive. Since layout methods are typically highly specialised, they cannot readily be integrated with the basic framework. Instead, their use has to be confined to a dedicated output phase in which highly specialised algorithms can be used.

15.3 Formalisation of Diagrammatic Evaluation

We now take a closer look at the constraint-based extensions in the context of picture transformations. We have already said that a picture in which all relations are interpreted can no longer contain any inconsistencies. Unfortunately, there are cases where *transient* inconsistencies are unavoidable. We will illustrate how such transiently inconsistent states occur during the diagrammatic evaluation of visual mathematical expressions.

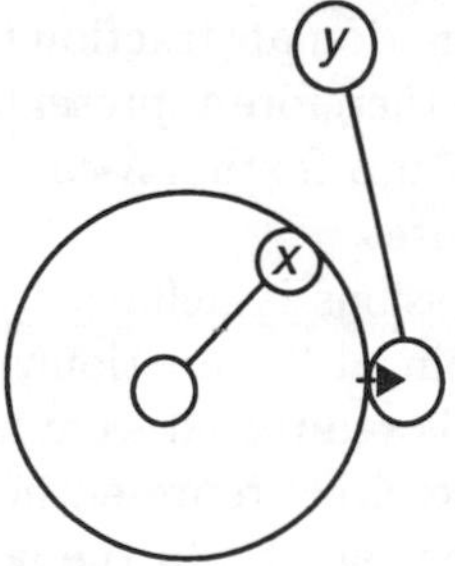

Figure 15.3. The VEX expression $(\lambda x.x)y$.

When formalising evaluation rules or consequence relations for visual mathematical notations, we often have to manipulate several graphical objects in a single step. In a rule-based approach such transformations must often be expressed in several rules each of which transforms only a part of the object set concerned. Intermediate transformation results, in which some of the objects are already modified, while others still have their original form or properties, may well be meaningless in terms of the original diagram notation, because the notation as such considers the whole transformation as an atomic step. In terms of the underlying constraint system, such transitory states may even be inconsistent. Semantically this is not a problem, since we know that at the end of the transformation step a consistent state will be restored. However, a CLP derivation cannot work with an inconsistent state, even if we can guarantee that it is only transient. In consequence, a *transaction concept* is required, where the application of some set of transformation rules is regarded as an atomic transformation. Upon entering and exiting the transaction, the diagrams must be consistent, but within the transaction intermediate states are allowed to relax the constraints as long as a well-defined state is restored at the end of the transaction. We will use the visual λ-calculus VEX to illustrate this point.

A complete discussion of VEX is beyond the scope of this chapter and can be found in [6]. A VEX expression consists only of circles and lines. Textual labels may be used to improve readability, but they do not have any semantic meaning. In VEX an isolated circle represents a variable. A functional abstraction is represented by enclosing its body completely in another circle which does not touch any circles that belong to the body. The formal parameters of an abstraction are given as circles touching the abstraction circle from the inside.[2] Lines which connect circles are used to declare the identity of two graphical objects. A special rule for variables says that their corresponding circle has to be depicted on the level of graphical inclusion on which the variable is free. This graphical instance is called the root of a variable. Copies of the variable may appear in arbitrary positions of the expression and are connected to the root by identity lines. A function application is depicted by letting the applied function and the operand touch from the outside and by drawing an arrow from the function abstraction to the actual parameter. The VEX expression in Fig. 15.3 therefore represents the expression $(\lambda x.x)y$. A complete translation of VEX into textual λ-calculus can be specified in only eight simple picture logic clauses.

Evaluation of VEX expressions is defined by graphical transformations. Informally speaking, the graphical β-reduction rule for VEX is defined in the following way: (1) the arrow is redirected from the functional abstraction to the formal parameter; (2) the circle representing the functional abstraction is removed; (3) the formal parameter and the actual parameter are merged;

[2] We are using a pure version of VEX here that allows only abstractions over single parameters.

and (4) variable links are shrunk accordingly. The order of these steps is irrelevant. The complete β-reduction of $(\lambda x.x)y$ is given in Fig. 15.4.

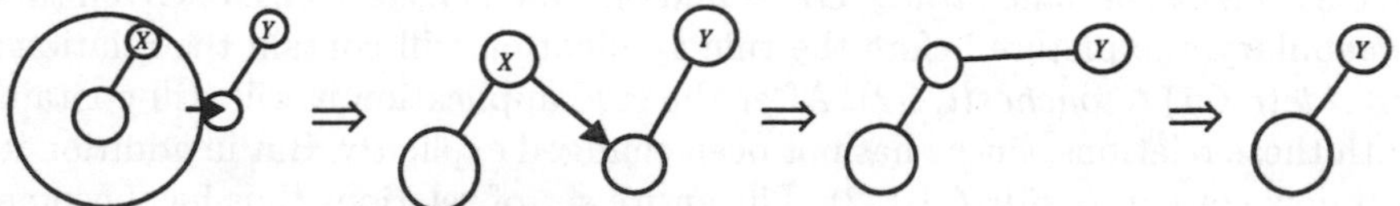

Figure 15.4. VEX β-reduction.

Here we can observe two interesting facts: (1) when performing the steps of a β-reduction separately, the intermediate states are not *semantically* well-defined diagrams; and (2) intermediate steps obtained by the manipulation of single objects could even generate *geometrically* ill-defined pictures. Let us look at the second case more closely. Assume we had chosen a different order of steps and we are merging the formal parameter with the actual parameter before removing the circle of the functional abstraction. This is basically done by removing the formal parameter and substituting the actual parameter in its place. A naive approach to solve this problem is given by the rule in Fig. 15.5. Remember that the two rectangles labelled B and F are not picture objects but the context variables for background and frame.

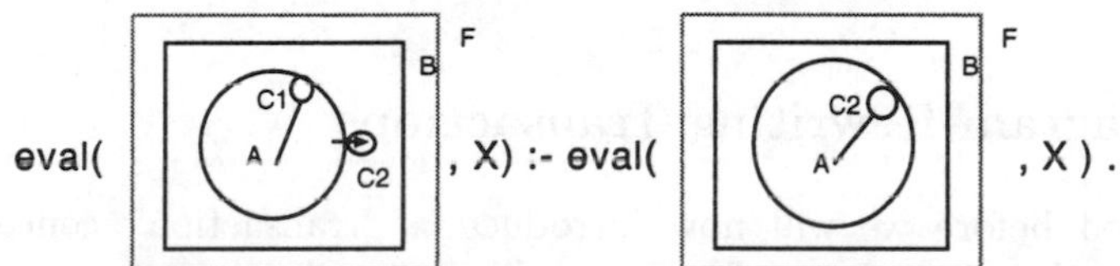

Figure 15.5. A naive approach to define β-reduction.

We apply this rule to the VEX diagram given in Fig. 15.3. Since picture unification automatically maintains the context relations between picture elements, the line from the root node y to the actual parameter will automatically follow the merging, because the actual parameter $C2$ to which it is connected is moved. Thus the pictures before and after the application of this rule are given by Fig. 15.6.

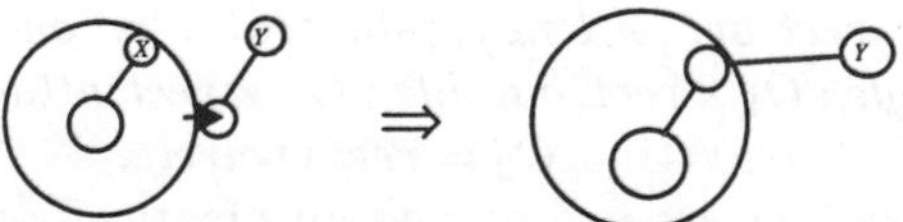

Figure 15.6. Diagram instances before and after rule application.

What happens to the line connecting the root node y and the actual parameter? Call this line l. A reasonable picture vocabulary for VEX contains the spatial relations $touches : line \times circle$, $inside : line \times circle$, $inside : circle \times circle$, $outside : line \times circle$ and $outside : circle \times circle$. Given this vocabulary, the picture before the rule application will contain the relations $outside(l, C1) \wedge touches(l, C2)$. After the rule application it will still contain both these relations, since l has not been changed explicitly. But in addition it will now contain $inside(C1, C2)$. The entire set of relations thus has become geometrically inconsistent: $outside(l, C1) \wedge touches(l, C2) \wedge inside(C1, C2)$. If the spatial relations are interpreted geometrically, and we have argued above that they have to be interpreted, this geometric inconsistency will be detected by the constraint solver and the derivation will be rejected. The rule is therefore not applicable.

Note that this is only a local conflict that would be resolved at the end of the transformation once the functional abstraction circle is removed. While one can find workarounds to avoid generating this particular inconsistency, this is not always possible in the general case. For a meaningful way of specifying diagram transformation, it must be possible to manipulate spatial constraints individually. In a rule-based approach this necessarily introduces the risk of producing inconsistent intermediate states. More importantly, the above rule expresses quite precisely and directly the transformation we have in mind so that it would be preferable if we can extend our framework to accommodate such transformations.

15.4 Diagram-Rewriting Transactions

As outlined before we will now introduce a "transaction" concept that can handle the kind of transformation discussed above by temporarily relaxing the spatial constraints. For the sake of a more concise discussion we will use a simplified example diagram language *boxes* in the following. The sentences of *boxes* consist of rectangles and circles. Circles are connected to a single "parent" rectangle by a line, and circles sharing the same parent may be interconnected by arrows. The desired transformation of *boxes* is to move all circles into their parent boxes while maintaining their interconnections. Thus the picture on the left-hand side of Fig. 15.7 is a sentence of *boxes* and the right-hand side is the desired transformation of this sentence. As above, a naive approach to formalise the transformation consists only of the two clauses in Fig. 15.8 with the vocabulary $(\{circle, rect, arrow, line\}, \{startsat : arrow \times circle, endsat : arrow \times circle, inside : O_1 \times rect, outside : O_1 \times rect, attached : line \times O_2\})$, where $O_1 = circle \cup arrow \cup line$, $O_2 = rect \cup circle$.

Since context relations are maintained automatically we do not need to worry about handling the arrows at all. Just swapping the boxes inside of their parent box, the arrows are following them automatically, since the *startsat*

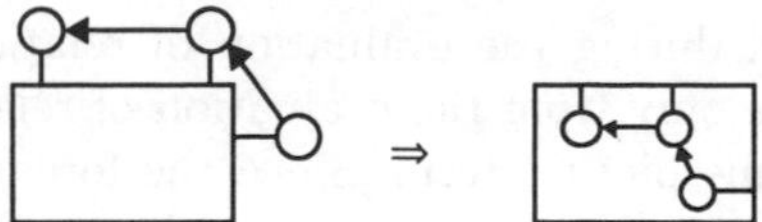

Figure 15.7. The example language.

and *endsat* relations are maintained. However, we are facing the same problem as above. When moving the circles one by one the first rule application generates the state given in Fig. 15.9. Thus the picture contains the original relations $outside(A, R) \wedge endsat(A, C)$ and the new relation $inside(C, R)$ which are contradictory.

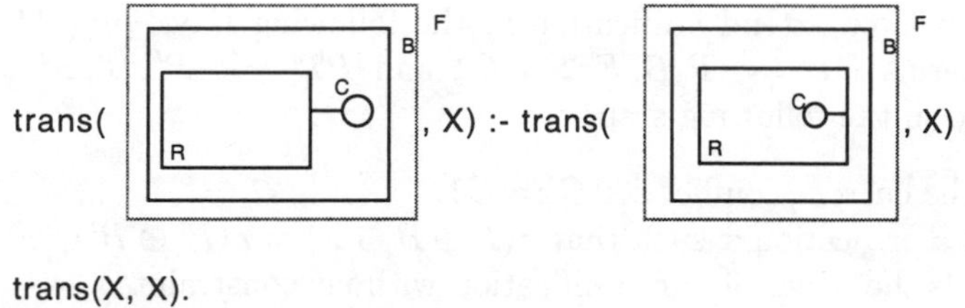

Figure 15.8. A first approach to formalise the *boxes* transformation.

The basic idea for the transactions is to define an extension of picture unification which can succeed despite of conflicts. Of course, unification must remain a consistent operation. Therefore the conflicts have to be identified and have to be resolved. This can be done by letting the unification compute the conflict set and by having it remove these conflicts from the picture terms. The conflict set must explicitly be yielded as a result of the unification so that the remainder of the transaction can resolve the conflicts.

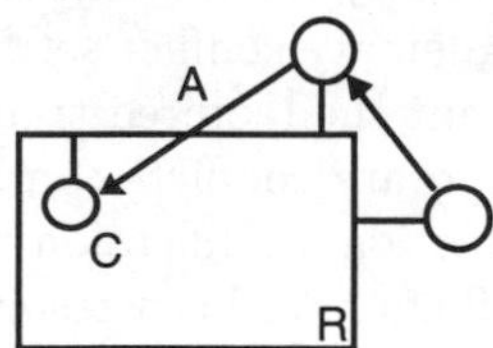

Figure 15.9. Intermediate state in the *boxes* transformation.

In order to achieve this we extend the notation of picture terms. Let P be some picture term, then the term $P\&C$ denotes the picture term and its associated conflict set C. Unification of two picture terms without conflict sets remains the same as above, but it behaves differently with conflict sets. First we have to analyse where conflicts can be detected: (1) Conflicts can only be detected during the evaluation of relation interpretations. (2) No conflicts can arise only from the evaluation of relations directly given in the foreground of some picture term. Since the foreground has a consistent geometric interpretation (it was given as a picture in the first place!), the relations in the foreground must have a consistent spatial interpretation. (3) Conflicts therefore have to be detected during the interpretation of relations in the background or frame.

If a conflict between two relations $r_1 \wedge r_2$ occurs and neither of them is in the foreground we have to define which of them is preferred, i.e., will be kept in the picture term. The less preferred relation will be put into the conflict set. This is achieved by defining a partial order over the relation types in the vocabulary so that r_1 is preferred over r_2 if $r_1 < r_2$. For the purpose of our example it is sufficient to define ***startsat, endsat*** $<$ ***inside, outside.***

We can now extend unification in the following way: $unif(PT, PT')$ for picture terms $PT =< P, B, F > \&C1$ and $PT' =< P', B', F' > \&C2$ is computed in the following steps:

1. Set the current conflict set $C := C1$.
2. Find a projection π such that $\pi(P \oplus B \oplus F) = \pi(P' \oplus B' \oplus F')$. This is the usual picture unification without constraints.
3. Evaluate the interpretations attached to all the relations in P'.
4. For every interpreted relation r in B', in the sequence defined by their partial order, evaluate its interpretation. If the evaluation fails, a conflict is detected. In this case let $B' := B' - \{r\}$ and $C := C + \{r\}$.
5. Repeat the last step with F' in place of B'.
6. If $C2$ is a variable or $C = C2$ then succeed with $C2 := C$.
7. Otherwise fail.

Note that unification with conflict sets has become a directed operation. New conflicts can only be introduced in PT' but not in PT. The conflict set is an aggregative structure and $C2$ will contain all conflicts from $C1$ plus the new conflicts that were caused by the unification. A conflict-free picture can still be enforced by using an empty conflict set for the second argument. We now have simple means to put the required transaction concept into action: during the transaction temporary conflicts can be gathered in a conflict set and at the end of the transaction a unification with an empty conflict set is used to ensure that all conflicts have been resolved and a consistent picture state has been restored.

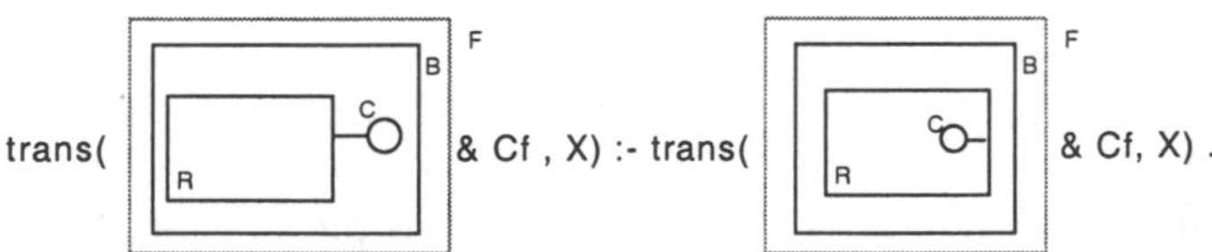

Figure 15.10. Revised transformation rule for *boxes*.

Let us now look at the transformation of our example language *boxes* again. If we extend the first of the above transformation clauses with conflict sets as shown in Fig. 15.10, it is easy to verify that after it has exhaustively been applied the required transformation is achieved, but the conflict set contains the relations $outside(A, B)$ for all arrows A and their respective parent boxes B. Thus the original second clause, which has an empty conflict set, is not applicable. It would be possible to simply drop the conflict set when calling the second clause, but any control would be lost over whether the conflict set indeed contains only the anticipated conflicts. We therefore introduce a special predicate *resolve* for controlled conflict resolution. Its first and fourth arguments are picture terms together with their conflict sets, the second argument is a picture term describing a conflict item, and the third argument is a picture term describing how the conflict is resolved. *resolve* matches all relations which are given in the second argument with the conflict set of the first argument. If the match is successful it removes the matched relations from the conflict set of the first argument and adds all relations given in the third argument to the picture term of the first argument. The modified picture term and the modified conflict set are returned in the fourth argument. *resolve* fails, if the match of the relations in the second argument with the conflict set of the first argument fails. Using this predicate we can replace the original productions as shown in Fig. 15.11.[3] *resolve* is used to move all arrows with *outside* conflicts from the outside of the rectangle to the inside. The new production set will now perform the intended transformation. The derivation would still fail as required if additional conflicts occurred that have not been defined in the specification. In this way global consistency can be guaranteed.

As we can see from the abstracted example we now have the means to define diagram transformations like those that occur in the β-reduction of the visual λ-calculus VEX. The combination of constraint-based extensions and resolution of transient conflicts therefore makes picture logic applicable to the definition and evaluation of the diagrammatic mathematical notations we were looking at.

15.5 Implementation

Basic picture logic was implemented as a fully interactive graphical system consisting of a Lisp-based front end and a Prolog-based back end that serves as a compiler and a runtime environment. Diagrammatic input can be given with an object-oriented graphics editor or it can be sketched on a pen tablet. Prototyped versions of the constraint-based extensions have been implemented in CLP(RI), but are not fully integrated with the interactive environment.

[3] The notation "P1:Picture-Term" used in the second clause introduces the variable $P1$ as a shorthand for the depicted term.

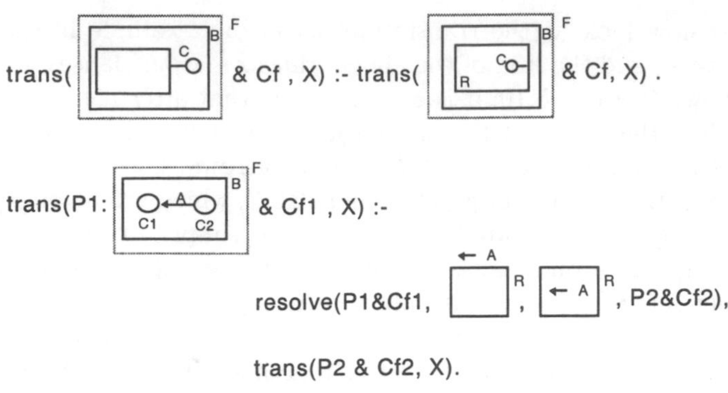

trans(Pic & nil, Pic).

Figure 15.11. Resolving anticipated conflicts after the transformation.

Since the underlying unification mechanism utilises full graph unification (implemented by set partitioning), high execution costs are incurred. Particular attention has to be paid to the fact that picture unification is a non-deterministic operation, since it is impossible to define a unique most general unifier. While full picture unification is desirable for use as a specification language, it is actually not utilised in most picture logic programs. In the cases we have explored, it is sufficient if unification deterministically yields a single unifier. Significant speed improvements have been achieved by introducing committed choice unification which implements this restriction. Since normal constraint logic programming languages are unable to delete constraints, we have to simulate this by meta-programming techniques which require frequent re-evaluation of constraints. A better approach would be based on more recent constraint solvers that allow to delete constraints [25]. This integration promises significant performance improvements.

15.6 Related Work and Conclusions

A logic approach to diagram specification based on description logic extended by concrete domains is presented in [26]. While this is a powerful method for the specification and analysis of diagrams, it is not aimed at transformation or animation. In [27] a spatial logic based on Clarke's point calculus [28] is presented and a specification of the visual programming language Pictorial Janus is given, but only single execution steps can be specified, since the framework does not have an underlying notion of state or sequence. This approach does not lend itself easily to an implementation, since it is based on full first-order logic and point sets.

Logic-based approaches to visualisation are either oriented towards generating layouts for static pictures [29] or based on procedural notions of algorithm animation [30], whereas we are aiming at animation as an automatic side effect of the declarative specification of diagram languages.

There is a large body of work on the formal specification of visual languages based on grammars [31], but grammars appear unsuitable as a basis of reasoning, because they have limited expressiveness and are aimed at one-step processing tasks like parsing. Dynamic transformations are difficult to capture in such approaches. However, in [32,33] we have shown how multidimensional grammars can formally be mapped into a constraint logic framework. This promises to be a basis for the integration of grammar-based and logic-based approaches.

We have presented picture logic, a visual logic language for handling diagrams. Its major distinguishing features are that it uses the expressive power of visualisation within the specification formalism itself and that the same specification framework can be used for specification, translation, transformation, and animation of diagrams. We have in particular looked at the evaluation and animation of visual mathematical expressions as a special application area and have defined constraint-based extension of picture logic that allow handling of this domain. In the same way consequence relations in visual logic notations like Peirce's α-β-calculus or Shin's extended Venn diagram systems should be formalisable. We will continue to investigate the usage of picture logic for the specification of such visual notations and hope that it will prove useful as a general meta-language.

References

1. Nelsen, R.B. (1993). Proofs without words. Washington, DC: Mathematical Association of America.
2. Allwein, G. and Barwise, J. (Eds) (1996). Logical reasoning with diagrams. New York: Oxford University Press.
3. Shin, S.-J. (1995). The logical status of diagrams. Cambridge, UK: Cambridge University Press.
4. Brown, J.R. (1999). Philosophy of mathematics: An introduction to the world of pictures. London: Routledge.
5. Glasgow, J., Narayanan, N.H. and Chandrasekaran, B. (Eds) (1995). Diagrammatic reasoning. Menlo Par, CA: AAAI Press / Cambridge, MA: MIT Press.
6. Citrin, W., Hall, R. and Zorn, B. (1995). Programming with visual expressions. In IEEE workshop on visual languages. Darmstadt, Germany. Los Alamitos, CA: IEEE Computer Society Press, pp. 294–301.
7. Keenan, D. (1995). To dissect a mockingbird: A graphical notation for the lambda calculus with animated reduction. Technical report, Smalltalk Computing.
8. Bricken, W.M. (1988). An introduction to boundary logic with the losp deductive engine. Research report, University of Washington.
9. Harel, D. (1988). On visual formalisms. Communications of the ACM 31(5):514–530.

10. Kahn, K.M. and Saraswat, V.A. (1990). Complete visualizations of concurrent programs and their execution. In IEEE Workshop on visual languages, Los Alamitos, CA: Skokie, IL. IEEE Computer Society Press, pp. 7–15.
11. Hammer, E. (1996). Representing relations diagrammatically. In G. Allwein and J. Barwise (Eds), Logical reasoning with diagrams. New York: Oxford University Press.
12. Hammer, E. and Danner, N. (1996). Towards a model theory of diagrams. In G. Allwein and J. Barwise (Eds), Logical reasoning with diagrams. New York: Oxford University Press.
13. Tanaka, T. (1991). Definite clause set grammars: A formalism for problem solving. Journal of Logic Programming 10:1–17.
14. Marriott, K. (1994). Constraint multiset grammars. In IEEE symposium on visual languages, St Louis, MO. Los Alamitos, CA: IEEE Computer Society Press.
15. Ferrucci, F., Pacini, G., Tortora, G., Tucci, M. and Vitiello, G. (1991). Efficient parsing of multidimensional structures. In IEEE workshop on visual languages. Kobe, Japan. Los Alamitos, CA: IEEE Computer Society Press, pp. 105–110.
16. Bolognesi, T. and Latella, D. (1989). Techniques for the formal definition of the g-lotos syntax. In IEEE workshop on visual languages, Rome. Los Alamitos, CA: IEEE Computer Society Press, pp. 43–49.
17. Meyer, B. (1992). Pictures depicting pictures: On the specification of visual languages by visual grammars. In IEEE workshop on visual languages, Seattle, WA. Los Alamitos, CA: IEEE Computer Society Press, pp. 41–47.
18. Meyer, B. (1993). Logic and the structure of space: Towards a visual logic for spatial reasoning. In D. Miller (Ed.), International symposium on logic programming, Vancouver. Cambridge, MA: MIT Press.
19. Meyer, B. (1994). Visual logic languages for spatial information handling (in German). Doctoral thesis, FernUni Hagen.
20. Older, W. and Vellino, A. (1990). Extending prolog with constraint arithmetic on real intervals. In Canadian conference on electrical and computer engineering, Ottawa.
21. Borning, A., Maher, M., Martindale, A. and Wilson, M. (1989). Constraint hierarchies and logic programming. In G. Levi and M. Martelli (Eds), International conference on logic programming. Cambridge, MA: MIT Press, pp. 149–164.
22. Wilson, M. and Borning, A. (1993). Hierarchical constraint logic programming. Journal of Logic Programming 16(3,4):277–318.
23. Jampel, M., Freuder, E. and Maher, M. (Eds) (1996). Over-constrained systems. Berlin: Springer.
24. DiBattista, G., Eades, P., Tamassia, R. and Tollis, I.G. (1999). Graph drawing. Upper Saddle River, NJ: Prentice Hall.
25. Borning, A., Marriott, K., Stuckey, P. and Xiao, Y. (1997). Solving linear arithmetic constraints for user interface applications. In UIST'97: ACM symposium on user interface software and technology, Banff, Canada, October.
26. Haarslev, V. (1998). A fully formalized theory for describing visual notations. In K. Marriott and B. Meyer (Eds), Visual language theory. New York: Springer, pp. 261–292.
27. Gooday, J.M. and Cohn, A.G. (1996). Using spatial logic to describe visual programming languages. Artificial Intelligence Review 10:171–186.
28. Clarke, B.L. (1981). A calculus of individuals based on "conncection". Notre Dame Journal of Formal Logic 23(3):204–218.
29. Kamada, T. and Kawai, S. (1991). A general framework for visualizing abstract objects and relations. ACM Transactions on Graphics 10(1).

30. Takahashi, S., Miyashita, K., Matsuoka, S. and Yonezawa, A. (1994). A framework for constructing animations via declarative mapping rules. In IEEE symposium on visual languages. St Louis, MO, pp. 314–322.
31. Marriott, K., Meyer, B. and Wittenburg, K. (1998). A survey of visual language specification and recognition. In K. Marriott and B. Meyer (Eds), Visual language theory. New York: Springer, pp. 5–86.
32. Meyer, B. (1999). Constraint diagram reasoning. In Principles and practice of constraint programming: CP'99. Alexandria, VA, October. New York: Springer.
33. Meyer, B. (2000). A constraint-based framework for diagrammatic reasoning. In Applied artificial intelligence 14(4):327–344.

16. A Topological Framework for Modelling Diagrammatic Reasoning Tasks

Jean-Louis Giavitto
Erika Valencia

In this chapter we propose to model some diagrammatic reasoning tasks within a topological framework. We only focus on problems that require a specific diagrammatic reasoning procedure to be solved rather than on reasoning on diagrams. To model such cognitive tasks, we propose to represent the problem with topological objects and to model the diagrammatic operations performed on them as topological operations. The idea underlying this proposition is that *combinatorial algebraic topology* is an adequate and unifying framework to specify and analyse diagrammatic representations and reasoning. To illustrate such a proposal, we present here three applications of this topological framework: the first concerns a categorisation problem, the second deals with hierachy restructuring, and the last one is the ESQIMO system for simple intra-domain analogy solving in unsupervised IQ tests.

16.1 Introduction

Several issues are addressed in the fields of diagrammatic reasoning (see [14] for an excellent introduction), e.g., visual formalism [19], diagrammatic inference [17, 28], diagrammatic approach of logic [5, 33], qualitative physics [11], cognitive issues [1, 13, 14], logical formalisation of spatial relationships [6, 15, 18, 29].

The last example accounts for the search of a formal theory of diagrammatic representations. A unique conceptual framework cannot encompass simultaneously all the issues investigated in the field of diagrammatic reasoning. However, it is possible to develop a formal framework to describe the basic objects and processes that are specific to it.

We are interested in reasoning on a problem with an internal diagrammatic representation of knowledge. More precisely, we are not interested in reasoning *on* diagrams but *with* (some kind of) diagrams on *various* kinds of problems. We will not enter into the debate concerning the distinction of internal and external diagrammatic representations (Cf. for instance the discussion papers of TWD'98 [34]). We only propose to explore a representational system in which we develop conceptual models of cognitive tasks. The formal framework that we explore here comes from combinatorial algebraic topology, and will be called CAT in this chapter.

Instead of rephrasing well-known diagrammatic applications in the CAT framework, we have found it more illuminating to present a topological modelling of cognitive tasks that can be diagrammatically solved (they involve lattice graph and geometric configuration) but that have not received until now a specific diagrammatic treatment.

The first application is a categorisation problem. The second one concerns the taxonomic reasoning and the problem of restructuring ontologies. The third application, more widely presented, has raised the development of the ESQIMO system [37, 38] for solving analogies in unsupervised IQ tests.

16.2 Algebraic Topology for Knowledge Representation

We were guided towards topological tools for several reasons. First, we are interested in diagrammatic reasoning as the use of spatial relationships such as neighbourhood, border, dimension, path, hole, ... , to represent and structure knowledge.

Although geometry studies these relations, we are not interested in the continuous and metric structure of geometrical objects. The primitive objects and relations involved in diagrams have a finite and discrete nature. For instance, a graph involves edges represented as line segments. A line segment has a continuous nature but this is irrelevant for the graph structure: the precise shape of the edges does not matter, only the connection implied between two nodes does. The same remark holds for *Venn diagrams*, *statecharts* and symbolic maps where it is only the configuration of finite sets of objects that is relevant. When metric aspects turn out to be important, they are often restricted to represent *partial order* relationships: A is bigger that B, C is closer from D than E, path F is shorter that path G, etc.

Moreover, we cannot restrict diagrams to plane geometry. For example, the realisability of a Venn diagram representing an arbitrary predicate requires working in a 3-dimensional space [29]. Path equivalence depends on the underlying structure of space (e.g. all closed paths are equivalent on the plane, but not on a torus). So we have to consider general spatial structures in *many* dimensions.

Hence, if we neglect quantitative diagrammatic representations (like barchart, geological survey map, etc.) we can focus on *n-dimensional combinatorial algebraic topology*. Algebraic topology develops the application of algebraic tools to topological problems. Such an approach is very attractive because we are particularly interested in the development of "constructive" objects, i.e. objects that can be tractable by a computer program.

16.2.1 Simplicial Complexes (SC)

Simplicial complexes are topological abstract structures that generalise the notion of *graph* [21,25]. Indeed, all complexes of dimension less than two are graphs. We find it interesting to consider some spatial properties of graphs and then generalise them to many dimensions to express more information. Simplicial complexes are the abstract objects that realise this generalisation. The following definition is standard in algebraic topology.

Definition 16.2.1. An *abstract simplicial complex* [21, 25] is a couple (V, K) where V is a set of elements called vertices of the complex and K is a set of finite parts of V such that if $s \in K$, then all the non-empty parts $s' \subseteq s$ belong also to K. The elements of K are called abstract simplices. The dimension of a simplex s is equal to $Card(s) - 1$. The dimension of the complex is the dimension of its biggest simplex.

All p-complexes with $p < 2$ are graphs. Indeed, graphs are composed of edges and vertices of dimension one as shown in Fig. 16.1(b). Simplicial complexes are particularly attractive to generalise semantic networks by keeping the possibility to express hierarchies as in a relational graph [23].

Definition 16.2.2. Let $\alpha = (\sigma_0, \sigma_1, \ldots, \sigma_n)$ be a sequence of simplices belonging to a complex K. The sequence α is called a *polygonal n-chain* of origin σ_0 and end σ_n if for all couples (σ_i, σ_{i+1}), $\sigma_i \cap \sigma_{i+1} \neq \emptyset$. The dimension of α is the smallest dimension of $\sigma_i \cap \sigma_{i+1}$.

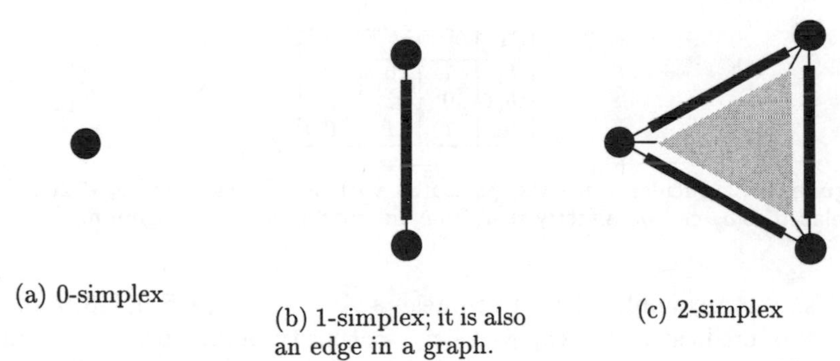

(a) 0-simplex

(b) 1-simplex; it is also an edge in a graph.

(c) 2-simplex

Figure 16.1. Geometrical representation of p-simplices for p varying from 0 to 2.

A p-simplex s is noted: $s = \langle v^0 v^1 \ldots v^p \rangle$, where $v^i \in V$. Fig. 16.1 shows the geometrical representation of 0, 1 and 2-simplices. We say that two simplices σ_1 and σ_2 are q-connected if there is a polygonal chain of dimension q that connects σ_1 with σ_2. Any p-simplex is p-connected to itself with a 0-chain.

Now, we propose to use simplicial complexes to represent knowledge.

16.2.2 Representing Binary Relations with Simplicial Complexes: Q-analysis

Atkin has proposed to represent binary relations with simplicial complexes: the **Q-Analysis** [3, 4, 26]. Q-Analysis has been used to model traffic [27], interactions between agents [8 (Ch. 8), 9, 32], position analysis at chess [2] and social relationships [3, 8, 16].

Let Λ be the incidence matrix of a binary relation $\lambda \subset A \times B$. Let $a \in A$, and the set B_a of elements $b_i \in B$ such that $(a, b_i) \in \lambda$. The set B_a can be directly read from Λ, as the a-column (see Fig. 16.2).

We represent the elements b_i of B_a as vertices and a as a simplex build on these vertices. The dimension of the simplex S_a representing a depends on the number of vertices in B_a.

The whole matrix Λ can then be represented as a simplicial complex containing all the simplices representing each element $a_i \in A$; we note it $K_A(B, \lambda)$ (see Fig. 16.3(a)).

Likewise, we represent Λ^{-1} with the dual simplicial complex $K_B(A, \lambda^{-1})$. In this case, the elements a_i are taken as vertices and the elements b_i are represented as simplices (see Fig. 16.3(b)). We say that $K_A(B, \lambda)$ and $K_B(A, \lambda^{-1})$ are conjugates; they contain the same information but present it in a different and complementary way.

We extended Q-Analysis to allow the representation of sets of predicates as a simplicial complex too [35]. We take a set of predicates $P = \{p_1, p_2, \ldots, p_n\}$ and represent the binary relation $\mu \subset A \times P$ such that $(a_i, p_j) \in \mu$ if $p_j(a_i)$ holds.

λ	a_1	a_2	a_3
b_1	1	0	0
b_2	0	1	1
b_3	1	1	0

Figure 16.2. Incidence matrix associated with λ. The elements $b_i \in B$ that are λ-related to a_j can be directly read from the matrix as the a_j-column.

Take, for example, the set of integers $A = \{1, 2, 3, 4, 5, 6, 7, 8, 9, 10\}$ and the set of predicates $P = \{p_1, p_2, p_3, p_4\} = \{\text{parity}, \text{oddity}, \text{primality}, \text{multiple of 3}\}$. The incidence matrix of μ is then obviously the one given in Fig. 16.4(a). We can represent the dual complex of μ, each element $a_i \in A$ being a simplex built with vertices $p_i \in P$. This dual representation enlightens the fact that elements 4, 8 and 10 have exactly the same representation when taking these few predicates.

A representation based upon simplicial complexes associates the same simplex to elements of A that cannot be distinguished. In other words, two elements will be separated only if there is at least one predicate that allows the differentiation. The same situation occurs with the dual complex.

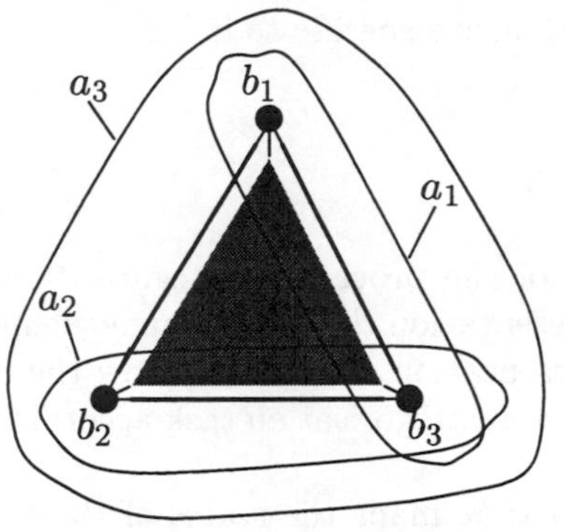

(a) Simplicial representation of λ taking b_i as vertices and a_i as simplices. We have $\lambda(a_1) = \{b_1, b_2\}$. So we represent a_1 as a 1-simplex, b_1 and b_2 being its two vertices.

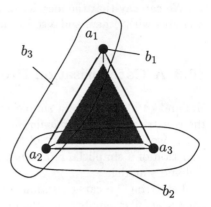

(b) Dual simplicial representation of λ taking a_i as vertices and b_i as simplices.

Figure 16.3. Simplicial representation of the binary relation λ using column (a) or row (b) of the incidence matrix as simplices.

μ	p_1	p_2	p_3	p_4
1	0	1	0	0
2	1	0	1	0
3	0	1	1	1
4	1	0	0	0
5	0	1	1	0
6	1	0	0	1
7	0	1	1	0
8	1	0	0	0
9	0	1	0	1
10	1	0	0	0

(a)

(b)

Figure 16.4. Incidence matrix (a) associated with μ in the numbers example and dual complex (b) associated with $\mu \subset \{1, 2, 3, 4, 5, 6, 7, 8, 9, 10\} \times \{p_1, p_2, p_3, p_4\}$, where we can see that the integers 4, 8 and 10 are identical with respect to these criteria.

Two simplices that have a smaller k-simplex in common are said to share a k-face. In terms of representation, it means that *they have k features in common.* As Freska emphasised, we call here for the use of discriminating features rather than for precise characterisation in terms of a universally applicable reference system [12].

We can say that the identity of an element is represented by the features it shares with others and also by those that are specific to it [20].

16.3 A Categorisation Problem

Holland et al. [22] give a simple model of the process of categorisation for the construction of a *homomorphic* representation that maps many elements of the world to one element of the representation. We present now the construction of a simplicial representation by a categorisation task according to Holland's model.

Let C be the categorisation function that maps the states of the world onto a smaller number of categories. The categorisation is made with the detection of the states of the world through detectors. Let $d_1, \ldots, d_n$ be binary detectors, which can take the value 0 if they are off or 1 if they are on.

When a state S_1 is perceived, the detectors take value 0 or 1. We can represent the values of the detectors for this state by the vector $V_1 = (V_1^{d_1}, \ldots, V_1^{d_n})$ of length n, where $V_1^{d_1}$ is the value of d_1 and so on. Then V_1 represents the state S_1.

We can now consider many successive states $S_1, \ldots, S_p$ and their encoding into binary vectors $V_1, \ldots, V_p$ of length n. If we write the vectors representing this list of states, we construct the matrix of Fig. 16.5 of the relationship ν between the detectors and the states.

ν	d_1	...	d_n
V_1	0/1	...	0/1
...	...	...	...
V_p	0/1	...	0/1

Figure 16.5. Incidence matrix of the relationship ν between the n detectors of a system and a succession of p states detected and encoded as vectors V_i.

Starting from this matrix, we can now build a simplicial representation of the states encoded. Indeed, we build the matrix by writing the lines V_i corresponding to the encoding of each state S_i, but we can now see that each detector has a representation as a column of the matrix. Thus, each detector that detects a particular feature can be represented as a simplex. The representation of the whole matrix as a complex and its dual representation will show the categories extracted through this perception. Indeed, two states indistinguishable by the detectors will be represented as equivalent.

16.3.1 Analysing the *Little Red Riding Hood* Tale

We now illustrate this construction model with a concrete example. We try to extract an ontology from the perception of the successive states that describe the *Little Red Riding Hood* tale.

Objects	Encoding
Red	alive, good, small
Humans	alive, good
Wolf	alive, animal, small
Trees	alive, place, several, bad
House	place
Basket	small
Give	motor, exterior
Sleep	motor, good
Eat	motor,interior, good
Walk	motor
Talk	motor, exterior

(a) Detectors encoding the objects of the world.

Detectors activated
1. Red, Mother, talk, house
2. Red, Mother, give, basket, house
3. Red, walk, tress, basket
4. Red, Wolf, talk, trees, basket
5. Red, walk, trees, basket
6. Wolf, walk, trees
7. Grandma, sleep, house, bed
8. Grandma, Wolf, talk, house, bed
9. Wolf, eat, house, bed
10. Red, Wolf, talk, house, basket, bed
11. Wolf, eat, house, bed, basket

(b) *Little Red Riding Hood* told in 11 scenes or states of the world.

Figure 16.6. The world and the story of the *Little Red Riding Hood*. Objects are perceived only through their attributes and not *per se*. A scene is perceived as a whole. For instance, scene one of the story corresponds to the activation of detectors: alive, good, small motor, exterior, place. The categorisation task builds a hierarchy of objects (ontology) from the sequence of perceived scenes.

To represent the objects of the *The Little Red Riding Hood*, we chose for example the detectors *alive*, *animal*, *good*, *bad*, *place*, *small*, *several*, *motor*, *exterior*, *interior*, which encode the principal characters, objects and concepts of the story according to Fig. 16.6(a). These encodings are of course arbitrary, but the important thing here is that we have a finite number of detectors that can encode the states of the world and that allow the distinction between different objects of the world.

Note that this analysis is held at a naive level. The story can be told, for example, in the 11 states of the world presented Fig. 16.6(b), also called images.

Several strategies are possible to extract the simplices considered as categories; we implemented two [35] in the `Mathematica` [39] programming language (from here, all sentences typeset in `typewriter style` are Mathematica expressions). The first one extracts concepts *incrementally* from the first image to the last one. This means that a base of categories is extracted from the first image. Then if this base is not sufficient to express the second image as a linear combination of the simplices of the base, we add to the base the

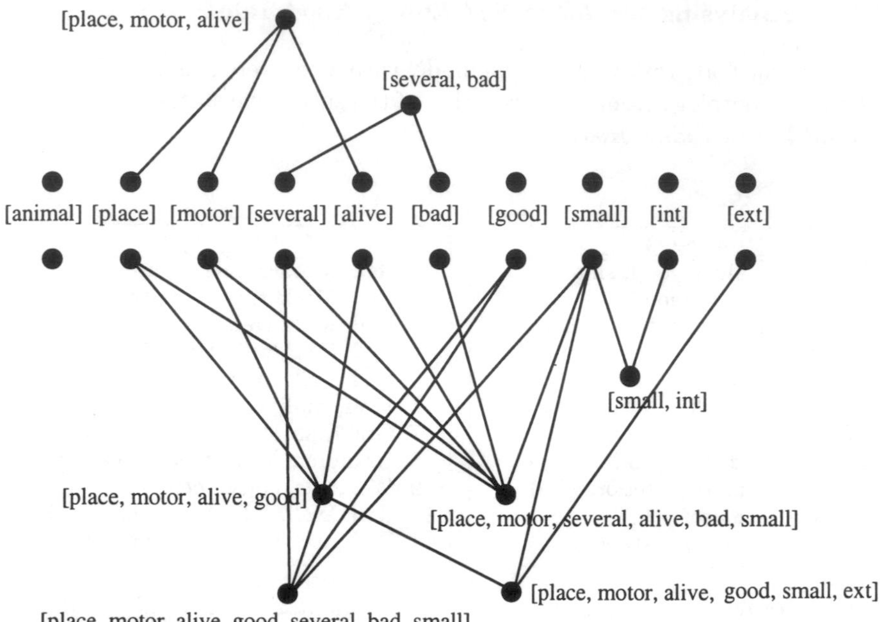

Figure 16.7. Two ontologies extracted from the Little Red Riding Hood story. The ontology extracted instantaneously is represented top-down in the higher part of the figure; the one extracted incrementally is represented bottom-up in the lower part of the figure.

simplices necessary to express it and so on. This is done with a Mathematica function, which we called `Incremente`.[1] We illustrate briefly its functioning with an abstract example before using it on the images of the tale.

Let us take the abstract sequence `l` of images:

```
l = {{1, 2, 3}, {3, 4}, {5, 6}, {1},
     {1, 2, 3, 4}, {3, 4, 5, 6}, {1, 2, 3}}
```

where an image is written between `{ }` and the story itself, composed of images, is also written between `{ }`. When we ask for the *incremental* base, we get the simplices:

```
Incremente[l]
cSimplex[{1, 2, 3}, {3, 4}, {5, 6}, {1}]
```

The base is expressed with simplices (a list with head `cSimplex[]`). The first element of the base is the first image itself, since it is perceived alone with no "history", and thus no base to express it. When the second, third and fourth images are perceived, they are also entirely added to the base since they are

[1] The function names are in French.

necessary to express themselves. But then, these simplices are sufficient to express the last three images as intersections and unions of the previous ones.

The other strategy builds immediately a taxonomy from the 11 images detected as a whole. This means that we get a minimal basis necessary to express all the categories. This strategy is implemented with the function `SimplexBase`. If we take the same succession of images `l`, we will not get the same base:

```
SimplexBase[l]
cSimplex[{1},  {2},  {3},  {4},  {5,6}]
```

where only the objects 5 and 6 cannot be distinguished since they appear together each time they appear in an image. For all the other objects, there is an image (a state) that makes possible their distinction by the detectors.

The incremental and instantaneous ontologies extracted from the 11 Little Red Riding Hood images are given in Fig. 16.7, where we can only see the maximal simplices represented in a Hasse diagram.[2] In this representation, each point is a simplex and the vertices represent inclusion relationships. The concepts represented at level n are ontologically precedent to the ones represented at level $n+1$ (the atomic simplices of the inner layers are at level 0).

16.4 Inheritance Restructuring

We present now an algorithm for *inheritance hierarchy restructuring* proposed in [31] in the field of object-oriented programming. The aim of this algorithm is to infer or restructure the inheritance hierarchy of classes to achieve a smaller, consistent data structure and better code reuse. We chose this example because it is simple to explain and well formalised. The CAT framework provides a concise and clear language to specify this algorithm and exhibit its diagrammatic nature.

We will call *features* any property, behaviour, instance variable or method that can be used for the description of objects. A class corresponds to the description of a type of objects sharing a set of features. Using inheritance to specify classes, we express explicitly the hierarchy relationships between the classes.

Moore [31] proposes an algorithm, called IHI, to infer automatically the inheritance hierarchy from the flat description of objects by their features. In the computed hierarchy, there must be a class corresponding to each concrete object (see Fig. 16.8). Further criteria must be specified to constrain the possible hierarchies:

[2] The extraction of categories from the Little Red Riding Hood story is being rethought more rigorously and applied to the analysis of hypertext structures. See the web sites `http://www.lami.univ-evry.fr/~giavitto/UTopoIa` and `http://www.limsi.fr/Individu/erika` for future developments.

1. Every feature should appear in only one class (maximal sharing of features between classes).
2. Minimal number of classes.
3. All inheritance links that are consistent with the objects' structure must be present.
4. The number of explicit inheritance links must be minimised.
5. The concrete objects should correspond to leaves of the inheritance hierarchy tree.

These criteria together are sufficient to specify a unique solution as shown by Moore in [31].

The problem of inferring a hierarchy from a set of concrete objects can now be rephrased into the CAT framework. We represent the features by vertices, and the classes by different simplices built with the vertices corresponding to the features that define the class. The inheritance relation of classes in the hierarchy is then simply modelled as the inclusion relation of the simplices. Finally, the inheritance graph is the minimal complex containing all the representations of the classes.

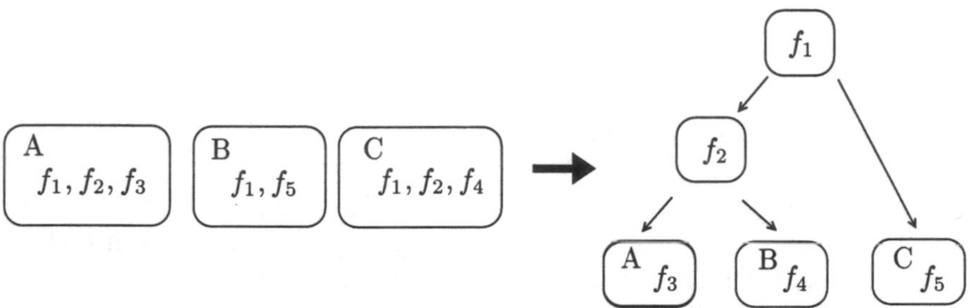

Figure 16.8. From a collection of concrete objects to an inheritance hierarchy. A set of concrete objects (A, B and C) is given on the left together with their features'set. A possible hierarchy that accepts this set of objects is given on the right. The features linked to a class are the set of features defined for this class merged with the features inherited (recursively) from the parent classes. A class without name is called an *abstract class* in object-oriented programming terminology [7] and corresponds to an internal node of the inheritance graph (hierarchies with multiple inheritance will be graphs rather than trees).

The five criteria used by Moore to constrain the class hierarchy are now *topological constraints* that have a simple and intuitive meaning. The corresponding topological constraints are respectively:

1. Every feature appears in a distinguished simplex.
2. A minimal number of simplices are distinguished.
3. The third property is automatically achieved within our translation.
4. A class inherits from the maximal classes it contains.
5. Concrete objects are simplices of maximal dimension.

The problem of inferring an inheritance hierarchy is now simply to find simplices satisfying the previous properties in the complex made by the concrete objects.

16.5 Analogy Solving with the ESQIMO System

We explore now the possibility of a topological representation to support the analogy [38]. The analogy solving between a source and a target domain is modelled as a topological transformation of the representation of the source into the representation of the target in some underlying abstract space of knowledge representation.

The task is to answer a typical IQ test by giving an element called D such that it completes a four-term analogy with three other given elements A, B and C:

"Find D such that it is to C what B is to A".

This kind of analogy solving has already been studied by Evans [10], but in our work the solution has to be built from scratch since no set of possible solutions is given to choose from. We call these kind of problems ***non-supervised*** IQ tests. This four-term analogy solving is usually decomposed into four steps [10]:

- Find the possible relations R_{AB} between A and B.
- Find the possible relations R_{AC} between A and C.
- Apply R_{AB} to C only on a domain determined with R_{AC}.
- Verify the symmetry by applying R_{AC} to B.

16.5.1 Diagrammatic Representation of the problem

Usually, IQ tests are given in terms of geometrical elements so that they can express many different properties at the same level and still stay simple. We chose a geometrical universe similar to the one investigated in [40] of 12 basic elements $E = \{e_1, \dots, e_{12}\}$, as shown in Fig. 16.9(a). These elements are all the possible combinations of the seven properties (or predicates): $P = \{p_1, \dots, p_7\} =$ { Roundness, Squareness, Triangleness, Whiteness, Darkness, Bigness, Smallness }.

These two sets are the only knowledge used by ESQIMO to solve the tests. We can represent this knowledge with a simplicial complex $K(\Omega)$ or its conjugate $K'(\Omega)$ (see Fig. 16.9(b)) by representing the binary relation $\lambda \subset A \times P$ such that $(a_i, p_j) \in \lambda$ if $p_j(a_i)$ holds.

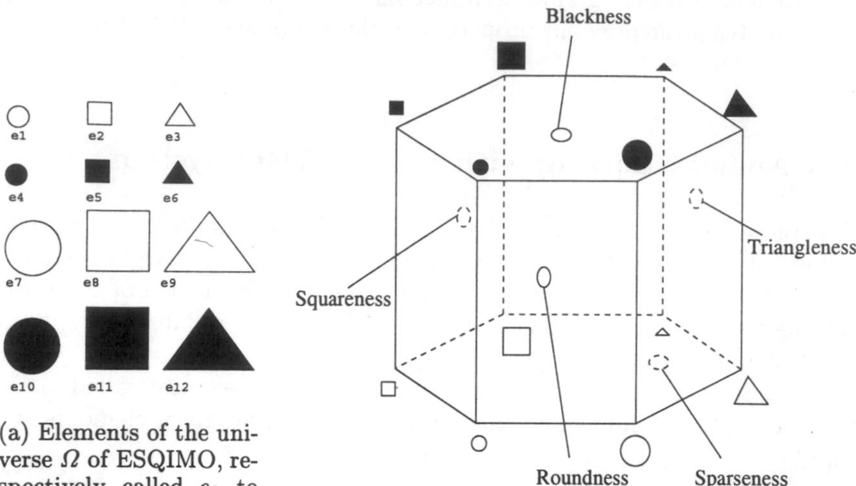

(a) Elements of the universe Ω of ESQIMO, respectively called e_1 to e_{12} starting from the top left element.

(b) A 2-D view of the dual complex $K'(\Omega)$. The elements of E are the vertices and the properties $p_i \in P$ are simplices of $K'(\Omega)$. Notice that the 6-simplex representing the property of blackness is normally 5-dimensional.

Figure 16.9. Elements manipulated by ESQIMO and their representation as a simplicial complex.

16.5.2 Algorithm Based on an SC Representation

When a problem is presented, each figure A, B and C is composed of one or more elements $e_i \in E$. Each element e_i can be represented as a simplex of $K(\Omega)$. Its vertices are the properties p_j such that $p_j(e_i)$ holds. Thus, a simple figure (composed of only one element) will be represented as a simplex and a composed figure (more than one element) will be represented with a set of simplices (a complex if we consider the subparts of each simplices).

The problem is now to find a relation between the (set of) simplex(ices) representing A and the (set of) simplex(ices) representing B and apply it to the (set of) simplex(ices) representing C.

Case of Simple Figures. In the case of simple figures, the transformation T_{AB} is seen as a polygonal chain from S_A to S_B in $K(\Omega)$. An elementary step linking S_i to S_{i+1} in a chain is then viewed as an elementary transformation $T_{S_i,S_{i+1}}$. A polygonal chain from S_A to S_B is then a transformation of A into B given by: $T_{S_l,S_B} \circ \cdots \circ T_{S_A,S_1}$.

If there are several chains, then we say that there are several possible relations between A and B. We can select a *best* solution giving a higher priority to polygonal chains that are short and of higher dimension, to choose a transformation that requires fewer steps and that preserves more properties.

To apply T_{AB} to S_C we have to extend the domain of T_{AB}, and so extend T_{AB} to T'_{AB} such that $T'_{AB}(S_C) = S_D$ and $T'_{AB}(S_A) = S_B$. T' is then a *simplicial application* [21]. There are different possible strategies to determine the domain of $S(C)$ on which we can apply T_{AB}, and we implemented three of them, presented in [36] (unfortunately, we do not have enough space to develop them here).

Case of Composed Figures. For composed figures, the transformations can be of several types: destruction, creation, metamorphosis, division, junction (as in the changes introduced by Hornsby [20]). We first pair the simplices of S_A with those of S_B and look for transformations between the simplices of each pair. The transformation T_{AB} is then the parallel application of the transformation found for each pair. There are many possible pairings leading to different solutions or to the same solution [36]. The only constraint we need is that all the vertices and faces of S_B are paired with vertices from S_A.

16.5.3 Examples of Analogy Solving with ESQIMO

We give three examples of IQ test solving with ESQIMO in Fig. 16.10. In the first example, we ask ESQIMO to solve the IQ test with the call of the function `Resolve` with the pairing parameters `App2` and `AppApp2` (they are pairing strategies) as shown below (for more details see [36]). The three given figures A, B and C are defined in terms of e_i elements of E. As seen in the first example of Fig. 16.10, A is composed of a white small circle plus a white small square.

```
A={e1,e2};  B={e7,e5};  C={e3,e1};
Resolve[A,B,C,App2,AppApp2]
```

Here, A is a composed figure; its representation corresponds to the set of simplices $S_A = \{e_1, e_2\} = \{\langle p_1, p_4, p_7\rangle, \langle p_2, p_4, p_7\rangle\}$. Likewise, the representations of B and C are, respectively, $S_B = \{e_7, e_5\} = \{\langle p_1, p_4, p_6\rangle, \langle p_2, p_5, p_7\rangle\}$ and $S_C = \{e_3, e_1\} = \{\langle p_3, p_4, p_7\rangle, \langle p_1, p_4, p_7, \rangle\}$.

`App2` is a strategy for the pairing between the set of simplices of A and the set of simplices of B that minimises the changes of properties between paired simplices. Here, this strategy gives the following pairings: $(e_1 \rightarrow e_7)$ and $(e_2 \rightarrow e_5)$. ESQIMO provides output about intermediate results such as pairings. For each pairing, an elementary transformation is proposed depending on the heuristic used, which is another parameter (that is internally settled until now [36]). We call them respectively T_1 for $(e_1 \rightarrow e_7)$ and T_2 for $(e_2 \rightarrow e_5)$. Intuitively, T_1 is a transformation that changes e_1 into e_7. However, in general, this can be done in many ways and one specific transformation has to be chosen. Then, the pairing strategy `AppApp2` is used to apply these elementary transformations to the elements of the set of simplices representing C. Here, AppApp2 proposes to apply one transformation in parallel to each simplices representing C: $T_1(e_3) // T_2(e_1)$. The corresponding output, produced by ESQIMO, is:

```
Par[Domain[1,Seq["D-elem"[SmallQ->0,BigQ->1]],{e3}],
    Domain[2,Seq["D-elem"[WhiteQ->0,BlackQ->1]],{e1}]]]
```

where Par means a parallel application and Seq a sequential application of the elementary transformation described in terms of change of properties (or predicates). For instance, "D-elem"[SmallQ->0,BigQ->1] is the elementary transformation that "increases the size" of a figure (i.e., "lose Smallness and gain Bigness"). This transformation is applied to e_3 in parallel with the transformation that darkens a figure which is applied to e_1. Finally, the solution is composed of two elements represented by the simplices $S_D = \{\langle p_3, p_4, p_6\rangle, \langle p_1, p_5, p_7\rangle\} = \{e_9, e_4\}$ (see Fig. 16.10, first example); the corresponding output is:

```
Choose[{e9,e4}]
```

Throughout the solving process, ESQIMO uses the prefix Choose in all its outputs. That is because many different solutions are possible and acceptable for psychological plausibility. ESQIMO can compute many solutions in parallel without selecting a *best* one; in that case there are many solutions that the user can Choose at the end.

Many choices made in ESQIMO's algorithm can be discussed. In fact, they can be seen as additional strategies parameterising the ESQIMO kernel. For example:

- The description of the properties of each figure in terms of predicates can be a problem for properties such as position. We could give each possible position a predicate that could be true or false.
- The way we associate a transformation to a given polygonal chain is not unique. In particular, our transformations could be called 0-degree since they preserve the minimum of topological properties along a chain. The next step consists in pairing higher-order structures between the sets of simplices.
- The way we determine the domain of S_C on which to apply T_{AB} can also lead to different strategies depending on whether we consider only the intersection between S_A and S_C or the whole S_C.
- The measure of satisfaction to select a *best* solution is here to take the shorter and wider polygonal chain between the two complexes. Other measures of satisfaction can be tested.

Furthermore, note that our formalisation of IQ test problems does not depend on their geometrical nature. Indeed, only the representational level is based on topology, while the objects manipulated by the system could have been non-geometrical. We could, for example, try ESQIMO on verbal IQ tests more like in the Copycat system [24].

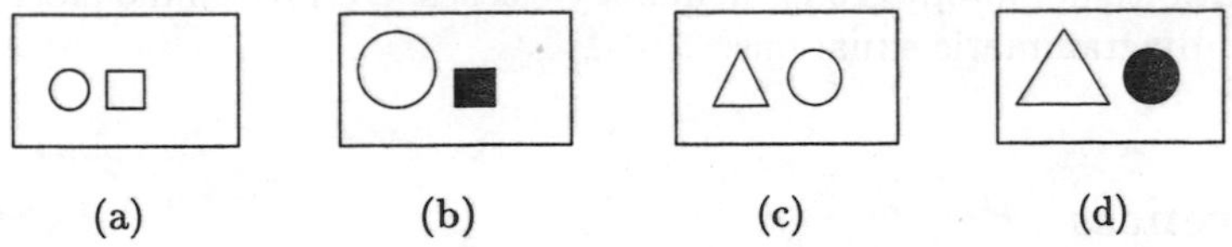

The first element becomes bigger and the second becomes black.

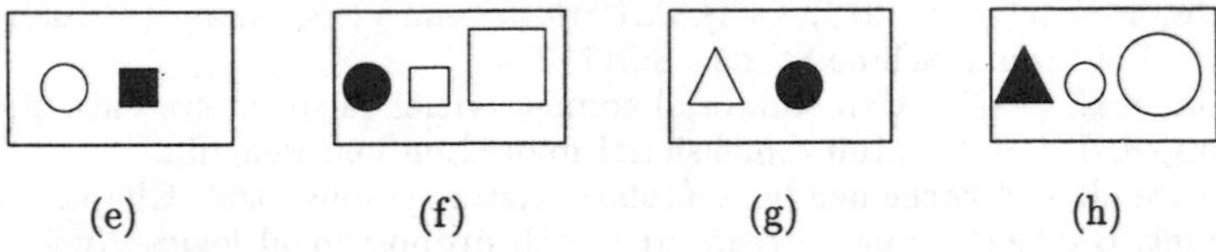

The first element becomes black and the second becomes white, is duplicated and one of the duplicates is bigger.

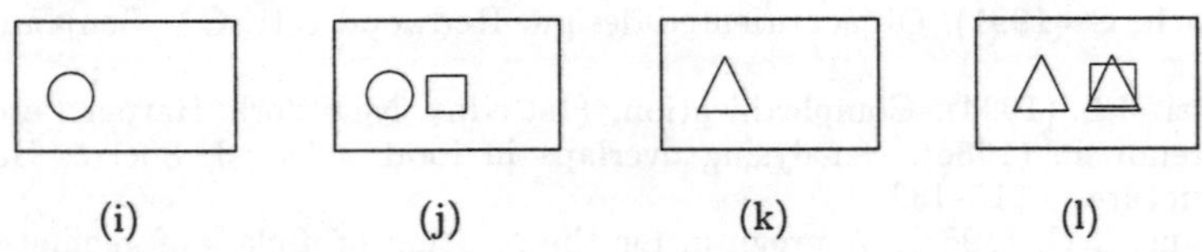

The first element is duplicated and one duplicate is squared. When squared, the property of triangleness is not taken off; this then creates an unstable solution, called a monster.

Figure 16.10. Three examples of analogy resolution with ESQIMO. The last two examples are not detailed in the text. They involve the duplication of a simplex.

Even if the ESQIMO system can be considered as very simple, we are convinced that a topological representational structure is well adapted to support analogy modelling. We find the results presented here already surprisingly satisfying with respect to the simplicity of the underlying machinery and this motivates further investigation.

16.6 Conclusions

The idea developed in this chapter is that *combinatorial algebraic topology* is an adequate and unifying framework to specify and analyse diagrammatic representations and diagrammatic reasoning.

It is important to notice that we have only used *elementary* CAT notions: simplicial complexes generalise the concept graph and polygonal chains extend the concept of path. These two notions have an immediate and intuitive meaning, even in higher dimensions, and are obviously diagrammatic. Future work must include the use of further CAT constructions (like simplicial applications, homotopy group, homology classes, etc.) to handle more sophisticated diagrammatic situations.

References

1. Arnheim, R. (1969). Visual thinking. Berkeley, CA: University of California Press.
2. Atkin, R.H. et al. (1976). CHAMP positional chess analyst. International Journal of Man–Machine Studies 8:517–529.
3. Atkin, R.H. (1977). Combinatorial connectivities in social systems. Springer.
4. Atkin, R.H. (1981). Multidimensional man. London: Penguin.
5. Barwise, J. and Etchemendy, J. (1995). Heterogenous logic. Chapter in [14].
6. Bennet, B. (1994). Spatial reasoning with propositional logics. In P. Torasso, J. Doyle and E. Sandewall (Eds), Proceedings of the fourth international conference on principles of knowledge representation and reasoning (KR94). San Francisco: Morgan Kaufmann.
7. Booch, G. (1991). Object oriented design. Redwood City, CA: Benjamin Cummings.
8. Casti, J.L. (1994). Complexification. (1st edn). New York: Harper Perennial.
9. Doreian, P. (1986). Analysing overlaps in food webs. J. Soc. & Biological Structures 9:115–139.
10. Evans, T.G. (1968). A program for the solution of a class of geometric analogy intelligence-test questions. In Semantic information processing. Cambridge, MA: MIT Press, Ch. 5, pp. 271–353.
11. Forbus, K.D. (1995). Qualitative spatial reasoning framework and frontiers. Chapter in [14].
12. Freska, C. (1997). Spatial and temporal structures in cognitive processes. In C. Freska, M. Jantzen and R. Valk (Eds), Foundations of computer science, Vol. 1337 of LNCS. Berlin: Springer.
13. Gardner, H. (1983). Frames of mind. New York: Basic Books, (Ch. 7).
14. Glasgow, J., Narayanan, N.H. and Chandrasekaran, B. (1995). Diagrammatic reasoning: Cognitive and computational perspectives. Cambridge, MA: AAAI Press/MIT Press.
15. Gotts, N.M. (1994). How far can we "C"? defining a "doughnut" using connection alone. In P. Torasso, J. Doyle and E. Sandewall (Eds), Proceedings of the fourth international conference on principles of knowledge representation and reasoning, Bonn, May. San Mateo, CA: Morgan Kaufmann, pp. 246–257.
16. Gould, P. (1980). Q-analysis, or a language of structure: An introduction for social scientists, geographers and planners. International Journal of Man–Machine Studies 13:169–199.
17. Grigni, M., Papadias, D. and Papadimitriou, C. (1995). Topological inference. In Proceedings of the internation joint conferennce on artificial intelligence (IJCAI'95), Montreal, August. San Mateo, CA: Morgan Kaufmann, pp. 901–907.
18. Hammer, E. (1997). Logic and visual information. Journal of Logic, Language, and Information 6(2):213–216.
19. Harel, D. (1995). On visual formalism. Chapter in [14].
20. Hornsby, K. and Egenhofer, M.J. (1997). Qualitative representation of change. Chapter in [30].
21. Henle, M. (1994). A combinatorial introduction to topology. New York: Dover.

22. Holland, J.H., Holyoak, K.J., Nisbett, R.E. and Thagard, P.R. (1986). Induction – Processes of inference, learning, and discovery. Cambridge, MA: MIT Press.
23. Hirtle, S.C. (1997). Representational structures for cognitive space: Trees, ordered trees and semi-lattices. Chapter in [30].
24. Hofstadter, D.R. (1984). The Copycat Project: An experiment in nondeterminism and creative analogies. AI memo 755, MIT Artificial Intelligence Laboratory, Cambridge, MA.
25. Hocking, J.G. and Young, G.S. (1988). Topology. New-York: Dover.
26. Johnson, J. (1991). The mathematics of complex systems. In J. Johnson and M. Loomes M. (Eds), The mathematical revolution inspired by computing. Oxford: Oxford University Press, pp. 165–186.
27. Johnson, J. (1991). The dynamics of large complex road systems. In J. Griffiths (Ed.), Transport Planning and Control. Oxford: Oxford University Press, pp. 165–186.
28. Lindsay, R.K. (1995). Images and inferences. Chapter in [14].
29. Lemon, O. and Pratt, I. (1997). Spatial logic and the complexity of diagrammatic reasonning. Graphics and Vision 6(1):89–109.
30. Lukose, D., Delugach, H., Keeler, M., Searle, L. and Sowa, J. (1997). Spatial information theory. Berlin: Springer, Vol. 1257 of LNCS.
31. Moore, I. (1996). A simple and efficient algorithm for conferring inheritance hierarchies. In TOOLS 96. TOOLS Europe. Englewood Cliffs, NJ: Prentice-Hall.
32. Pimm, J., Lawton, J. and Cohen, J. (1991). Food web patterns and their consequence. Nature 350:669–674.
33. Shin, S.-J. (1991). An information-theoretic analysis of valid reasoning with Venn diagrams. In J. Barwise et al. (Eds), Situation theory and its applications, Part 2. Cambridge, UK: Cambridge University Press.
34. Electronic forum of the Thinking With Diagram Workshop. Discussion papers TwD'98. `http://www.mrc-cbu.cam.ac.uk/projects/twd/discussion-papers`
35. Valencia, E. (1997). Un modèle topologique pour le raisonnement diagrammatique. Rapport pour le DEA Sciences Cognitives, Université de Paris-Sud, LIMSI, France. See also `http://www.lami.univ-evry.fr/~giavitto/UTopoIa`.
36. Valencia, E. (1998). Hitch hiker's guide to Esqimo. Research report RR 1173, LRI, ura 410 CNRS, Université Paris-Sud, Orsay, France, May.
37. Valencia E. and Giavitto, J.-L. (1998). Algebraic topology for knowledge representation in analogy solving. In C. Rauscher (Ed.), European conference on artificial intelligence (ECAI'98), Brighton, UK, 23–28 August. Christian Rauscher, pp. 88–92.
38. Valencia, E., Giavitto, J.-L. and Sansonnet J.-P. (1998). ESQIMO: Modelling analogy with topology. In F. Ritter and R. Young (Eds), Second European conference on cognitive modelling (ECCM2), Nottingham, UK, 1–4 April. Nottingham University Press, pp. 212–213.
39. Wolfram, S. (1988). Mathematica. Redwood City, CA: Addison-Wesley.
40. Weber, S.H. and Stolcke, A. (1990). L_0: A testbed for miniature language acquisition. International Computer Science Institute.

17. Multiple Readings of Peirce's Alpha Graphs

Sun-Joo Shin

The system of Existential Graphs (henceforth, EG), consisting of three parts – Alpha, Beta and Gamma – was invented by Charles S. Peirce at the dawn of modern logic. Out of these three, Alpha and Beta systems are proven to be sound and complete deductive systems which are equivalent to sentential and first-order languages.[1] However, logicians have strongly preferred symbolic systems to EG. The following two complaints against EG explain logicians' choice: first, reading off Peirce's graph is not easy; second, EG's inference rules are not as intuitive as the inference rules of natural deductive systems. Many have also believed that EG lacks the kind of visual power present in a system like Euler diagrams.[2] This chapter will show that if we take full advantage of the visual features of EG, we can disarm these criticisms of the system.[3]

17.1 Overview of the Chapter

The first section, after a brief introduction of Alpha graphs, summarises the traditional method of reading these graphs, and points out the main problems of this reading. In the second section, I uncover an important visual feature of the Alpha system and present a new reading method – one which gives us *directly* a more useful reading of the graphs. In the third section, I revive another important visual feature the system, called "scroll", which was discussed by Peirce and mentioned in other literature but has never been developed fully. Based on the visual features discussed in the above sections, that is, cut, juxtaposition as conjunction, juxtaposition as disjunction, and scroll, another reading method of the Alpha system is introduced. I show that this method is more *natural* than the other two examined in the first and second sections. In the fourth section, the inference rules of the Alpha system are restated in a natural way and reinterpreted in a more natural way

[1] [6, (pp. 139–151)], [11, (pp. 124–137)].
[2] For this aspect of the Euler system, refer to [1–3].
[3] In this chapter, I will focus on Peirce's Alpha system to demonstrate this claim, and refer to [8] for the application of this method in EG's Beta part.

based upon the visual features discovered in the previous section. In the fifth section, I make a stronger claim for the Alpha system: this graphic system is more efficient than symbolic languages for certain purposes.

17.2 "Endoporeutic" Reading

The Alpha system has two kinds of primitive vocabulary: sentential symbols and cut. Sentential symbols, A_1, A_2, ... , represent propositions, and a cut represents negation. Simple examples illustrate how to read off Alpha graphs:

Example 17.2.1. Let S stands for the proposition "It is sunny", and W "It is windy".

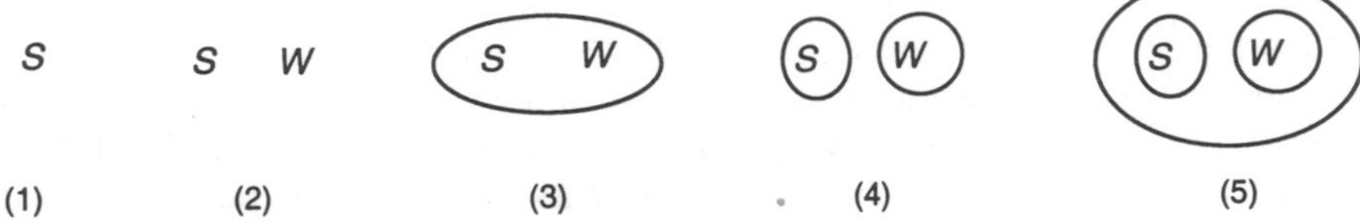

Graph (1) means "It is sunny", (2) "It is sunny and windy", (3) "It is not the case that it is sunny and windy", (4) "It is neither sunny nor windy", and (5) "It is not the case that it is neither sunny nor windy", which is the same as "It is sunny or windy".

A cut corresponds to a negation symbol and juxtaposition to a conjunction symbol in a propositional language. Therefore, intuitively, this system sounds similar to a sentential language which has only two kinds of connectives, i.e. conjunction and negation.[4] As we will see soon, every existing work on the Alpha system has so far been built on this intuition.[5] Hence, it should not be surprising that Alpha graphs are considered difficult to read off, since a sentence using only conjunction and negation usually needs many parentheses, and, accordingly, is cumbersome to use.

I will show that the relation made between Alpha graphs and sentences with only conjunction and negation is not only misleading but has prevented us from using this graphical system fully. The important differences between these two systems will be discussed in detail later in the paper.

It has long been believed that Peirce's Alpha system has only two kinds of syntactic operation: cut and juxtaposition. Hence, the meaning of a graph has been obtained through two stages: (i) translate a cut into a negation and a juxtaposition into a conjunction to get a sentence with nested negation and conjunction symbols, and (ii) if the result looks complicated, simplify it by adopting an additional connective to get an equivalent sentence.

[4] Hence, this graphical system, like the sentential symbolic system with conjunction and negation symbols, must be truth-functionally complete.

[5] Refer to [1, (Ch. 8)], [6, 9, 11].

Example 17.2.2. For example, the following graph is translated into the sentence "$\neg(\neg P \wedge \neg(\neg Q \wedge R))$":

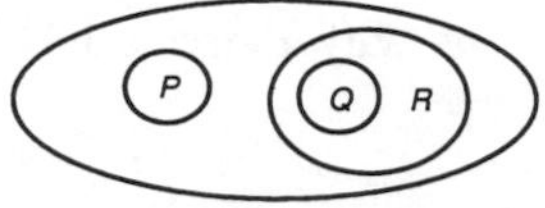

Next, DeMorgan's law is used to get a simpler-looking sentence, "$(P \vee (\neg Q \wedge R))$". The impression has endured that Peirce's system has fewer syntactic devices, and therefore, in most cases, Peirce's graphs are bound to be read indirectly. This indirect reading method was what Peirce originally had in mind, as the following passage shows:

> The interpretation of existential graphs is *endoporeutic*, that is proceeds inwardly; so that a nest sucks the meaning from without inwards unto its centre, as a sponge absorbs water.[6]

When we consider negation and conjunction to be the basic relations of a graph, the question of how to proceed, i.e. whether outwardly or inwardly, is crucial, since the negation of conjunctions is different from the conjunction of negations. No challenge has been made against this *endoporeutic* reading method.

I claim that this reading method has prevented us from benefiting from the visual power of the system. That is, this method does not reflect visually clear facts in the system. For example, in the graph of Example 17.2.2 it is truc that both P and R are evenly and Q is oddly enclosed. However, the endoporeutic reading does not reflect directly this visually clear fact at all, and what is worse, it leaves the impression that this visual fact is misleading. To put this criticism in a more general way: this method forces us to read a graph only in one way. Hence, we are supposed to read this graph in the following way only: This graph is a cut of [the juxtaposition$_1$ of (a cut of P and$_1$ a cut of [the juxtaposition$_2$ of a cut of Q and$_2$ R])]. However, as we will see soon, there are other possible readings.

By interpreting cuts as negation and juxtaposition as conjunction, the meaning of a graph usually turns out to be a negation of a sentence which consists of several conjuncts, each of which is again a negation of a sentence, etc. These nested negations and conjunctions add only unnecessary complication, just as a formula which uses only negation and conjunction symbols is more cumbersome than a formula which has the same meaning but uses disjunction and conditional symbols as well. A similar problem arises when we translate a sentential formula into a graph. Given a simple sentence, "$(P \vee Q)$", we need to change this sentence to a more complicated looking sentence, "$\neg(\neg P \wedge \neg Q)$", since it has been believed that only two kinds of syntactic devices exist in the Alpha system.

[6] [5, (Ms 650)], quoted by [6, (p. 39, n. 13)].

When we interpret what is visually represented or when we represent information visually, if we do not have a direct way to carry out either task, then this visual representation system cannot be said to be efficient or useful. This is one of the main reasons why Peirce's EG have not been used much as a deductive system.

17.3 "Negation Normal Form" Reading

In this section, I identify overlooked visual properties that recognise more syntactic distinctions in EG than on the traditional approach. I do not introduce new syntactic devices into EG, but only observe significant visual differences already present in EG. The reading method presented below will always give us a translation in negation normal form. I call this translation method NNF reading.

I suggest that the following two features be interpreted *directly*: (i) the visual distinction between what is asserted in an evenly enclosed area and what is asserted in an oddly enclosed area, and (ii) the visual fact that some juxtapositions occur in an evenly enclosed area and some in an oddly enclosed area. Several definition are in order:

Definition 2.1. Let X be a subgraph of a Peircean Alpha graph.[7] Then,

X is in an *E-area* iff X is enclosed by an even number of cuts, and
X is in an *O-area* iff X is enclosed by an odd number of cuts.

Definition 2.2. Let X and Y be disjoint subgraphs of a given graph. Then,

the juxtaposition of X and Y is an *E-jux* iff
X and Y are juxtaposed in an E-area, and
the juxtaposition of X and Y is an *O-jux* iff
X and Y are juxtaposed in an O-area.

Definition 2.3. graph D is a *simple graph* iff D is a sentence letter, a single cut of a sentence letter, an empty space or an empty cut.[8]

Definition 2.4. x is a *simple* iff x is a sentence letter, $\neg y$ (for a sentence letter y), $\top$, or $\bot$.

[NNF Reading]

The following function reads off a simple graph into a simple.

[7] Throughout the chapter, "subgraph" means the same as Peirce's "partial graph". For Peirce's meaning of partial graph see [6, (p. 33)]. For a formal definition of subgraphs, see [1, (p. 98)]. Hammer used the term "subgraphs".

[8] By empty space, we mean a blank sheet of paper. By empty cut, we mean a cut with a blank inside.

(a) Let A_i be a *token* of a letter in graph D. Then,
$f(A_i) = A_i$
$f([A_i]) = \neg A_i$.[9]
(b) $f(\emptyset_{sp}) = \top$.[10]
(c) $f([\ \ ]) = \bot$.

Now, we extend this function f to $\overline{f}$ to obtain translations of Alpha graphs.

1. $\overline{f}(G) = f(G)$ if G is a simple graph.
2. $\overline{f}([[G]]) = \overline{f}(G)$.
3. $\overline{f}(G_1 \ldots G_n) = \overline{f}(G_1) \wedge \ldots \wedge \overline{f}(G_n)$.
4. $\overline{f}([G_1 \ldots G_n]) = \overline{f}([G_1]) \vee \ldots \vee \overline{f}([G_n])$.

The first clause reads off the visual fact whether a sentence symbol is in an E-area or in an O-area to obtain a simple. The second clause erases any double cut of a graph. The third and the fourth clauses read off E-jux and O-jux to obtain conjunction and disjunction respectively. Unlike with the endoporeutic reading discussed in the previous section, this reading does not yield any nested negation and conjunction. A negation, if any, gets in only as a simple by the first clause.[11]

Let me illustrate how these clauses work through the following examples.

Example 17.3.1. Let us return to Example 17.2.2 of the previous section. We have seen the following graph is translated into "$\neg(\neg P \wedge \neg(\neg Q \wedge R))$" by the traditional reading method:

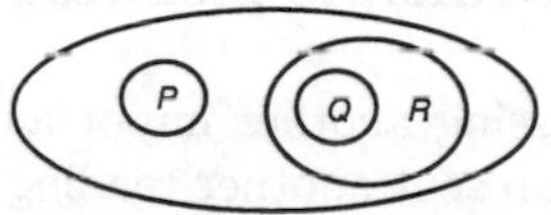

Our function $\overline{f}$ defined above is applied in the following way:

$$
\begin{aligned}
\overline{f}(\,[\,[P]\ \ [\,[Q]\ R\,]\,]) &= \overline{f}([\,[P]\,]) \vee \overline{f}([\,[\,[Q]\ R\,]\,]) && \text{by clause 4}\\
&= \overline{f}(P) \vee \overline{f}([Q]\ R) && \text{by clause 2}\\
&= \overline{f}(P) \vee (\overline{f}([Q]) \wedge \overline{f}(R)) && \text{by clause 3}\\
&= f(P) \vee (f([Q]) \wedge f(R)) && \text{by clause 1}\\
&= P \vee (\neg Q \wedge R) && \text{by def. of } f
\end{aligned}
$$

---

[9] Following Peirce's linear notation [4, (4.378)], for a graph G, we write "$[G]$" as a single cut of G.

[10] We write "$\emptyset_{sp}$" for an empty space and "[ ]" for an empty cut.

[11] Therefore, there is no clause for $\overline{f}([G])$, which is different from the endoporeutic reading.

Example 17.3.2. Let us test the following complicated looking graph which I chose from Roberts' examples:[12]

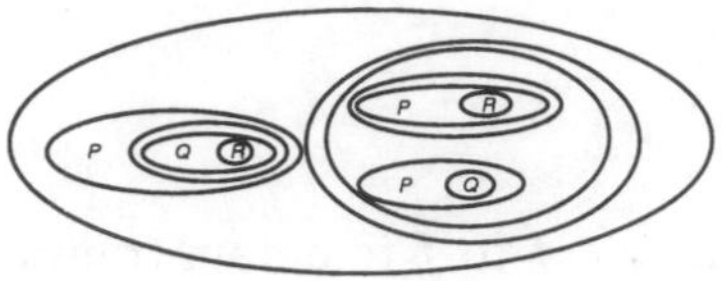

According to our new reading method, a double cut may be erased any time to get the following:

$$\begin{array}{ll}
\overline{f}([\,[\,PQ\,[R]\,]\;\;P\,[R]\;\;[P\,[Q]]\,]) & \\
= \overline{f}([\,[\,PQ\,[R]\,]]) \vee \overline{f}([P]) \vee \overline{f}([[R]]) \vee \overline{f}([\,[P\,[Q]]\,]) & \text{by clause 4} \\
= \overline{f}(\,PQ[R]) \vee \overline{f}([P]) \vee \overline{f}(R) \vee \overline{f}(P\,[Q]) & \text{by clause 2} \\
= (\overline{f}(P) \wedge \overline{f}(Q) \wedge \overline{f}([R])) \vee \overline{f}([P]) \vee \overline{f}(R) \vee (\overline{f}(P) \wedge \overline{f}([Q])) & \text{by clause 3} \\
= (f(P) \wedge f(Q) \wedge f([R])) \vee f([P]) \vee f(R) \vee (f(P) \wedge f([Q])) & \text{by clause 1} \\
= (P \wedge Q \wedge \neg R) \vee \neg P \vee R \vee (P \wedge \neg Q) & \text{by def. of } f
\end{array}$$

17.4 Algorithm for Multiple Readings

In this section, after discussing another important visual feature, I will incorporate Peirce's idea to present another reading algorithm, which gives us more flexibility in understanding this system.

When Peirce presented his EG, he provided the required conventions and rules of transformation. These conventions tell us how graphs should be interpreted. That is, they correspond to an informal semantics of the system. There are conventions on how to interpret juxtaposition and cut. These two conventions have been adopted in most studies of EG, as discussed in the first section of this paper. It is interesting to notice that Peirce had a separate third convention for material implication. Let me introduce Roberts' presentation of this convention:

> C4 [Convention 4][13] concerns the way in which EG is to express the conditional proposition. ... How shall we graph "If P then Q"? In

[12] [6, (p. 46)].

[13] According to Roberts' rearrangement of Peirce's conventions, the convention for material implication became convention 4, while it was the third convention according to Peirce. For Peirce's numberings, see [4, (4.436-7)]. Footnote is mine.

> order to assert it we must place it on SA [sheet of assertion, i.e. a sheet of paper on which a graph is drawn]. But since "If P then Q" asserts neither P nor Q, we must be careful not to scribe them on SA. We get what we want by means of what Peirce called a "scroll" — "two closed lines one inside the other" (Ms 450, p. 14), like this:
>
> 
>
> ... Suppose now that we place the graph Q in the innermost circle, and the graph P in the outermost compartment, obtaining this graph:
>
> 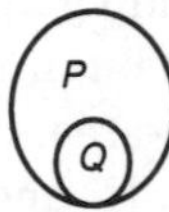
>
>
> Note that we have succeeded in diagramming both P and Q, yet not on the surface of SA itself. And we agree to express in this way the conditional proposition *de inesse*: If P then Q. ... Here then is C4 [Convention 4]: ***The scroll is the sign of a conditional proposition de inesse*** (that is, of material implication) (Ms 450, p. 14).[14]

An interesting question is why Peirce had this third convention while the rest of the conventions are enough to get a correct reading of EG's Alpha system. Clearly, Peirce's intention was to get a direct reading of a scroll, rather than taking a detour through nested cuts and juxtapositions. Quite surprisingly, the importance of this scroll convention has not been discussed. One of the main reasons for this neglect is closely related to a general direction which most research of EG has taken: to understand EG only theoretically, but not for practical purposes. However, when we put EG into practical use as a deductive representation system, this reading adds much more convenience and naturalness not only to the existing reading method but to our understanding of Peirce's transformation rules. The following example will illustrate this point.

Example 17.4.1. Reading off cut, conjunctive juxtaposition, and scroll, we know easily that graph 1 represents tautology, graph 2 corresponds to modus ponens reasoning, and graph 3 modus tollens:

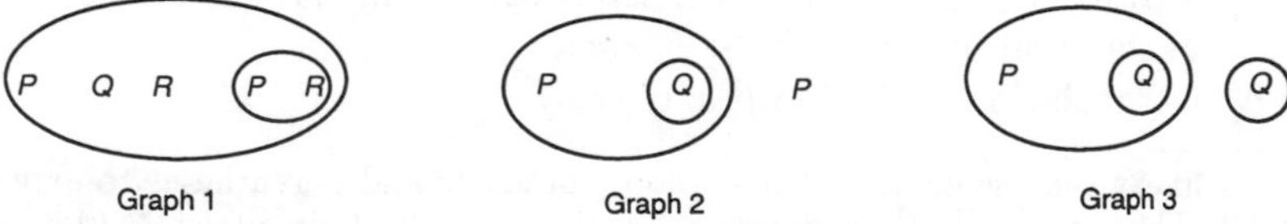

[14] See [6, (pp. 33–35)].

The first graph represents $(P \wedge Q \wedge R) \rightarrow (P \wedge R)$, the second $(P \rightarrow Q) \wedge P$, and the third $(P \rightarrow Q) \wedge \neg Q$.

We put Peirce's convention for scroll in a more general way:

Reading a scroll. Let X and Y be subgraphs, and suppose that each of them is translated into α and β, respectively. Then, the following subgraph with a scroll is translated into $(\alpha \rightarrow \beta)$:

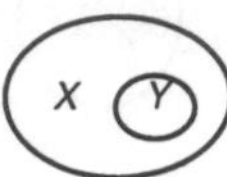

Now, we have three choices for reading graphs: (i) endoporeutic reading: to proceed reading inwardly, by reading off cut and conjunctive juxtaposition, (ii) NNF reading: to get a translation in negation normal form, by making a distinction between E-area and O-area, and between E-jux and O-jux, and (iii) to add the reading of scroll to (i) and (ii).

Example 17.4.2. The following graph is translated into several different but logically equivalent formulas, depending upon which method we take:

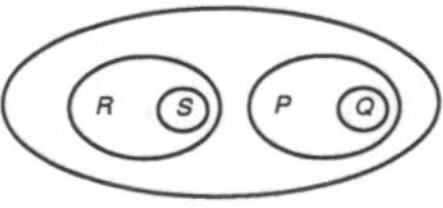

1. $\neg(\neg(R \wedge \neg S) \wedge \neg(P \wedge \neg Q))$ by (i)
2. $(R \wedge \neg S) \vee (P \wedge \neg Q)$ by (ii)
3. $(R \rightarrow S) \rightarrow (P \wedge \neg Q)$ by (iii)

This flexibility corresponds to the fact that each reader might perceive a given graph differently from another. And the following algorithm is defined to reflect this aspect of the graphical representation.

Multiple readings. Let X and Y be subgraphs.

1. If X is a sentence letter, then its translation is X.
2. If a translation of X is α, then a translation of $[X]$ is $\neg\alpha$.
3. If a translation of X is α and a translation of Y is β, then
   a) a translation of XY is $(\alpha \wedge \beta)$,
   b) a translation of $[XY]$ is $(\neg\alpha \vee \neg\beta)$,
   c) a translation of $[X\ [Y]]$ (i.e. scroll with X in the outside cut and Y in the inner cut)[15] is $(\alpha \rightarrow \beta)$, and
   d) a translation of $[[X]\ [Y]]$ is $(\alpha \vee \beta)$.

[15] For a linear representation, Peirce adopts brackets and parentheses to express a scroll. That is, "$[X\ [Y]]$" is expressed as "$[X(Y)]$" by Peirce (see [4, (4.378)]).

As its name suggests, this reading algorithm allows us to translate one and the same graph into more than one sentence. That is, this method does not guarantee a unique translation. Does this cause any problem? I claim that this is not a problem at all. On the contrary, this flexibility renders the Multiple Readings method the most natural reading method among those proposed.

In what sense is this reading method more natural than the other two algorithms we discussed in the previous sections, i.e. endoporeutic and NNF readings? Each of these two previous readings forces the reader to read off a graph only *one* way, which guarantees a unique translation of the graph. On the other hand, in many cases, this is against the natural way of perceiving a graph. For example, if no instruction is given, not everybody would perceive a graph from outside inwards (the only perception which the endoporeutic reading allows us to adopt), and not everybody would pay attention to E-jux or O-jux visual features (one used by the NNF reading). We sometimes mix these two different kinds of perception, or sometimes a scroll might catch our eyes first, etc. A graph, unlike a language in a linear system, can be perceived in more than one way depending upon which way the reader happens to carve up the given graph. It is this actual practice of perceiving a graph that the Multiple Readings method aims to implement. Hence, this reading method allows each reader to interpret a graph in a way which is the most natural to him. This flexibility is lacking both in the traditional method of the first section and the NNF reading method of the second section.

For example, the graph in Example 17.4.2 may be carved up in many more different ways. Let me illustrate only a few of them, by highlighting the cuts which catch the reader's eyes first:

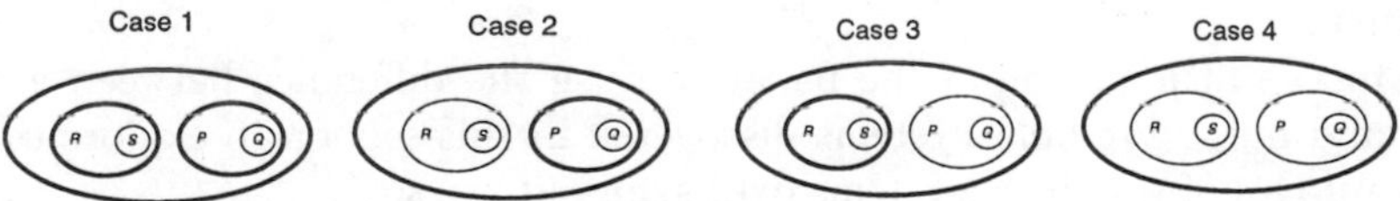

The first case is where someone perceives this whole graph as $[[X]\ [Y]]$, where $X = R[S]$ and $Y = P[Q]$. On the other hand, someone might see a scroll first to perceive this whole graph as $[X\ [Y]]$, where $X = [R\ [S]]$ and $Y = P\ [Q]$, or as $[[X]\ Y]$, where $X = R\ [S]$ and $Y = [P\ [Q]]$. These correspond to cases 2 and 3, respectively. Or, some might perceive this graph as $[X]$, where $X = [R\ [S]]\ [P\ [Q]]$. This case is where the outermost cut catches the reader's eye to provide a different pattern.

How the reader perceives a given graph determines which is the most convenient and the most natural reading to him. The algorithm of Multiple Readings presented in this section covers all these possible ways of carving up one and the same graph, and allows the reader to choose whichever is the most intuitive and, therefore, the most natural, to him. This flexibility corresponds to the fact that readers might perceive a given graph differently.

Accordingly, I claim that the reading method reflecting this intuition for the Peircean graphs is more *natural* than either the endoporeutic or the NNF reading.

It is interesting to notice that this flexibility issue does not arise in a linear symbolic language. On the contrary, a symbolic system is very careful to prevent any multiple readings of a formula since it would yield ambiguity. Sentential languages are defined so that each sentence may have one and only one derivational history, and the semantics is built on this unique history. For example, a string "$P \wedge Q \vee R$" is not well-formed, and, accordingly, we do not bother with its semantics. In order to secure unique readability, parentheses or prefix notations have been adopted so that one and only one way of parcelling up a sentence may be available.

Neither the endoporeutic nor the NNF reading violates this principle of symbolic languages. Each of these readings defines the well-formed graphs so that each graph may have one and only one way of being composed. Each algorithm is based on this uniquely defined inductive syntactic history. As examined above, there is no theoretical defect in either method.

However, I claim that there is no need to keep the unique readability principle in this graphical system as long as we have flexible reading methods, and, therefore, that both the endoporeutic and the NNF readings blindly followed the practice of symbolic languages without any necessity. As seen in the above example, one and the same Alpha graph may be carved up in multiple ways. Then, it would be quite unnatural to prevent the reader from carving up a graph in any other way than a particular reading algorithm can handle. Hence, the Multiple Readings method is practically more desirable for graphical systems. This is a prime example to demonstrate that different approaches should be taken depending on the nature of a representation system.

In the fifth section of the paper, we use the difference between graphical and linear symbolic systems discovered in this section to explore certain advantages of graphical systems over symbolic ones.

17.5 Transformation Rules

With more visual features than are traditionally recognised, we made the reading of graphs easier, and now I claim that inference rules can be stated with a more specific symmetry and can be understood more intuitively.

Peirce's inference rules may be summarised as follows:[16]

R1: The rule of erasure. *Any evenly enclosed graph may be erased.*

R2: The rule of insertion. *Any graph may be scribed on any oddly enclosed area.*

[16] For more discussions of Peirce's rules I mention here, refer to [6, (pp. 40-45)].

R3: The rule of iteration. *If a graph P occurs on SA or in a nest of cuts, it may be scribed on any area not part of P, which is contained by {P}.*[17]

R4: The rule of deiteration. *Any graph whose occurrence could be the result of iteration may be erased.*

R5: The rule of the double cut. *The double cut may be inserted around or removed (where it occurs) from any graph on any area.*

The way Peirce states the rules keeps certain symmetry, that is, erasure versus insertion, and iteration versus deiteration. However, the symmetries built into Peirce's inference rules are not fine-grained enough to be useful. The only symmetry these rules exhibit is between writing and erasing, which occurs in any kind of manipulations of signs, whether symbols or graphs. Peirce is perfectly aware that this is the only kind of symmetry present in EG: "All our transformations are analysed into insertions and omissions".[18]

Unlike symbolic systems, the Alpha system does not introduce new syntactic objects for conjunction or disjunction, which is a strong point of graphical systems over symbolic ones.[19] Instead, these two different connectives are represented in terms of a very clear visual feature, that is, whether a juxtaposition between two subparts of a graph takes place in an area enclosed by an even number or by an odd number of cuts. Therefore, we could provide another kind of symmetry by relying on the visual distinction between an E-area and an O-area which we discussed in the second section.

Let's recall that a natural deductive system is natural because the rules are presented in terms of the history of the formation of a formula. However, this graphical system should not pursue this specific kind of naturalness for the following reason: A unique derivational history of a graph does not seem to be necessary when a unique reading method is not desirable, as discussed at the end of the previous section.[20] In a graphical system like EG, visual facts (rather than a history of syntactic composition) should be implemented in the system in order to achieve naturalness. It is a crucial visual feature of EG whether something is written or erased in an E-area or in an O-area. Therefore, I suggest a new version of the Alpha rules to implement the inference rules of the Alpha system around the following basic visual features: what we may *draw* or *erase* in which area, either in an *E-area* or in an *O-*

[17] "For any graph P, let '{P}' denote the place of P" [6, p. 38].

[18] See [4, (4.380)].

[19] For more discussion of this issue, refer to [7, (Ch. 6)].

[20] For more detail, refer to [7].

area.[21] This will make the rules *natural* for a different reason from why a natural deductive system is natural.

[Rules of transformation][22]

1. In an E-area, say, area a,
   a) we may *erase* any graph, and
   b) we may *draw* graph X, if there is a token of X
      either (i) in the same area, i.e. area a
      or (ii) in the next-outer area from the area a.[23]
2. In an O-area, say, area a,

[21] Richard White also rewrote Peirce's permissions (i.e. Peirce's rules) to make the symmetry more specific, by using "even introduction", "even elimination", "odd introduction", and "odd elimination" [10, (p. 353)]. However, his project starts with the conviction that "'iconic' notation is logically inessential" [10, (p. 352)]. Hence, he transforms Alpha graphs into linear expressions, which does not make the meanings of inference rules understood easily. White seems to believe that graphical notations are confusing, but does not realise that linear expressions with many parentheses, not having visual clarity, are even more confusing. For instance, in one of his examples, ([A([B(C)])][([A(B)][(A[C])])]), it is not easy to see how parentheses and brackets are matched [10, (p. 353)].

[22] The symmetries in the first two rules can be summarised as follows:

	E-area	O-area
Erase	Anything	X if there is another X either in the same area or in the next-outer area
Draw	X if there is X either in the same area or in the next-outer area	Anything

[23] I adopt Roberts' notations here. Roberts says "It is convenient to inject some order into the areas of graphs. For any graph P, let {P} denote the place of P. And let the relation symbol $\supseteq$ be defined as follows: {B} *is enclosed by every cut that encloses* {A} *if and only if* {A} $\supseteq$ {B}. The sign $\supseteq$ may be read 'contains' " [6, (p. 38)]. With these notations, Roberts defined the identity of the areas: "If $\{P\} \supseteq \{Q\}$ and $\{Q\} \supseteq \{P\}$, then $\{P\} = \{Q\}$, i.e., P and Q are scribed on the same area" [6, (n.11, p. 38)].) Then, we give the following definition:

Definition 4.1. Area b is the *next-outer area* from area a if and only if (i) $a \subseteq b$, (ii) $a \neq b$, and (iii) there is no other area between a and b.

For example, in the following, P is in the next-outer area from $\{Q\}$:

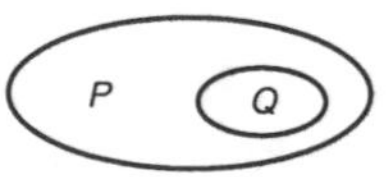

a) we may *erase* graph X, if there is another token of X
either (i) in the same area, i.e. area a
or (ii) in the next-outer area from the area a, and

b) we may *draw* any graph.

3. A double cut may be erased or drawn around any part of a graph.

Now, we will see what these rewritten rules mean. The soundness of Peirce's Alpha system has not been understood at an intuitive level as the soundness of a natural deductive system has. A main reason for this difference is that the inference rules for a natural deductive system are understood easily through obvious correspondence with the meaning of each connective, but there has been no attempt at a similar understanding of Peirce's inference rules. With the traditional method of reading graphs based on using nested negations and conjunctions, for example, Peirce's iteration rule is hard to understand. I will show that understanding our reformulated inference rules is much easier if we apply all the visual features discussed in the previous sections.

The first inference rule has two clauses: Clause 1(a) allows us to erase anything in an E-area. As demonstrated in the second section, the juxtaposition in an E-area (we called it "E-jux") corresponds to conjunctive information. Clearly, erasing a conjunct should be valid. Clause 1(b) says we may draw X in an E-area when one of the two conditions is satisfied. This case requires some examination, since adding a conjunct is not a valid step.

First, as I will show presently, this rule is quite different from the $\wedge$-introduction rule of symbolic logic. The rule of $\wedge$-introduction allows us to conjoin previously obtained formulas, that is, combining more than one piece of information. However, none of the Alpha rules deals with more than one graph at the same time. That is, the inference is done from one *single* graph to another *single* graph. Rule 1(b)(i) corresponds to the following inference in symbolic logic: $\alpha \Longrightarrow (\alpha \wedge \alpha)$. In the case of symbolic logic we do not need this copy rule separately, since the reiteration and $\wedge$-introduction rules together do this job. The second condition of 1(b), i.e. *if* a token of X is in the next-outer area from the area where we want to draw another token of X, is rather complicated. Since this token is in the next-outer area from an E-area, it must be in an O-area. Then, we know that we are dealing with a graph with the following scroll as its subpart:[24]

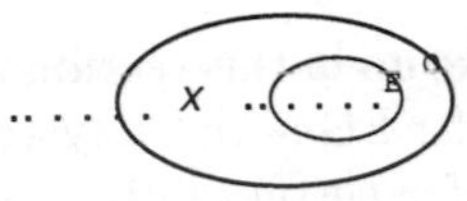

Recalling that a scroll represents a conditional proposition, this subgraph represents a piece of conditional information with X as an antecedent. Therefore, it is a valid step to infer that X occurs as a consequent as well. Thus,

[24] Dots express other possible subparts of a graph.

rule 1(b), i.e. the drawing rule in an E-area, is valid but different from the $\wedge$-introduction rule.

There are two main reasons why it is much easier to see the validity of this rule than the validity of Peirce's iteration rule. First, our rule 1(b) cuts off the redundant part of the iteration rule, and spells out only the necessary part.[25] Second, we make use of the visual feature discussed in the third section, that is, scroll.

Rule 2(b) allows us to draw any graph in an O-area. In the second section we saw that O-jux represents disjunctive information. Hence, this rule corresponds to adding a disjunct, which is a valid step. Rule 2(a) is not the same as the $\vee$-elimination rule of propositional logic. For the same reason as why there is no similar rule in EG to $\wedge$-introduction, there is no single inference rule in this system that has the same function as the $\vee$-elimination rule of symbolic logic: The Alpha system manipulates one single graph to another single graph. The first clause of 2(a) is the same as the following inference in symbolic logic: $(\alpha \vee \alpha) \Longrightarrow \alpha$. This inference is obtained by both the reiteration and $\vee$-elimination rules in first-order logic. The second clause says that we may erase token X in an O-area if another token of X is in the next-outer area to this O-area. That is, in the following, we may transform the graph on the left side to the graph on the right side:[26]

According to the reading algorithm of the third section, if α is a translation of X and β a translation of Y, then the part of the graph on the left side will be translated into $(\alpha \wedge (\neg\alpha \vee \neg\beta))$.[27] Then, we may infer from this to $(\alpha \wedge \neg\beta)$, which is the translation of the part of the graph on the right side and is obtained by erasing the token X of the inner O-area. Therefore, this is a valid rule.

The third rule, i.e. the erasure and drawing of a double cut, is clearly valid, just as erasing or adding a double negation is.

17.6 Visual Efficiency

One last challenge we will take up in this section is to explore whether there is any good reason for us to prefer EG's Alpha system to standard propositional languages. I will claim that for certain purposes there are good reasons for

[25] Peirce's iteration and deiteration rules partially overlap with the insertion and erasure rules.

[26] Dots express other possible subparts of a graph.

[27] This reading is obtained by applying clauses 3 and 4 in that order.

the preference of the Alpha system over sentential languages, without making any general judgement about these two different kinds of systems.

First, I suggest that we reverse the traditional relationship between formulas and graphs: while we have understood graphs in terms of formulas, we will here attempt to understand formulas in terms of graphs. An algorithm from sentences to graphs is presented below:

Let *SS* be a set of sentence symbols, *WFF* be a set of formulas of sentential logic and $\mathcal{G}$ be a set of the Alpha graphs. We define function $\mathcal{K}$ as follows:

$$\mathcal{K} : SS \longrightarrow \mathcal{G}, \text{ where } \mathcal{K}(A_i) = A_i$$

Then, we extend this function to $\overline{\mathcal{K}}$ as follows:

$$\overline{\mathcal{K}} : WFF \longrightarrow \mathcal{G}, \text{ where}$$

1. If α is a sentence symbol, then $\overline{\mathcal{K}}(\alpha) = \mathcal{K}(\alpha)$.
2. $\overline{\mathcal{K}}((\neg\beta)) =$

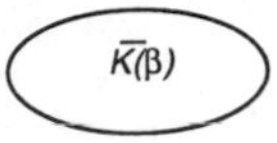

3. $\overline{\mathcal{K}}((\beta \wedge \gamma)) =$

4. $\overline{\mathcal{K}}((\beta \vee \gamma)) =$

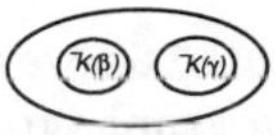

5. $\overline{\mathcal{K}}((\beta \rightarrow \gamma)) =$

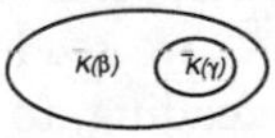

We will briefly discuss two kinds of applications for this algorithm, $\overline{\mathcal{K}}$: one is to find logical equivalence among formulas, and the other is to transform a formula into a negation normal form.

We observe that different formulas can sometimes be translated into one and the same graph, which is consistent with our previous observation that one and the same graph can sometimes be translated into more than one formula. When more than one syntactically different formulas are translated into one and the same graph, we know that these formulas are logically equivalent since the Alpha system is an unambiguous representation system. We thus establish the following simple proposition:

Simple Proposition. Let α and β be formulas and let G be the graph of α and G' be the graph of β.[28] If *either* G and G' is the same graph *or* G

[28] That is, $\overline{\mathcal{K}}(\alpha) = G$ and $\overline{\mathcal{K}}(\beta) = G'$.

and G' is the same except that we may transform one to another by erasing or drawing a double cut, then α and β are logically equivalent.

There are several uses we can make out of this proposition. In order to test logical equivalence among different sentential formulas, we either set up a truth table or see whether we can deduce one from another and vice versa. We can now add one more method to this list, that is, to use Peirce's Alpha system.

Example 17.6.1. The two formulas, $P \to (Q \to R)$ and $(P \wedge Q) \to R$, are logically equivalent, since we get the following graphs for each formula:

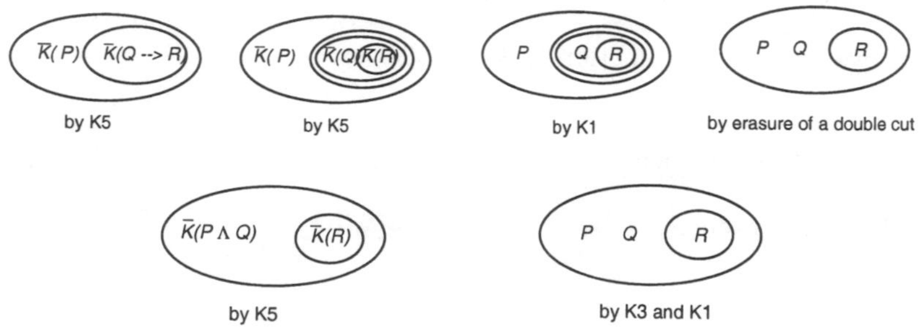

Of course, our method is not complete since getting the same graph (or, the same except double cuts) is a sufficient but not a necessary condition for logical equivalence. We get two different graphs for two logically equivalent formulas, e.g. $(P \wedge \neg P)$ and $(Q \wedge \neg Q)$.[29]

Combining the simple proposition stated above and the method presented in the second section, we can use Alpha graphs to transform a formula into a formula in negation normal form. That is, after translating a given formula into an Alpha graph by the algorithm $\overline{\mathcal{K}}$, we read the graph by the reading method of the second section, i.e. NNF reading, to get a formula in negation normal form.

Example 17.6.2. We would like to change the formula $\neg((P \vee Q) \wedge \neg R)$ into a formula in negation normal form. According to our algorithm $\overline{\mathcal{K}}$, we obtain the following graph:

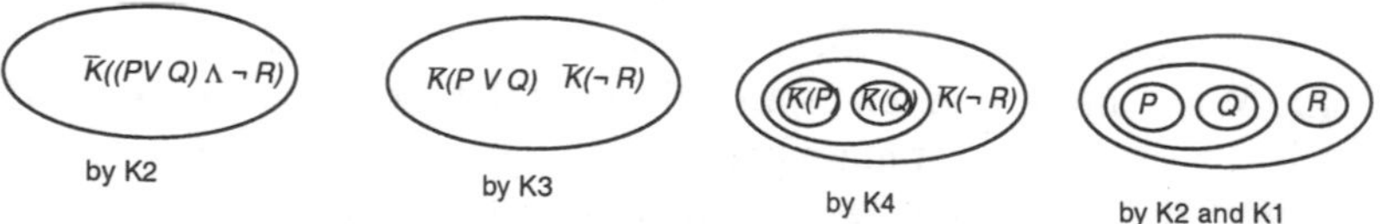

[29] There is a way to get around this problem, by introducing substitution among sentence letters and subgraphs.

Now, we read this Alpha graph according to NNF reading to get the formula $((\neg P \wedge \neg Q) \vee R)$, which is a formula in negation normal form for $\neg((P \vee Q) \wedge \neg R)$.

An important point is that in the Alpha system we *see* logical equivalence. It is more efficient to see that the resulting graphs are the same graph than to find a deduction sequence from one formula to the other and vice versa. Accordingly, some laws in propositional logic, e.g. DeMorgan's laws and distributive laws, can be presented clearly in this graphic system. Also, in the case of obtaining a negation normal form sentence, we obtain this formula by *reading off* the graph of the given formula, rather than transforming it by using inference rules. Again, this is more efficient than finding a deduction sequence.

The efficiencies we have discussed so far can be seen as a more general advantage of EG over symbolic systems: EG is efficient in that only two kinds of syntactic devices, i.e. cut and juxtaposition, allow us to express negative, conjunctive, disjunctive and conditional propositions, while propositional languages adopt different syntactic devices for each connective. The fewer syntactic devices a deductive system has, the fewer inference rules it requires, and accordingly, the simpler the search for a deduction becomes. However, a symbolic system with only two connectives, though complete, is quite inconvenient. On the other hand, EG's Alpha system has fewer syntactic devices than propositional languages, but without suffering from the inconvenience of a symbolic system with only two connectives.

Acknowledgements

I wish to acknowledge financial support from the Faculty Research Program and the Institute for Scholarship in the Liberal Arts of the University of Notre Dame.

References

1. Hammer, E. (1995). Logic and visual information. Stanford, CA: Center for the Study of Language and Information.
2. Hammer, E. and Shin, S-J. (1996). Euler and the role of visualization in logic. In J. Seligman and D. Westerstaåhl (Eds), Logic, language and computation, CSLI, Lecture notes 58: 271-286. Stanford, CA: Center for the Study of Language and Information.
3. Hammer, E. and Shin, S-J. (1998). Euler's visual logic. History and Philosophy of Logic, 19(1): 1-29.
4. Peirce, C.S. (1931–1958). Collected papers of Charles Sanders Peirce. C. Hartshorne and P. Weiss (Eds), Cambridge, MA: Harvard University Press.
5. Peirce, C.S. Manuscripts in Houghton Library, Harvard University.
6. Roberts, D. (1973). The existential graphs of Charles S. Peirce. The Hague: Mouton.

7. Shin, S-J. (1994). The logical status of diagrams. New York: Cambridge University Press.
8. Shin, S-J. (2000). Reviving the iconicity of beta graphs. In M. Anderson, P. Cheng, and V. Haarslev (Eds), Theory and application of diagrams 58-73. Berlin: Springer.
9. Sowa, J. (1984). Conceptual structure: Information processing in mind and machine. Reading, MA: Addison-Wesley.
10. White, R. (1984). Peirce's alpha graphs: The completeness of propositional logic and the fast simplification of truth-function. Transactions of the Charles S. Peirce Society 20: 351-361.
11. Zeman, J. (1964). The graphical logic of C.S. Peirce. PhD thesis, University of Chicago.

18. On Automating Diagrammatic Proofs of Arithmetic Arguments

Mateja Jamnik
Alan Bundy
Ian Green

Theorems in automated theorem proving are usually proved by formal logical proofs. However, there is a subset of problems which humans can prove by the use of geometric operations on diagrams: so-called diagrammatic proofs. Insight is often more clearly perceived in these proofs than in the corresponding algebraic proofs; they capture an intuitive notion of truthfulness that humans find easy to see and understand. We are investigating and automating such diagrammatic reasoning about mathematical theorems. Concrete, rather than general diagrams are used to prove particular concrete instances of the universally quantified theorem. The diagrammatic proof is captured by the use of geometric operations on the diagram. These operations are the "inference steps" of the proof. An abstracted schematic proof of the universally quantified theorem is induced from these proof instances. The constructive ω-rule provides the mathematical basis for this step from schematic proofs to theoremhood. In this way we avoid the difficulty of treating a general case in a diagram. One method of confirming that the abstraction of the schematic proof from the proof instances is sound is proving the correctness of schematic proofs in the meta-theory of diagrams. These ideas have been implemented in the system, called DIAMOND, which is presented here.[1]

18.1 Introduction

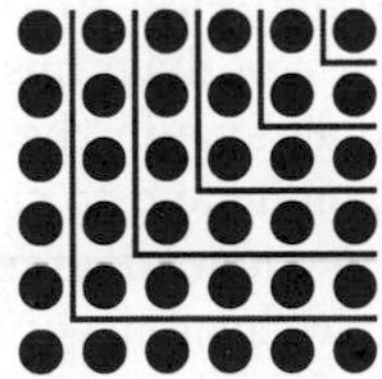

$$1 + 3 + 5 + \cdots + (2n - 1) = n^2$$

[1] This is a reprint of an article first published in the *Journal of Logic, Language and Information* 8(3):297–321, 1999, Jamnik et al, reprinted with kind permission of Kluwer Academic Publishers.

It requires only basic secondary school knowledge of mathematics to realise that the diagram above is a proof of a theorem about the *sum of odd naturals*.

It is an interesting property of diagrams that allows us to "see" and understand so much just by looking at a simple diagram. Not only do we know what theorem the diagram represents, but we also understand the proof of the theorem represented by the diagram and believe it is correct.

Is it possible to simulate and formalise this sort of diagrammatic reasoning on machines? Or is it a kind of intuitive reasoning particular to humans that mere machines are incapable of? Roger Penrose claims that it is not possible to automate such diagrammatic proofs.[2] We are taking his position as a challenge and are trying to capture the kind of diagrammatic reasoning that Penrose is talking about so that we will be able to emulate it on a computer.

The importance of diagrams in many domains of reasoning has been extensively discussed by Larkin and Simon [10], who claim that "a diagram is (sometimes) worth *ten* thousand words". The advantage of a diagram is that it concisely stores information, explicitly represents the relations among the elements of the diagram, and it supports a lot of perceptual inferences that are very easy for humans. Diagrams have been extensively used in the history of mathematics to *aid informal mathematical reasoning*. The use of diagrams in explanations of theorems and proofs of geometry dates back to Ancient Greece, and the time of Aristotle and Euclid. Thus it is surprising perhaps that more recently, starting with the invention of formal axiomatic logic in the sense of Frege, Russell and Hilbert, diagrams have been denied a *formal* role in theorem proving. It is generally thought by logicians that diagrams have no accepted syntax nor semantic theory in a logical formalism which would make them rigorous enough to be used in formal proofs. Only very recently, in the last two decades, have there been efforts to fill this gap and investigate whether and how diagrams can be used in formal proofs. For instance, the investigation of Pierce's existential graphs in [16], the work on GROVER in [2], the work on Hyperproof in [3], the introduction of computational models for interpreting Euler's circles in [17], the analysis of the use of Venn diagrams as a formal system in [14], and the formalisation of a logical theory of Venn diagrams in [6].

Our work contributes in some sense to the effort in the research from the formal perspective on the use of diagrams, especially that of automated reasoning systems which use diagrams in the reasoning process. Our aim is to formalise diagrammatic reasoning and to show that diagrams can be used for proofs. In this paper we show how diagrams can be used for proofs in a formal system. We look into how theorems of mathematics can be expressed as diagrams for some concrete values, i.e., ground instantiations of a theorem. The initial diagram is manipulated using some geometric operations. The se-

[2] Roger Penrose presented his position in the lecture at the International Centre for Mathematical Sciences in Edinburgh, in celebration of the 50th anniversary of UNESCO on 8 November 1995.

quence of geometric operations on a diagram represents the inference steps of a diagrammatic proof. Such a concrete proof instance is called an example proof. The set of all available operations defines the proof search space. A general pattern is extracted from these proof instances, and is captured in a recursive program. This recursive program constitutes a general diagrammatic proof for the universally quantified theorem. An existing technique in logic, namely the constructive ω-rule, justifies the step from schematic proofs to theoremhood.

We also aim to investigate the relation between formal algebraic proofs and more "informal" diagrammatic proofs. Usually, theorems are *formally* proved with the use of inference steps which often do not convey an intuitive notion of truthfulness to humans. The inference steps of a formal symbolic (as opposed to diagrammatic) proof are statements that follow the rules of some logic. The reason we trust that they are correct is that the logic has been previously proved to be sound. Following and applying the rules of such a logic guarantees that there is no mistake in the proof. We want to have such a guarantee in our proof system, and moreover, to gain an insight into the proof. Ultimately, the entire process of diagrammatically proving theorems will illuminate the issues of formality, rigour, truthfulness and power of diagrammatic proofs.

We implemented a diagrammatic proof system called DIAMOND, which automates such diagrammatic reasoning and applies it to problem solving in mathematics. The user interactively constructs example proofs by choosing an initial diagram which represents the theorem, and then applies diagrammatic operations to build a proof. DIAMOND then automatically extracts a general pattern from these instances, and captures it in a recursive program.

First, we list some of the theorems and their diagrammatic proofs. These help us define our problem domain. Second, we present DIAMOND's architecture, some operations required, the abstraction mechanism employed, and indicate how to verify the abstracted proof. Next, we report on some of our results and discuss future work. Then, we discuss some of the related diagrammatic reasoning systems. Finally, we conclude by summarising the main points of this paper.

18.2 "Diagrammatic" Theorems

We are interested in mathematical theorems that admit diagrammatic proofs. In order to clarify what we mean by diagrammatic proofs we first list some example theorems. Then, we introduce a taxonomy for categorising these examples in order to be able to characterise the domain of problems under consideration.

18.2.1 Examples

Pythagoras' Theorem. Pythagoras' theorem states that the square of the hypotenuse of a right-angle triangle equals the sum of the squares of its other two sides. Here is one of the many different diagrammatic proofs of this theorem, taken from [11, (p. 3)]:

$$a^2 + b^2 = c^2$$

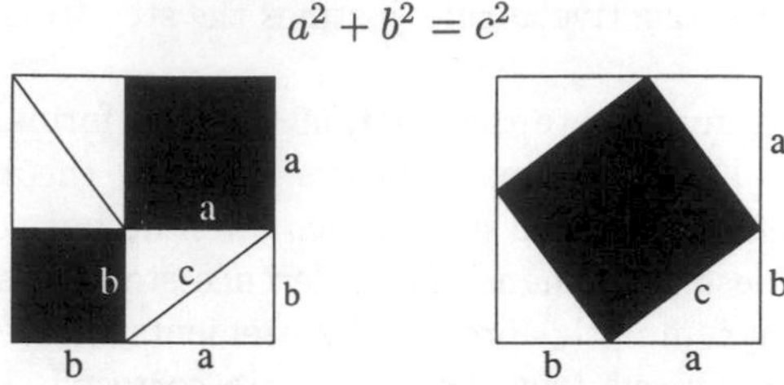

The proof consists of first taking any right-angle triangle, completing a bigger square by joining to it identical triangles and squares along its sides, and then rearranging the triangles in a bigger square.

Sum of Odd Naturals. This example is also taken from [11, (p. 71)]. The theorem about the *sum of odd naturals* states the following:

$$1 + 3 + \cdots + (2n - 1) = n^2$$

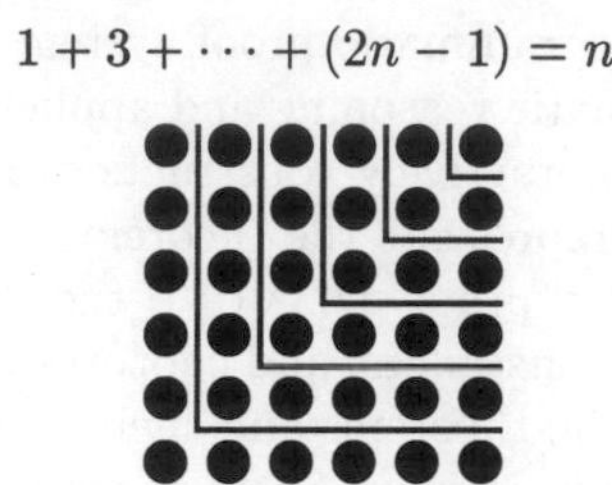

Note the use of parameter n. If we take a square we can cut it into as many ells (which are made up of two adjacent sides of the square) as the magnitude of the side of the square. Note that one ell is made out of two sides, i.e., $2n$, but the shared vertex has been counted twice. Therefore, one ell has a magnitude of $(2n - 1)$, where n is the magnitude of the square.

Geometric Sum. This example is also taken from [11, (p. 118)]. A theorem about a geometric sum of $\frac{1}{2^n}$ states the following:

$$\tfrac{1}{2} + \tfrac{1}{4} + \tfrac{1}{8} + \cdots = 1$$

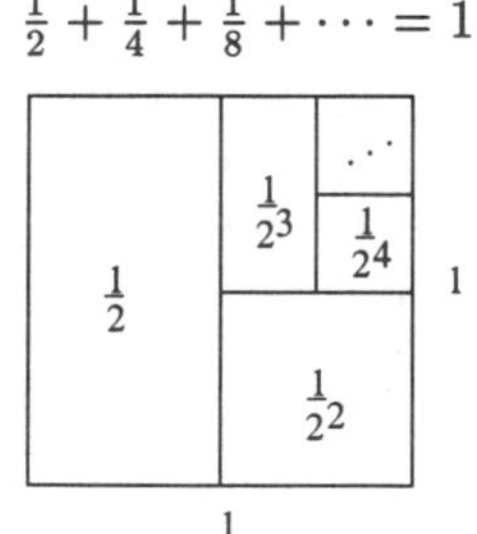

Note the use of ellipsis in the diagram. Take a square of unit magnitude. Cut it down the middle. Now, cut one half of the previously cut square into halves again. This will create two identical squares making up a half of the original square. Take one of these two squares and continue doing this procedure indefinitely.

18.2.2 Classification

From the analysis of the examples that we presented above, and many others (see [8, 11]), three categories of proofs can be distinguished:

Category 1: Non-inductive theorems. Usually, there is only one representative diagram for all instances of the theorem. There is no need for induction to prove the general case: proofs are not schematic. Simple geometric manipulations of a diagram prove the individual case. Abstraction is required to show that this proof will hold for all a, b. Theorems are of continuous space. Example theorem: *Pythagoras' theorem.*

Category 2: Inductive theorems with a parameter. A diagram is a representative of a particular instance of a theorem. Proofs are schematic: they require induction for the general diagram of magnitude n (a concrete diagram cannot be drawn for this instance). An alternative method can sometimes be used to capture the generality of the proof. Theorems are of discrete space. Example theorem: *sum of odd naturals.*

Category 3: Theorems whose proofs are inherently inductive: for each individual concrete case of the diagram they need an inductive step to prove the theorem. Every particular instance of a theorem, when represented as a diagram, requires the use of abstractions to represent infinity. Theorems are of continuous space. Example theorem: *geometric sum.*

18.2.3 Problem Domain

We choose mathematics as our domain for theorems since it allows us to make formal statements about the reasoning, proof search, induction, generalisations, abstractions and such issues. Having introduced the examples and their categorisation, which is by no means exhaustive, we are now able to further restrict our domain of mathematical theorems.

First, we narrow down the domain to a subset of theorems that can be represented as diagrams without the need for abstraction (e.g., the use of ellipsis, as in the above example theorem for *geometric sum*). Conducting proofs and using abstractions in diagrams is problematic, since it is very difficult to keep track of these abstractions while manipulating the diagram during the proof procedure.

Second, we consider diagrammatic proofs that require induction to prove the general case (i.e., Category 2 above). Namely, diagrams can be drawn only for concrete situations and objects. We cannot draw, for example, an

$n \times n$ square without some abstraction device, e.g., ellipsis. Our challenge is to find a mechanism for extracting a general proof that does not require using abstractions in diagrams.[3] The generality of the proof will be captured in an alternative way (by using the constructive ω-rule – see Section 18.2.4).

Third, to date we consider theorems of natural number arithmetic only. DIAMOND is designed to prove examples of Category 2, where diagrams represent natural numbers. A more formal definition of diagrams is given in Section 18.3.4. We may extend diagrammatic theorem proving for examples of Category 1 as well.

One of the possibilities for future work is to consider a need for a more formal problem domain definition.

18.2.4 Constructive ω-Rule

As mentioned above we use the constructive ω-rule to prove theorems of Category 2. Siani Baker in [1] did some work on the constructive ω-rule and schematic proofs for theorems of arithmetic. Here, we explain the idea behind constructive ω-rule and schematic proofs and how they can be applied to diagrammatic proofs.

Schematic Proof. Schematic proofs use the constructive ω-rule which is an alternative to induction. The constructive ω-rule allows inference of the sentence $\forall x P(x)$ from an infinite sequence $P(n)$ $n \in \omega$ of sentences.

$$\frac{P(0), P(1), P(2), \ldots}{\forall n.P(n)}$$

where "if each $P(n)$ can be proved *in a uniform way* (from parameter n), then conclude $\forall n P(n)$". The criterion for uniformity of the procedure of proof using the constructive ω-rule is taken to be the provision of a general schematic proof, namely the proof of $P(n)$ in terms of n, where some rules R are applied some function of n (i.e., $f_R(n)$) times (a rule can also be applied a constant number of times). Let the proof of $P(n)$ be captured using a recursive function $\mathsf{proof}(n)$. Now, $\mathsf{proof}(n)$ is schematic in n, since we applied some rule R n times. The following procedure summarises the essence of using the constructive ω-rule in schematic proofs:

1. Prove a few special cases (e.g. $P(2)$, $P(16)$, ...).
2. Abstract (guess) $\mathsf{proof}(n)$ (e.g. from $\mathsf{proof}(2)$, $\mathsf{proof}(16)$, ...).
3. Prove that $\mathsf{proof}(n)$ proves $P(n)$ by meta-induction on n.

The general pattern is extracted (guessed) from the individual proof instances by (learning type) inductive inference. By meta-mathematical induction we mean that we introduce system META such that for all n:

[3] Note that [2] formalises the use of abstractions, however, in the domain of well founded relations.

$$\vdash_{\text{META}} \mathsf{proof}(n) : P(n)$$

where ":" stands for "is a proof of". Baker used PA_ω (i.e., Peano arithmetic with ω-rule) for the system META [1]. The meta inductive rule is defined as follows:

$$\frac{\vdash_{\text{META}} \mathsf{proof}(0) : P(0) \quad \mathsf{proof}(r) : P(r) \vdash_{\text{META}} \mathsf{proof}(s(r)) : P(s(r))}{\vdash_{\text{META}} \forall n\ \mathsf{proof}(n) : P(n)}$$

This essentially says that by using the rules on $P(s(n))$ we can reduce it to $P(n)$. For more information, see [1].

Diagrams and Schematic Proofs. We claim that we can extend Baker's work on schematic proofs to our diagrammatic proofs so that the generality of the diagrammatic proof is embedded in the schematic proof. Thus, we eliminate the need for abstractions in diagrams, and can extract a general schematic proof from manipulations on concrete diagrams.

The diagrammatic schematic proof starts with a few particular concrete cases of the theorem represented by the diagram. The diagrammatic procedures (i.e., operations) on the diagram are performed next, capturing the inference steps of the diagrammatic proof. This step corresponds to the first step of the schematic proof procedure given in the previous section (18.2.4).

The second step is to abstract the operations involved in the schematic proof for n. Note that the generality is represented as a recursive program which specifies a sequence of diagrammatic procedures (operations) that are used on a diagram, and not as a general representation of a diagram. More precisely, the basic idea is to consider proofs for $n+1$ which can be reduced to proofs for n (or conversely, such proofs for n which can be extended to proofs for $n+1$ by adding to them some additional sequence of operations). The difference between the proof for $(n+1)$ and the proof for n, i.e., the additional sequence of operations in the proof for $(n+1)$ with respect to the proof for n, is referred to as the step case of the abstracted schematic proof.

The last step in the schematic proof procedure is to prove by meta-induction that the abstracted diagrammatic schematic proof is indeed correct. One way of proving the correctness of schematic proofs is to create a theory of diagrams that models the processes in a diagrammatic reasoning system and prove correctness there. A formal definition of a diagrammatic proof of an arithmetic statement and the correctness of this diagrammatic proof will be discussed in Section 18.3.4.

Schematic Diagrammatic Proof for the Sum of Odd Naturals. Now we can structure the diagrammatic proofs in a more formal way. Here we list the proof for the theorem about the *sum of odd naturals* as a sequence of steps that need to be performed on the diagram:

1. Cut a square into n ells, where an ell consists of two adjacent sides of the square.

2. For each ell, continue splitting from an ell pairs of dots at the end of two adjacent sides of the ell until only 1 dot is left (note that for each ell of magnitude n, we will have $n-1$ pairs of dots plus another dot which is a vertex of the two adjacent sides, i.e., $2(n-1)+1$).

Identifying the operations (i.e., geometric manipulations) that were required to prove the theorem will help us define a large repertoire of such operations which will be used in the diagrammatic proofs. The generality of the proof is captured by the use of the constructive ω-rule, by which we take a few special cases of the diagram (say squares of magnitudes 15 and 16), and find the general pattern of the proof that will hold for each case (e.g., the schematic proof given above).

18.3 DIAMOND System

The diagrammatic proof system DIAMOND is an embodiment of some of the ideas presented in this paper. DIAMOND stands for **Dia**gram**m**atic Reas**on**ing and **D**eduction.

Clearly, an important issue in the development of DIAMOND is the *internal* representation of diagrams and operations on them. It was George Pólya who was first to advise us on the importance of knowledge representation [12, 13]. Simon argued Pólya's point further in [15] by stating that solving a problem means representing it so that the solution becomes trivial, or at least transparent. In automated reasoning it is difficult to see how to use this advice, since there is normally only one representation scheme for the problem which is available to the system. In DIAMOND we choose a representation which we hope captures the intuitiveness, rigour and simplicity of human reasoning with diagrams. We aim to represent diagrams in a way which enables a theorem prover to prove theorems using diagrams.

In DIAMOND we use a mixture of Cartesian and topological representations. DIAMOND uses a primitive notion of a diagram: a dot. All other *elementary* and *derived* diagrams (e.g., rows, columns, ells, frames, squares, triangles, rectangles) are composed in various ways out of dots. The advice of Pólya about alternative representations can readily be used in DIAMOND. Namely, diagrams can be represented in a variety of different ways. For instance, a square is represented as: a sequence of rows; a sequence of columns; a concentric sequence of circumferences, each of which is called a frame; a nested sequence of ells; a sequence of four similar squares; a matrix; a sequence of diagonals.

The choice of the representation that DIAMOND uses is important. Most of the proofs that DIAMOND proves require some kind of recursive decomposition of a diagram. Each alternative representation makes available a different form of recursive decomposition. For more information on the choice of internal representation for diagrams in DIAMOND see [8].

The architecture of DIAMOND consists of two parts. The *diagrammatic component* forms and processes the diagram. It is the interface between DIAMOND and the user. The *inference engine* deals with the diagrammatic inference steps. It processes the operations on the diagram. An important submodule is the abstraction mechanism which is used to extract general schematic proofs from example proofs.

The rest of this section presents the operations used to construct proofs, the structure of proofs and the abstraction mechanism used in DIAMOND.

18.3.1 Geometric Operations

Geometric operations (also referred to as manipulations or procedures) capture the inference steps of the proof. Thus, a sufficiently large number of such operations, which are then available to the user in the search for the proof, needs to be identified and formalised. Since we are not generating, i.e., discovering diagrammatic proofs, but rather we are trying to understand them, we can expect from the user to input these operations. To date, a small number of such operations have been implemented and are available to the user.

DIAMOND is targeted to prove theorems of discrete arithmetic. Diagrams are a way of representing natural numbers. The interest lies in the effect on the numbers that diagrams represent after an operation has been applied on the diagrams. Thus, the operations join and split diagrams apart in various ways. Some operations are just simple ones (e.g., split a row from a square), and some are more complicated ones (e.g. decompose a square into a sequence of rows). Hence, DIAMOND distinguishes between two types of operations: *atomic* and *composite*.[4]

- *Atomic operations* are basic one-step operations that can be combined into more complex operations. Examples of such operations are: rotate, translate, cut, split, join, remove and insert a segment. To date, there are 14 atomic operations implemented in DIAMOND.
- *Composite operations* are more complex, typically recursive operations, composed from simple atomic ones. One can think of them as tactics in automated reasoning. Composite operations are defined in terms of decomposition of different recursive representations of diagrams. Depending on the theorem at hand, the diagram is viewed using a particular representation, which enables one to use a particular recursive composite operation. Ideally, the internal representation of the diagram is pertinent to the composite operation that is being carried out on it. Such a representation would render an operation very easy to apply. It would be just a simple decomposition of the representation of a diagram. Examples of such operations

[4] A complete list of operations can be found in [8], however, a more formal definition of operations as part of the diagrammatic theory can be found in Section 18.3.4.

are: recursive decomposition of a square into rows, or columns, or cells, or frames.

In the example of the theorem for the *sum of odd naturals* the proof consists of the following operations: lcut and split_ends.

18.3.2 Constructing a Proof

DIAMOND's example proof consists of a sequence of applications of geometric operations on a diagram. The abstraction is then carried out automatically, if any such abstraction exists for the two example proofs given.[5] DIAMOND expects the example proofs to be formulated in a particular way, where the order of operations in the user's formulation of the example proofs is crucial. Both example proofs are expected to be given with the same order of operations, but with some extra operations in the case of the proof of $(n+1)$ for some particular n.

Consider the example for the *sum of odd naturals.* The step cases for proofs for $n=4$ and $n=3$ look as follows:

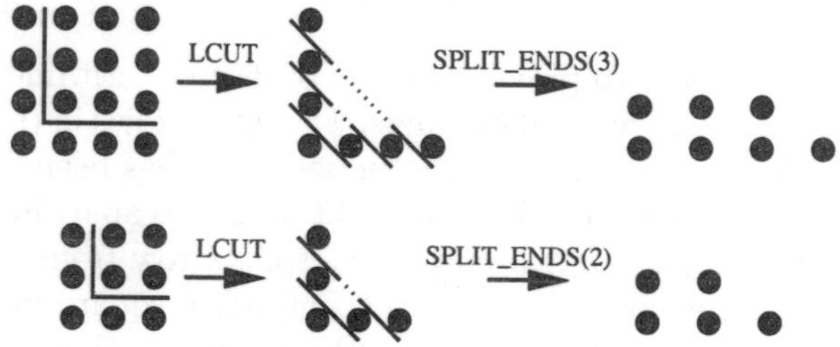

The aim is to recognise automatically the structure of the proof from a linear sequence of applications of operations, so that the example proofs for n and $n+1$ can be reformulated in the general case into the following:

$$\begin{aligned} \mathsf{proof}(n) &= \mathcal{A}(n), \mathcal{A}(n-1), \ldots \mathcal{A}(1), \mathcal{B} \\ \mathsf{proof}(n+1) &= \mathcal{A}(n+1), \mathcal{A}(n), \mathcal{A}(n-1), \ldots \mathcal{A}(1), \mathcal{B} \end{aligned}$$

where for each n, $\mathcal{A}(n)$ is a step case consisting of a sequence of applications of some operations and $\mathcal{B}$ is a base case for $n=0$. Alternatively, we seek this recursive reformulation:

$$\begin{aligned} \mathsf{proof}(n+1) &= \mathcal{A}(n+1), \ \ \mathsf{proof}(n) \\ \mathsf{proof}(0) &= \mathcal{B} \end{aligned}$$

[5] If the proof contains a case split for say, even and odd integers, and the two example proofs given are for two different cases, then DIAMOND cannot abstract from them. However, DIAMOND recognises that the example proofs were given for different cases, and requests the user to supply another example proof for each case, in order for it to be able to abstract. This will be further explained in Section 18.3.3.

Note that proof(0) is often an empty list of operations, because often no diagram is defined for $n = 0$, i.e., a diagram which consists of no dots.

A further issue that we are investigating currently is to relax the requirement for a particular ordering of operations in formulating example proofs. Sets with partial ordering could be used as an alternative.

18.3.3 Abstraction

Given some example proofs DIAMOND needs to abstract from them, so that the final diagrammatic proof is not only for the cases of specific n's, but holds for all n. Such a schematic proof is a general program which specifies the applications of some operations, where the number of application of each operation is dependent on n or is a constant.

We distinguish between two types of example proofs: *destructor*, i.e., the example proofs which are formulated so that the base case operations are performed last (in a sense, the initial diagram is "destructed" by the application of operations down to a trivial diagram, forming the proof along in this way); and *constructor*, i.e., the base case operations are performed first, followed by the step case operations. In DIAMOND we arbitrarily choose to use destructor schematic proofs.

A proof that has the same structure for all n, i.e., one recursive function defines a complete proof for all n, is called a 1-homogeneous proof. Proofs can be c-homogeneous; then there are c cases of the proof. For instance, when there are two cases of a proof, one for odd and one for even natural numbers, then there need to be two recursive functions defining each case of the proof – the proof is 2-homogeneous. We say that if all concrete instances of the proof (for instances of numbers that "equal modulo c") have the same structure and can be abstracted, then the proof is c-homogeneous. If there are c cases, then there are c different abstracted proofs, one for each case. We seek the smallest complete recursive definition of a proof, i.e., c potentially different schematic proofs, if there are c cases. The following theorem and corollary will help us define what we mean by the smallest complete proof:

Theorem 18.3.1. If a proof is c-homogeneous, then it is also (kc)-homogeneous for every natural number $k > 0$.

The immediate consequence of Theorem 18.3.1 is:

Corollary 18.3.1. If a proof is *not* c-homogeneous, then it is also *not* f-homogeneous for every factor f of c.

In a c-homogeneous proof we will denote by $\mathcal{B}_r$ a base case for a branch of numbers which give remainder r when divided by c. $\mathcal{B}_r$ is actually a proof for the smallest natural number that gives remainder r when divided by c.

A schematic proof is defined to be the smallest complete proof if there is no other f-homogeneous proof obtainable from a c-homogeneous proof for any factor f of c, and all f schematic proofs for f cases are defined.

The general representation of a destructor proof is formalised as follows – let: $n = kc + r$ where $c = \textit{number of cases}$ and $r < c$, and $i \geq 1$. Then the recursive definition of a general proof is:

$$\mathsf{proof}(ic + r) = \mathcal{A}_r(ic + r),\ \ \mathsf{proof}((i-1)c + r)$$
$$\mathsf{proof}(r) = \mathcal{B}_r$$

where $\mathcal{A}_r$ is a step case and $\mathcal{B}_r$ is a base case for a class of proofs where $n \equiv r(mod\ c)$. "," denotes concatenation of sequences of operations: "do operations of $\mathcal{A}(n+1)$, then $\mathsf{proof}(n)$". The formalisation of abstracted proof for *constructor* proofs is symmetric to the one given above.

Abstracting for all Linear Functions. As mentioned above, we aim to recognise the particular recursive structure of the given example proofs. More precisely, we want to extract the step case $\mathcal{A}$ and the base case $\mathcal{B}$ of the proof and then abstract them for all n. The general methodology employed for doing this can be demonstrated as:

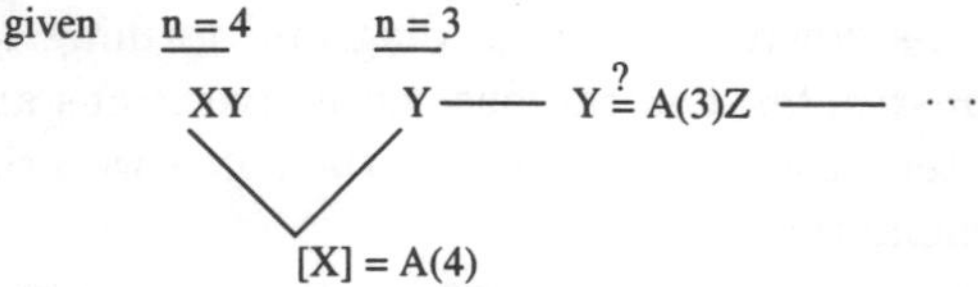

The first step of the abstraction algorithm is to extract the difference between the two given example proofs for n_1 and n_2 ($n_1 > n_2$), where $c = n_1 - n_2$, in the hope that this, when abstracted, will be the step case $\mathcal{A}$ of the proof. This is done by commutative and associative matching[6] which detects and returns the difference between the two example proofs. Now we have a concrete step case of the proof. This difference consists of a few operations op_k each applied x_{k,n_1} times for some natural k.

To make a step case general, we need to find the dependency function between every x_{k,n_1} and n_1. This demands identifying a function of n_1, which would give a specific x_{k,n_1}, i.e., $f_k(n_1) = x_{k,n_1}$ for some k and n_1. DIAMOND assumes that the dependency is linear: $an+b$. This is a heuristically adequate choice. Thus, let us write for each op_k a linear equation $an_1+b = x_{k,n_1}$, where n_1 and x_{k,n_1} are known.

The subsequent stage of the abstraction is to extract the next step case from the rest of the example proof for the corresponding new n (i.e., n_2). If successful, continue extracting step cases for the corresponding n's from the rest of the proof until only the base case is left.

Since we are dealing with inductive proofs, it is expected that every step case of a proof will have the same structure, i.e., will consist of the same sequence of application of operations, but a different number of times. Thus, we could in the same way as above for every operation op_k write a linear

[6] Using commutative and associative matching reduces the sensitivity to the order of proof steps [8].

equation $an_2 + b = x_{k,n_2}$. However, the number x_{k,n_2} of applications of a particular operation op_k in the next step case is not known. A possible value of x_{k,n_2} is acquired by counting the number x' of times every operation op_k of the initial step case occurs in the rest of the proof. The actual value of the number of occurrences of each operation could be any number from 0 to x'. Thus, we do branching for all such values and thus we have:

$$an_1 + b = x_{k,n_1}$$
$$an_2 + b = x_{k,n_2}$$

where n_1, n_2, x_{k,n_1} and x_{k,n_2} are known, so the equations can be solved for a and b, and x_{k,n_2} takes values from 0 to x'. This results in several possible potential abstractions of the step case. The aim is to eliminate those that are impossible. After checking if step cases for all n down to the base case are structurally consistent one hopes to be left with at least one possible abstraction of the example proofs. The step case is rejected when the sequence of operations in the subsequent step cases is impossible, i.e., the functions were wrong. This normally occurs when the dependency function gives a negative number of applications of a particular operation, when the calculated sequence is not identical to the rest of the example proof, or when there is no integer solution to our equations. Usually, there will be only one possible abstraction of the two given example proofs.

The example proof for the *sum of odd naturals* is abstracted into the following step case and base case:

$$\mathcal{A}(n) = [(\mathsf{lcut}, 1), (\mathsf{split\_diagonal\_ends}, n-1)]$$
$$\mathcal{B} = [\,]$$

where the function in parentheses indicates the number of times that the operations are applied for each particular n.

***f*-Homogeneous Proof.** Assume two example proofs for the *sum of odd naturals* (the example proof would consist of making n lcuts, and then showing that each ell consists of an odd number of dots). If the user supplies two example proofs for values of n and $n+1$, for some concrete n, then there is no problem, so DIAMOND will abstract normally and determine that the proof is 1-homogeneous. However, should the user supply proofs for n and $n+2$ for some concrete n, the first stage of abstraction would determine that the step case consists of two lcuts. However, a complete recursive function for abstraction requires a step case to consist of one lcut only.

DIAMOND checks this by trying to split the step case into a further f structurally identical sequences of operations, for all factors f of c in order to obtain an f-homogeneous proof. If the method fails, then there is no such f-homogeneous further abstraction of the step case $\mathcal{A}(n)$. If the method succeeds, and DIAMOND finds a new abstraction of the step case, call this $\mathcal{A}'(n)$, then it also needs to find a new base case $\mathcal{B}'_{r'}$ where the previous r

for c was such that $n = kc + r$ and $r < c$, and the new r' is now such that $n = kf + r'$ and $r' < f$.

18.3.4 Correctness of the Schematic Proof

The last stage of extracting a diagrammatic proof is to check that the *guessed* general schematic proof is indeed correct. To prove that the schematic proof is correct we need to show in some meta-theory that proof(n) uniformly proves $P(n)$ for all n, i.e., it gives a proof tree with $P(n)$ at its root, and axioms at its leaves. This requires reasoning about proofs, i.e., meta-level reasoning. A meta-level proof using general diagrams would be an obvious method for verifying our schematic proof. However, such a meta-level proof would reintroduce the need for abstractions (e.g. ellipsis) of diagrams, which we are trying to avoid.

One way of overcoming this problem is to define diagrams and operations in a theory of diagrams where we can express abstract diagrams symbolically rather than diagrammatically. In this theory we can verify schematic proofs by defining the notion of applicability of a posited proof. Given that a particular theorem is expressed as an equality, its schematic proof is correct if applying the operations specified in the schematic proof of the diagrammatic representation of the left-hand side of the theorem results in the diagrammatic representation of the right-hand side of the theorem. There are two conditions that need to be satisfied. The first condition is that there is an appropriate diagrammatic representation available for the conversion of the theorem into its diagrammatic representation. The second condition is that the operations of the schematic proof are defined on those diagrams.

Before we can state the definition of the correctness property of schematic proofs, we need to formalise the machinery which will enable us to model the processes of a diagrammatic proof. Therefore, we need to formally define diagrams, operations on them, and the applicability of operations of a schematic proof.

Diagrams. Diagrams in the theory are defined to be of object type. Some examples of the different kinds of object names in the theory are: row, column, ell, frame, square, rectangle, and triangle.

Diagrams of the theory model natural numbers. DIAMOND's primitive notion of a concrete diagram, a dot, is represented in the theory as the natural number 1. Objects are introduced via a constructor function, diagram, which takes the name of the type of diagram and the list of parameters of its magnitude. Thus, the type of constructor function diagram is name $\times$ pnat list $\rightarrow$ object. So, for instance, a square of magnitude 4 is expressed in the theory as diagram(square,[4]). All elementary and derived concrete diagrams are expressed using a primitive object dot, hence in the theory they can be expressed using a constructor function, the object name and some parameter representing a natural number for the magnitude of the diagram.

Constant $\emptyset$ denotes a null diagram, or in other words an empty diagram. We define that any diagram that is of 0 magnitude is an empty diagram (note that $a \in b$ denotes that a natural number a is an element of a list b; thus the type of infix $\in$ is: $\mathsf{pnat} \times \mathsf{pnat\ list} \rightarrow \mathsf{boolean}$):

$$0 \in s \rightarrow \mathsf{diagram}(x, s) \equiv \emptyset \tag{18.1}$$

Note also that all triangles are equilateral.[7] Here are some examples of diagrams: diagram(row,n), diagram(column,n), diagram(square,n), diagram(ell,n).

Operators. This section gives the operators available in the theory. First, we write diagrammatic equality using $\stackrel{d}{=}$ which denotes that two lists of diagrams are identical. Here is the definition of $\stackrel{d}{=}$:

$$\mathsf{X} \stackrel{d}{=} \mathsf{Y} \longleftrightarrow \forall \mathsf{d}.\ \mathsf{count}(\mathsf{d}, \mathsf{X}) = \mathsf{count}(\mathsf{d}, \mathsf{Y})$$

where the function count can be defined by:

$$\begin{aligned} \mathsf{count}(\mathsf{d}, [\,]) &= 0 \\ \mathsf{count}(\mathsf{d}, \mathsf{d} :: \mathsf{D}) &= 1 + \mathsf{count}(\mathsf{d}, \mathsf{D}) \\ \mathsf{d} \neq \mathsf{e} \rightarrow \mathsf{count}(\mathsf{d}, \mathsf{e} :: \mathsf{D}) &= \mathsf{count}(\mathsf{d}, \mathsf{D}) \end{aligned}$$

Diagrammatic equality $\stackrel{d}{=}$ is a larger relation than an arithmetic equality $=$, because it has all the properties of $=$, i.e., reflexivity, symmetry, transitivity and substitution properties, plus an additional one – the order of elements in a list does not matter. Therefore, two lists of diagrams, X and Y, are diagrammatically equal, $\mathsf{X} \stackrel{d}{=} \mathsf{Y}$, even if the orders in which the diagrams are listed in both lists differ.[8]

We now define some operators that introduce the existence of several diagrams (note that the data type pnat stands for non-negative natural number of Peano arithmetic): "@" is append on lists, "::" and nil are list constructors (concatenation of elements onto a list, and an empty list), $\otimes$ is an infix operator which introduces a combination of a number of identical lists of diagrams, and $\biguplus$ denotes a collection of diagrams of increasing magnitudes which are all of the same kind – it is analogous to $\sum$ for summation of integers. Here is the recursive definition of $\biguplus_{i=a}^{b}$ for all $a \leq b$:[9]

[7] It is hard to represent discrete triangles that are of any magnitude, i.e., the sides are of different and any magnitudes. Triangles are represented in a discrete space. Hence, they appear to be right-angle triangles, despite the fact the all the sides of any triangle are of equal discrete magnitude. Were we to extend DIAMOND to prove theorems of real arithmetic (see Section 18.4), then there would be a need for continuous space, and therefore a scope for triangles of any magnitude.

[8] Note that our definition of diagrammatic equality of lists is equivalent to bag equality. The order of the elements in a bag does not matter.

[9] Note that to simplify the notation we write $\biguplus_{i=a}^{b} D(i)$ instead of $\biguplus(a, b, \lambda i.D(i))$.

$$\biguplus_{i=a}^{a} \mathsf{diagram}(\mathsf{name}, f(i)) \stackrel{d}{=} [\mathsf{diagram}(\mathsf{name}, f(a))] \quad (18.2)$$

$$a \leq b \rightarrow \biguplus_{i=a}^{b+1} \mathsf{diagram}(\mathsf{name}, f(i)) \stackrel{d}{=} \biguplus_{i=a}^{b} \mathsf{diagram}(\mathsf{name}, f(i))@ [\mathsf{diagram}(\mathsf{name}, f(b+1))] \quad (18.3)$$

Note that f is some function which generates a list of natural numbers for a given number i. This list denotes the parameters of a magnitude of a diagram.

Operations. Diagrammatic operations are represented via a function op : opname × object list → object list. We give here a definition of one operation only, but there are many more operations defined in the theory – see [8].

$$\mathsf{op}(\mathsf{lcut}, \mathsf{diagram}(\mathsf{square}, n) :: \mathsf{D}) \stackrel{d}{=} [\mathsf{diagram}(\mathsf{square}, n-1), \mathsf{diagram}(\mathsf{ell}, n)]@\mathsf{D} \quad (18.4)$$

One_Apply and Apply. Here we define what it means to apply an operation on a diagram several times. We use a function apply and function one_apply. Let:

$$\mathsf{one\_apply}(0, \mathsf{opnm}, \mathsf{D}) \stackrel{d}{=} \mathsf{D} \quad (18.5)$$

$$\mathsf{one\_apply}(n+1, \mathsf{opnm}, \mathsf{D}) \stackrel{d}{=} \mathsf{op}(\mathsf{opnm}, \mathsf{one\_apply}(n, \mathsf{opnm}, \mathsf{D})) \quad (18.6)$$

$$\mathsf{apply}([\,], \mathsf{D}) \stackrel{d}{=} \mathsf{D} \quad (18.7)$$

$$\mathsf{apply}((\mathsf{opnm}, x) :: \mathsf{opss}, \mathsf{D}) \stackrel{d}{=} \mathsf{apply}(\mathsf{opss}, \mathsf{one\_apply}(x, \mathsf{opnm}, \mathsf{D}) \quad (18.8)$$

Equations. Here we give a theorem which will be needed. (18.9) is provable from (18.7) and (18.8). Its proof is not given here, but can be found in [8].

$$\mathsf{apply}(\mathsf{ops}, \mathsf{D} :: \mathsf{D}_s) \stackrel{d}{=} \mathsf{apply}(\mathsf{ops}, [\mathsf{D}])@\mathsf{D}_s \quad (18.9)$$

Conversion relation dmap. Let dmap denote a relation between a particular class of statements of arithmetic and their equivalent diagrammatic expressions in the theory of diagrams. The equivalence is defined to be over the *size* of the diagram. The size of a diagram is defined to be the number of counters (dots) in the diagram, i.e., the natural number that the diagram represents. dmap takes two arguments, an arithmetic expression and a list of diagrams which could collectively represent this expression. Hence, the type of the relation dmap is pnat × object list. Here are some general conversions:

$$\mathsf{dmap}(0, [\,]) \quad (18.10)$$

$$\mathsf{dmap}(n^2, [\mathsf{diagram}(\mathsf{square}, [n])]) \quad (18.11)$$

$$\mathsf{dmap}(2n-1, [\mathsf{diagram}(\mathsf{ell}, [n])]) \quad (18.12)$$

$$\mathsf{dmap}(\textstyle\sum_{j=a}^{b} f(j), \biguplus_{j=a}^{b} \mathsf{D}_j) \text{ such that } \forall j, a \leq j \leq b, \mathsf{dmap}(f(j), [\mathsf{D}_j]) \quad (18.13)$$

We have now formalised enough machinery to be able to define the correctness property of a schematic proof.

Definition 18.3.1 (Correctness of Schematic Proofs). proof is a correct schematic proof of a particular conjecture $\forall n \quad L(n) = R(n)$ if for all n there exist two lists of diagrams D and E such that $\mathsf{dmap}(L(n), \mathsf{D})$ and $\mathsf{dmap}(R(n), \mathsf{E})$, and

$$\mathsf{apply}\,(\mathsf{proof}\,(n),\ \mathsf{D}) \stackrel{d}{=} \mathsf{E}$$

It is possible to prove the property in Definition 18.3.1 only if $L(n)$, $R(n)$ and proof are known, i.e., for a specific case of a conjecture and a schematic proof. Knowing $L(n)$ and $R(n)$ allows us to infer some conversion relations which specify two lists of diagrams D and E. This satisfies the first part of Definition 18.3.1. In the next section we prove the correctness of a schematic proof for a particular conjecture at hand.

Proof of Correctness of Schematic Proofs for an Example. Here we prove the property given in Definition 18.3.1 for an example of a schematic proof of a theorem about the *sum of odd naturals*. The theorem is stated as $n^2 = \sum_{i=0}^{n}(2i-1)$. The schematic proof of this theorem is given as:[10]

$$\mathsf{proof}(0) = [\,] \tag{18.14}$$

$$\mathsf{proof}(n+1) = [(\mathsf{lcut}, 1)],\ \mathsf{proof}(n) \tag{18.15}$$

The proof of correctness of a schematic proof for this particular example requires induction on n. The base case for $n = 0$ is trivial, since by (18.10) no operations are applied to an empty diagram list which results in [ ]. We consider a step case of induction.

Step case:

Hypothesis: for n

Using (18.11) notice $\mathsf{dmap}(n^2, [\mathsf{diagram}(\mathsf{square}, [n])])$,
hence let $\mathsf{D} = [\mathsf{diagram}(\mathsf{square}, [n])]$.
Using (18.13) and (18.12) notice $\mathsf{dmap}(\sum_{i=0}^{n}(2i-1), \biguplus_{i=0}^{n} \mathsf{diagram}(\mathsf{ell}, [i]))$,
hence let $\mathsf{E} = \biguplus_{i=0}^{n} \mathsf{diagram}(\mathsf{ell}, [i])$.

$$\mathsf{apply}(\mathsf{proof}(n), [\mathsf{diagram}(\mathsf{square}, [n])]) \stackrel{d}{=} \biguplus_{i=0}^{n} \mathsf{diagram}(\mathsf{ell}, [i])$$

Conclusion: for $n + 1$

Similarly to the hypothesis, D and E are converted for $n + 1$.

---

[10] For the brevity of presentation we take a simpler version of the schematic proof which does not include the operation split_ends.

$$\mathsf{apply}(\mathsf{proof}(n+1), [\mathsf{diagram}(\mathsf{square}, [n+1])]) \stackrel{d}{=} \biguplus_{i=0}^{n+1} \mathsf{diagram}(\mathsf{ell}, [i])$$

$$\mathsf{proof}(n+1) = [(\mathsf{lcut}, 1)],\ \mathsf{proof}(n) \Downarrow$$

$$\mathsf{apply}(((\mathsf{lcut}, 1), \mathsf{proof}(n)), [\mathsf{diagram}(\mathsf{square}, [n+1])]) \stackrel{d}{=} \biguplus_{i=0}^{n+1} \mathsf{diagram}(\mathsf{ell}, [i])$$

$$(18.8) \Downarrow$$

$$\mathsf{apply}(\mathsf{proof}(n), \mathsf{one\_apply}(1, \mathsf{lcut}, [\mathsf{diagram}(\mathsf{square}, [n+1])])) \stackrel{d}{=} \biguplus_{i=0}^{n+1} \mathsf{diagram}(\mathsf{ell}, [i])$$

$$(18.6) \text{ and } (18.5) \Downarrow$$

$$\mathsf{apply}(\mathsf{proof}(n), \mathsf{op}(\mathsf{lcut}, [\mathsf{diagram}(\mathsf{square}, [n+1])])) \stackrel{d}{=} \biguplus_{i=0}^{n+1} \mathsf{diagram}(\mathsf{ell}, [i])$$

$$(18.4) \Downarrow$$

$$\mathsf{apply}(\mathsf{proof}(n), [\mathsf{diagram}(\mathsf{square}, [n]), \mathsf{diagram}(\mathsf{ell}, [n+1])]) \stackrel{d}{=} \biguplus_{i=0}^{n+1} \mathsf{diagram}(\mathsf{ell}, [i])$$

$$(18.9) \Downarrow$$

$$\mathsf{apply}(\mathsf{proof}(n), [\mathsf{diagram}(\mathsf{square}, [n])]) @[\mathsf{diagram}(\mathsf{ell}, [n+1])] \stackrel{d}{=} \biguplus_{i=0}^{n+1} \mathsf{diagram}(\mathsf{ell}, [i])$$

$$(\text{RHS of hypothesis}) \Downarrow$$

$$\biguplus_{i=0}^{n} \mathsf{diagram}(\mathsf{ell}, [i]) @[\mathsf{diagram}(\mathsf{ell}, [n+1])] \stackrel{d}{=} \biguplus_{i=0}^{n+1} \mathsf{diagram}(\mathsf{ell}, [i])$$

$$(18.3) \Downarrow$$

$$\biguplus_{i=0}^{n+1} \mathsf{diagram}(\mathsf{ell}, [i]) \stackrel{d}{=} \biguplus_{i=0}^{n+1} \mathsf{diagram}(\mathsf{ell}, [i])$$

■

18.3.5 Arithmetic Conjecture and Diagrammatic Proof

Definition 18.3.1 makes no claims about the link between a schematic proof and the theoremhood of a conjecture $\forall n\ L(n) = R(n)$. We still need to consider the possibility of a *correct* schematic proof of a *false* conjecture. To establish that the conjecture is true when proved by a schematic proof, an explicit algebraic link between them needs to be defined. We establish this link via the *size of diagrams.* We first define the size of a diagram, and later, in Theorem 18.3.2, we state the theorem concerning the algebraic correctness of a schematic proof for a given conjecture.

Let us denote the size of the diagram D by $|\mathsf{D}|$. Here is a definition for the size of a diagram:

Definition 18.3.2 (Size of Diagrams). The size of a list of diagrams is equal to the value of the arithmetic expression that it represents: if $\mathsf{dmap}(e, \mathsf{D})$ then $|\mathsf{D}| = e$.

Note that the type of $|\ |$ is: $\mathsf{object\ list} \rightarrow \mathsf{pnat}$. Using the property of size defined in Definition 18.3.2 on formulae from (18.10) to (18.13), we have the following:

$$|[\,]| = 0 \tag{18.16}$$

$$|[\mathsf{diagram}(\mathsf{square}, [n])]| = n^2 \tag{18.17}$$

$$|[\mathsf{diagram}(\mathsf{ell}, [n])]| = 2n - 1 \tag{18.18}$$

$$\left| \biguplus_{j=a}^{b} \mathsf{D}_j \right| = \sum_{j=a}^{b} |[\mathsf{D}_j]| \tag{18.19}$$

Apart from being diagrammatically correct, we want every schematic proof to be *algebraically correct* as well. A schematic proof is algebraically correct if the sizes of the diagrams representing both sides of the proposition after the operations of the schematic proof have been applied are the same. Theorem 18.3.2 states the property of algebraic correctness for any schematic proof.

Theorem 18.3.2 (Algebraic Correctness of Schematic Proofs). For all instances of a schematic proof P and for all pairs of lists of diagrams D and E, a schematic proof P is algebraically correct if and only if

$$\mathsf{apply}\,(\mathsf{P}, \mathsf{D}) \stackrel{d}{=} \mathsf{E} \longrightarrow |\mathsf{D}| = |\mathsf{E}|$$

The proof of Theorem 18.3.2 is straightforward by appealing to the properties of diagram size invariance under applications of multiple operations. The lemmas about these properties and the proof of Theorem 18.3.2 are not given here, but can be found in [8].

There is one last theorem needed in the formalisation of diagrammatic theory which will allow us *to prove* theorems of arithmetic using *diagrammatic* proofs. We state in Theorem 18.3.3 the property concerning the diagrammatic provability of arithmetic arguments.

Theorem 18.3.3 (Diagrammatic Provability of Arithmetic Conjecture). A conjecture $\forall n\ L(n) = R(n)$ is diagrammatically provable if and only if for all n there exist two lists of diagrams D and E such that $\mathsf{dmap}(L(n), \mathsf{D})$ and $\mathsf{dmap}(R(n), \mathsf{E})$, and

$$|\,\mathsf{D}\,| = |\,\mathsf{E}\,| \longrightarrow L(n) = R(n)$$

The proof of Theorem 18.3.3 is trivial by the definition of size of a list of diagrams given in Definition 18.3.2.

Diagrammatic Provability for an Example. We consider now an example of an arithmetic conjecture and prove it diagrammatically using a schematic proof that DIAMOND extracts. Let the arithmetic conjecture be

$$\forall n\; n^2 = \sum_{i=0}^{n} 2i - 1$$

and the schematic proof proof that DIAMOND extracted be as defined in (18.14) and (18.15). Here are the reasoning steps of the proof:

1. Appealing to Theorem 18.3.3 we can discharge the conjecture by:
   - using (18.11) notice $\mathsf{dmap}(n^2, [\mathsf{diagram}(\mathsf{square}, [n])])$, hence let $\mathsf{D} = [\mathsf{diagram}(\mathsf{square}, [n])]$,
   - using (18.13) and (18.12) notice $\mathsf{dmap}(\sum_{i=0}^{n}(2i-1), \biguplus_{i=0}^{n} \mathsf{diagram}(\mathsf{ell}, [i]))$, hence let $\mathsf{E} = \biguplus_{i=0}^{n} \mathsf{diagram}(\mathsf{ell}, [i])$,

   and proving for all n

$$|\,[\mathsf{diagram}(\mathsf{square}, [n])]\,| = \left|\, \biguplus_{i=0}^{n} \mathsf{diagram}(\mathsf{ell}, [i]) \right| \tag{18.20}$$

2. Appealing to Theorem 18.3.2 and $\mathsf{proof}(n)$ that DIAMOND extracted, we can discharge the expression in (18.20) by proving for all n

$$\mathsf{apply}\,(\mathsf{proof}(n), [\mathsf{diagram}(\mathsf{square}, [n])]) \stackrel{d}{=} \biguplus_{i=0}^{n} \mathsf{diagram}(\mathsf{ell}, [i]) \tag{18.21}$$

3. Finally, notice that we have already proved (18.21) in Section 18.3.4.

■

18.4 Results and Further Work

DIAMOND is implemented in Standard ML of New Jersey, Version 109.[11] The code is available upon request to the first author.

[11] Standard ML of New Jersey (SML/NJ) is a compiler and programming environment for the Standard ML programming language. SML/NJ is publicly available via the Internet on the following site: `http://cm.bell-labs.com/cm/cs/what/-smlnj/index.html`

The entire process of interactive construction of proofs, automatic abstraction from example proofs, and automatic verification of schematic proofs in the theory of diagrams have been implemented in DIAMOND. In the evaluation of DIAMOND we distinguished between a development and a test set of theorems which we proved using DIAMOND. The development set of theorems included three theorems: *the sum of odd naturals*, i.e., $n^2 = \sum_{i=0}^{n} 2i - 1$, *the sum of all naturals*, i.e., $\frac{n(n+1)}{2} = \sum_{i-0}^{n} i$, and *an odd triangular sum*, i.e., $Tri_{2n+1} = Tri_{n+1} + 3Tri_n$, where Tri_i is an i-th triangular number [11]. The test set included 26 theorems. Some proofs of these theorems are reported more elaborately in [8]. All of these theorems contribute to the significant range and depth of theorems proved using DIAMOND. For more information, the reader is referred to [8].

We want to relax the restriction currently imposed on the formulation of example proofs. Our abstraction mechanism can deal with a linear sequence of operations. This sequence is in fact a linearisation of some partially ordered sequence of operation. We want an abstraction mechanism which would be sensitive to partially ordered sequences of operations.

There is also a possibility of allowing non-linear dependency functions in general schematic proofs: e.g. exponential or polynomial function.

Some recognition and generalisation of diagrams using abstractions could be an interesting issue to consider. This requires some formalisation of abstractions (e.g. ellipsis) in diagrams.

There is a possibility to extend DIAMOND's problem domain from natural number arithmetic to geometry or even further to a different field such as hardware verification. Extending our problem domain to geometry would enable us to prove theorems of Category 1 which are usually geometric theorems of continuous space. These do not require induction, hence there would be no need for DIAMOND's abstraction mechanism. The generality is embedded in the use of continuous space and diagrams of general magnitude. The existing operations and the formalisation of schematic proofs can be used. Additional operations for moving diagrams in various directions would need to be implemented. For more information on the possible extension of DIAMOND to other problem domains, see [8].

DIAMOND is an interactive proof checker. A long-term goal is to design an automated theorem prover capable of discovering diagrammatic proofs.

18.5 Related Work

Several diagrammatic systems such as the Geometry Machine [5], Diagram Configuration model [9], GROVER [2] and Hyperproof [3] have been implemented in the past and are of relevance to our system. Additional information about issues in reasoning with diagrams can be found in [4], which is a good reference for demonstrating how extensive and important this field is.

One of the first systems to use a diagram in proving theorems was Gelernter's Geometry Machine [5]. The diagram in the geometry machine has two roles. Its *negative* role is to reject hypotheses (subgoals) that are not true in the diagram. In this way the search space is pruned. The *positive* role of the diagram is to shorten the inference paths by assuming various facts that are obvious in the diagram as true.

Koedinger and Anderson [9] implemented a geometry problem solver called the Diagram Configuration (DC) model. The key feature of the system is that it organises its data in perceptual chunks, called diagram configurations. These are analogical to key features of diagrams that humans recognise when they inspect a diagram. Therefore, during the process of generating a solution path, DC infers the key steps first, and ignores along the way the less important features of the diagram, i.e., the less important inference steps.

"&"/GROVER, developed in [2], is an automated reasoning system which uses information from a diagram to guide proof search. Its problem domain are theorems of well-founded relations. It consists of the "&" automated theorem prover, based on the sequent calculus for Zermelo set theory, and GROVER, which is the diagram-interpreting component of the system. It passes information extracted from the diagram and translated into logical formulae in the language of "&" to the "&" theorem prover. These formulae are then used as additional hypotheses to the main proof of the conjecture. GROVER considers only subgoals that are known to be true in the diagram.

The main common feature of these three systems is their use of diagrams. The diagram is used to model algebraic statements, and the system uses these models for heuristic guidance while searching for an *algebraic* proof. Thus, the basic underlying reasoning process is non-diagrammatic. In contrast, proofs in DIAMOND are explicitly constructed by operations on diagrams, thus the inference steps of the proof are entirely diagrammatic.

Perhaps a more closely related system to DIAMOND is Hyperproof [3]. Hyperproof reasons about the blocks world. It is an educational tool to teach principles of logic reasoning and proof construction. Hyperproof is an interactive tool for proof checking, as opposed to an automated theorem prover. It is a heterogeneous logic system, because it models inferencing between different kinds of representation. Unlike traditional systems for first-order logic, which use sentential representation, Hyperproof uses sentential and diagrammatic (graphical) representation. It uses a diagram for a concise representation of a complex system aiming to aid human reasoning. The user can take advantage either of conventional sentential inference rules or diagrammatic inference rules. It differs from DIAMOND in that Hyperproof's diagrammatic inference rules deduce from a diagram to a sentential formula or vice versa. It does not have diagrammatic inference rules between two diagrams, as is the case in DIAMOND. Moreover, in Hyperproof the sentential inference rules (as well as diagrammatic description of a situation) are essential to construct a proof.

Closer to our work is work done by Siani Baker [1] described in Section 18.2.4 on the constructive ω-rule, whereby she exploits the uniform structure of inductive proofs to abstract from example proofs. The main difference between Baker's work and ours is the problem domain. Baker implemented the use of the constructive ω-rule for proving arithmetic theorems. Our domain, on the other hand, is diagrammatic theorems. Furthermore, Baker's motivation was to use schematic proofs for theorems that require a cut rule in the inductive proof, otherwise the proof cannot be carried out automatically. Schematic proofs avoid the need for a cut rule. On the other hand, we use the constructive ω-rule in order to justify the automatic provision of general arguments about theorems and their proofs from particular instances.

18.6 Conclusion

One of the aims in the research reported here is to see whether it is possible to automate the use of diagrams in formal proofs. The hope is that automating the "informal" diagrammatic reasoning of humans will shed light on the issues of formality, informality, rigour and "intuitive" understanding of the correctness of diagrammatic proofs. We have made good progress in exploring this important and difficult area. In particular, we have an explicit handle on abstraction. We showed that diagrams *can* be used for *formal* proofs. We presented, as an example, a diagrammatic reasoning system, DIAMOND, which supports interactive construction of diagrammatic proofs. DIAMOND applies diagrammatic reasoning to problem solving in mathematics. The user proves concrete examples of a theorem, and the system automatically abstracts these instances to give a general schematic proof which we hope holds for all n. In DIAMOND we have the logical machinery (meta-theory, constructive ω-rule) to subsequently justify that the schematic proof does indeed prove the original theorem.

Acknowledgements

The research reported in this paper was supported by an Artificial Intelligence Department Studentship from the University of Edinburgh and a Slovenian Scientific Foundation supplementary studentship for the first author, and by EPSRC grant GR/L/11724 for the other two authors. The copyright reprint permission fee to Kluwer was financed by the School of Computer Science, University of Birmingham.

References

1. Baker, S., Ireland, A. and Smaill, A. (1992). On the use of the constructive omega rule within automated deduction. In A. Voronkov (Ed.), International

conference on logic programming and automated reasoning – LPAR 92, St. Petersburg. Lecture Notes in Artificial Intelligence No. 624, Berlin: Springer, pp. 214–225.

2. Barker-Plummer, D. and Bailin, S.C. (1997). The role of diagrams in mathematical proofs. Machine Graphics and Vision 6(1):25–56.
3. Barwise, J. and Etchemendy, J. (1991). Visual information and valid reasoning. In W. Zimmerman and S. Cunningham (Eds), Visualization in teaching and learning mathematics. Washington, DC: Mathematical Association of America, pp. 9–24.
4. Chandrasekaran, B., Glasgow, J. and Narayanan, N.H. (Eds) (1995). Diagrammatic reasoning: Cognitive and computational perspectives. Cambridge, MA: AAAI Press/MIT Press.
5. Gelernter, H. (1963). Realization of a geometry theorem-proving machine. In E. Feigenbaum and J. Feldman (Eds), Computers and thought. New York: McGraw Hill, pp. 134–52.
6. Hammer, E.M. (1995). Logic and visual information. Stanford, CA: Center for the Study of Language and Information.
7. Jamnik, M., Bundy, A. and Green, I. (1997). Automation of diagrammatic reasoning. In M.E. Pollack (Ed.), Proceedings of the 15th IJCAI, Vol. 1. International joint conference on artificial intelligence. San Mateo, CA: Morgan Kaufmann, pp. 528–533. Also published in the Proceedings of the 1997 AAAI fall symposium. Also available from Edinburgh as DAI Research Paper No. 873.
8. Jamnik, M. (1999). Automating diagrammatic proofs of arithmetic arguments. Unpublished PhD thesis. Available from Edinburgh University.
9. Koedinger, K.R. and Anderson, J.R. (1990). Abstract planning and perceptual chunks. Cognitive Science 14:511–550. Reprinted in Chandrasekaran, B., Glasgow, J. and Narayanan, N.H. (Eds) (1995). Diagrammatic reasoning: Cognitive and computational perspectives. Cambridge, MA: AAAI Press/MIT Press, pp. 577–625.
10. Larkin, J.H. and Simon, H.A. (1987). Why a diagram is (sometimes) worth ten thousand words. Cognitive Science 11:65–99. Reprinted in Chandrasekaran, B., Glasgow, J. and Narayanan, N.H. (Eds) (1995). Diagrammatic reasoning: Cognitive and computational perspectives. Cambridge, MA: AAAI Press/MIT Press, pp. 69–109.
11. Nelsen, R.B. (1993). Proofs without words: Exercises in visual thinking. Washington, DC: Mathematical Association of America.
12. Pólya, G. (1945). How to solve it. Princeton, NJ: Princeton University Press.
13. Pólya, G. (1965). Mathematical discovery (2 vols). New York: Wiley.
14. Shin, S.J. (1995). The logical status of diagrams. Cambridge, UK: Cambridge University Press.
15. Simon, H.A. (1996). The sciences of the artificial (3rd edn). Cambridge, MA: MIT Press.
16. Sowa, J. (1984). Conceptual structures: Information processing in mind and machine. Reading, MA: Addison-Wesley.
17. Stenning, K. and Oberlander, J. (1995). A cognitive theory of graphical and linguistic reasoning: Logic and implementation. Cognitive Science 19:97–140.

19. On the Practical Semantics of Mathematical Diagrams

Dave Barker-Plummer

Sidney C. Bailin

This chapter describes our research into the way in which diagrams convey mathematical meaning. Through the development of an automated reasoning system, called &/GROVER, we have tried to discover how a diagram can convey the meaning of a proof. &/GROVER is a theorem-proving system that interprets diagrams as proof strategies. The diagrams are similar to those that a mathematician would draw informally when communicating the ideas of a proof. We have applied &/GROVER to obtain automatic proofs of three theorems that are beyond the reach of existing theorem-proving systems operating without such guidance. In the process, we have discovered some patterns in the way diagrams are used to convey mathematical reasoning strategies. Those patterns, and the ways in which &/GROVER takes advantage of them to prove theorems, are the focus of this chapter.

19.1 Introduction

Diagrams and visual images play an essential role in both the comprehension and communication of mathematical proofs. We contend that this role is to make the content of the proof "real" rather than formal. Diagrams are used to represent the objects and relations to which a proof refers. When successfully used, the validity of a proof can be "seen" in the diagram rather than justified as a step-by-step application of formal rules. We suggest that visualisation distinguishes "following" a proof from "seeing" it to be true. In the former case, the proof is not fully assimilated, and thus, we might argue, not fully understood.

What distinguishes the full comprehension of a proof from just following the individual steps? The difference concerns the interpretation of the mathematical language: whether it is understood as a purely formal system of formulae, rules, and inferences, or whether it points to something that, however abstract, is real in the world of the mathematician.

Visualisation, then, is a means by which mathematics sheds its purely formal character and takes on meaning. As such, it is a key aspect not just of mathematical learning but also of mathematical discovery. Diagrams, in turn, are a vehicle for communicating the visualised images. Far from being an expendable aid, diagrams play an essential role in the communication of mathematical meaning.

This chapter summarises our research into the way in which diagrams convey mathematical meaning. Through the development of an automated reasoning system, called &/GROVER, we have tried to discover how a diagram can convey the meaning of a proof. &/GROVER is a theorem-proving system that takes a conjecture to be proved and also a diagram intended to represent the essence of the proof. &/GROVER interprets the diagram as a strategy for performing a detailed formal proof. The diagram focuses &/GROVER's attention on the relevant facts at each stage of the proof.

&/GROVER consists of two parts: GROVER, the diagram processor which is the subject of this paper, and an underlying theorem prover, called &. The diagram processor constructs a strategy on the basis of information extracted from the diagram; & is then called upon to prove the subgoals in this strategy.

Three non-trivial theorems which we have proved fully automatically using the &/GROVER system are: the Diamond Lemma, a theorem from the theory of well-founded relations; the Multiple Peaks Theorem, a generalisation of the Diamond Lemma; and the Schröder–Bernstein Theorem, a theorem from the theory of functions.

In working with these theorems we have discovered a number of heuristics that appear to play a significant role in the interpretation of a mathematical diagram. The heuristics concern the identification and ordering of steps in the proof strategy, and the determination of relevant facts to be used at each stage of the proof.[1]

19.2 How can a Proof be Seen?

The basic hypothesis underlying our work is that visualisations are partial models of the world to which a proof refers.

A visualisation of a theorem consists of exemplars of the patterns asserted to hold. When we prove a universal statement of the form $\forall x.A(x)$, for example, we typically say something like "let c be an arbitrary x", and then proceed to demonstrate $A(c)$. If A is an existential formula of the form $\exists y.B(x,y)$ then we might construct a y for which $B(c,y)$ holds.

We hypothesise that the diagram illustrating a proof is a trail of the instantiations performed along the way: the objects themselves, together with a representation of the relevant facts about them. These facts are mathematical assertions composed of primitive or defined relations between the objects, and logical operators such as conjunction, disjunction, negation, and implication. For GROVER then *a diagram represents a set of facts concerning the*

[1] A more detailed description of the work presented here, including more detail on all of the three proofs mentioned above, can be found in [2].

properties of, and relations between, exemplar objects that are identified in the course of a proof.

19.3 Example of a Diagram-Based Proof: The Diamond Lemma

The Diamond Lemma, a theorem in the theory of well-founded relations, states that a *well-founded* relation that is *locally confluent* is also *globally confluent.*[2] The definitions of these terms are as follows:

- The *domain* of a relation R is the set of all elements that are related by R to some other element, that is, all a such that for some b, either $R(a, b)$ or $R(b, a)$.
- A relation R is *well founded* (WF_R) if there are no infinite R-chains, that is, no infinite series of elements $a, b, c \ldots$ such that $R(a, b)$, $R(b, c)$, ...
- A relation R is *locally confluent* (LC_R) if and only if for any three elements a, b, and c in the domain of R, if $R(a, b)$ and $R(a, c)$, then there is an element d such that $R(b, d)$ and $R(c, d)$.
- The *transitive closure* of R is the relation R^* such that $R^*(a, c)$ if and only if there is an R-chain from a to c, that is, a series $b_1, b_2, \ldots b_n$ such that $R(a, b_1)$, $R(b_1, b_2)$, $\ldots R(b_{n-1}, b_n), R(b_n, c)$.
- The relation R is *globally confluent* (GC_R) if and only if its transitive closure R^* is locally confluent.

The standard proof of this theorem uses a diagram that begins as the upper half of the diamond in Fig. 19.1 and is elaborated in steps, eventually yielding the element h. The proof begins by assuming that arbitrary elements a, b, and c have been selected with $R^*(a, b)$ and $R^*(a, c)$. Since $R^*(a, b)$, there is an R-chain from a to b and therefore there is an element d that is the first element of this chain. Similarly, there is an e one step along an R-chain from a to c.

Now the local confluence of R is invoked to deduce that there is an element f which completes the small diamond shown in Fig. 19.1.

The next step of the proof uses *transfinite induction*, which is a technique for proving properties about the domain of a well-founded relation.[3] Transfinite induction states that, in order to prove a property $P(a)$ for all elements a in the domain of a well-founded relation R, it suffices to show that the property "climbs up" R. That is, it suffices to show that for every x in the

[2] A more detailed description of this proof is given in [1].

[3] There is another diagrammatic proof of the theorem as we have stated it, but it requires mathematical (as opposed to transfinite) induction, and may in fact be more complicated for an automatic prover. The proof (and diagram) that we use is required when local confluence is defined more broadly, to assert the existence of an R* relation between b and d and between c and d.

domain of R, $P(x)$ holds if P holds for every y "lower than x" (i.e. with the property that $R(x,y)$.

Transfinite induction is applied by observing that e is "lower" than a: we can, therefore, assume the theorem to hold when e is the upper vertex of an R^* diamond. We now have the upper half of an R^* diamond with vertex e, the other elements being f and c. Although e and f are illustrated as related by R, not R^*, we can see that there is an R-path from e to f with no intermediate elements (the degenerate case), and therefore $R^*(e,f)$ holds.

Applying the theorem to the half-diamond with vertices e, f and c, we obtain an element g such that $R^*(f,g)$ and $R^*(c,g)$.

The next step is to observe that there is an R-path from d to g, passing through f. Thus, $R^*(d,g)$ holds even though it is not explicitly noted in Fig. 19.1.

The R-path from d to g provides another opportunity to apply transfinite induction. This time we observe that d is "lower" than a and therefore that the half-diamond with vertices d, b, and g can be completed with an element h as shown in Fig. 19.1.

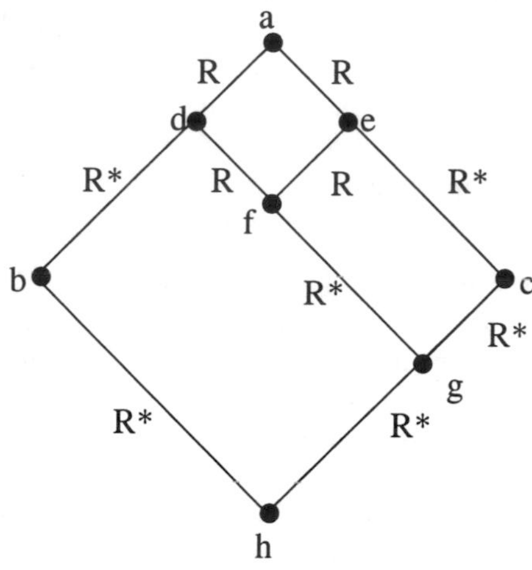

Figure 19.1. Completion of the proof of the Diamond Lemma.

Finally, observing in Fig. 19.1 that there is an R-path from c to h (through g), we see that the theorem has been successfully proven.

19.4 Diagrams as Staged Observations: The Existential Solve Heuristic

Existential solve is a heuristic procedure that we use to infer the trail of existence proofs implicit in a diagram. We first developed *existential solve* by considering the proof of the Diamond Lemma. We then discovered that it plays an essential role in the proofs of two other difficult theorems, which are described in this paper.

Existential solve implements the reasoning described in the proof of the Diamond Lemma above. The goal of the heuristic is to construct a sequence of lemmas, each of which proves the existence of (or "solves" for) one existential object in the diagram.[4] The key point is that the objects are solved for sequentially, and that the heuristic is responsible for determining the order in the sequence.

Solving for an existential object means proving the existence of an object that has the properties asserted in the diagram. This is not as obvious a process as it might seem, however, because some properties may involve other objects which may not have been solved for yet. We therefore need the notion of a *defining* property for an object. A *defining property* for an existential object $\mathbf{x}$ is a formula whose variables consist only of existential objects that have already been solved for, universal objects, and $\mathbf{x}$ itself. The procedure determines a succession of existential objects, on the basis of these *defining* properties.

At the beginning of the proof of the Diamond Lemma there are two existential objects with defining properties: $\mathbf{d}$ and $\mathbf{e}$. When there is more than one candidate, *existential solve* chooses the existential object whose defining properties, taken together, contain the most other objects (universals and previously solved for existentials). The rationale for this criterion is that a greater number of objects in the properties means, in some sense, more information, or greater constraint, and thus a stronger definition. If there are ties when this criterion is applied, *existential solve* proves the existence of the remaining candidates in (logical) parallel.

Existential solve organises the existential objects in the diagram into a partial order by repeatedly applying the criteria just described. With each selection of the next object to be solved for, that object becomes available to appear in the defining properties of other objects. Eventually, every existential object will have at least one defining property, and the ordering process is then complete.

19.4.1 Existential Solve in the Diamond Lemma

To see how *existential solve* works in the Diamond Lemma, we apply it to the diagram in Fig. 19.1. The following formulae are explicitly represented in the diagram:

$$R(a,b)\ R(a,c)\ R(a,\mathbf{d})\ \ R(a,\mathbf{e})\ \ R^*(\mathbf{d},b)\ R^*(\mathbf{e},c)$$
$$R(\mathbf{d},\mathbf{f})\ R(\mathbf{e},\mathbf{f})\ R^*(c,\mathbf{g})\ R^*(\mathbf{f},\mathbf{g})\ R^*(b,\mathbf{h})\ R^*(\mathbf{g},\mathbf{h})$$

[4] Throughout this paper we write existential objects in bold face.

All of the objects are existential except a, b, and c, which are identified as universal in the hypothesis of the theorem.

In the first pass of *existential solve*, the existential objects with potentially defining properties are **d**, **e**, **g**, and **h**. The defining properties of **d** are $R(a, \mathbf{d})$ and $R^*(\mathbf{d}, b)$ and the defining properties of **e** are $R(a, \mathbf{e})$ and $R^*(\mathbf{e}, c)$, each set containing two universal objects.

The only defining property of **g** at this stage is $R(c, \mathbf{g})$, and the only one for **h** is $R(b, \mathbf{h})$. Since each of these contains only one universal, **g** and **h** are ruled out at this stage. There is no way to break the tie between **d** and **e**, so the order in which they are solved for is randomly chosen.

In the next pass, d and e may appear in the defining properties of other objects, so the object **f** has the defining properties: $R(d, \mathbf{f})$ and $R(e, \mathbf{f})$. The presence of the two previously solved for objects, d and e, means that **f** now wins out over **g** and **h**, each of which still has only one defining property containing only one other object.

In the next pass, **g** has the defining properties: $R(f, \mathbf{g})$ and $R(c, \mathbf{g})$. Now **g** wins over **h** because its defining properties contain two other objects, f and c, while **h** still has only one defining property, containing one other object.

In the final pass, **h** has the defining properties: $R(b, \mathbf{h})$ and $R(g, \mathbf{h})$ and this marks the end of the trail.

19.5 Diagrams as Elisions of Infinitely Many Observations

Recall our view of diagrams as partial models of the world to which a proof refers. Diagrams are finite, while mathematical worlds are typically infinite. While a theorem may quantify over an infinite range of objects (as in "for every integer $i \ldots$"), a diagram expressing the theorem will focus on an arbitrary example in that range (as in "let i_0 be an arbitrary integer").

When a mathematical argument relies on the implicit performance of an arbitrary number of calculations or operations, rigorous presentation of the argument must be based on inference rules that permit such reasoning. The most common of such rules are various forms of *induction*.[5]

When GROVER detects the presence of ellipses in a diagram, it tries to determine whether a finite (but arbitrarily long) series is being represented, and hence whether mathematical induction should be applied. If the objects connected by the ellipses are labelled similarly except for numerical (integer) subscripts, GROVER interprets this to indicate a situation requiring mathematical induction. When GROVER recognises a diagram calling for mathematical induction, it decomposes the diagram into two simpler diagrams, one

[5] These are not the only rules that permit such reasoning: others include the Axiom of Choice and its many equivalents.

for the base case and one for the step case, using the ellipses to determine where the separation should occur.

GROVER's assumption in performing this decomposition is that each of the resulting diagrams will contain enough information to prove its part of the theorem. In particular, the diagram for the step case must not only express the desired conclusion (e.g., that the property $P(x)$ holds for $x = n + 1$), but it must also express the inductive hypothesis (i.e., that $P(n)$ holds) which will be used to derive the conclusion $P(n + 1)$. GROVER verifies this as part of the more general process of matching the diagram with the corresponding conjecture. This process was described in [1].

19.5.1 The Multiple Peaks Theorem

To see how the interpretation of ellipses works, we present the proof of the Multiple Peaks Theorem. In Fig. 19.2, R^* refers to the *transitive closure* of a relation R. The theorem states that, if R is globally confluent then an object h can be found so that the figure can be completed along the dotted lines, i.e., $R^*(b_0, h)$ and $R^*(b_{k+1}, h)$.

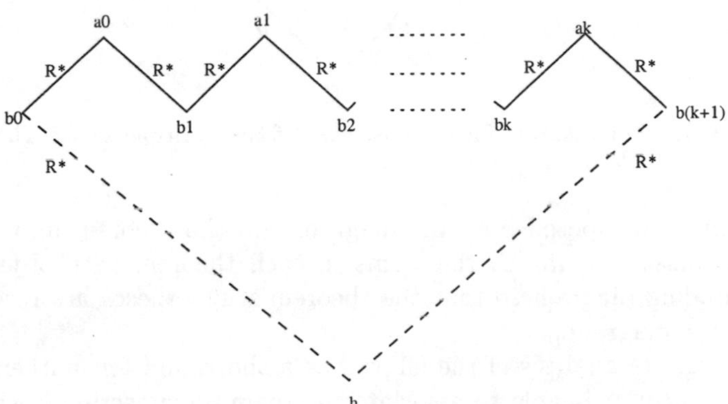

Figure 19.2. Graphical statement of the Multiple Peaks Theorem.

Most notable about this theorem is the fact that the number of "peaks", represented by the variable n, is arbitrary.

The proof is a straightforward application of mathematical induction. The base case ($n = 0$) follows immediately from the global confluence of R. The step case ($n = k + 1$) follows from the inductive hypothesis, which gives us the existence of an h_k such that $R^*(b_0, h_k) \wedge R^*(b_{k+1}, h_k)$.

Since we also have, from the assumptions of the theorem, that

$$R^*(a_{k+1}, b_{k+1}) \wedge R^*(a_{k+1}, b_{k+2})$$

we infer, from the transitivity of R^*, that $R^*(a_{k+1}, h_k)$. We therefore use the global confluence of R to get the existence of an h_{k+1} such that

$$R^*(h_k, h_{k+1}) \wedge R^*(b_{k+2}, h_{k+1})$$

and then, from the transitivity of R^* again, infer that $R^*(b_0, h_{k+1})$.

As in the proof of the Diamond Lemma, the transitivity of R^* is automatically inferred by & and applied where needed.

GROVER interprets each ellipsis in Fig. 19.2 as representing a *sequence* $t_1 \ldots t_m$ of objects. Since the objects in one of the sequences are existential, GROVER infers that their existence is to be proven by mathematical induction. GROVER therefore replaces Fig. 19.2 with Figs 19.3 and 19.4, and the theorem itself is decomposed into a base case and a step case by applying &'s mathematical induction tactic.

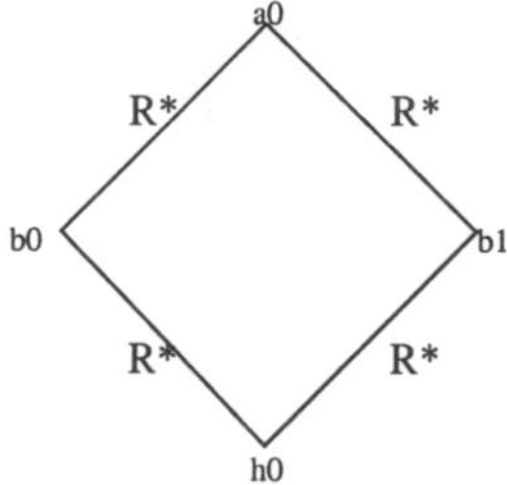

Figure 19.3. The diagram for the base case of the Multiple Peaks Theorem.

Having decomposed both the diagram and the theorem into two parts, GROVER must now match the terms in each theorem with objects in the corresponding diagram so that the theorem's hypotheses are recognised as facts in the diagram.

Through its analysis of the ellipses as a shorthand for mathematical induction, GROVER is able to associate the diagram subscript k with the induction variable n in the theorem. Completing the association process is complicated, however, by a discrepancy in representation between the theorem and the diagram. The theorem (in its original form as well as in the step case) does not contain the universal variables $a_0, a_1 \ldots a_k, a_{k+1}$ and $b_0, b_1, b_2 \ldots b_k, b_{k+1}, b_{k+2}$ but rather two universal variables a and b, which are applied as functions to an index variable i. In order to complete the association, therefore, GROVER must establish the correspondence between the variables a and b in the theorem and the instantiated terms in the diagram.

GROVER solves this problem using the idea of *spanning hypotheses* – hypotheses of the form $\forall x.(x \leq n+1 \rightarrow A)$ where $n+1$ is associated with a diagram *spanning limit*, which is the subscript of the final term of a sequence in the diagram. GROVER replaces each spanning hypothesis with the instantiated formulae $A[t/x]$ for all *spanning instances* t, which are the diagram

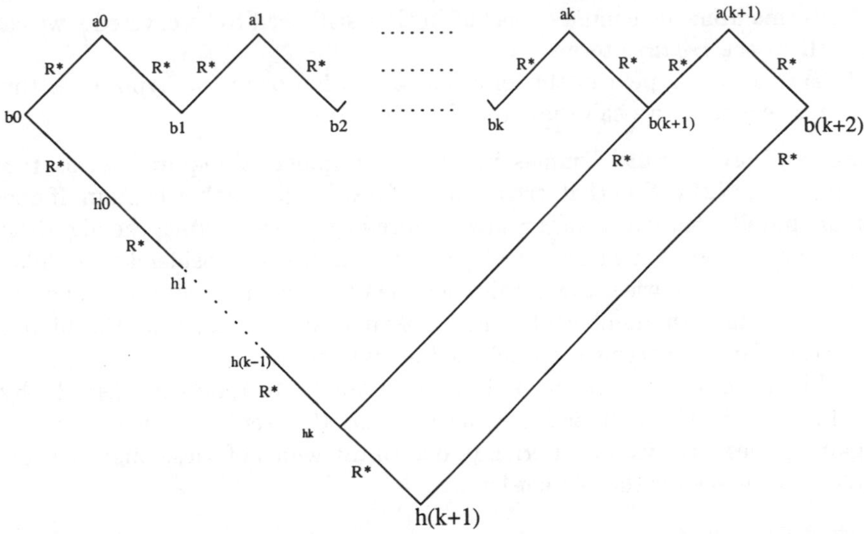

Figure 19.4. The diagram for the step case of the Multiple Peaks Theorem.

objects participating in one of the diagram sequences. GROVER is then able to match the hypotheses of both the base case and step case theorems with facts in their respective diagrams.

19.5.2 Focus of Attention: Choosing Relevant Hypotheses

A diagram fact that has been proven from the conjecture's hypotheses is available as a hypothesis during any individual step of the proof strategy. Furthermore, the conclusion of any previous step in the strategy is available as a hypothesis in subsequent steps. Not all of these potential hypotheses are necessarily useful, however, and in order to facilitate &'s search for a proof, GROVER tries to keep the hypotheses to a minimum. GROVER determines the relevancy of other facts by comparing the terms found in the current lemma to those found in the potential hypotheses. The objective is to find hypotheses that, taken together, mention all of the terms found in the lemma's conclusion. We call this a process of *covering* all of the lemma's terms.

To determine the hypotheses for a given lemma, a heuristic algorithm examines the preceding lemmas to see whether any of them can contribute to "covering" the current lemma's terms. The algorithm proceeds backwards, examining the most recent lemmas first and then, if necessary, moving on to the earlier lemmas. As this process continues, the set of terms that still need to be covered shrinks.

A measure of relevancy is provided by defining two classes of terms in the current lemma:

1. Terms from the lemma's conclusion that still need to be covered – we call these the *required* terms.
2. Terms that appear in the lemma's conclusion or in the hypotheses thus far selected – we call these the *desired* terms.

GROVER sorts parallel lemmas by (1) the number of required terms they contain, and (2) within that, the number of desired terms they contain. If none of the parallel lemmas contains any required or desired terms, the algorithm proceeds to the next latest set of parallel lemmas to consider as candidate hypotheses. Otherwise, the parallel lemmas that come out best in the sort – i.e., the highest number of required terms, and within that the highest number of desired terms – are selected as hypotheses.

When the process described above is complete GROVER considers the hypotheses of the theorem, and the diagram facts that represent them. GROVER again applies a relevancy criterion to determine which of these might be suitable hypotheses for the current lemma.

Example: Choosing Relevant Hypotheses in the Multiple Peaks Theorem. To understand how the procedure we have just described helps to prune hypotheses, we consider the final lemma step of the Multiple Peaks Theorem, which is the theorem's conclusion: $\exists h.(R^*(b_0,h) \wedge R^*(b_{k+2},h))$. We back up to the preceding lemma, which is $R^*(h_k,\mathbf{h_{k+1}}) \wedge R^*(b_{k+2},\mathbf{h_{k+1}})$. This lemma contains the required term b_{k+2}, but the required term b_0 still needs to be covered, so we back up to the parallel lemmas

$$R^*(b_0,\mathbf{h_0}) \wedge R^*(b_1,\mathbf{h_0}) \qquad \text{and} \qquad R^*(b_0,\mathbf{h_k}) \wedge R^*(b_{k+1},\mathbf{h_k})$$

Both lemmas contain the required term b_0, so we must look to the desired terms in order to break the tie. The $\mathbf{h_k}$ goal wins because it contains the desired term $\mathbf{h_k}$ while the $\mathbf{h_0}$ goal contains no other desired term.

19.6 Diagram Idioms: Visualisation and Abstraction

In this section we will describe the process by which we move from the diagram to a collection of formulae which it represents. This is a crucial step in GROVER's automatic processing of the diagram.

One of the key components of &/GROVER is a graphical editor called DEGAS. DEGAS is a rather conventional graphical editor, with tools allowing the drawing of lines, ellipses, and rectangles, and for attaching labels to these objects. The most important feature of DEGAS for GROVER is that it is able to save the diagram in the form of a *geometry facts file* (G-file).The G-file is a generic textual representation of the diagram structure, irrespective of any semantics that we associate with the diagram.

19.6.1 Interpreting the Diagram

When presented with a diagram, GROVER must interpret it as representing facts that are expected to follow from the hypotheses of the current theorem. We have developed a small expert system for carrying out this task. The rules of the expert system are intended to capture the usual practice of mathematical diagrams.

The interpretation of the diagram is divided into two parts: a local analysis, and a global analysis. The local analysis phase produces atomic formulae from the spatial and explicit relationships in the diagram, and writes them to a *logic file* (L-file). This is described in [2]; we will focus here on the global analysis phase which detects larger constructions in the diagram.

19.6.2 Global Analysis: Verify Logic

The result of the local analysis of the G-file is a collection of atomic formulae, which are implicitly conjoined. We call this representation a *Logic File* (L-file). Diagrams can represent more complex structures than a flat collection of atomic formulae, however. These structures are detected in an analysis of the L-file which we call *verify logic*. *Verify logic* is only activated once the G-file representation has been completely interpreted as an L-file, so it is an operation on logical formulae. In principle, the same processing could be performed on the G-file representation, or interleaved with the *geometry to logic* phase. From an implementor's point of view, however, it is simpler to wait until the L-file representation is complete before looking for higher-level structures.

The global analysis is implemented as a collection of "critics", each of which looks for specific conditions that might hold within the diagram, and modifies the logical representation appropriately. For example, one of the critics implemented in GROVER is the `definition by cases` critic.

The `definition by cases` critic is triggered by the presence of two equalities in the L-file of the form $x = t_1, x = t_2$, where x is an existential object, and t_1, t_2 are arbitrary terms involving only universal objects. It is a general feature of diagrams that distinct tokens represent distinct objects (token referentiality, see [3]), and therefore such a pair of equalities presents a puzzle on the face of it. One explanation is that the diagrammer is attempting to assert $t_1 = t_2$, but the role of x is then unexplained. The `definition by cases` critic attempts to gather evidence that the existential object x is being defined by cases, as under some circumstances being equal to t_1 and under other disjoint circumstances being equal to t_2. If such evidence can be found, the equalities $x = t_1$ and $x = t_2$ are replaced by the critic with the more complex formulae: $P \to x = t_1 \wedge Q \to x = t_2$, where P and Q are possibly complex formulae representing the two alternative conditions.

19.6.3 Example: The Schröder–Bernstein Theorem

The `definition by cases` critic has a crucial role in the Schröder–Bernstein Theorem, a theorem from the theory of functions which concerns the way in which the "size" of sets can be measured. The Schröder–Bernstein Theorem states that if there is a one–one function (an *injection*) from the set A *into* the set B, and a one–one function from B into A, then there is a *bijection* between the two sets, i.e., a one–one function from A *onto* B.

$\forall f, g, A, B.Injection(f, A, B) \wedge Injection(g, B, A) \rightarrow \exists h.Bijection(h, A, B)$

An intuitive proof of the Schröder–Bernstein Theorem would proceed as follows: The bijection h must be some combination of of f and g^{-1}, i.e., for each $a \in A$, $h(a)$ will be either $f(a)$ or $g^{-1}(a)$. The problem is therefore to define a partition of A into sets A_1 and A_2 so that h behaves like f for members of A_1 and g^{-1} on members of A_2. Since h is to be a bijection, every $b \in B$ will have to be in $range(h)$. Therefore, if b is not in $range(f)$, then $h^{-1}(b)$ must be in A_2. So A_2 contains $g^{-1}(B - range(f))$. Moreover, A_2 must be closed under $g \circ f$, because if $a \in A_2$ then $h(a) = g^{-1}(a)$, so $h(a)$ cannot be $f(a)$ unless $f(a) = g^{-1}(a)$. Therefore, unless $f(a) = g^{-1}(a)$, $f(a)$ must be "hit" under h by some other element of A, which can only be $g(f(a))$. So let A_2 be the smallest set containing $g^{-1}(B - range(f))$ and closed under $g \circ f$, and let A_1 be $A - A_2$.

The diagram illustrating this strategy (Fig. 19.5) contains objects **h**, $\mathbf{A_1}$, $\mathbf{A_2}$ and **C** whose existence must be proved, and in addition it represents the *definition* of these objects.

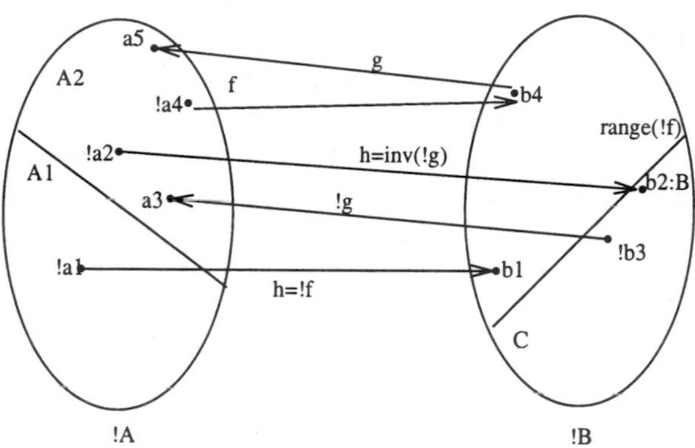

Figure 19.5. The diagram for the Schröder–Bernstein Theorem.

The two arrows defining the function **h** in the diagram, one arrow labelled $\mathbf{h} = f$ and the other labelled $\mathbf{h} = g^{-1}$, are recognised by the `definition by cases` critic as indicating a definition of the function **h** by cases. The critic looks at the source points of the respective arrows, to determine whether they indicate that the function **h** is defined to be f on some subset of its domain, and g^{-1} on the other subset of the domain.

Three other critics are needed in the diagrammatic proof of the Schröder–Bernstein Theorem. The **function chains** critic looks for information concerning items at the end points of arrows, in order to construct appropriate assertions concerning the relationships between objects at the ends of these arrows. The diagram of Fig. 19.5 contains universal objects a_1, a_2, a_4 and b_3, whose role in the diagram is to serve as starting points for function arrows. Since these are universal objects, they are exemplars for arbitrary objects with the same properties that they themselves exhibit. The function chains critic generalises the formulae containing these objects to universal formulae.

b_3, for example, is a member of **C** which is mapped by the function g onto some member of A_2. Rather than view this structure as three distinct formulae, $b_3 \in \mathbf{C}$, $\langle b_3, \mathbf{a_3} \rangle \in f$ and $\mathbf{a_3} \in \mathbf{A_2}$, we recognise that the geometric structure is intended to represent that every member of **C** is mapped by g to some member of $\mathbf{A_2}$.

The function chains critic examines the L-file for formulae which match this pattern, constructing the appropriate generalisations of the specific formulae. The part of the diagram which is significant for this step is shown in Fig. 19.6.

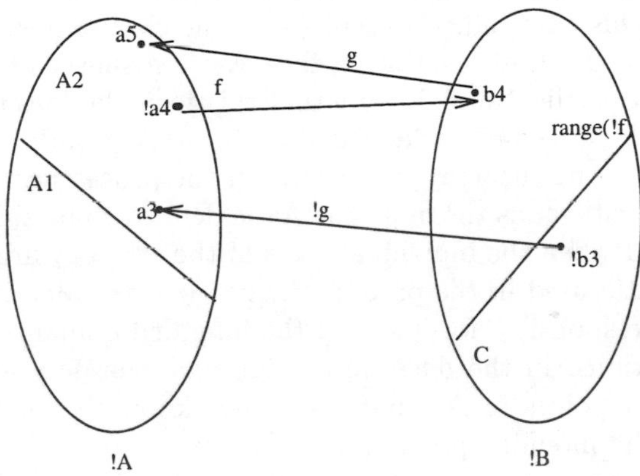

Figure 19.6. Function chains.

The result of applying the function chains critic to the formulae just mentioned is the new formula:

$$\forall b3.(b3 \in \mathbf{C} \rightarrow \forall x.(\langle b3, x \rangle \in g \rightarrow x \in \mathbf{A_2}))$$

The same critic notes that formulae $a_4 \in A_2$, $\langle a_4, b_4 \rangle \in f$, $\langle b_4, a_5 \rangle \in g$ and $a_5 \in A_2$ indicate that an arbitrarily chosen element in $\mathbf{A_2}$ maps under $g \circ f$ back into $\mathbf{A_2}$.

The formulae involving a_4, b_4 and $\mathbf{a_5}$ have the same structure, except that this represents a chain of function applications. Again, the chain beginning with the universal object is traversed, and the properties of the beginning and end points of the chain examined. The result is a universal formula which asserts that all start points with the same properties as the exemplar are mapped by the same chain, to end points with the same properties as *its* exemplar.

The result of applying this critic to the chain is:

$\forall a4.(a4 \in \mathbf{A_2} \rightarrow \forall x, y.(((\langle a4, x\rangle \in f \wedge \langle x, y\rangle \in g) \rightarrow y \in \mathbf{A_2})))$

These formulae capture the intent of the larger structure in the diagram, by aggregating facts recognised as forming a pattern into an appropriate compound formula.

On the basis of the formulae derived by the function chains critic, the `Closure` critic recognises that $\mathbf{A_2}$ contains the image under g of $B - range(f)$ and is closed under the composition $g \circ f$ and therefore that $\mathbf{A_2}$ is (probably) intended to be the closure of the given base set under the composition of g and f. The crucial part of the diagram for this critic is coincidentally identical to the part relevant to the function chains critic, so consult Fig. 19.6.

The choice to consider $\mathbf{A_2}$ as the closure rather than some superset of the closure is heuristic, but we believe that this is generally likely to be the intention, particularly when no additional information about the set is available, as in this case. The choice of $\mathbf{A_2}$ as the closure means that we will add a formula to the L-file indicating that $\mathbf{A_2}$ is a subset of all non-empty sets with the properties 1 and 2. Note that this formula does not imply that the set A_2 itself enjoys properties 1 and 2 above. A proof of this fact must be constructed by the theorem prover later in the processing.

The closure critic adds the hint that $\mathbf{A_2}$ is defined to be this intersection. This hint is used when the individual goals of the strategy are constructed.

The final critic used in the proof is the `generalise domain and range` critic, which is responsible for inferring the intended domains and ranges of *Function* assertions. In the diagram of Fig. 19.5, the only arrows labelled by g have target points in $\mathbf{A_2}$, but we do not know that g's range is just $\mathbf{A_2}$. Indeed in the intended proof, g is an injection into A. The `generalise domain and range` critic examines the diagram looking at all of the target points of arrows sharing the same label. Having identified these end points the critic identifies the largest graphical object containing all of these end points, and asserts this as the set into which the function maps. This results in the L-file formula $Function(g, B, \mathbf{A_2})$ being replaced by $Function(g, B, A)$, and $Function(f, A, range(f))$ by $Function(f, A, B)$.

Like the `closure` critic, the action of the `generalise domain and range` critic can be undesirable. It may over-generalise, since for example the intended range of g may indeed have been $\mathbf{A_2}$, or under-generalise, since the intended range of the function may in fact not appear as an object in the diagram, but may contain the inferred range. Experience with other diagrams will determine which of these cases is the most likely to occur, and the diagram cues that we may use to determine the likely intended values for the domains and ranges of sets.

19.7 Other Approaches to Graphical theorem-proving

We are aware of work in graphical theorem-proving by Gelernter, Jamnik, Barwise and Etchemendy, and Pastre. We try to identify both the similarities and differences of our work with these other approaches.

19.7.1 Gelernter's Geometry Machine

The concept of graphical theorem-proving was introduced by Gelernter in his Geometry Machine ([4–6]) GROVER resembles the Geometry Machine (GM) in the following respects: the diagram is used as a model of the goal to be proven, and the diagram suggests constructions of terms that are needed in the intended proof.

The principal difference between the Gelernter approach and GROVER concerns the way in which the graphical information is used. In GM, the diagram is consulted in order to guess bindings that will prove the current subgoal. It is assumed that if the instantiated subgoal is true in the diagram, it may be provable. GROVER offers a different form of guidance: the advice takes the form of specifying the subgoals themselves. Thus with GROVER the high-level structure of the proof is determined by the diagram.

19.7.2 Jamnik's DIAMOND

Jamnik has described a theorem-proving system called DIAMOND which uses only graphical inference rules to construct a proof [7, 8]. The user of DIAMOND draws a diagram that represents an instance of the theorem to be proved (many of Jamnik's examples are drawn from [9]). The user then edits the diagram using graphical inference rules, i.e. operations that preserve relevant semantic properties of the diagram (for example, the number of objects present in the diagram). Once a proof of a specific instance of a theorem has been generated, the system automatically generalises the proof of the instance to the proof of the general case, and if successful thereby proves the general theorem.

DIAMOND differs from the approach that we adopt by utilising an entirely graphical inference system. Proofs in DIAMOND involve exactly one representation, namely a diagrammatic one. DIAMOND's proof generalisation routines, which operate at the meta-level, are syntactic operations on the structure of the proof of the instance of the theorem, and these are sentential in style. This is exaclty the reverse of the approach that we take in &/GROVER. In our system, the formal proof is carried out in an entirely sentential representation, and it is the meta-information, about how to carry out the proof, that is carried by the diagram.

19.7.3 Barwise and Etchemendy's Hyperproof

Barwise and Ethemendy's team have developed a system called Hyperproof which allows the user to reason about blocks worlds both graphically and with formulas.

Hyperproof is a proof checker, rather than an automatic prover such as GROVER. The user can invoke either standard logical inferences on formulas, or graphical inferences on the diagrams. For example, the user can perform an operation that splits a diagram into two alternative diagrams, each with more information than the original one (proof by cases), and an operation that merges two diagrams into a single diagram containing the information common to both. The graphical inference rules supported by Hyperproof are formulated at a fairly low level, comparable to those of first-order logic. Nevertheless, the graphical inferences serve to elide what would otherwise be large blocks of logical inferences.

Unlike both DIAMOND and &/GROVER, Hyperproof has no meta-level. The diagrams and sentences participate equally in the proof, representing the conjecture to be proved. But the proof is constructed entirely by the user, and any strategic information lies entirely within her head.

19.7.4 Pastre's DATTE

Pastre has described a theorem prover that uses diagrams to aid the proof of theorems [10]. This work is quite different from ours. The diagram in Pastre's theorem prover, DATTE, is an internal representation of the formulae that the theorem prover is currently manipulating rather than something that the user provides to guide the prover, as in GROVER.

The major difference between GROVER and DATTE is that, in GROVER, the user provides the diagram as guidance for the prover, and the system views it not only as a representation of formulae, but also as a proof strategy. DATTE's diagram represents only those formulae that are known to be true at a particular point in the proof.

19.8 Conclusions

We have described various issues that arise when an automated system tries to interpret a diagram as a mathematical proof. In our investigation of three theorems whose proofs require different techniques – transfinite induction, mathematical induction, and set theory, respectively – we found that a common element is the decomposition of the proof into a series of existence proofs;

the diagram suggests the conditions to be proven in "solving" for successive objects. The diagram also suggests the degree of relevance of each previously solved for object to the current existence proof, thus providing a tractable set of hypotheses to be used in each lemma. Finally, patterns in the diagram may suggest higher-order abstractions that are crucial in proving the theorem.

Our goal is to develop a system that will foster the development of proofs by students of mathematics and even by working mathematicians. By raising the level of the conversation to the types of abstractions contained in diagrams, a theorem-proving system could serve as a kind of surrogate colleague with whom ideas are tested and the implications of different constructs explored.

References

1. Barker-Plummer, D. and Bailin, S.C. (1992). Proofs and pictures: Proving the diamond lemma with the GROVER theorem proving system. In Working notes of the AAAI symposium on reasoning with diagrammatic representations, Stanford, CA, 25–27 March.
2. Barker-Plummer, D. and Bailin, S.C. (1997). The role of diagrams in mathematical proofs. Machine Graphics and Vision 6(1):25–56.
3. Barwise, J. (1993). Heterogeneous reasoning. In G. Allwein and J. Barwise (Eds), Working papers on diagrams and logic. Indiana University Logic Group, pp. 1–13.
4. Gelernter, H. (1963). Realization of a geometry theorem proving machine. In E. Feigenbaum and J. Feldman (Eds), Computers and thought. New York: McGraw Hill.
5. Gelernter, H., Hansen, J.R. and Loveland, D.W. (1963). Empirical explorations of the geometry theorem proving machine. In E. Feigenbaum and J. Feldman (Eds), Computers and thought. New York: McGraw Hill.
6. Gilmore, P. (1970). An examination of the geometry theorem proving machine. Artificial Intelligence 1:171–187.
7. Jamnik, M., Bundy, A. and Green, I. (1997). Automation of diagrammatic proofs in mathematics. In B. Kokinov (Ed.), Perspectives on cognitive science, Vol. 3. Sofia: NBU Press, pp. 168–175. Also available as Department of Artificial Intelligence Research Paper No. 835.
8. Jamnik, M., Bundy, A. and Green, I. (1999). On automating diagrammatic proofs of arithmetic arguments. Journal of Logic, Language and Information 8(3):297–321. Also available as Department of Artificial Intelligence Research Paper No. 910.
9. Nelson, R.B. (1993). Proofs without words. Number 1 in Classroom Resource Materials. Washington, DC: The Mathematical Association of America.
10. Pastre, D. (1977). Automatic theorem proving in set theory. Technical report, University of Paris (VI).

20. EnE Sentences and Local Extent in Diagrams

Norman Foo

Most diagrams, even if they can be interpreted as embeddings in infinite objects, are nevertheless finite in extent if they are to be presented conventionally. In many applications where diagrams are used to explain, instruct, communicate, cogitate or conjecture, this finiteness implies local extent. By this we mean that even if they are used to reason about potentially unbounded domains or constructs, the fact that they can be used at all suggests that only local properties are being examined. We are interested in how the features of such diagrams can be described in logic, and how diagram manipulations that represent actions can be justified. In particular, we establish a correspondence between a substructure construction which formalises local extent and a class of sentences preserved under extension from, and reduction to, this substructure. The sentence class is called EnE because it has the form of successively nested existential and negated existential subsentences, and appear to cover most of the applications so far encountered. The hope is that this understanding can be used to mark out the local regions of diagrams that can be safely isolated for manipulations.

20.1 Introduction

Recent papers [4, 5] examined the use of diagrams in reasoning about blocks world and link lists respectively. In both domains simple actions were considered and state transitions were represented in diagrammatic form. Some features that appear to be typical of such diagrams are:

1. They are generic or schematic, hence can be parametised.
2. They represent local regions of concrete or abstract domains.
3. All changes due to actions are confined to such regions.
4. Objects (and their attributes) are conflated with icons.

On reflection, none of this is surprising, nor even peculiar to diagrams. In programming languages, for instance, each instruction (or even procedure) or rule execution normally changes only a finite number of things (values, pointers, bindings, etc.). The genericity of procedures is taken for granted. In visual programming iconic tokens play a central role. That such properties are also inherent in many uses of diagrams probably says a lot about the psychology of procedural thinking. While in programming languages these features are there by design, the fact that diagrams have to be finitely displayed[1] largely constrains them to have these features. If we can understand in rigorous terms what these features actually mean, we may be able to use such diagrams more confidently to explain, instruct, communicate, cogitate or conjecture. This chapter is an attempt to do this. Some of it provides a different perspective on work already done; some are extensions and generalisations of prior results; the remainder connects model theory with the idea of local tests and effects.

20.2 The Role of Logic

There is strong evidence from cognitive psychology that human processing of diagrams has many aspects, not the least of which is the fact that some form of immediate comprehension, called "free rides" by Shimojima [10] (see [8] for a recent discussion), is involved. Moreover, as Barwise and Etchemendy [1] have pointed out, there is in general no interlingua between logic and diagrams that is fully adequate. As this chapter addresses the ostensibly perceptual feature of local extent – humans presumably have an intuitive grasp of which objects are locally relevant to given objects in the context of an application – it is legitimate to ask why logic should be invoked.

There are at least two justifications for using logic (more precisely, model theory) to examine local extent. The first is well known, and it is the analogy of non-monotonic logics that are used to explicate common-sense reasoning. The roles of such logics are (i) to realise common-sense reasoning in computational models, and (ii) to classify the various nuances of common-sense approaches by reference to different modes of inference. Likewise, there may well be various nuances of local extent that can be classified and given computational realisations using logical models. The second is less traditional, but perhaps more persuasive from the viewpoint of applications. If it is possible to give simply declarative and realisable accounts of reasoning about local extent – and logic may be one way to do so – then interfaces that display diagrams can have local regions automatically constructed and highlighted. This would be a congenial aid in diagrammatic reasoning. Conversely, if a local region were to be user-selected, a logic could be used to confirm its validity. Logic is therefore used to explain and sanction intuition.

[1] Infinite objects or extent are usually represented by a finite ellipsis convention.

20.3 Basic Definitions

For completeness we recapitulate some basic definitions and establish notation in this section. A model theory text such as [2] should be consulted for more details.

Definition 1 Suppose $\mathcal{L}$ is a first-order language with equality but no function symbols. An $\mathcal{L}$-structure $\mathcal{U}$ is a triple $\langle \mathcal{A}, \mathcal{R}, \mathcal{C} \rangle$ where $\mathcal{A}$ is a non-empty set called the domain of $\mathcal{U}$, $\mathcal{R}$ is a map that assigns relations of the appropriate arity on $\mathcal{A}$ to predicate symbols in $\mathcal{L}$, and $\mathcal{C}$ is a map that assigns elements of $\mathcal{A}$ to constant symbols in $\mathcal{L}$.

The relationship between diagrams and structures has been much discussed. If diagrams are regarded as alternatives to textual forms conventional to first-order logic (terms, formulas, etc.) then they should both be interpretable into structures. If so, it should be the case that there are *calculi* of diagrams as well, and indeed a number have been actively researched [11, 13]. We accept the broad thrust of this view, and provide a syntactic condition for confining diagrammatic reasoning to "small" local pieces of potentially unbounded diagrams. Diagram calculi can then operate safely on these pieces.

Definition 2 (Sub-structure) If $\mathcal{U}_1$ and $\mathcal{U}_2$ are $\mathcal{L}$-structures with domains $\mathcal{A}_1$ and $\mathcal{A}_2$ respectively such that $\mathcal{A}_1 \subseteq \mathcal{A}_2$, and the relation and constant assignment maps of $\mathcal{U}_1$ are that of $\mathcal{U}_2$ restricted to $\mathcal{A}_1$, then $\mathcal{U}_1$ is a substructure of $\mathcal{U}_2$, and conversely $\mathcal{U}_2$ is an extension of $\mathcal{U}_1$.

Sub-structures will be the meanings of sub-diagrams. Sub-diagrams can arise in a variety of ways. In systems like Xfig, a widely used picture-drawing program, it is possible to "border" a region containing drawn objects. It is also possible to "select" non-contiguous objects. These are typical sub-diagrams whose objects give rise to substructures (when there are no function symbols). It may also be necessary to add to such initially chosen objects other objects to form domains of sub-structures, as will be the case below.

20.4 Localness and Genericity

This section summarises work already completed using new insights and vocabulary. The paper [4] used STRIPS and a blocks world to illustrate how localness and genericity can be formalised, and suggested how action invariants can be conjectured from these features. The paper [5] continued this work in the domain of linked lists but used logic programs as the underlying semantics for the diagrams. Both relied on confining the precondition of actions to formulas that had a particular syntax. We generalise this syntax and provide the intuitive reasons for it.

20.4.1 Genericity

Suppose we want to execute an action A on some objects (they need not be physical – linked lists are an example). Typically this action actually belongs to a *class* of similar actions of which this one is an instance. Say, for this instance, the objects are $a_1, \ldots, a_n$. If so, it is natural to regard A here as really the instantiated form of a parametised or generic action $A[x_1, \ldots, x_n]$, with the substitutions a_i for x_i. This is the essence of *genericity*, and it sanctions the following methodology. If we were to reason about the *specific* action $A[a_1, \ldots, a_n]$ – let us call it A for short – and come to certain conclusions, so long as we did not use any specific properties of the names $a_1, \ldots, a_n$, we can *generalise* the conclusions by lifting these names to the variables $x_1, \ldots, x_n$. This is a version of the theorem on Generalisation from Constants [12]. If we believe that our diagrams are generic, then this is precisely what we are doing, and exactly the caveat that must be observed. This imposes the following design constraint on any system of diagrammatic reasoning that permits lifting of names to variables in the construction of "templates": the variables must be "tagged" with the used properties of the names from which they were lifted. This is not so onerous as it might sound, for it is satisfied by systems in which icons are strongly typed. From these observations the informal notion of a template boils down to the isomorphism of sub-structures constructed from the names of the action. This construction is considered in the next sub-section.

20.4.2 Local Extent

A generic action $A[x_1, \ldots, x_n]$ can act on a situation or state only if a precondition is satisfied. For instance, a node can be inserted into a particular location of an ordered linked list only if its value is in between those of the two nodes sandwiching the location; likewise a block can be moved on top of another only if they both do not have other blocks on them. We can test for such preconditions procedurally or diagrammatically, but can also express them declaratively in logic. It will be assumed that these are mutually translatable, so we will use logic to explain our intuitions about local extent. Let the precondition of the action above be expressed by the formula $\phi[x_1, \ldots, x_n]$. Then for any specific action $A[a_1, \ldots, a_n]$ the corresponding precondition is the sentence $\phi[a_1, \ldots, a_n]$. The names $a_1, \ldots, a_n$ refer to the objects of the action, but in general *there may be more objects which the action may influence.* For instance, in the blocks world the precondition of a $move(a, b)$ action, meaning move block a to the top of block b, may have the precondition $\exists x\ on(a, x) \wedge \exists y\ on(b, y) \wedge \neg\exists u\ on(u, a) \wedge \neg\exists v\ on(v, b)$. If this is satisfied, there will be instantiations for the variables x and y – yes, we are interpreting the existential quantifiers substitutionally, but diagrams are presumably constructive. These will normally figure in the postcondition of the action, and hence any diagram that displays the effects of the action will show the objects that are names for the x and y. So, if we wish to reason about the action it is necessary to use not only the names in it but also other names implied by its precondition. In this example the extra names were pulled in via the existential quantifiers. We may describe this informally as a kind of *domain expansion* or *closure* with respect to the instantiated action parameters and the predicates that occur in its precondition. The definition below is a generalisation of the one that appeared in [4] (that was closure of extent 1; see below).

Definition 3 (Closure) Given a set S of constants in $\mathcal{U}$, the 1-expansion of S by n-ary predicate P in $\mathcal{U}$, denoted $closure(\mathcal{U}, S, P, 1)$, is
$\bigcup\{\{a_1, \dots, a_n\} \mid \text{some } a_i \text{ is in } S, \text{and } \mathcal{U} \models P(a_1, \dots, a_n)\}$.
Inductively, the j+1-expansion of S by P, $closure(\mathcal{U}, S, P, j+1)$, is
$\bigcup\{\{a_1, \dots, a_n\} \mid \text{some } a_i \text{ is in } closure(\mathcal{U}, S, P, j), \text{and } \mathcal{U} \models P(a_1, \dots, a_n)\} \cup closure(\mathcal{U}, S, P, j)$.
The notation $closure(\mathcal{U}, S, P_1, \dots, P_k, n)$ is the obvious generalisation to several predicates in which any of them can be used at any stage.

The last parameter in the closure notation is the *extent* of the closure. It will be omitted if the context is clear. By a slight abuse of notation we will identify the sub-structure of $\mathcal{U}$ (restricted to the predicates $P_1, \dots, P_k$)) with $closure(\mathcal{U}, S, P_1, \dots, P_k, j)$.

This is really a model-theoretic construction inspired by a metric introduced by Gaifman [6]. He wanted a metric between points in a domain to be measured by the cost of connecting them via the available predicates. Thus, if only the predicate P was considered, he decreed distance 0 for the equality predicate, then distance 1 between any two points c and d such that $\mathcal{U} \models P(\dots, c, \dots, d, \dots)$. Higher distances n are defined inductively to reflect the n-closure above. More precisely, if we denote the distance between c and d by $G(c, d)$, then:

Lemma 1 $d \in closure(\mathcal{U}, \{c\}, P, n)$ iff $G(c, d) \leq n$.

Observation 1 Membership in a closure can be computed efficiently using Dijkstra's shortest path algorithm.

Notation 1 When we wish to denote either the positive or the negated form of a formula ϕ without specifying which is the case, we will write $\pm\phi$. In the definition of EnE sentences below, by $\bar{x}$ we mean a sequence of variables, say $x_1, \dots, x_k$. By $\exists\bar{x}$ we mean a sequence of existential quantifiers, say $\exists x_1, \dots, \exists x_k$. When we write $P(\bar{x}, \bar{y})$ we mean that predicate P has free variables among the sequences $\bar{x}, \bar{y}$ in no particular order.

Definition 4 (Primitive EnE Sentences) A primitive EnE sentence[2] of width n based on $\bar{c}$ is $\pm\exists\bar{x}_1\phi_1(\bar{x}_1)$ where:

[2] More generally, the P_i's here can be conjunctions of atoms.

$\phi_1(\bar{x}_1)$ is $P_1(\bar{c}, \bar{x}_1) \wedge \pm\exists\bar{x}_2\phi_2(\bar{x}_2)$;
$\phi_2(\bar{x}_2)$ is $P_2(\bar{x}_1, \bar{x}_2) \wedge \pm\exists\bar{x}_3\phi_3(\bar{x}_3)$;
$\vdots$
$\phi_n(\bar{x}_n)$ is $\pm\exists\bar{x}_n P_n(\bar{x}_{n-1}, \bar{x}_n)$

The collection of such sentences is denoted by EnE(S, n) where S is the set of constants from $\bar{c}$. Informally, a primitive EnE$(S, 1)$ sentence is one that has a prefix of the form $\exists x$ or $\neg\exists x$ followed by an atom in which there is at least one constant from S. A primitive EnE(S, n) formula is so defined that its ground instances are conjunctions of ground literals with n-chains (see Definition 6 as arguments. In Section 20.5 we exhibit examples of EnE formulas in special domains of application.

Definition 5 (EnE Sentences) An EnE(S, n) sentence is a Boolean combination of primitive EnE(S, j) sentences where $j \leq n$.

The connection between such sentences and closure is the key to local extent.

Proposition 1 (EnE Preservation) Suppose ϕ is an EnE(S, n) sentence with predicates $P_1, \ldots, P_k$. Then $\mathcal{U} \models \phi$ iff $closure(\mathcal{U}, S, P_1, \ldots, P_k, n) \models \phi$.

This proposition (see Observation 4 for a strengthening), whose proof is given in the Appendix, is the basis for local extent if we confine preconditions (or any property of interest) to EnE sentences. Observe that unless the set of ground atoms involving S that satisfies $\mathcal{U}$ is sparse, there is little to be gained from considering the closure rather than the entire structure. In many applications, not only is this set sparse but many interesting properties are expressible as EnE sentences. The informal paraphrase of this proposition is that for sparse relations and EnE sentences it suffices to consider the local extent marked out by the closure of the given constants.

20.5 Examples

In this section we illustrate the above ideas with two examples. Both examples rely on EnE Preservation to justify the confinement of interest to the local extent when reasoning about potential actions.

20.5.1 Floor Maps

We consider floor maps, much like those in [9], and assume that the predicates *person(p), in_room({o,p},r), next_to(r,s), connected_to(r,s)* describe the topology in logic. We have indicated the types as follows: o are objects, p are people, r, s are rooms. The action is to locate a room not occupied by a person but is next to one with a fax machine. As there are two such machines, named $f1$ and $f2$, we assume the truth of the formula $fax = f1 \vee fax = f2$. The precondition sentence is expressible as $\exists r(in\_room(fax, r) \wedge \exists s(next\_to(r, s) \wedge \neg\exists p(person(p) \wedge in\_room(p, s))))$ which is an EnE($\{fax\}$,3) sentence with respect to the mentioned predicates. The closure of *f1,f2* will pull in the names (constants) of following items: all rooms with a fax machine, all rooms next to these, then all people who are in these latter rooms. The order in which they are pulled in reflects the increasing width j of the closure. Indeed, once this highly intuitive idea is grasped, the proof of the EnE Preservation proposition is easy. These remarks are illustrated in Fig. 20.1. The longer dashed region corresponds to one instantiation of the EnE($\{fax\}$,3) sentence. The room $r1$ is picked up as one next to the room with $f1$. Other objects, including the person in the other room next to the $f1$ room, are also picked up and used in testing the precondition. The shorter dashed region corresponds to the other instantiation of the sentence. The total local extent with respect to *fax* is the union of these two regions.

Now, consider the predicate $connected\_to(r, s)$ which did not figure in the precondition. Is that the kind of property that is easily represented in diagrams, and how do they figure in closure and EnE sentences? A hint can be obtained from a Prolog-like definition of this predicate in terms of the $next\_to(r, s)$ predicate:

$$connected\_to(r, s) \leftarrow next\_to(r, s)$$
$$connected\_to(r, s) \leftarrow next\_to(r, v) \wedge connected\_to(v, s)$$

In the fixed-point semantics of logic programs this says that the relation $connected\_to(r, s)$ is the transitive closure of $next\_to(r, s)$, and this is known to be inexpressible in first-order languages as it is a global property. For the usual diagrams though, because of local extent, connectedness is usually quickly apparent and indeed a "free ride". In Fig. 20.1 for instance, if we add a hallway to connect the rooms, the predicate is visually trivial. These observations merit further investigation in diagrammatic reasoning, and are part of our research program.

20.5.2 Linked List Insertion

It is common experience in teaching or designing algorithms that diagrams are used to explain or test ideas and operations. For algorithms that work on potentially unbounded data structures, the diagrams drawn are nevertheless *finite*, with finite conventions for representing infinite completions. Here we will consider an example first addressed in [5] as another illustration of how EnE sentences capture the local nature of diagrams used to explain the insertion of a new node into a sorted linked list data structure.

Let us assume that the predicates and their intended meanings in this ontology are: $node\_value(n, v)$ – the node n has value v in its value field; $node\_link(n1, n2)$ – the node $n1$ has a next-node pointer $n2$ in its link field;

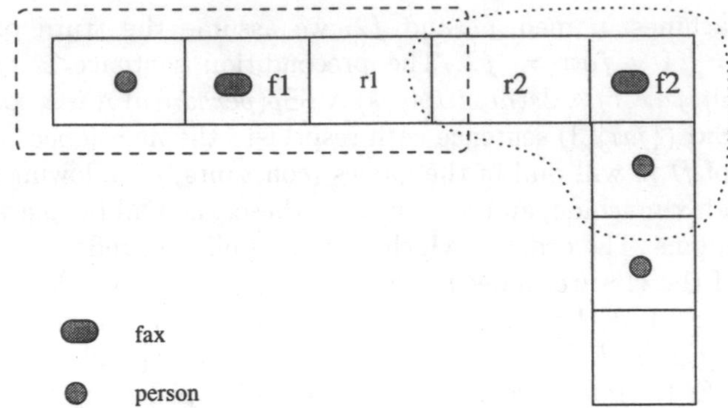

Figure 20.1. A floor plan showing the closure in dashed boundaries.

and a binary ordering predicate $v1 < v2$ – the value $v1$ is less than the value $v2$. The action is to insert the node n with value v into an existing ordered list. Figure 20.2 shows the situation.

The EnE formula for linked list insertion is a precondition formalisable by: $\exists v1 \exists v2 (v1 < v < v2 \wedge node\_value(n, v) \wedge \exists n1 \exists n2 (node\_value(n1, v1) \wedge node\_value(n2, v2) \wedge node\_link(n1, n2)))$.

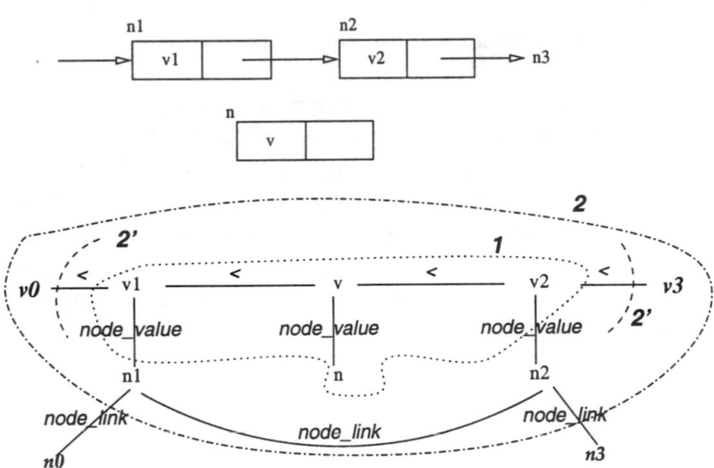

Figure 20.2. Node insertion and its local extent.

The upper part of Fig. 20.2 is the conventional diagram for explaining the beginning of the node insertion algorithm. The part below corresponds to the local closures of widths 1 and 2 (boundaries are labelled accordingly), starting from the initial constant v. The structure is representable as a graph since most of its predicates are binary, and each edge is labelled by the predicate that connects its two object arguments.

The important point to observe is that the local extent displayed in the lower diagram was constructed from the EnE formula above. However, this local extent is larger than the usual node-cell form diagram in the upper part, having attracted the values $v0$ and $v1$ (from nodes preceding and succeeding $n1$ and $n2$ respectively). This identified local extent is not wrong (it is *safe*) as it certainly suffices for reasoning about the operation or action, and it is reassuring that Proposition 1 guarantees this correctness. But it would be more satisfactory if we can achieve the smaller closure implied by common practice. This is indeed possible. The reason for the sometimes larger than necessary local extent is that the definition of EnE formulas is a little more generous than it need have been. Unless we want to preserve the Gaifman metrical properties of the closure (there are often good reasons for wanting this), we need not insist that at every stage of the closure the same set of predicates be used. If this non-uniformity is accepted, the closure can be re-computed according to a more stringent EnE formula, resulting in the local extent indicated by $2'$, which excludes the values $v0$ and $v1$, and is faithful to the diagram in the upper part of Fig. 20.2.

20.6 Cut and Paste

What we can do after determining a local extent is to use it for *cut and paste* operations. That is, the action specification should permit us to construct a "replacement diagram" (local state update) that can be used to supplant the original one. Under what circumstances can this be done?

Foo et al. [4] used STRIPS [3] as the action specification because of its simplicity and provided a partial answer to the above question. The *postcondition* of an action in STRIPS are two lists, the add- and the delete-lists, which are ground atoms that are added and deleted respectively from the prior state to yield the new state. The assumption is that all states are Herbrand models. With STRIPS the following property is usually satisfied – every constant in the postcondition is already mentioned in the precondition. The add- and delete-lists are certainly expressible as EnE sentences, so it follows that the closure structure constructed from the precondition suffices for the postcondition, and consequently a Preservation proposition for the post-condition EnE sentence also holds with the same closure domain. This is the essence of the "paste" part of state update.

Applied to the two examples above, if the the add- and the delete-lists of the intended actions are constrained as above, then any change in state of the floor plan or the linked list will take place only *within* the local extents indicated in the figures.

This does not answer the more general question. As a number of action theories admit causal propagation rules, even fixed point constructions, we expect that more elaborate constructions will be necessary for closure. But as this is ongoing work we hope to report progress later.

20.7 Conclusion

It has become evident in our work that finite model theory plays an important role in the model theory of diagrams and can be used to validate proposed calculi. Our work in the use of diagrams to convey state change has raised some interesting issues. What is the status of closures that have pictorially disjoint sub-diagrams? These are common in illustrating, say, two relevant local portions of a large diagram in which the "in-between" parts are irrelevant. Logically they correspond to having two far-apart constants as starting points for closure chains. Other issues are the following. What is the best way to illustrate recursion in algorithms? Can ellipses in, say, infinite series, be given model theoretic meanings and reasoned about? Does the idea of "telescoping" diagrams have a semantics in local extent? How do causal rules affect the construction of local extent? What kinds of diagram expansions are needed for capturing action ramifications? Is there a notion of interpolation between diagrams? We believe our report here is a pointer to some of the techniques that may be fruitful.

Acknowledgements

This research was supported in part by an Australian Research Council Special Investigator Award.

References

1. Barwise, J. and Etchemendy, J. (1995). Heterogeneous logic. In J.I. Glasgow, N.H. Narayanan and B. Chandrasekaran (Eds), Diagrammatic reasoning: Cognitive and computational perspective. Cambridge, MA: MIT Press, pp. 209–232.
2. Chang, C.C. and Keisler, H.J. (1973). Model theory. Amsterdam: North-Holland.
3. Fikes, R.E. and Nilsson, N.J. (1971). STRIPS: A new approach to the application of theorem proving to problem solving. Artificial Intelligence 2:189–208.
4. Foo, N., Nayak, A., Pagnucco, M., Peppas, P. and Zhang, Y. (1997). Action localness, genericity and invariants in STRIPS. In Proceedings of the fifteenth international joint conference on artificial intelligence, IJCAI'97, Nagoya, August. San Mateo, CA: Morgan Kaufmann, pp. 549–554.
5. Foo, N. (1998). Diagrammatic reasoning about linked lists. In Lee, H.Y. and Motoda, H. (Eds), Proceedings of the fifth Pacific Rim international conference on artificial intelligence, PRICAI'98: Topics in Artificial Intelligence, LNAI v. 1531. Berlin: Springer, pp. 565–574.
6. Gaifman, H. (1982). On local and nonlocal properties. In J. Stern (Ed.), Logic colloquium '81. Amsterdam: North-Holland, pp. 105–135.
7. Glasgow, J.I., Narayanan, N.H. and Chandrasekaran, B. (Eds) (1995). Diagrammatic reasoning: Cognitive and computational perspectives. Cambridge, MA: MIT Press.

8. Gurr, C.A. (1998). Theories of visual and diagrammatic reasoning: Foundational issues. In Proceedings of the AAAI fall symposium on visual and diagrammatic reasoning. Orlando, FL: AAAI Press, pp. 3–12.
9. Myers, K. and Konolige, K. (1995). Reasoning with analogical representations. In J.I. Glasgow, N.H. Narayanan and B. Chandrasekaran (Eds), Diagrammatic reasoning: Cognitive and computational perspectives. Cambridge, MA: MIT Press, pp. 273–301.
10. Shimojima, A. (1996). Operational constraints in diagrammatic reasoning. In J. Barwise and G. Allwein (Eds), Logical reasoning with diagrams. New York: Oxford University Press.
11. Shin, S.J. (1994). The logical status of diagrams. Cambridge, UK: Cambridge University Press.
12. Shoenfield, J. (1967). Mathematical logic. Reading, MA: Addison-Wesley.
13. Sowa, J. (1984). Conceptual structures. Reading, MA: Addison-Wesley.

Appendix

This appendix contains the proof of the main proposition in the chapter. It uses subsidiary propositions which may be independently interesting.

Notation 2 The sub-structure $closure(\mathcal{U}, S, P, n)$ is abbreviated as $C(\mathcal{U}, S, P, n)$. When the language is fixed, so the set of predicates is unambiguous, we write $C(\mathcal{U}, S, n)$ for $C(\mathcal{U}, S, \mathcal{P}, n)$ where $\mathcal{P}$ is the set of predicates. Also, if S is a singleton $\{a\}$, we write $C(\mathcal{U}, a, n)$ for $C(\mathcal{U}, S, n)$.

Lemma 2 $a_n \in C(\mathcal{U}, a_0, n)$ iff there is a sequence of predicates (not necessarily distinct) $P_1, \ldots, P_n$ such that the ground atoms below satisfy the conditions:

$$\mathcal{U} \models P_1(\ldots, a_0, \ldots, a_1, \ldots)$$

$$\mathcal{U} \models P_2(\ldots, a_1, \ldots, a_2, \ldots)$$

$$\vdots$$

$$\mathcal{U} \models P_n(\ldots, a_{n-1}, \ldots, a_n, \ldots)$$

.

Definition 6 The sequence $\langle a_0, \ldots, a_n \rangle$ in the above lemma is called an n-chain.

Corollary 1 $\langle a_0, \ldots, a_k \rangle$ and $\langle a_k, \ldots, a_n \rangle$ are chains iff $\langle a_0, \ldots, a_n \rangle$ is a chain.

Definition 7 An n-chain $\langle a_0, \ldots, a_n \rangle$ is P-extendable if in addition to the conditions for an n-chain there is a predicate P and constant a_{n+1} such that $\mathcal{U} \models P(\ldots, a_n, \ldots, a_{n+1}, \ldots)$.
An n-chain that is not P-extendable for some predicate P is an n-P-cecum. In this case, for arbitrary position of the variable x, $\mathcal{U} \models \neg\exists x P(\ldots, a_n, \ldots, x, \ldots)$.

Corollary 2 $\langle a_0, \ldots, a_k \rangle$ is a chain and $\langle a_k, \ldots, a_n \rangle$ is an $n - k - P$-cecum iff $\langle a_0, \ldots, a_n \rangle$ is an $n - P$-cecum.

Observation 2 An n-P-cecum may well be extendable by some other predicate Q, so that $\langle a_0, \ldots, a_n \rangle$ is only a "dead-end" as far as predicate P is concerned.

From the preceding definitions and lemmas, the next corollary is immediate.

Corollary 3 For any n-chain or n-cecum $\langle a_0, \ldots, a_n \rangle$, $\mathcal{U} \models \langle a_0, \ldots, a_n \rangle$ iff $C(\mathcal{U}, a_0, n) \models \langle a_0, \ldots, a_n \rangle$.

Definition 8 A positive EnE sentence is one in which there is no negated quantifier. A tail-negative EnE sentence is one in which the only negated quantifier is the deepest nested.

Observation 3 The next two remarks are entailed by the standard semantics of existential quantifiers. If α is a positive EnE sentence, then α is equivalent to a prenex sentence which has the same form as α except that all the existential quantifiers have been moved outwards to the prefix. Likewise, if α is a tail-negative EnE sentence, it is equivalent to a "near-prenex" sentence in which all but the deepest (the only negated) existential quantifier has moved outwards to the prefix.

The following propositions are easy consequences of this observation.

Proposition 2 $\langle a_0, \ldots, a_n \rangle$ is an n-chain iff $\mathcal{U} \models \alpha$ where α is equivalent to a positive EnE(a,n) sentence.

Proposition 3 $\langle a_0, \ldots, a_n \rangle$ is an n-P-cecum iff $\mathcal{U} \models \alpha$ where α is equivalent to a tail-negative EnE(a,n+1) sentence with P as the deepest nested predicate.

The principal result of this paper, viz., Proposition 1, is a direct consequence of the next proposition and Corollary 3.

Proposition 4 Every EnE(a,n) sentence is equivalent to a disjunction of sentences from the class of positive EnE(a,n) and tail-negative EnE(a,k) sentences where $k < n$.

Proof. If the EnE(a,n) sentence is positive the assertion is trivially true. So, suppose it is not positive. For simplicity of exposition, let us assume that the predicates occurring in the sentence are binary – the proof structure for the general case is similar. We use strong induction on n. If $n = 1$, then the sentence is $\neg\exists x P(a, x)$ for some predicate P, say. Then $\mathcal{U} \models \neg\exists x P(a, x)$ iff $\langle a \rangle$ is a *1-P-cecum*. Assume the assertion for for at most n, consider an EnE(a,n+1) sentence in which the first occurrence of a negated existential is $\neg\exists x_k(P_k(x_{k-1}, x_k) \wedge \pm\exists x_{k+1}\phi(x_k, x_{k+1}))$, which is equivalent to $\forall x_k(P_k(x_{k-1}, x_k) \rightarrow \pm\exists x_{k+1}\phi(x_k, x_{k+1}))$. Then there is a $k - 1$ chain $\langle a_0, \ldots, a_{k-1} \rangle$ and either (i) $\mathcal{U} \models \neg\exists x_k P_k(a_{k-1}, x_k)$ or (ii) for some a_k $\mathcal{U} \models P_k(a_{k-1}, a_k)$. Case (i) holds iff $\langle a_0, \ldots, a_{k-1} \rangle$ is a k–P_k–*cecum*, hence is equivalent to a tail-negative EnE(a,k) sentence. Case (ii) holds if there is some a_k such that $\mathcal{U} \models P_k(a_{k-1}, a_k)$, which implies that $\mathcal{U} \models \pm\exists x_{k+1}\phi(x_k, x_{k+1})$, where this is by assumption an EnE(a,n-k) sentence. By the induction hypothesis $\pm\exists x_{k+1}\phi(x_k, x_{k+1})$ is equivalent to a disjunction of sentences from the two classes above. But by Propositions 2 and 3 each such disjunct holds iff their corresponding chains or ceca hold. Moreover $P_k(a_{k-1}, a_k)$ corresponds to a chain $\langle a_{k-1}, a_k \rangle$, so each of the chains or ceca appended to it is, by Corollary 1 also a chain or cecum. Hence, using Propositions 2 and 3 again, the EnE(a, n+1) sentence is equivalent to a disjunction of sentences from the two classes, so completing the induction.

Observation 4 From Proposition 1, if formula α is EnE, then it is preserved under expansion from and restriction to a local closure. There is the question whether any formula with this property is equivalent to an EnE formula. This is so if we can add inequality to the the predicate set. The reason is that with this addition, it is possible to say in EnE form that exactly k distinct points satisfy some predicate argument. With this expressive power, we can characterise up to isomorphism all closures of a given width.

This observation says that EnE sentences with inequality are essentially precise ways to describe the reachability set of a finitely branching graph, where only certain labelled edges matter. It is therefore not surprising that in most applications "test conditions" are expressible as EnE sentences, if the vertices of the graph model objects and edges model relations.

21. Implementing Euler/Venn Reasoning Systems

Nik Swoboda

This chapter proposes an implementation of a Euler/Venn reasoning system using directed acyclic graphs(DAGs) and shows that this implementation is correct with respect to a slightly modified version of the mathematical model of Euler and Venn reasoning proposed by Shin and Hammer. This DAG system will be presented as an independent diagrammatic reasoning system that could easily be implemented. In proving the correctness of this alternative method of thinking about Euler/Venn reasoning it will also be shown that the proposed system preserves or inherits the soundness and completeness properties of the mathematical model of the Euler/Venn system. These results showing relationships between classes of Euler/Venn diagrams and the proposed DAGs then allow us to conclude that Euler/Venn diagrams can be represented as discrete objects and be employed in mechanical reasoning systems.

21.1 Introduction

In the following study, we will look at an implementation of a Euler/Venn mechanical reasoning system and show that this implementation captures the essential properties[1] of a system similar to the Shin/Hammer mathematical Euler and Venn systems as given in [2–4]. To do this, we will first look at a modified version of the Shin/Hammer mathematical system that is associated with Euler/Venn diagrams. Then a second diagrammatic system representing Euler/Venn reasoning, one lending itself naturally to implementation, will be proposed using DAGs[2], and the relations between this system and the formal mathematical system associated with Euler/Venn diagrams will be explored. It will be argued that this second representation is in fact true to the formal mathematical model of Euler/Venn reasoning and thereby preserves the properties of being sound and complete.

[1] One system *captures the essential properties* of another system if there is a translation or mapping between them that preserves deductive and semantic relations.

[2] A DAG is a directed acyclic graph.

21.2 Formal Specification of Mathematical System

The mathematical formalisation of the diagrammatic language of Euler/Venn, EV_F, is defined to be the three-tuple $\langle \Gamma, \Delta, \Sigma \rangle$, with Γ as the set of grammatical or well-formed formulas, Δ the deductive system, and Σ the semantics of the system. EV_F is defined to be a traditional Venn system with Euler-like extensions (see below). While this treatment was inspired by and is quite similar to that found in [3], there are a number of important differences that should be noted, the most important of which include that the grammar presented here adds more well-formed diagrams, and that the system's semantics have been changed to accommodate these new diagrams. As the result of having a modified semantics and more well-formed diagrams, two new inference rules are introduced to maintain the completeness of the system.

21.2.1 The Vocabulary

1. Rectangles: Each rectangle denotes the domain of discourse to be represented by the diagram.
2. Closed curves: A countably infinite set $C_1, C_2, C_3 \ldots$ of uniquely labelled closed curves. Each closed curve must not intersect itself. Each of these curves is taken to represent the set which corresponds to its label.
3. Shading: The shading of any region denotes that the set represented by that region is empty.
4. $\otimes$: A countably infinite set $\otimes_1, \otimes_2, \otimes_3, \ldots$ of individual constants.
5. Lines: Lines are used to connect individual constants $\otimes_n$ of the same n, in different regions to illustrate the uncertainty of which set contains that constant.

21.2.2 Γ: The Mathematical Grammar

Notion of Region. A *region* is any area of a diagram completely enclosed by lines of that diagram. Any region of the diagram completely enclosed by a closed curve is referred to as a *basic region.* A *minimal region* is any region which is not the combination of other regions. A complete Euler/Venn diagram having n basic regions will contain 2^n minimal regions. Other diagrams with Euler-like features, expressing set containment, will contain fewer minimal regions. This lack of certain minimal regions in these diagrams is taken to mean that the set that the missing region represents is empty. Since this "lack of a region" is meaningful we will refer to these regions as *missing regions.* For example in Fig. 21.1 the region denoting the intersection of C and the complement of B is missing. The following set theoretic operations on regions will be allowed:

1. $\cup$ The union of two regions is the region containing both of those regions.

2. $\cap$ The intersection of two regions is the region that is common to both regions.
3. $\subset$ One region is the subset of another if that region is entirely contained within the other.
4. $-$ The difference of two regions is the regions of the first not contained by the second.
5. $\overline{r}$ The complement of a region is the region not contained in that region but still within the rectangle of the diagram.

Formation Rules. Formation rules for well-formed diagrams V_{EV_F} of EV_F:

1. Any diagram containing only a Rectangle is a member of V_{EV_F}.
2. If $V \in V_{EV_F}$ then:
   a) V with the addition of any closed curve C with unique label N completely within the rectangle of V so that the regions intersected by C are split into at most two new regions, is a member of V_{EV_F}.[3]
   b) V with the addition of a $\otimes_n$ of a new n within any region of a closed curve of V is a member of V_{EV_F}.
   c) V with the shading of any enclosed region is a member of V_{EV_F}.
   d) If V contains a certain $\otimes_n$ then the result of adding another $\otimes_n$ to any region not containing $\otimes_n$ and then connecting the two of them together with a line is a member of V_{EV_F}.
3. No other diagram is in V_{EV_F}.

Well-Formed

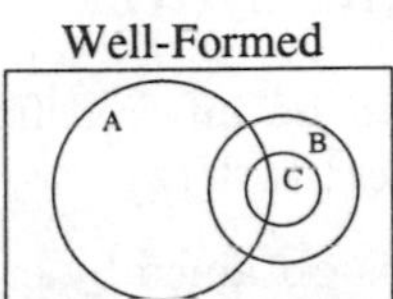

Not Well-Formed

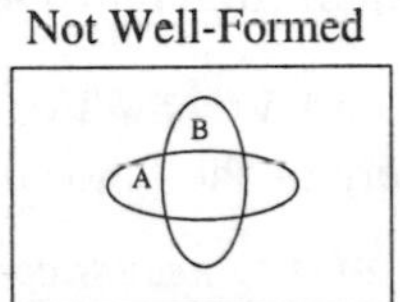

Figure 21.1. Examples of diagrams of EV_F.

Notion of a Tag. By a *label* in a diagram, any symbol labelling some curve of the diagram will be meant. Given the set $\{L_1, \ldots, L_n\}$ of labels of V, a *tag* is a subset of $\{L_1, \overline{L_1}, \ldots, L_n, \overline{L_n}\}$ containing at most one of L_i and $\overline{L_i}$ for each i. A tag τ is said to be *complete* if for each label L_i of V, either $L_i \in \tau$ or $\overline{L_i} \in \tau$. A tag is said to be *positive* if none of its elements are of the form $\overline{L_i}$. A positive tag containing one element will be referred to as a *basic tag.*

Thus for each basic region labelled L there will be a tag $\{L\}$ corresponding to it and tag $\{\overline{L}\}$ corresponding to the complement of that region. Then given two regions tagged with τ_1 and τ_2 the tag for the intersection of those regions

[3] This grammatical stipulation while more general than that used in [3] is still not as general as one might like. But it can be argued that any Euler/Venn diagram can be expressed under this restriction.

will be $\tau_1 \cup \tau_2$. We then see that the complete tags correspond exactly to all the "potential" minimal regions of the diagram. These intuitions are then made precise by the following definition:

Definition 21.2.1. Tag Assignment Function. Given a diagram $V \in V_{EV_F}$ containing curves labelled $L_1, \ldots, L_n$, the function $region_V$ from the tags of V's labels to the regions of V will be defined as follows:

1. For each basic region r labelled L in V $region_V(\{L\}) = r$ and $region_V(\{\overline{L}\}) = \overline{r}$.
2. If $region_V(\tau_1) = r_1$ and $region_V(\tau_2) = r_2$ and $\tau_1 \cup \tau_2$ is a tag, then if the region $r_1 \cap r_2$ is missing in V(it is not represented in the diagram) then $region_V(\tau_1 \cup \tau_2) = \emptyset$ otherwise $region_V(\tau_1 \cup \tau_2) = r_1 \cap r_2$.[4]

It is important to point out that not every region has a tag, but rather only regions that are the intersection of basic regions and the complements of basic regions.

Notion of Counterpart. Given two diagrams V and V' we will say that region r of V and r' of V' are *counterparts* if there is a tag τ such that $region_V(\tau) = r$ and $region'_V(\tau) = r'$. *Counterparts agree with respect to shading and* $\otimes$ *sequences* in two diagrams when for any two regions that are counterparts one is shaded iff the other is shaded, and one contains a link of a $\otimes_n$ sequence iff the other contains a $\otimes_n$ link of the same n.

21.2.3 Δ: The Mathematical Deductive System

Given diagrams V and V' of EV_F, V' can be inferred from V if V' is the result of applying any of the following rules[5] to V:

1. *Erasure of part of a* $\otimes$ *sequence* – V' is obtained by erasing a $\otimes_n$ of a $\otimes$ sequence of V where that $\otimes_n$ falls within a shaded region and provided that the possibly split $\otimes$ sequence is rejoined by a line if necessary.
2. *Extending a* $\otimes$ *sequence* – V' is the result of adding a new $\otimes_n$ link to a $\otimes$ sequence of V in a minimal region not already containing a link of that sequence.
3. *Erasure* – V' is obtained from V by erasing:
   a) an entire $\otimes$ sequence;
   b) the shading of a region;
   c) a closed curve (and possibly redrawing the remaining curves to keep the diagram well-formed) if the removal does not cause any counterpart regions to disagree with regard to shading or containment of links of a $\otimes$ sequence.

[4] Here the issue of missing regions is addressed for the system to be able to deal correctly with Euler type diagrams illustrating set inclusion.

[5] Please note that the rules of *Adding shaded regions* and *Removing shaded regions* are the above-mentioned new rules.

4. *Introduction of a new curve* – V' is the result of adding a new curve to V in such a way that V' is well-formed, the other labels of V are left undisturbed, and all counterparts agree with respect to shading and containment of links of a $\otimes$ sequence.
5. *Inconsistency* – V' of any form can obtained from V if V contains a region that is both shaded and contains all the links of some $\otimes$ sequence.
6. *Adding shaded regions* – V' is the result of adding a new minimal (but not basic) region corresponding to a missing region in V provided that this new region is shaded and is drawn so that the region is contained within the basic regions to whose intersection it is intended to correspond.
7. *Removing shaded regions* – V' is the result of removing a shaded minimal but not basic region of V. To emphasise the fact that the region has been removed the lines enclosing the now non-existing region should be smoothed into curves, and the remaining curves should be spaced out to remove points of unintended intersection.

Unification – V' can be inferred from diagrams V_1 and V_2 if it is the case that:

1. The set of labels of V' is the union of the labels of V_1 and V_2.
2. Counterparts in both V' and V_1 and V' and V_2 agree with respect to shading and containment of a link of a $\otimes$ sequence.

The following are figures to illustrate the use of the system's two new rules.

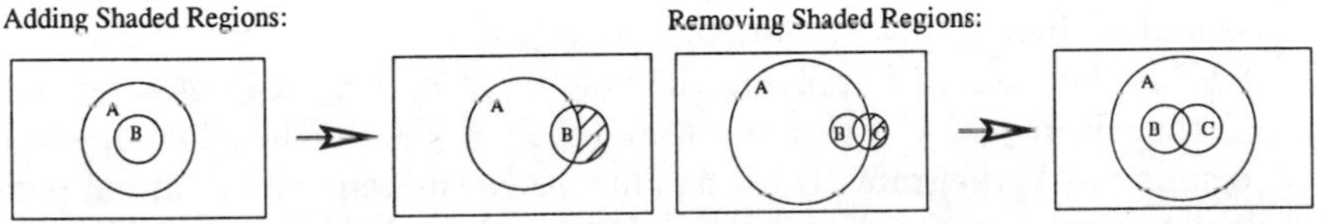

A diagram V is provable from the set of diagrams $\mathcal{V}$ in EV_F, written as $\mathcal{V} \vdash_{EF_V} V$, if there is a sequence of diagrams $V_1 \ldots V_n$ where V_n is equal to V and all $V_1 \ldots V_n$ are either members of $\mathcal{V}$ or the result of applying one of the above rules of inference to a prior diagram in the sequence.

21.2.4 Σ: The Mathematical Semantics

The semantics of the system is given by the assignment of a domain to the diagram and subsets of this domain to each basic region of the diagram. Formally this assignment is the pair (U, f) where U is the domain and f is a function associating a subset of U with each basic region. Basic regions of the same label are assigned the same subset of U by f. Let $\mathcal{L}$ be a finite set of labels.

Proposition 21.2.1. (Extension of Hammer [3].) If (U, f) is an assignment of U to basic regions then there is a unique set assignment (U, g) to minimal regions. (Note that each basic region and minimal region have a unique tag.) Given this (U, g) and (U, f), there is a unique model (U, I) assigning subsets U to tags of $\mathcal{L}$ s.t. it appropriately extends both (U, f) and (U, g).

Diagram V is *true* in model $M = (U, I)$ of EV_F iff for every tag τ constructed from V's labels, if $region_V(\tau)$ is shaded or missing then $I(\tau) = \emptyset$, and if $region_V(\tau)$ completely contains a $\otimes$ sequence then $I(\tau) \neq \emptyset$. When this is the case $M \models_{EV_F} V$ will be written. With $\mathcal{V} \cup \{V\}$ a set of diagrams, V is a *logical consequence* of $\mathcal{V}$ in EV_F iff every model which makes all of $\mathcal{V}$ true in EV_F also makes V true. This is written as $\mathcal{V} \models_{EV_F} V$.

21.2.5 Soundness and Completeness of EV_F

Theorem 21.2.1. Soundness of EV_F (extension of Hammer [3]). For every set of diagrams $\mathcal{V} \cup \{V\}$, if $\mathcal{V} \vdash_{EV_F} V$ then $\mathcal{V} \models_{EV_F} V$.

Proof Sketch. It suffices to show that the two new rules of inference preserve soundness; this plus Hammer's Soundness proof will demonstrate the soundness of EV_F.

1. If V' is the result of applying the rule of *Adding a shaded region* to V, then $V \models_{EV_F} V'$. Suppose that $(U, I) \models_{EV_F} V$ then for all minimal regions r not existing in V $I(r) = \emptyset$. Thus since the newly added region is shaded then $I(r) = \emptyset$, and $(U, I) \models_{EV_F} V'$.
2. If V' is the result of applying the rule of *Removing a shaded region* to V, then $V \models_{EV_F} V'$. Suppose that $(U, I) \models_{EV_F} V$ then for all shaded regions r in V diagram $I(r) = \emptyset$. Thus since the removed minimal region does not exist in the diagram then $I(r) = \emptyset$, and $(U, I) \models_{EV_F} V'$.

Theorem 21.2.2. Finite Completeness of EV_F (extension of Shin [4]). For every finite set of diagrams $\mathcal{V} \cup \{V\}$, if $\mathcal{V} \models_{EV_F} V$ then $\mathcal{V} \vdash_{EV_F} V$.

Proof Sketch. For this proof, Hammer's completeness proof found in [3] will be greatly relied upon. First all diagrams in $\mathcal{V}$ are extended to Venn diagrams and put into the set $\mathcal{V}'$, through the repeated application of the *Adding shaded regions* inference rule. The same is done to V extending it to V'. From soundness and the transitivity of $\models_{EV_F}$ it is concluded that $\mathcal{V}' \models_{EV_F} V'$. Now Hammer's completeness result will be used to show that $\mathcal{V}' \vdash_{EV_F} V'$. We now only need to apply the rule of *Removing shaded regions* to show $V' \vdash_{EV_F} V$. Hence $\mathcal{V} \vdash_{EV_F} V$.

To further clarify the above proof the following diagram has been provided.

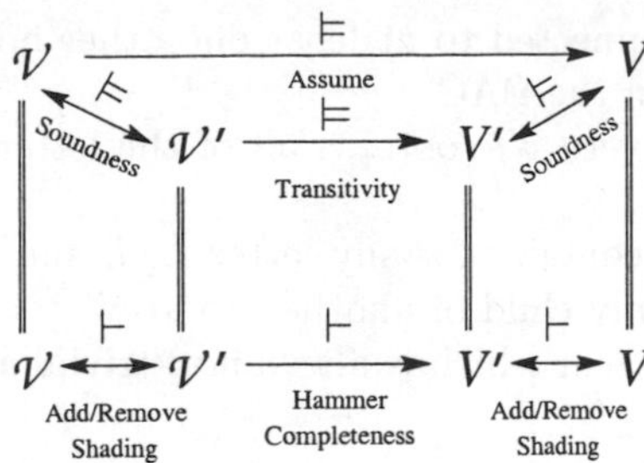

21.3 Formal Specifications of the DAG Implementation

The implementation of the diagrammatic language of Euler/Venn, EV_I, is defined to be the three-tuple $\langle \mathsf{G}, \mathsf{D}, \mathsf{S} \rangle$.

21.3.1 The Vocabulary of the Implementation

1. Nodes: Each node represents a set that can be expressed as the intersection of one or more of the sets represented by the diagram. Associated with each node are a number of attributes: a name, whether the set is empty (shading), and whether the set possibly contains any individual constants ($\otimes_n$). Regions are named so that the set that they represent is exactly the intersection of the sets associated with the letters of its name. Likewise a node can also be thought of as a region, not necessarily minimal, of the diagram.
2. Directed edges: Directed edges connect nodes A and B, leading from A to B, expressing that their associated sets, $S(A)$ and $S(B)$, are such that $S(A)$ *covers*[6] $S(B)$. Likewise the edge relation can also be thought of in terms of region containment.

In the sections to follow the natural meaning of the predicates parent, child, ancestor, and descendent will be used.

21.3.2 G: The Grammar of the Implementation

Formation rules for proper DAGs D_{EV_I} of EV_I:

1. Any DAG containing only one node (usually named U) and no edges is a member of D_{EV_I}.
2. If $D \in D_{EV_I}$, then D with the addition of one or more new nodes $N_1, \ldots, N_n$ adhering to the following stipulations is a member of D_{EV_I}:

[6] "A *covers* B" iff $B \subsetneq A$ and there is no C such that $B \subsetneq C$ and $C \subsetneq A$.

   a) Every N_i is connected to at least one other node N', and does not cause a cycle in the DAG.
   b) The name of every N_i contains all of the letters of the names of its parents.
   c) No N_i's name contains for any letter L, L and $\overline{L}$.
   d) No N_i is the only child of another node N'.
   e) No N_i has a parent which is also a non-trivial ancestor of another of its parents.
   f) Every leaf contains L or $\overline{L}$ for every letter occurring in the DAG.
3. If $D \in D_{EV_I}$, then D with the following modifications is member of D_{EV_I}:
   a) Some node along with all of its descendents are shaded.
   b) Some terminal node and all of its ancestors contain a $\otimes_n$.
4. No other DAG is a member of D_{EV_I}.

Notion of Region. A *region* is represented by a node of the DAG. A region is referred to as a *basic region* if its label contains a letter with no bar, not contained in the labels of any of its parents. Likewise a region is the *complement of a basic region* if its label contains a letter with a bar, not contained in the labels of any of its parents. This new letter is referred to as the region's *identifying letter*. A *minimal region* is any terminal node of the DAG. As before the set theoretic operations $\cup$, $\cap$, $\subset$, $-$, $\overline{r}$ will be allowed on regions.

Notion of a Tag. Using a notion of tag as that defined above a second tag assignment function will be defined.

Definition 21.3.1. Tag Assignment Function. Given a DAG $D \in D_{EV_I}$ containing basic regions with identifying letters $L_1, \ldots, L_n$, the function $region_D$ from the tags of these identifying letters to the regions of D will be defined as follows:

1. For each basic region r with the identifying letter L in D $region_D(\{L\}) = r$ and $region_D(\{\overline{L}\}) = \overline{r}$.
2. If $region_D(\tau_1) = r_1$ and $region_D(\tau_2) = r_2$ and $\tau_1 \cup \tau_2$ is a tag, then if the region $r_1 \cap r_2$ is missing in D then $region_D(\tau_1 \cup \tau_2) = \emptyset$ otherwise $region_D(\tau_1 \cup \tau_2) = r_1 \cap r_2$.

Notion of Counterpart. Given two DAGs D and D' we will say that region r of D and r' of D' are *counterparts* if there is a tag τ such that $region_D(\tau) = r$ and $region'_D(\tau) = r'$.

21.3.3 D: The Deductive System of the Implementation

Given DAGs D and D' of EV_I, D' can be inferred from D if it is the case that D' is the result of applying any of the following rules to D.

1. *Erasure of part of a $\otimes$ sequence* – D' is obtained by removing a $\otimes_n$ of a $\otimes$ sequence from a minimal region provided that this minimal region is shaded. The $\otimes_n$ is also removed from its ancestors not having a different descendent also containing a $\otimes_n$ of the same n.
2. *Extending a $\otimes$ sequence* – D' is the result of adding a new $\otimes_n$ link to a $\otimes$ sequence of D in a minimal region not already containing a link of that sequence. The same link is added to all of that node's ancestors.
3. *Erasure* – D' is obtained from D by erasing:
   a) An entire $\otimes$ sequence, removing all $\otimes_n$'s of a certain n occurring in any node of the DAG.
   b) The shading of a minimal region, and the shading of any of its parents not having all shaded descendants.
   c) A basic region and all regions containing that region's identifying letter or its complement, provided that the removal does not cause any counterpart regions to disagree with regard to shading or containment of links of a $\otimes$ sequence.
4. *Introduction of a new curve* – D' is the result of adding a new basic region to D as specified by the Inductive Construction Technique (defined below) and the other labels of D are left undisturbed and all counterparts agree with respect to shading and containment of links of a $\otimes$ sequence.
5. *Inconsistency* – Any D' can be obtained from D if D contains a region that is both shaded and all of the links of some $\otimes$ sequence are contained in it and its descendants.
6. *Adding shaded regions* – D' is the result of adding a missing minimal region in D as specified by the Direct Construction Technique (defined below) and provided that this minimal region is shaded and is not a basic region.
7. *Removing shaded regions* – D' is the result of removing a minimal but not basic region that is shaded from D and rearranging the DAG as specified by step 4 of the Direct Construction Technique.

Unification – D' can be inferred from DAGs D_1 and D_2 if it is the case that:

1. The set of basic regions of D' is the union of the basic regions of D_1 and D_2.
2. Counterparts in both D' and D_1 and D' and D_2 agree with respect to shading and containment of a link of a $\otimes$ sequence.

A DAG D is provable from the set of DAGs $\mathcal{D}$, written as $\mathcal{D} \vdash_{EV_I} D$, if there is a sequence of DAGs $D_1 \ldots D_n$ where D_n is equal to D and all $D_1 \ldots D_n$ are either members of $\mathcal{D}$ or the result of applying one of the above rules of inference to a prior DAG in that sequence.

21.3.4 S: The Semantics of the Implementation

The semantics of the system is given by the assignment of a domain to the root of the DAG, and subsets of this domain to each basic region of the

DAG. Formally this assignment is the pair (U, f), U being the domain and f being a function associating with each basic node a subset of U. Nodes of the same identifying letter are assigned the same subset of U. Once again Proposition 21.2.1 is used to establish that given (U, f) there is a unique assignment to minimal regions (U, g) and a unique model assigning subsets of U to tags (U, I) appropriately extending them both.

Diagram D is *true* in model $M = (U, I)$ of EV_I iff for every tag τ of D, if $region_D(\tau)$ is shaded or missing then $I(region_D(\tau)) = \emptyset$, and if $region_D(\tau)$ and its descendants (if any) contain an entire $\otimes$ sequence then $I(region(\tau)) \neq \emptyset$. When this is the case $M \models_{EV_I} D$ will be written. With $\mathcal{D} \cup \{D\}$ a set of diagrams, D is a *logical consequence* of $\mathcal{D}$ in EV_I iff it is true in every model which makes all of $\mathcal{D}$ true in EV_I. This is written as $\mathcal{D} \models_{EV_I} D$.

21.4 Relationships Between $\langle \Gamma, \Delta, \Sigma \rangle$ and $\langle \mathsf{G}, \mathsf{D}, \mathsf{S} \rangle$

21.4.1 Relationship Between Γ and G

This section explains the grammatical relation between the formal mathematical representation of a Euler/Venn diagram and its corresponding DAG. It will be shown that there is a translation process that results in a bijection between classes of isomorphic Euler/Venn diagrams and DAGs. This translation process will be given in two forms: one inductive and the other direct. Each method is needed to explain algorithms used in the deductive system of the implementation (namely the rules of *Introduction of a new curve*, *Adding shaded regions*, and *Removing shaded regions*).

Translating Euler/Venn Diagrams into DAGs, Inductive Construction. Knowing that a Euler/Venn diagram V can be constructed by a sequence of adding circles in a certain way to an empty diagram, it suffices to define the translation technique inductively on this sequence.[7]

1. Base: The empty diagram is the DAG with one node U and no edges.
2. Induction: When adding circle A to an existing Euler/Venn diagram V' and its corresponding DAG D' proceed as follows:
   a) Identify the region Y that covers A and the region X that covers $\overline{A}$. Add nodes YA and $X\overline{A}$ to D' directly below Y and X with an edge from Y to YA and from X to $X\overline{A}$.[8]
   b) Identify all regions represented by nodes in the DAG crossed by YA and $X\overline{A}$ (previously known as A and $\overline{A}$ respectively.)[9] For each of

[7] If given an already constructed Euler/Venn diagram this sequence can be arbitrarily chosen. The order of the placement of the circles on the page makes no difference as long as the two resulting diagrams are equivalent.

[8] In the case that Y is U leave out the U from the names of the nodes to make our DAGs easier to read.

[9] From the grammar Γ it is known that each of these regions is crossed once, creating two new minimal regions.

these crossed regions W, add to D' WYA and $WX\overline{A}$ with edges from W to each of them and edges from YA to WYA and from $X\overline{A}$ to $WX\overline{A}$. Then any duplicate letters that might occur in an individual name are removed.

c) Starting with the top of the DAG, determine if any region is now covered by or covers one of the newly created regions Z, all regions whose name contains A or $\overline{A}$. Assume that it is W that now covers Z. Remove the edge leading to W from its previously covering parent, connect this old parent to Z and draw a new edge from Z to W.[10] Appropriately rename each of these nodes W to include the letter(s) of their new parent Z not previously in W. After doing this the new letters are added to each of W's descendents, also adding new edges to their possible new parents. This is done to retain only edges between covering regions.

3. Final step: To finish the construction shading and $\otimes$ sequence information needs to be included. Shade all minimal regions of the DAG which correspond to shaded minimal regions of the diagram. Likewise add a $\otimes_n$ to a minimal region of the DAG if its corresponding region of the diagram has a link of a $\otimes$ sequence of a certain n. Then starting at the bottom of the DAG and working up shade any node having all shaded children and put an $\otimes_n$ into any node having any children with a $\otimes_n$ of a certain n.

Example of Inductive Construction.

1. The empty diagram.

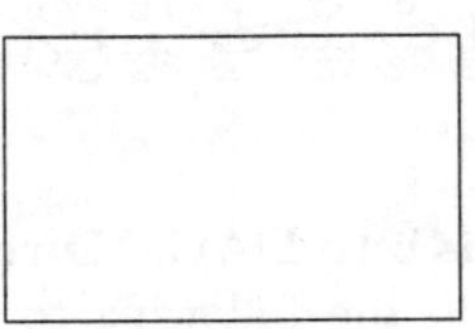

2. After adding the set A.

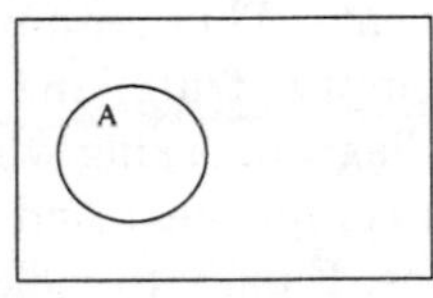

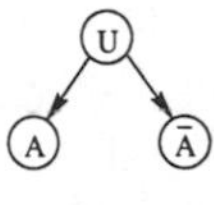

[10] Here since it is known that there is a unique covering parent so at most one edge needs to be changed for each node.

3. After adding the set B. B and $\overline{B}$ are added below V and both A and $\overline{A}$ are split.

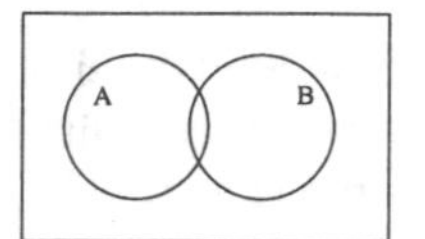

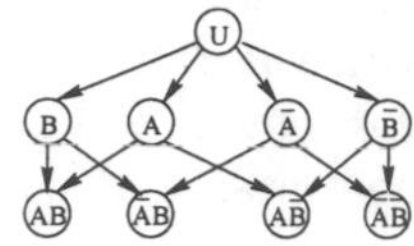

4. After adding the set C. C is added below $\overline{A}$ and renamed $\overline{A}C$, $\overline{C}$ is added below V. B, $\overline{A}$, $\overline{B}$, $\overline{A}B$, and $\overline{A}\,\overline{B}$ are split. Duplicates $\overline{A}C$, $\overline{A}BC$, and $\overline{A}\,\overline{B}C$ are removed.

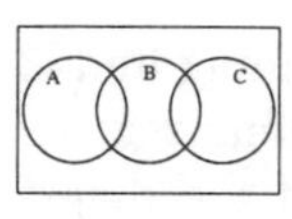

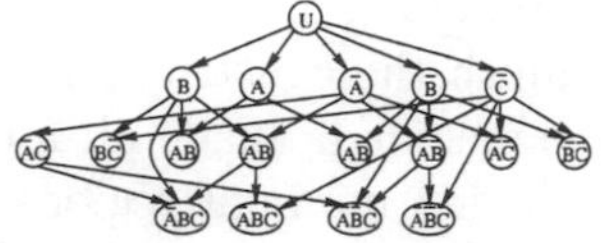

5. Reorder, A becomes $A\overline{C}$, AB becomes $AB\overline{C}$, $A\overline{B}$ becomes $A\overline{B}\,\overline{C}$, BC becomes $\overline{A}BC$, and $\overline{B}C$ becomes $\overline{A}\,\overline{B}C$. Edges from $\overline{C}$ to $\overline{A}\,\overline{B}\,\overline{C}$ and from B to $\overline{A}BC$ are removed and appropriately replaced. Finally remove duplicates $\overline{A}BC$ and $\overline{A}\,\overline{B}C$.

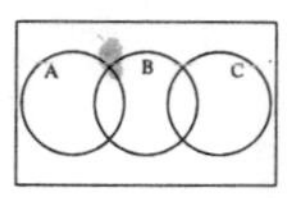

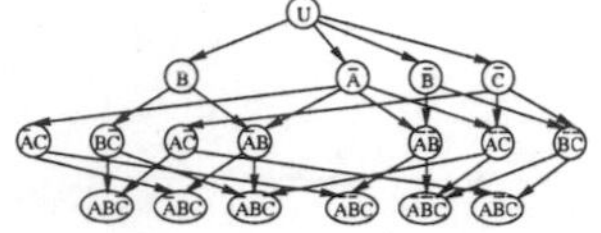

Translating Euler/Venn Diagrams into DAGs, Direct Construction. Given any Euler/Venn diagram V do the following to translate it into a corresponding DAG D:[11]

1. First identify all minimal regions of the Euler/Venn diagram V. To each of these regions associate a name which contains the letters of all of the sets of the diagram or their complements. Thus for any minimal region R start by naming the region Λ (the empty string) and iteratively look at each of the sets represented in the diagram asking whether the minimal region is a subset of that set or its complement and concatenating the appropriate letter to its current name. Finally to make the names easier to read alphabetise the letters of the name.

[11] Note that the intermediate DAGs used in this construction may not be well-formed in terms of the above grammar G.

2. Start by adding to an empty DAG one node appropriately named for each minimal region of the Euler/Venn diagram V.
3. Let N be the number of sets in the Euler/Venn diagram; thus the name of each of the minimal nodes consists of exactly N letters.[12] For each of these minimal nodes add to the diagram as their parents nodes with names consisting of N choose $N-1$, sometimes written as $\binom{N}{N-1}$, letters from each of their names, while being careful not to duplicate any node. Thus for each minimal region with a name of N letters construct parents for that node all of which have names of length $N-1$ and are subsets of the name of the minimal region. (See Fig. 21.2 for further clarity.) If a duplicate occurs connect that minimal region to the already existing node. Continue this process for each of the nodes of $N-1$ letters and so on, until only nodes consisting of one letter are added. Finally add the node U as the parent of each of these nodes consisting of only one letter.

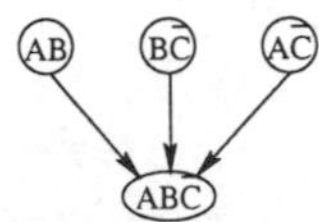

Figure 21.2. Example of rule 3 for one minimal node.

4. Starting with the minimal nodes and working up the DAG, eliminate all nodes that only have one child, and following this delete any nodes with no descendent minimal regions. If a node only has one child it is the union of one region thus equal to its child, and if it has no descendent minimal regions it is null.
5. Lastly shade and add $\otimes_n$'s to the minimal nodes of the DAG and then the entire DAG as done in the last step of the Inductive Construction.

Example of Direct Construction.

1. After the first two steps, a partial DAG with only minimal regions.

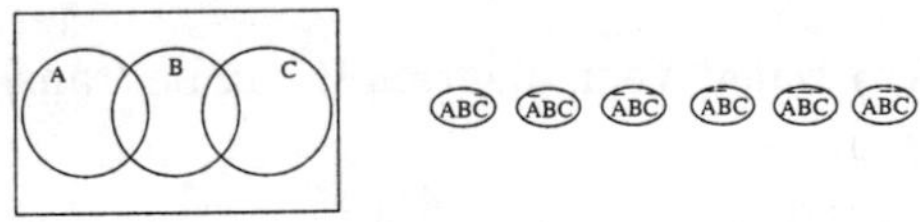

2. After the first iteration of step 3, adding nodes $\{AB, B\overline{C}, A\overline{C}, \overline{A}B, BC, \overline{A}C, \overline{A}\,\overline{C}, \overline{A}\,\overline{B}, \overline{B}C, \overline{B}\,\overline{C}, A\overline{B}\}$.

[12] Each minimal region has a name of length N from part 1 of the current construction.

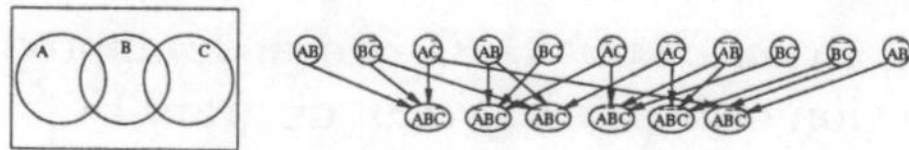

3. After the second iteration of step 3, adding nodes $\{C, B, A, \overline{A}, \overline{B}, \overline{C}\}$.

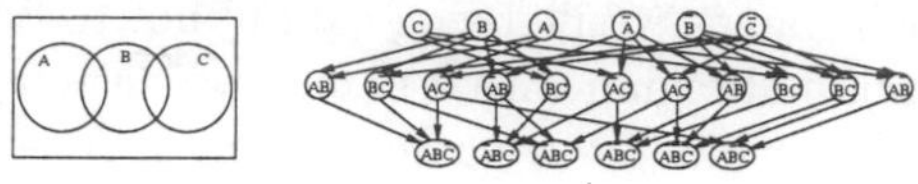

4. After the third iteration of step 3, adding node $\{V\}$.

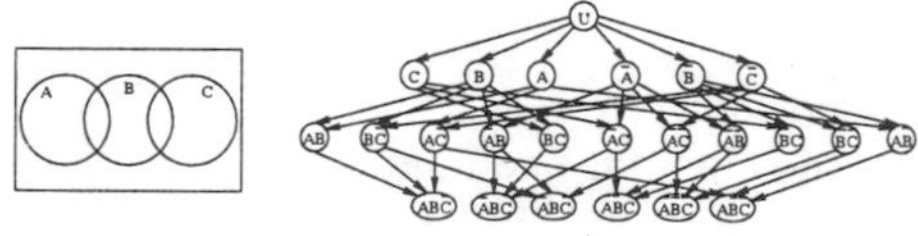

5. Final diagram after step 4, removing nodes $\{AB, BC, \overline{B}C, C, A\}$.

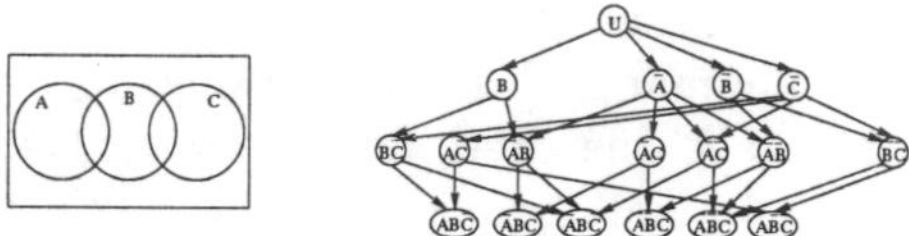

21.4.2 The Capturing of Essential Properties

Lemma 21.4.1. Any Euler/Venn diagram V can be translated into at least one DAG D.

This is direct from either of the above construction techniques.

Lemma 21.4.2. Any Euler/Venn diagram V can be translated into a unique DAG D.

Proof Sketch. First we notice that the regions of a Euler/Venn diagram that have tags can be ordered using the subset relation into a partial order. Also any partial order or poset can be described uniquely up to isomorphism by its comprising covers relations. We then shade the nodes of the poset if that region is empty and put a $\otimes_n$ in the node if it contains a link of the $\otimes$ sequence of the same n. We now have that for each Euler/Venn diagram there is a unique characterising poset. By Lemma 21.4.1 and looking closely at the above construction technique it is seen that the DAG being constructed is, with the directed edges interpreted as spatial relations, just this poset. This can be shown inductively, focusing on the inductive construction technique. For the base case we look at the empty diagram; this has by definition a unique DAG. Assume Euler/Venn diagram V has a unique DAG and show that V with the addition of one set has exactly one new node for each new region and that the edge relation preserves the covers ordering. Here by rule 2(a) two nodes are added, one for the new circle and another for its complement, so that there are at least two new regions: one corresponding to the new set and the other to its negation. Due to the grammar, each region crossed by the new set is divided into two new regions, and by rule 2(b) exactly those nodes are added to the DAG. Hence exactly the right number of nodes are being added. Finally by rule 2(c) the new DAG is reordered to preserve the covering ordering. Trivially it is noted that the DAG and the diagram both have the same nodes shaded and containing links of $\otimes$ sequences, since the same rules are used to shade the poset and the DAG. Thus for all Euler/Venn diagrams our translation results in a unique DAG.

Lemma 21.4.3. Each DAG D is the translation of a unique class of isomorphic Euler/Venn diagrams.

Proof Sketch. First we observe that each DAG has a unique set of terminal nodes with shading and $\otimes$ information corresponding to the minimal regions of the diagram. We next realise that each class of isomorphic Euler/Venn diagrams is characterised by a unique set of minimal regions with shading and $\otimes$ information.[13] Hence any DAG is the translation of a unique class of isomorphic Euler/Venn diagrams.

Lemma 21.4.4. The translation of a single Euler/Venn diagram into its corresponding DAG by the inductive and non-inductive techniques stated above results in two equivalent DAGs.

Theorem 21.4.1. For any set of Euler/Venn diagrams $\mathcal{V} \cup V \subset D_{EV_F}$, there exists a unique corresponding set of DAGs $T(\mathcal{V} \cup V) \subset D_{EV_I}$ such that: $\mathcal{V} \vdash_{EV_F} V$ iff $T(\mathcal{V}) \vdash_{EV_I} T(\{V\})$ and $\mathcal{V} \models_{EV_F} V$ iff $T(\mathcal{V}) \models_{EV_I} T(\{V\})$.

The proof of this theorem uses the above Lemma 21.4.2 and Lemma 21.4.3 and then demonstrates the close relationship between the deductive and semantic systems of EV_F and EV_I.

[13] This can be seen from the fact that all of the diagrams in one isomorphism class can be shown to be equivalent to a Venn diagram with the shading of certain minimal regions and with $\otimes_n$'s in certain minimal regions. Thus it is by either the shaded or unshaded regions, since one is the complement of the other, and which minimal regions contain $\otimes_n$'s that the class is characterised.

21.5 Soundness and Completeness of EV_I

Theorem 21.5.1. Soundness. For every set of DAGs $\mathcal{D} \cup D$, if $\mathcal{D} \vdash_{EV_I} D$ then $\mathcal{D} \models_{EV_I} D$.

Proof.
Given $\mathcal{D} \vdash_{EV_I} D$ we know that $T^{-1}(\mathcal{D}) \vdash_{EV_F} T^{-1}(\{D\})$ from Theorem 21.4.1. From this it is concluded that $T^{-1}(\mathcal{D}) \models_{EV_F} T^{-1}(\{D\})$ from the soundness of EV_F. Lastly, again using Theorem 21.4.1, $\mathcal{D} \models_{EV_I} D$ is concluded. □

Theorem 21.5.2. Completeness. For every set of DAGs $\mathcal{D} \cup D$, if $\mathcal{D} \models_{EV_I} D$ then $\mathcal{D} \vdash_{EV_I} D$.

Proof.
Given $\mathcal{D} \models_{EV_I} D$ we know that $T^{-1}(\mathcal{D}) \models_{EV_F} T^{-1}(\{D\})$ from Theorem 21.4.1. From this it is concluded that $T^{-1}(\mathcal{D}) \vdash_{EV_F} T^{-1}(\{D\})$ from the completeness of EV_F. Lastly, again using Theorem 21.4.1, $\mathcal{D} \vdash_{EV_I} D$ is concluded. □

Acknowledgements

I would like to thank Jon Barwise, Gerard Allwein, Eric Hammer, and Kathi Fisler for their helpful comments on earlier drafts of this work. Special thanks also goes to the US Department of Education, whose Grant # P200A502367 provided support for this research.

References

1. Allwein, G. and Barwise, J. (Eds) (1996). Logical reasoning with diagrams. Oxford University Press.
2. Hammer, E. and Danner, N. (1996). Towards a model theory of Venn diagrams. In Allwein and Barwise [1], pp. 109–127.
3. Hammer, E. (1995). Logic and visual information. CSLI Publications and FoLLI.
4. Shin, S.-J. (1996). Situation-theoretic account of valid reasoning with Venn diagrams. In Allwein and Barwise [1], pp. 81–108.

22. Visual Spatial Query Languages: A Semantics Using Description Logic

Volker Haarslev
Ralf Möller
Michael Wessel

We present a first treatment dealing with the semantics of visual *spatial* query languages for geographic information systems using a suitable description logic. This decidable space logic is described and its usefulness for geographic information systems is exemplified. The logic supports the specification of a semantics, reasoning about query subsumption and about applying default knowledge.

22.1 Introduction

For accessing spatial databases or geographic information systems (GIS), different query specification techniques have been proposed. For instance, the visual spatial query system VISCO developed in our group [9,13] can be used to query a spatial database (GIS) in a *visual* way. In contrast to conventional textual query systems the user is not required to learn a complicated textual query language in order to effectively use an information system. Users can query the database by drawing diagrammatic representations of what is to be retrieved from the spatial information system. However, experiences with the current VISCO system indicate that in the context of VISCO (and query systems in general), the specification of queries in a GIS still could be made easier by advances in research areas combining spatial and terminological reasoning with visual language theory.

In this chapter we discuss the application of a new logic-based formalism to specifying the semantics of visual spatial queries. To the best of our knowledge this is the first proposal utilising an expressive and decidable spatial logic for this task. The formalism can be used to define the semantics of visual spatial queries, to reason about query subsumption, and to deal with multiple worlds or query completion with the help of default reasoning. Examples for these kinds of reasoning are discussed in this chapter. Our formalism is based on the description logic $\mathcal{ALCRP}(\mathcal{D})$ [6, 7] offering mechanisms for integrating so-called *concrete domains* and on a recent extension for default reasoning [11].

We wish to emphasise that the work on VL theory presented in this chapter truly extends our previous research as summarised in [3], where we used a logic that is more expressive than $\mathcal{ALCRP}(\mathcal{D})$ since it allows *qualified number restrictions* but also is less expressive than $\mathcal{ALCRP}(\mathcal{D})$ since it has no *defined roles*. In [4] we made the first proposal for using $\mathcal{ALCRP}(\mathcal{D})$ for reasoning about visual representations. This proposal was extended in [5] by considering the semantics of visual spatial query languages. In [8] we integrated default reasoning. This chapter summarises our previous work and extends it by using so-called ABox patterns for describing n-ary queries.

22.2 The Description Logic $\mathcal{ALCRP}(\mathcal{D})$

This section gives a brief introduction to the description logic $\mathcal{ALCRP}(\mathcal{D})$ and to description logics (DLs) in general, summarising the notions important for this chapter. We refer to [1, 14] for more information about description logics.

Many DLs can be viewed as subsets of first-order predicate logic. However, it is important to note that particular DLs are only considered as practical if they are based on *sound, complete* and *terminating* reasoning algorithms, i.e. the *decidability* of a DL is of utmost importance. Of course, this is a major distinction to reasoning with general first-order predicate logic.

DLs are based on the ideas of structured inheritance networks [2]. In a DL a factual world consists of named individuals and their relationships that are asserted through binary relations. Hierarchical descriptions about sets of individuals form the terminological knowledge. Descriptions (or terms) about sets of individuals are called *concepts* and binary relations are called *roles*. Descriptions consist of identifiers denoting concepts, roles, and individuals, and of description constructors. For any individual x the set $\{\mathsf{y}|\mathsf{r}(\mathsf{x},\mathsf{y})\}$ is called the set of *fillers* of the role r. A role which may have at most one filler is referred to as a *feature*.

For instance, consider the following description used in our GIS scenario with the intended meaning "a cottage that is enclosed by a forest" that contains concept names (e.g. cottage), role names (e.g. is_g_inside), and constructors (e.g. $\sqcap$ and $\exists$).

$$\textbf{cottage\_in\_forest} \doteq \mathsf{cottage} \sqcap \exists\, \mathsf{is\_g\_inside} \,.\, \mathsf{forest}$$

22.2.1 Terminology

In this section, the language (syntax and semantics) for defining concepts and roles in $\mathcal{ALCRP}(\mathcal{D})$ is presented. $\mathcal{ALCRP}(\mathcal{D})$ is parameterised with a concrete domain which consists of a set of concrete objects and a set of predicates.

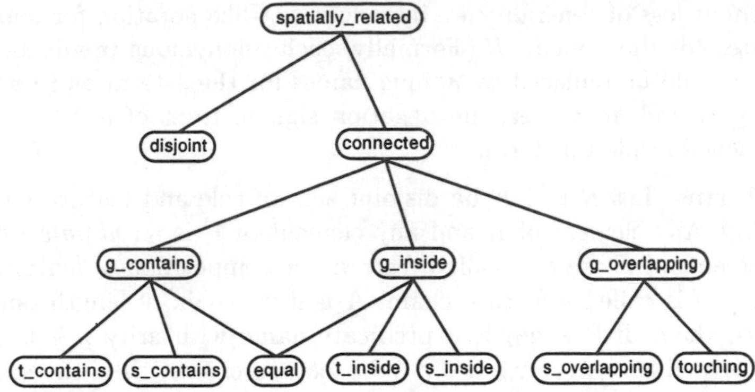

Figure 22.1. Subsumption hierarchy of spatial predicates.

Figure 22.2. Spatial relations between A and B. The inverses of t_contains and s_contains as well as the relation equal have been omitted.

Concrete Domains. A *concrete* domain $\mathcal{D}$ is a pair $(\Delta_\mathcal{D}, \Phi_\mathcal{D})$, where $\Delta_\mathcal{D}$ is a set called the domain, and $\Phi_\mathcal{D}$ is a set of predicate names. Each predicate name P from $\Phi_\mathcal{D}$ is associated with an arity n, and an n-ary predicate $\mathsf{P}^\mathcal{D} \subseteq \Delta_\mathcal{D}^n$. A concrete domain $\mathcal{D}$ is called *admissible* iff (1) the set of its predicate names is closed under negation (i.e. for any $\mathsf{P} \in \Phi_\mathcal{D}$ there exists a $\overline{\mathsf{P}} \in \Phi_\mathcal{D}$ denoting the negation of P) and contains a name $\top_\mathcal{D}$ for $\Delta_\mathcal{D}$ and (2) the satisfiability problem for finite conjunctions of predicates is decidable.

A concrete domain can be understood as a device providing a bridge between conceptual reasoning with abstract entities and (qualitative) constraint reasoning with concrete or symbolic data. In this chapter we use the admissible concrete domain $\mathcal{RS}_2$. It is the union of the domains $\mathcal{R}$ (over the set $\mathbb{R}$ of all real numbers with predicates built by first-order means from (in)equalities between integer polynomials in several indeterminates, see [12]) and $\mathcal{S}_2$ (over the set of all two-dimensional polygons with topological relations from Figs 22.1 and 22.2 as predicates, see [7]). The name "concrete domain" is in some sense misleading since it suggests that a concrete domain realises reasoning about "concrete" (e.g. numeric) data. This kind of reasoning is sometimes supported (e.g. in the domain $\mathcal{R}$) but in our application we mainly use concrete domains for reasoning about the satisfiability of finite conjunctions of predicates. For instance, the domain $\mathcal{S}_2$ qualitatively decides the satisfiability of conjunctions such as $\mathsf{touching}(I_1, I_2) \wedge \mathsf{contains}(I_2, I_3) \wedge \mathsf{touching}(I_1, I_3)$ without any notion of "concrete" polygons. This is a well-known example for a constraint satisfaction problem.

Without loss of generality we introduce a λ-like notation for anonymous predicates for the domain $\mathcal{R}$. Formally, each anonymous predicate and its negation could be replaced by unique names for the λ-term and its negated counterpart and, moreover, the negation sign in front of a λ-term can be safely moved inside this term.

Role Terms. Let R and F be disjoint sets of role and feature names, respectively. Any element of R and any element of F is an *atomic role term*. The elements of F are also called *features*. A composition of features (written $\mathsf{f}_1\mathsf{f}_2\cdots$) is called a feature chain. A feature chain of length one is also a feature chain. If $\mathsf{P} \in \Phi_{\mathcal{D}}$ is a predicate name with arity $n+m$ and u_1, ... ,u_n as well as v_1, ... ,v_m are $n+m$ feature chains, then the expression $\exists(\mathsf{u}_1,\ldots,\mathsf{u}_\mathsf{n})(\mathsf{v}_1,\ldots,\mathsf{v}_\mathsf{m}).\mathsf{P}$ (*role-forming predicate restriction*) is a *complex role term*. Let S be a role name and let T be a role term. Then $\mathsf{S} \doteq \mathsf{T}$ is a terminological axiom.

An example for using a role-forming predicate operator is the definition of a role is_g_inside for a corresponding topological predicate g_inside (see Section 22.2.1 for an explanation of the semantics). Intuitively speaking, this role holds for any pair of individuals (I_1, I_2) iff the spatial area (associated via the feature has_area) of I_1 is g_inside of the area of I_2 (see also Fig. 22.3 where this is illustrated for the predicate s_inside).

$$\textbf{is\_g\_inside} \doteq \exists(\mathsf{has\_area})(\mathsf{has\_area}).\mathsf{g\_inside}$$

Concept Terms. Let C be a set of concept names which is disjoint from R and F. Any element of C is an *atomic concept term*. If C and D are concept terms, R is an arbitrary role term or a feature, $\mathsf{P} \in \Phi_{\mathcal{D}}$ is a predicate name with arity n, and $\mathsf{u}_1,\ldots,\mathsf{u}_n$ are feature chains, then the following expressions are also concept terms: $\mathsf{C} \sqcap \mathsf{D}$ (*conjunction*), $\mathsf{C} \sqcup \mathsf{D}$ (*disjunction*), $\neg\mathsf{C}$ (*negation*), $\exists\mathsf{R}.\mathsf{C}$ (*concept exists restriction*), $\forall\mathsf{R}.\mathsf{C}$ (*concept value restriction*), and $\exists\mathsf{u}_1,\ldots,\mathsf{u}_\mathsf{n}.P$ (*predicate exists restriction*). Concept terms may also be written in parentheses.

We illustrate the notion of concept and role terms by extending the cottage example mentioned above.

$$\begin{aligned}&\textbf{normal\_cottage\_in\_forest} \doteq\\&\quad\mathsf{cottage\_in\_forest} \sqcap\\&\quad\exists\,\mathsf{has\_space}.\lambda_{\mathcal{R}}x.(x \geq 30 \wedge x < 70)\end{aligned}$$

The definition of normal_cottage_in_forest roughly has the intended meaning "something is a standard cottage in a forest if and only if it is a cottage located in a forest with 30–70 square metres of total space for the cottage". This definition also gives an example for a predicate exists restriction for the domain $\mathcal{R}$ using a feature has_space.

In order to ensure the decidability, we had to restrict possible combinations of concepts terms w.r.t. defined roles (e.g. a nested concept term with defined roles such as $\forall$ is_touching . $\exists$ is_g_inside . cottage is not allowed). Note that all examples in this chapter are restricted. However, this restrictedness criterion is beyond the scope of this chapter and is explained elsewhere [7].

Terminology. Let A be a concept name and D be a concept term. Then A $\doteq$ D and A $\sqsubseteq$ D are terminological axioms as well. A finite set of terminological axioms $\mathcal{T}$ is called a *terminology* or *TBox* if no concept or role name in $\mathcal{T}$ appears more than once on the left-hand side of a definition and, furthermore, if no cyclic definitions are present.

The previous examples already informally introduced concept axioms for defined concepts using the $\doteq$ operator. For convenience, we also allow the $\sqsubseteq$ operator for the definition of primitive concepts, i.e. their definition consists only of necessary conditions. The concept cottage is a good candidate for a primitive definition documenting that we omitted in our terminology other conditions that are not relevant for this modelling task. For instance, a cottage has to be at least a building: **cottage** $\sqsubseteq$ building.

Of course, there exist other description logics that allow more than one axiom for a particular concept name or even support generalised concept inclusions (implications) with arbitrary concept terms on the left- and right-hand side of terminological axioms. These axioms can be used as a powerful modelling tool but are currently not supported in $\mathcal{ALCRP}(\mathcal{D})$ w.r.t. decidability.

Semantics. An *interpretation* $\mathcal{I} = (\Delta_\mathcal{I}, \cdot^\mathcal{I})$ consists of a set $\Delta_\mathcal{I}$ (the abstract domain) and an interpretation function $\cdot^\mathcal{I}$. The sets $\Delta_\mathcal{D}$ (see above) and $\Delta_\mathcal{I}$ must be disjoint. The interpretation function maps each concept name C to a subset $\mathsf{C}^\mathcal{I}$ of $\Delta_\mathcal{I}$, each role name R to a subset $\mathsf{R}^\mathcal{I}$ of $\Delta_\mathcal{I} \times \Delta_\mathcal{I}$, and each feature name f to a partial function $\mathsf{f}^\mathcal{I}$ from $\Delta_\mathcal{I}$ to $\Delta_\mathcal{D} \cup \Delta_\mathcal{I}$, where $\mathsf{f}^\mathcal{I}(a) = x$ will be written as $(a, x) \in f^\mathcal{I}$. If $u = \mathsf{f}_1 \cdots \mathsf{f}_n$ is a feature chain, then $u^\mathcal{I}$ denotes the composition $\mathsf{f}_1^\mathcal{I} \circ \cdots \circ \mathsf{f}_n^\mathcal{I}$ of the partial functions $\mathsf{f}_1^\mathcal{I}, \ldots, \mathsf{f}_n^\mathcal{I}$. Let the symbols C, D, R, P, $\mathsf{u}_1, \ldots, \mathsf{u}_m$, and $\mathsf{v}_1, \ldots, \mathsf{v}_m$ be defined as above. Then the interpretation function can be extended to arbitrary concept and role terms as follows:

$$(\mathsf{C} \sqcap \mathsf{D})^\mathcal{I} := \mathsf{C}^\mathcal{I} \cap \mathsf{D}^\mathcal{I}, \quad (\mathsf{C} \sqcup \mathsf{D})^\mathcal{I} := \mathsf{C}^\mathcal{I} \cup \mathsf{D}^\mathcal{I}, \quad (\neg \mathsf{C})^\mathcal{I} := \Delta_\mathcal{I} \setminus \mathsf{C}^\mathcal{I}$$

$$(\exists\, \mathsf{R} \,.\, \mathsf{C})^\mathcal{I} := \{a \in \Delta_\mathcal{I} \mid \exists b \in \Delta_\mathcal{I} : (a, b) \in \mathsf{R}^\mathcal{I}, b \in \mathsf{C}^\mathcal{I}\}$$

$$(\forall\, \mathsf{R} \,.\, \mathsf{C})^\mathcal{I} := \{a \in \Delta_\mathcal{I} \mid \forall b : (a, b) \in \mathsf{R}^\mathcal{I} \Rightarrow b \in \mathsf{C}^\mathcal{I}\}$$

$$(\exists\, \mathsf{u}_1, \ldots, \mathsf{u}_n \,.\, \mathsf{P})^\mathcal{I} := \{a \in \Delta_\mathcal{I} \mid \exists\, \mathsf{x}_1, \ldots, \mathsf{x}_n \in \Delta_\mathcal{D} : (a, \mathsf{x}_1) \in \mathsf{u}_1^\mathcal{I}, \ldots, (a, \mathsf{x}_n) \in \mathsf{u}_n^\mathcal{I}, (\mathsf{x}_1, \ldots, \mathsf{x}_n) \in \mathsf{P}^\mathcal{D}\}$$

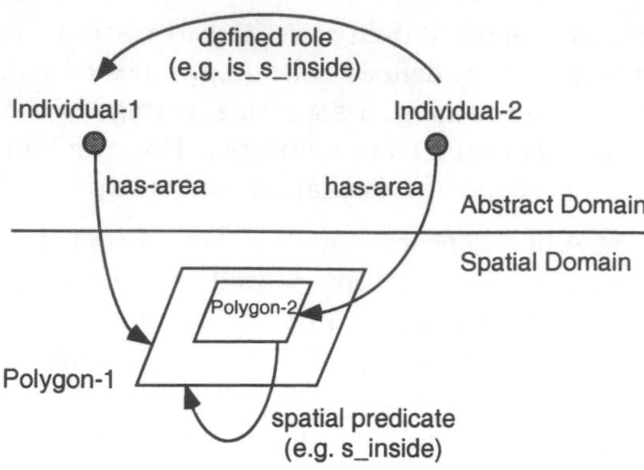

Figure 22.3. Relationship between abstract and spatial domain.

$$\begin{aligned}
&(\exists\,(\mathsf{u}_1,\ldots,\mathsf{u}_n)(\mathsf{v}_1,\ldots,\mathsf{v}_m)\,.\,\mathsf{P})^{\mathcal{I}} := \\
&\{(a,b) \in \Delta_{\mathcal{I}} \times \Delta_{\mathcal{I}} \mid \exists\, \mathsf{x}_1,\ldots,\mathsf{x}_n,\mathsf{y}_1,\ldots,\mathsf{y}_m \in \Delta_{\mathcal{D}} : \\
&\qquad (a,\mathsf{x}_1) \in \mathsf{u}_1^{\mathcal{I}},\ldots,(a,\mathsf{x}_n) \in \mathsf{u}_n^{\mathcal{I}}, \\
&\qquad (b,\mathsf{y}_1) \in \mathsf{v}_1^{\mathcal{I}},\ldots,(b,\mathsf{y}_m) \in \mathsf{v}_m^{\mathcal{I}}, \\
&\qquad (\mathsf{x}_1,\ldots,\mathsf{x}_n,\mathsf{y}_1,\ldots,\mathsf{y}_m) \in \mathsf{P}^{\mathcal{D}}\}
\end{aligned}$$

An interpretation $\mathcal{I}$ is a *model* of a TBox $\mathcal{T}$ iff it satisfies $\mathsf{A}^{\mathcal{I}} = \mathsf{C}^{\mathcal{I}}$ for all terminological axioms $\mathsf{A} \doteq \mathsf{C}$ in $\mathcal{T}$, and $\mathsf{A}^{\mathcal{I}} \subseteq \mathsf{C}^{\mathcal{I}}$ for $\mathsf{A} \sqsubseteq \mathsf{C}$ respectively.

Fig. 22.3 illustrates the idea behind the semantics of the role-forming predicate operator for the domain $\mathcal{S}_2$. The spatial predicates (e.g. g_inside) operate on concrete domain values (e.g. polygons) that are attached to corresponding abstract individuals via features. If a role (e.g. is_s_inside) is *defined* by a predicate (e.g. s_inside), then every pair $(\mathsf{p}_1, \mathsf{p}_2)$ of polygons that are fillers of has_area for two abstract individuals i_1 and i_2 is tested whether the binary predicate s_inside$(\mathsf{p}_2, \mathsf{p}_1)$ is fulfilled. In case of a successful test the role membership (e.g. is_s_inside) is also established for the abstract individuals i_1 and i_2, i.e. it holds is_s_inside$(\mathsf{i}_2, \mathsf{i}_1)$. This also applies for the opposite direction: if a role membership is asserted for a pair of abstract individuals, their associated concrete feature fillers are either established with the corresponding predicate or verified if concrete feature fillers already exist.

22.2.2 The Assertional Language

The *assertional language* of a DL is designed for stating concept or role memberships of named individuals that are used to describe the factual world. With respect to concrete domains we distinguish *abstract* and *concrete* individuals. Abstract individuals are elements of the abstract domain. Concrete individuals are elements of the concrete domain and are used as parameters for predicates. Both types of individuals can be feature fillers while only abstract individuals can be role fillers. For instance, in the VL domain abstract individuals can be used to represent geometric figures such as circles, rectangles, etc. that, in turn, represent *syntactic* elements of a query language. Additionally, abstract individuals can also be used to represent *semantic* elements (e.g. lake, estate, etc.) of a GIS. The descriptions might be associated via features with reals used to define geometric properties such as the diameter of a circle (lake), the width and height of a rectangle (estate), etc.

The set of assertions (ABox) has to comply with the definitions declared in the TBox. An *ABox* of $\mathcal{ALCRP}(\mathcal{D})$ is a finite set of assertions defined as follows.

Syntax. Let I_A and I_D be two disjoint sets of individual names for the abstract and concrete domain. If C is a concept term, R an atomic or complex role term, f a feature name, P a predicate name with arity n, a and b are elements of I_A and x, x_1, ... ,x_n are elements of I_D, then the following expressions are *assertional axioms*: a : C (*concept membership*), (a, b) : R (*role filler*), (a, x) : f (*feature filler*), (x_1,. . ., x_n) : P (*predicate membership*).

Semantics. For specifying the semantics of ABox assertions we have to extend the interpretation function $\mathcal{I}$. An *interpretation* for the assertional language is an interpretation for the concept language which additionally maps every individual name from I_A to a single element of $\Delta_\mathcal{I}$ and every individual name from I_D to a single element from $\Delta_\mathcal{D}$. We assume that the unique name assumption does not hold, that is $a^\mathcal{I} = b^\mathcal{I}$ may hold even if $a \neq b$.

$$\mathsf{a} : \mathsf{C} \text{ iff } \mathsf{a}^\mathcal{I} \in \mathsf{C}^\mathcal{I},\ (\mathsf{a},\mathsf{b}) : \mathsf{R} \text{ iff } (\mathsf{a}^\mathcal{I},\mathsf{b}^\mathcal{I}) \in \mathsf{R}^\mathcal{I}$$
$$(\mathsf{a},\mathsf{x}) : \mathsf{f} \text{ iff } \mathsf{f}^\mathcal{I}(\mathsf{a}^\mathcal{I}) = \mathsf{x}^\mathcal{I},\ (\mathsf{x}_1,\ldots,\mathsf{x}_n) : \mathsf{P} \text{ iff } \mathsf{x}_1^\mathcal{I},\ldots,\mathsf{x}_n^\mathcal{I} \in \mathsf{P}^\mathcal{D}$$

ABox Example. Using the cottage scenario the following ABox $\mathcal{A}$ illustrates the four different types of ABox assertions. Based on the semantics given above a $\mathcal{ALCRP}(\mathcal{D})$ reasoner will infer that c is a member of normal_cottage_in_forest. S_c and S_f denote the associated area polygon of c and f.

$$\mathcal{A} = \{\mathsf{c}:\mathsf{cottage}, (\mathsf{c}, 60):\mathsf{has\_space}, (\mathsf{c}, \mathsf{S_c}):\mathsf{has\_area}, \mathsf{f}:\mathsf{forest}, (\mathsf{f}, \mathsf{S_f}):\mathsf{has\_area}, (\mathsf{S_c}, \mathsf{S_f}):\mathsf{g\_inside}\}.$$

22.2.3 Reasoning Services and Complexity

The notion of a *model* (see above) is used to define the reasoning services that a DL inference engine has to provide, i.e. it proves for every concept specification that the following conditions hold:

- A term A *subsumes* a term B if and only if for every model $\mathcal{I}$, $B^\mathcal{I} \subseteq A^\mathcal{I}$.

- A term A is *coherent/satisfiable* if and only if there exists at least one model $\mathcal{I}$ such that $\mathsf{A}^{\mathcal{I}} \neq \emptyset$.
- Terms A and B are *disjoint* if and only if for every model $\mathcal{I}$, $\mathsf{A}^{\mathcal{I}} \cap \mathsf{B}^{\mathcal{I}} = \emptyset$.
- Terms A and B are *equivalent* if and only if for every model $\mathcal{I}$, $\mathsf{A}^{\mathcal{I}} = \mathsf{B}^{\mathcal{I}}$.
- An *ABox* $\mathcal{A}$ is *consistent* if and only if there exists a model $\mathcal{I}$ of $\mathcal{A}$.
- An individual a in an ABox $\mathcal{A}$ is a *member* of a concept C if and only if $\mathcal{A} \cup \{\mathsf{a}:\neg\mathsf{C}\}$ is not consistent.

The process of computing the direct subsumption relationships between all concept names in a TBox $\mathcal{T}$ is called *classification* and creates a concept taxonomy. The process of computing the direct concept membership for all individuals in an ABox $\mathcal{A}$ is called *realisation*.

The expressiveness and computational complexity of a particular DL depend on the variety of employed description constructors. Various complexity results for subsumption algorithms for specific description logics are summarised in [14]. The complexity for deciding satisfiability of $\mathcal{ALCRP}(\mathcal{D})$ concepts is NExpTime-complete [10].

The incoherence of a concept is illustrated by the following situation. We define a paradise cottage as a fishing cottage located in a mosquito-free forest, i.e. the forest is not spatially connected with a river.

$$\textbf{fishing\_cottage} \doteq \mathsf{cottage} \sqcap \exists\, \mathsf{is\_touching}\,.\,\mathsf{river}$$
$$\textbf{mosquito\_free\_forest} \doteq \mathsf{forest} \sqcap \forall\, \mathsf{is\_connected}\,.\,\neg\mathsf{river}$$
$$\textbf{paradise\_cottage} \doteq$$
$$\mathsf{fishing\_cottage} \sqcap \exists\, \mathsf{is\_g\_inside}\,.\,\mathsf{forest} \sqcap$$
$$\forall\, \mathsf{is\_g\_inside}\,.\,\mathsf{mosquito\_free\_forest}$$

However, a fishing cottage is defined as a cottage that touches a river. It follows that the forest containing a fishing cottage must also be spatially connected with this river. Obviously, the paradise cottage is only a dream that cannot exist in the real world. This is due to the intended semantics of the underlying spatial relations: A situation where a region r_1 (cottage) is *g_inside* another region r_2 (forest) and this region r_1 is also *touching* a third region r_3 (river) implies that r_2 is *connected* to r_3, i.e. $\mathsf{g\_inside}(\mathsf{r}_1,\mathsf{r}_2) \wedge \mathsf{touching}(\mathsf{r}_1,\mathsf{r}_3) \Rightarrow \mathsf{connected}(\mathsf{r}_2,\mathsf{r}_3)$. This implies that a paradise cottage cannot be located inside a mosquito-free cottage.

22.3 Semantics of Spatial Queries

The previous sections defined the description logic $\mathcal{ALCRP}(\mathcal{D})$ and demonstrated its usefulness for spatial reasoning. We introduced semantic entities such as buildings, cottages, forests, etc. These entities are suitable candidates for elements of visual spatial queries. In VISCO we assume that these

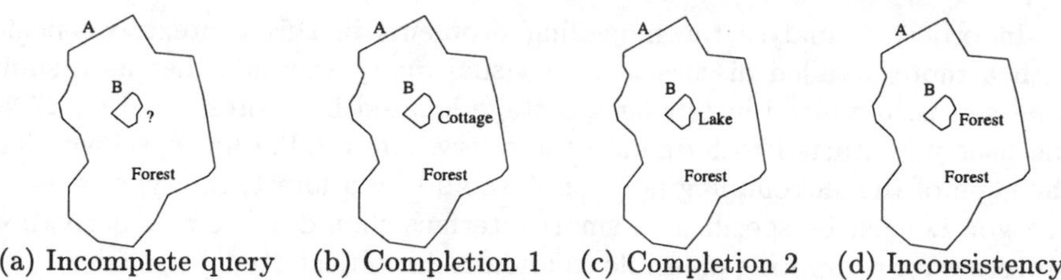

(a) Incomplete query (b) Completion 1 (c) Completion 2 (d) Inconsistency

Figure 22.4. Automatic completion of visual queries by application of default rules.

and other basic map objects are predefined in a GIS. Furthermore, spatial areas are defined by polygons. Map elements (e.g. polylines, polygons) are annotated with labels such as "forest", "building", "river" etc. that directly correspond to the semantic entities characterised above.

We imagine a VISCO application scenario for querying a GIS as follows. Instead of textually writing a complicated SQL query, a user simply draws a constellation of spatial entities that resemble the intended constellation of interest. Using thc basic vocabulary provided by the GIS the user has to annotate drawing elements by concept names (e.g. this polygon represents a forest). The parser of VISCO would analyse the drawing and create a corresponding ABox as semantic representation. Thus, the *semantics* of a query is defined by an ABox derived from a spatial constellation.

Sometimes it might be hard for users to fully specify a query. Therefore, a completion facility is needed to resolve semantic ambiguities or to complete underspecified information by using default rules for further specialisation. The next subsections describe the usefulness of spatial default reasoning and the query processing and reasoning process.

22.3.1 Completion of Queries

Default knowledge is used to make queries more precise if it can be applied in a consistent way. Due to space limitations we omit any discussions about the formal representation of default knowledge and its rules of inference. This is discussed elsewhere [11]. First of all, the process of formulating (visual) queries can be facilitated by *automatically completing* queries in a meaningful way, therefore reducing the number of mouse interactions. The process of selecting semantic concept descriptors for objects involved in a query (e.g. cottage, river, forest) can partly be automated by interpreting a partially specified query. For instance, in its current development stage, VISCO users can select concept descriptors from a list of over 300 predefined concepts. Thus, even a situation-adapted reduction of the complete list of possibilities to a suitable subset or an order relation for sorting groups of possible concept candidates would be very appropriate.

In order to analyse the modelling problems in this context, we begin with a more detailed discussion of a visual query example. Let us assume a person is interested in buying a cottage located in a forest. In Fig. 22.4a the user just started to formulate the query. After (s)he has specified that the type of the surrounding polygon A should be a forest, the type of small polygon B must be specified. A smart interface should use formal derivation processes for computing plausible candidates for object "type" specifications. For narrowing the set of possibilities we assume that two default rules are applicable: one is saying that the interior small polygon B could be a cottage (Fig. 22.4b) and another is stating that B could be a lake (Fig. 22.4c) if this does not lead to inconsistencies. Since an object can be either a lake or a cottage, there is no way to believe in both possibilities at a time. This kind of default rule interaction is a simple example demonstrating the necessity of considering different *possible worlds* which must be maintained by the reasoning system. Depending on the default rule being used to conclude new knowledge, different subsequent conclusions might be possible.

Other potentially active default rules might produce inconsistencies with the set of current assertions without providing a possibility of using multiple worlds to avoid inconsistencies. For example, if a default rule is applied that the small polygon B might as well be a forest (Fig. 22.4d), we would get a contradiction if an axiom (as part of our conceptual background knowledge) states that a forest can never contain another forest. Thus, in our query context, the latter default cannot be applied and, as a consequence of computing and appropriately interpreting the set of possible worlds, we can compose a situation-adapted menu for the graphical user interface and the user can select between meaningful concepts for object B. In our specific example, the menu will contain items for cottage and lake but not for forest.

If more than one possible world is computed, an intuitive criterion would be to select the world originating from a default with the more specific precondition or conclusion. E.g., in the query shown in Fig. 22.5a we would prefer a default concluding that the thin graphical object might be a "river flowing into a lake" (which might be a useful concept in our scenario) instead of a more general default concluding only that the object is an ordinary river.

The automatic augmentation of visual queries by conclusions of applied default rules can be seen as a specialisation process. Therefore, this process might not only be useful during the construction of a visual query, but also useful as a tool for query refinement after a query has been executed that yields too many results. In addition, not only conceptual information is important. In our GIS context we also have to consider the spatial relations between domain objects.

In the context of sketch-based visual querying, on the one hand it is sometimes useful to leave some spatial relations between graphical objects unspecified because they are unknown or simply because the user is not willing to specify them. On the other hand, in order to actually draw a

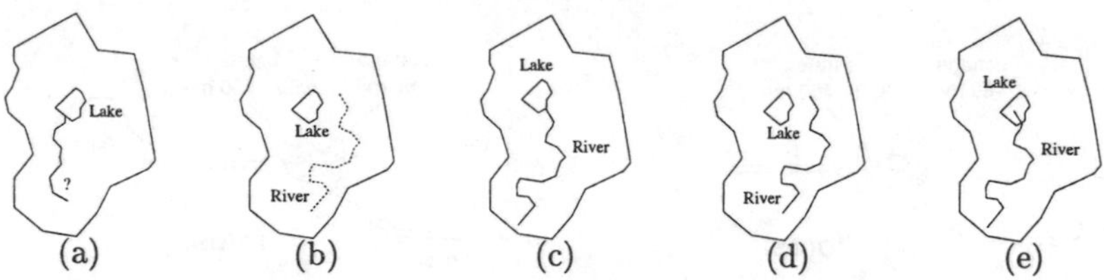

Figure 22.5. Scenarios for situation-adapted completion of queries (see text).

picture, the user *must* specify each spatial relation, even if it is just one of several possible (base) relations. The problem of how to specify "don't care relations" or "example relations" is well known and inherent in diagrammatic representations. It is similar to the problem of visualising visual disjunctions.

For example, in the query shown in Fig. 22.5b, we have a visible disjoint relation between the river and the lake. If we intended the river to be disjoint from the lake, the query-answering system would not find any rivers flowing into this lake. The problem is how can we specify that the river should be strictly inside the forest but leave the relation to the lake unspecified. As a possible solution to this problem, we could simply *ignore* each visible disjoint relation. But, with this interpretation, we can now no longer state a query searching for rivers *not* flowing into this specific lake, which might be a very useful concept. We propose the following solution. For objects like the river that are drawn with a specific drawing attribute such as dashing, the universal spatial relation to other objects (disjunction of all base relations) is asserted. Dashed objects introduce no spatial query constraints. However, in some cases this would usually not match the user's intention as there will be too many matches, i.e. the answer set will be too large. With the help of default knowledge we can automatically refine the query in a way that is appropriate according to the semantics of the objects involved in a query. So, we can guide the interpretation of spatial aspects by the help of conceptual background knowledge and application of defaults, yielding different hypotheses as possible worlds. A river flows into a lake or not, i.e. graphically both objects are either touching (see also Fig. 22.5c) or they are disjoint (see Fig. 22.5d). With respect to a lake, there are no other possibilities. In our world model a river never overlaps with a lake (see also Fig. 22.5e). This is assumed to be stated as an axiom as part of our general conceptual background knowledge. Besides defaults involving concept constraints we also have to take care of default rules with conclusions yielding new *relation* constraints. For instance, one rule could conclude the relationship touching, the other disjoint (see [11] for a discussion).

22.3.2 Reasoning about Visual Spatial Queries

We flesh out the scenario for the GIS query introduced above. The cottage should be located in a forest with a river in the immediate vicinity. The buyer

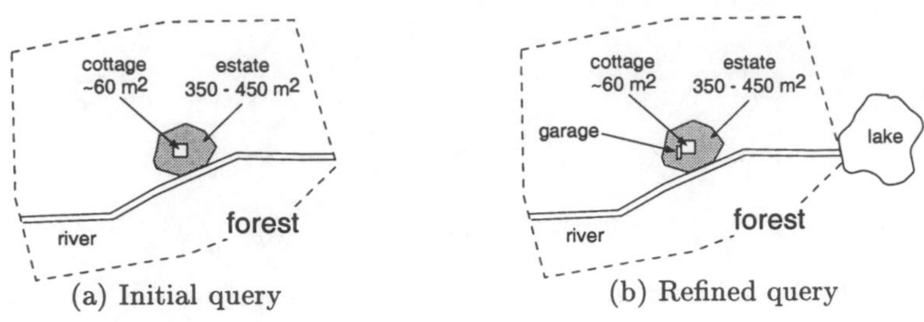

(a) Initial query (b) Refined query

Figure 22.6. Spatial sketches representing spatial queries.

also want a cottage that provides about 60 m^2 floor space. The estate itself should have about 400 m^2. Having these requirements in mind we sketch a query (see Fig. 22.6a) reflecting the intended spatial and geometric constraints. The parser translates the sketch to an equivalent ABox on the basis of a taxonomy containing concept descriptions for the spatial vocabulary of this GIS domain. We get the following Abox:

$$\begin{aligned}\mathcal{A}_0 = \{&\mathsf{c}:\mathsf{cottage} \sqcap \exists\, \mathsf{has\_space} \,.\, \lambda_{\mathcal{R}} x \,.\, (x > 40 \wedge x < 70),\\ &\mathsf{e}:\mathsf{estate\_area} \sqcap \exists\, \mathsf{has\_space} \,.\, \lambda_{\mathcal{R}} x \,.\, (x > 350 \wedge x < 450),\\ &\mathsf{r}:\mathsf{river}, (\mathsf{r},\mathsf{e}):\mathsf{is\_touching}, (\mathsf{c},\mathsf{e}):\mathsf{is\_g\_inside}, \mathsf{f}:\mathsf{forest}, (\mathsf{e},\mathsf{f}):\mathsf{is\_g\_inside}\}\end{aligned}$$

We use concept and role expressions as defined in the previous examples. The cottage is described by the individual c with a predicate-exists restriction asserting a floor space between 40 and 70 m^2. The cottage c has to be inside of an estate with a size between 350 and 450 m^2. As a simplification we assume that the river r has to touch the estate e that is inside of a forest f. Additionally, we assume the following new or revised concept definitions.

$$\mathbf{estate} \sqsubseteq \mathsf{spatial\_area} \sqcap \exists\, \mathsf{has\_space} \,.\, \lambda_{\mathcal{R}} x \,.\, (\top_{\mathcal{R}}(x))$$

$$\mathbf{estate\_in\_forest} \doteq \mathsf{estate} \sqcap \exists\, \mathsf{is\_g\_inside} \,.\, \mathsf{forest}$$

$$\mathbf{fishing\_cottage} \doteq \mathsf{cottage} \sqcap \exists\, \mathsf{is\_g\_inside} \,.\, (\mathsf{estate} \sqcap \exists\, \mathsf{is\_touching} \,.\, \mathsf{river})$$

The realising component of the $\mathcal{ALCRP}(\mathcal{D})$ reasoner will compute the following direct concept memberships of the individual c: normal_cottage_in_forest and fishing_cottage. The individual e will be the direct concept member of estate_in_forest. The other individuals r and f keep their asserted concept memberships.

In the following we employ a so-called *abstraction process* that can reduce particular ABoxes to corresponding ABoxes consisting of a single concept assertion representing the original query. The concept used in the assertion represents an abstraction of the ABox and, in turn, of the original query. It defines the *semantics* of this query and is used as a *query concept*. We assume that the GIS provides a query facility (implemented by the $\mathcal{ALCRP}(\mathcal{D})$ reasoner) that can retrieve all stored individuals that are members of a given query concept. Therefore, the concept mentioned in a reduced ABox is used to retrieve all "matching" individuals and to answer the query.

For instance, using the abstraction process we can replace ABox $\mathcal{A}_0$ by ABox $\mathcal{A}_1 = \{\mathsf{c} : \mathsf{cottage}_{\mathsf{c}_1}\}$. $\mathcal{A}_1$ uses the synthesised concept $\mathsf{cottage}_{\mathsf{c}_1}$. The concept $\mathsf{estate}_{\mathsf{e}_1}$ is only introduced to improve the readability of $\mathsf{cottage}_{\mathsf{c}_1}$.

$$\mathbf{estate}_{\mathsf{e}_1} \doteq \mathsf{estate} \sqcap \exists\, \mathsf{is\_g\_inside} . \mathsf{forest} \sqcap \exists\, \mathsf{is\_touching} . \mathsf{river}$$

$$\mathbf{cottage}_{\mathsf{c}_1} \doteq$$

$$\mathsf{cottage} \sqcap \exists\, \mathsf{is\_g\_inside} . \mathsf{estate}_{\mathsf{e}_1} \sqcap$$

$$\exists\, \mathsf{has\_space} . \lambda_{\mathcal{R}} x . (x > 40 \wedge x < 70)$$

The query concept $\mathsf{cottage}_{\mathsf{c}_1}$ is classified by the reasoner. The semantic validity of this query is automatically verified during classification, i.e. to check whether the query concept is coherent (see Section 22.2.3). For instance, if the forest f were required to be "mosquito-free" (see above), the $\mathcal{ALCRP}(\mathcal{D})$ reasoner would immediately recognise the incoherence of $\mathsf{cottage}_{\mathsf{c}_1}$. This information could be used by the spatial parser for generating an explanation to the user and for identifying the source of the contradiction.

Let us assume that the query with the concept $\mathsf{cottage}_{\mathsf{c}_1}$ returns more than 100 matches (i.e. individuals). The next step for the user might be to refine the query by adding more constraints.[1] One could add more requirements to the estate, e.g. we ask for a garage connected to the cottage. The extended sketch (see Fig. 22.6b) corresponds to the ABox $\mathcal{A}_2$ (ignoring the lake) by adding to ABox $\mathcal{A}_0$ new assertions: $\mathcal{A}_2 = \mathcal{A}_0 \cup \{\mathsf{g} : \mathsf{garage}, (\mathsf{c}, \mathsf{g}) : \mathsf{is\_touching}\}$. The abstraction process reduces ABox $\mathcal{A}_2$ to ABox $\mathcal{A}_3 = \{\mathsf{c} : \mathsf{cottage}_{\mathsf{c}_2}\}$ using the following synthesised concept description.

$$\mathbf{cottage}_{\mathsf{c}_2} \doteq \mathsf{cottage} \sqcap \exists\, \mathsf{has\_space} . \lambda_{\mathcal{R}} x . (x > 40 \wedge x < 70) \sqcap$$

$$\exists\, \mathsf{is\_g\_inside} . \mathsf{estate}_{\mathsf{e}_1} \sqcap \exists\, \mathsf{is\_touching} . \mathsf{garage}$$

$$\mathbf{cottage}_{\mathsf{c}_3} \doteq \mathsf{cottage}_{\mathsf{c}_1} \sqcap \exists\, \mathsf{is\_touching} . \mathsf{garage}$$

The $\mathcal{ALCRP}(\mathcal{D})$ reasoner recognises the taxonomic relationship that $\mathsf{cottage}_{\mathsf{c}_1}$ subsumes $\mathsf{cottage}_{\mathsf{c}_2}$. It can be rewritten as $\mathsf{cottage}_{\mathsf{c}_3}$ that even textually demonstrates the subsumption relationship.

Subsumption between query concepts immediately leads to subsumption between GIS queries and to its utilisation for query optimisation. In order to answer the refined query a query optimiser can benefit from the detected query subsumption and reduce the search space to the set of query matches already computed for the query concept from ABox $\mathcal{A}_1$. Note that these

[1] Of course, one of the most important criteria is the price of the estate. This is currently neglected due to the non-spatial nature of this part of the query (but see our proposal in Section 22.4).

query matches are members of the concept $\mathsf{cottage}_{\mathsf{c}_1}$. This type of query optimisation is an important aspect in applying description logics to database theory (see [1] for an introduction to these topics).

The benefits of computing a concept subsumption taxonomy can be even more subtle. Imagine a query from another user looking for a cottage located in a forest that is connected to a river. The ABox $\mathcal{A}_4$ is derived from the sketch. The abstraction process creates the following two concept definitions from ABox $\mathcal{A}_4$.

$$\mathcal{A}_4 = \{\mathsf{c}:\mathsf{cottage}, \mathsf{e}:\mathsf{estate\_area}, (\mathsf{c},\mathsf{e}):\mathsf{is\_g\_inside}, \mathsf{r}:\mathsf{river}, \mathsf{f}:\mathsf{forest}, (\mathsf{f},\mathsf{r}):\mathsf{is\_connected}, (\mathsf{e},\mathsf{f}):\mathsf{is\_g\_inside}\}$$

$$\mathbf{estate}_{\mathsf{e}_2} \doteq \mathsf{estate} \sqcap \exists\, \mathsf{is\_g\_inside}\,.\,(\mathsf{forest} \sqcap \exists\, \mathsf{is\_connected}\,.\,\mathsf{river})$$

$$\mathbf{cottage}_{\mathsf{c}_4} \doteq \mathsf{cottage} \sqcap \exists\, \mathsf{is\_g\_inside}\,.\,\mathsf{estate}_{\mathsf{e}_2}$$

The resulting ABox $\mathcal{A}_5 = \{\mathsf{c}:\mathsf{cottage}_{\mathsf{c}_4}\}$ uses the derived concepts. It turns out that the concept $\mathsf{cottage}_{\mathsf{c}_4}$ subsumes the other concepts $\mathsf{cottage}_{\mathsf{c}_i}$ although the concept descriptions are textually different. This is a rather complex proof based on the interaction between the spatial relations: $\mathsf{g\_inside}(\mathsf{e},\mathsf{f}) \wedge \mathsf{touching}(\mathsf{e},\mathsf{r}) \Rightarrow \mathsf{connected}(\mathsf{f},\mathsf{r})$.

The abstraction process works rather well for ABoxes containing no joins or cycles, i.e. the same individual is a filler of several roles or even related to itself through a cycle of role assertions. If joins or cycles are present in an ABox, it depends on the expressiveness of the description logic whether an ABox can be reduced to a single concept membership assertion. For instance, joins can be expressed by restricting the number of possible role fillers or by equality restrictions for feature fillers. As mentioned above, other DLs also support the definition of cyclic concepts that might be required to fully reduce some ABoxes. Due to unknown decidability results $\mathcal{ALCRP}(\mathcal{D})$ currently does not allow cyclic concepts or number restrictions. Therefore, in case of ABoxes with joins or cycles, we can only partially reduce these ABoxes. This is illustrated in Fig. 22.6b by adding a lake. The river has to flow into the lake and the same lake is touching the forest. This is an example for a join in a corresponding ABox. However, the reasoning with $\mathcal{ALCRP}(\mathcal{D})$ as described above is still valid and usable for query processing. Only the subsumption between ABox queries requires a more sophisticated approach.

22.4 Using ABox Patterns for n-ary Queries

We have demonstrated that the abstraction process of rewriting a query ABox to a concept term provides a means to specify the semantics of a visual query. Unfortunately, there are also some drawbacks with this approach. First of all, since the semantics of a concept description is only the set of individuals that satisfy the concept, no n-ary query results can be returned. This is not surprising, since concept descriptions correspond to first-order formulae with only one free variable. Thus, the abstraction process is only successfully applicable to ABoxes where it can select exactly one primary "target individual" from the query. The target individual remains as the single individual and the other individuals are represented by the query concept derived by the abstraction process. In the case of the cottage example, the target object of the query is the cottage the user is looking for. However, considering VISCO, the semantics of a VISCO query is a set of n-tuples, so one would need more than one free variable to fully specify the semantics of VISCO's query language which can handle aggregates.

As a solution to this problem we have developed so-called *ABox patterns*, which are ordinary $\mathcal{ALCRP}(\mathcal{S}_2)$ ABoxes that may – in addition to ordinary individuals – contain *variables*, e.g. $x?$, $y?$. Intuitively speaking, given a "GIS database ABox" $\mathcal{A}$ and a query ABox $\mathcal{Q}$ that contains variables, the *ABox pattern retrieval service* returns a set σ of *substitutions* which are mappings from variables in $\mathcal{Q}$ to ABox individuals in the database ABox $\mathcal{A}$ such that $\mathcal{A} \models_{\mathcal{T}} \sigma(\mathcal{Q})$ (w.r.t. the TBox $\mathcal{T}$).

$\mathit{abox\_pattern\_retrieval}(\mathcal{A}, \mathcal{Q}) := \{\sigma \mid \mathcal{A} \models_{\mathcal{T}} \sigma(\mathcal{Q})\}$.

$\sigma(Q)$ applies the given substitution σ to the query ABox $\mathcal{Q}$, replacing its variables with individuals from $\mathcal{A}$. As a subproblem, we need to decide the *ABox entailment problem*, e.g. the question whether $\mathcal{A} \models_{\mathcal{T}} \sigma(\mathcal{Q})$. This problem is decidable for $\mathcal{ALCRP}(\mathcal{S}_2)$ (see [11]). The substitutions can simply be enumerated as mappings from $\mathit{Vars}(\mathcal{Q}) \rightarrow \mathit{Individuals}(\mathcal{Q})^{\| \mathit{Vars}(\mathcal{Q}) \|}$ and applied to $\mathcal{Q}$ yielding $\sigma(\mathcal{Q})$. If $\sigma(\mathcal{Q})$ is entailed by $\mathcal{A}$, then the range (image) of σ is the n-ary query result. Note that computing the set of individuals that are members of a concept C is a special case of an ABox pattern retrieval.

$\mathit{concept\_members}(\mathcal{A}, \mathsf{C}) := \mathit{abox\_pattern\_retrieval}(\mathcal{A}, \{x?:\mathsf{C}\})$.

As an example, we reconsider our query where a user is looking for a cottage in a forest. Obviously, the price of the estate as well as the cottage and the forest itself are of interest, so instead of returning just the cottage the query should return triples such as $\langle \mathsf{cottage}, \mathsf{estate}, \mathsf{forest} \rangle$ which can be further inspected. The corresponding query ABox $\mathcal{Q}$ might be defined as

$\mathcal{Q} = \{x?:\mathsf{cottage}, y?:\mathsf{estate}, z?:\mathsf{forest}, (x?, y?):\mathsf{g\_inside}, (y?, z?):\mathsf{g\_inside}\}$.

The user can also refer to specific individuals, e.g.

$\mathcal{Q} = \{x? : \mathsf{cottage}, y? : \mathsf{estate}, (x?, y?) : \mathsf{g\_inside}, (y?, \mathsf{black\_forest}) : \mathsf{g\_inside}\}$, where black_forest is a specific database ABox individual. Obviously, joins can easily be specified. Suppose we are looking for two cottages within the same estate that are located at the same river:

$$\mathcal{Q} = \{x?:\mathsf{cottage}, y?:\mathsf{cottage}, x? \neq y?, z?:\mathsf{estate}, (x?, z?):\mathsf{g\_inside}, (y?, z?):\mathsf{g\_inside}, r?:\mathsf{river}, (x?, r?):\mathsf{touching}, (y?, r?):\mathsf{touching}\}.$$

Since the unique name assumption also does not hold for variables, we introduce an additional assertion $x? \neq y?$ in order to ensure that $\sigma(x?) \neq \sigma(y?)$. This simply constrains the substitution σ and has no impact on $\mathcal{ALCRP}(\mathcal{S}_2)$.

22.5 Conclusion

The formalism presented in this chapter can be used to define the semantics of visual spatial queries, to reason about query subsumption, and to deal with multiple worlds or query completion with the help of default reasoning. The proposed ABox pattern retrieval solves the problem with joins and/or cycles in ABoxes and supports n-ary query results. However, the price is an increased query execution complexity due to the more complex inference problem.

References

1. Borgida, A. (1995). Description logics in data management. IEEE Transactions on Knowledge and Data Engineering 7(5):671–682.
2. Brachman, R. and Schmolze, J. (1985). An overview of the KL-ONE knowledge representation system. Cognitive Science pp. 171–216.
3. Haarslev, V. (1998). A fully formalized theory for describing visual notations. In K. Marriott and B. Meyer (Eds), Visual language theory. Berlin:Springer, pp. 261–292.
4. Haarslev, V. (1998). A logic-based formalism for reasoning about visual representations (extended abstract). In Proceedings, AAAI workshop on formalizing reasoning with visual and diagrammatic representations, AAAI fall symposium series, 23–25 October. Orlando, FL: AAAI Press, pp. 57–66.
5. Haarslev, V. (1999). A logic-based formalism for reasoning about visual representations. Journal of Visual Languages and Computing 10(4):421–445.
6. Haarslev, V., Lutz, C. and Möller, R. (1998). Foundations of spatioterminological reasoning with description logics. In T. Cohn, L. Schubert and S. Shapiro (Eds), Proceedings, sixth international conference on principles of knowledge representation and reasoning (KR'98), Trento, Italy, 2–5 June, pp. 112–123.
7. Haarslev, V., Lutz, C. and Möller, R. (1999). A description logic with concrete domains and a role-forming predicate operator. Journal of Logic and Computation 9(3):351–384.
8. Haarslev, V., Möller, R. and Wessel, M. (1999). On specifying semantics of visual spatial query languages. In 1999 IEEE symposium on visual languages, Tokyo, Japan, 13–16 September. Los Alamitos, CA: IEEE Computer Society Press, pp.4–11.
9. Haarslev, V. and Wessel, M. (1997). Querying GIS with animated spatial sketches. In 1997 IEEE symposium on visual languages, Capri, Italy, 23-26 September. Los Alamitos, CA: IEEE Computer Society Press, pp. 197–204.
10. Lutz, C. (1999). On the complexity of terminological reasoning. Technical Report LTCS-99-04, RWTH Aachen.
11. Möller, R. and Wessel, M. (1999). Terminological default reasoning about spatial information: A first step. In Proceedings of COSIT'99, International conference on spatial information theory, Stade. Berlin: Springer, pp. 172–189.

12. Tarski, A. (1951). A decision method for elementary algebra and geometry. Berkeley, CA: University of California Press.
13. Wessel, M. and Haarslev, V. (1998). VISCO: Bringing visual spatial querying to reality. In 1998 IEEE symposium on visual languages, Halifax, Canada, 1–4 September. Los Alamitos, CA: IEEE Computer Society Press, pp. 170–177.
14. Woods, W. and Schmolze, J. (1992). The KL-ONE family. In F. Lehmann (Ed.), Semantic networks in artificial intelligence, Oxford: Pergamon Press, pp. 133–177.

Part IV

Applications of Diagrammatic Reasoning

Introduction

The following chapters address the computational and notational application of diagrammatic reasoning and diagrammatic representations in a variety of domains including human-computer interaction, diagram processing, case-based reasoning, sketch interpretation, diagram editing, novel representations for plans and understanding electricity, and grammars for diagrammatic languages.

From a computational perspective, the inception of diagrammatic reasoning research coincides with the birth of the first artificial intelligence applications of the late 1950s. Inspired by a comment made by Minsky that "a diagram might prove useful" to his work, Gelertner created the first diagrammatic reasoning program. His Geometry Machine (1959) proved theorems in plane geometry using a diagram to focus its backward-chaining search to only those branches whose premises held in the diagram, noting that if they did not hold in the specific they could not hold in general. He showed that using a diagram to focus the search cut the average branching factor of the search from 1000 to 5, a factor of 200! On the heels of this success, Evans followed with ANALOGY (1962), a program that attempted quite successfully to answer geometrical analogy questions from IQ tests in an effort to emulate human diagrammatic reasoning capabilities in this domain.

Rather surprisingly, these early successes did not immediately spawn further work in diagrammatic reasoning. This may be due in part to equally successful work being done at the time based on more generally applicable symbolic reasoning. It was not until Larkin and Simon's seminal paper "Why a Diagram is (Sometimes) Worth Ten Thousand Words" (1987) that Minsky's notion that a diagram could be useful as an effective search heuristic was once again proven insightful. Larkin and Simon showed that a large saving in the number of inferences required in a production system resulted when spatial information was provided to the system. It seemed that the time was now ripe for investigation of less traditional computational representations as this work was followed by a resurgence in diagrammatic reasoning research – work represented in this volume.

Although there has been some effort to repeat the successes of Gelertner and Larkin and Simon in gaining computational efficiency through use of diagrams in other arenas, it has proven to be more of a challenge than one might expect and, in fact, none of the work in this section is along these lines. On the contrary, though, work started by Evans attempting to emulate human diagrammatic reasoning capabilities via machine has experienced some success, some of which is represented in the chapters of this part.

The work by Anderson in Chapter 24 and Fish and McCartney in Chapter 25 might be considered to adhere closest to Evans' original vision in that it not only attempts to imbue a machine with some level of diagram understanding present in human beings but, in these attempts, exploits the spatial coherence of related collections of diagrams just as did Evans. Anderson argues that what he terms "diagram processing", the ability of a system to deal directly with diagrammatic input and produce cogent diagrammatic output, is as important as natural language processing and should be afforded equal research resources. He offers, as a step towards a full diagram processing agent, an example diagrammatic information system that is capable of responding to interesting queries posed to it in the domain of cartograms of the United States. Fish and McCartney investigate the synergy between diagrammatic reasoning and the artificial intelligence technique of case-based reasoning. They hypothesise that the understanding of a given diagram might be facilitated by comparing and contrasting this diagram to diagrams previously experienced. This use of similarity is the defining feature of case-based reasoning, making its marriage to diagrammatic reasoning particularly appropriate in this context. They offer, as an example of their concept, a program that uses case-based reasoning to understand diagrams in the domain of football plays. Such integration between diagrammatic reasoning and more traditional artificial intelligence techniques is crucial to the nascent diagrammatic reasoning field in that it exposes other communities to the field and folds the field into the mainstream.

The work of Anderson and Fish and McCartney differs fundamentally in how diagrams are represented by their systems. Where Anderson uses a low-level representation based on the tiling of a two-dimensional surface, Fish and McCartney use a more high-level representation that describes a diagram in terms of the concepts and structure present within it. This higher-level representation is in fact more in keeping with Evans' original representational scheme and, further, with the representational schemes of much of the work in diagrammatic reasoning.

Chapter 23, by Meyer, Marriott and Allwein, is a good overview of a number of application categories using high-level representational schemes that have developed within the diagrammatic reasoning community. They move towards a definition of a generic "intelligent diagrammatic environment" (IDE), a framework in which structured diagrams are used as a more natural means of communication between a user and a system. They discuss subtasks required by IDEs and identify four major IDE application areas: document understanding, diagramming support, diagrammatic query languages for spatial information systems and diagrammatic theorem provers. Such consolidation work is essential for uniting a very diverse community and is helpful in understanding the nature of diagrammatic reasoning.

The work of Stahovich in Chapter 26 represents sketches of diagrams using a high-level representational scheme as well. Although sketch understanding is discussed in conjunction with diagrammatic query languages in the work by Meyer, Marriott and Allwein, sketches within Stahovich's SketchIT program are not queries to a database of spatial information but, instead, are designs of mechanical devices considered as representative of whole families of designs with similar functionality. SketchIT uses domain knowledge to determine the behaviours of the various parts of a stylised sketch augmented with a state transition diagram describing the overall behaviour of the device and then produces new designs from this knowledge. This work advances human-computer interaction by permitting an engineer to communicate to a system via a more natural means.

As an example of an application in the category termed "diagramming support" – programs that automatically generate intelligent diagram editors from high-level specifications – in Chapter 28 Calder describes the design of Thistle, a diagram editor for the wide variety of diagrams found in the domain of linguistics. This work exploits formal language theory by treating diagrams as conforming to a grammar. Further, it is argued that the exploration of diagrammatic grammars is likely to bear fruit in future research. This assertion is seconded by Minas in Chapter 32, where the author offers hypergraph grammars as a means of specifying diagrammatic languages and describes DiaGen, a generator for creating graphical editors for specific diagram classes. Such formal approaches to diagrammatic reasoning are necessary to ground the field, providing a firm foundation for future work within it.

From a notational perspective, interest in diagrammatic reasoning and representations reaches back in time to when the first scratches were made in the dirt. A more formal interest can be traced back to the ancient Greeks and the beginnings of mathematics through Euler circles (1772), Venn diagrams (1894), and Pierce's existential graphs (1914). More recently, as evidenced by Part 3, interest in diagrams has turned towards a deeper reflection on their true benefits, limitations, and relationships to other formalisms. This deliberation has both elevated the status of diagrams in relation to formalisms thought to be more solidly grounded and debunked long-held intuitions concerning their perspicuity.

Three chapters in this part are concerned with application of this newly gained information in the form of new diagrammatic representations, for a variety of domains, that attempt to overcome shortcomings in representation schemes currently used in these domains. Ader, in Chapter 27, offers "functional notation" (f-notation) and its corresponding graphical representations (f-graphs) as a means to unambiguously diagrammatically capture the notions of context and time when planning research methodology. In

Chapter 30, Kosara, Miksch, Shahar and Johnson are concerned with the diagrammatic representation of standardised health care procedures and, finding traditional flowchart representations too limited for communicating these procedures, they describe a plan-representation language, Asbru, that overcomes the limitations of flowcharts and a corresponding visualisation tool, AsbruView, to diagrammatically realise plans represented in this language. Finally, as part of the continuing Representational Analysis and Design Project, a project whose goal is the development of principles for the design of effective representational systems. Cheng in Chapter 29 presents AVOW diagrams, a representational system designed to facilitate the understanding of electricity. AVOW diagrams are an example of Law Encoding Diagrams, a general class of diagrams whose properties are under investigation by the Project.

Chapter 31 presents a pair of case studies by Lucas and Cousin-Rittemard concerning diagrams within two disparate fields – linguistics and mechanics – and distils out commonalities between the representations that suggest general properties of diagrammatic communication.

As with Part 3, which addresses formalisation of diagrammatic reasoning, the following chapters only highlight a number of representative approaches to the computational and notational application of diagrammatic reasoning. It is clear that the application of diagrammatic reasoning and representations has only begun to scratch the surface of what promises to be a very fruitful endeavour, itself begun by scratches on a surface.

23. Intelligent Diagrammatic Interfaces: State of the Art

Bernd Meyer
Kim Marriott
Gerard Allwein

One important practical role for diagrammatic reasoning is to provide the foundation for intelligent diagrammatic environments (IDEs), in which structured diagrams are used as a means of natural visual communication between the human user and the system. The present chapter provides a characterisation of IDEs and gives an overview of the state of the art. Rather than attempting an exhaustive survey, we discuss the subtasks that have to be addressed in a general IDE framework and show how they are handled in representative systems from the following IDE application areas:

- Document understanding
- Diagramming support
- Diagrammatic query languages for spatial information systems
- Diagrammatic theorem provers

23.1 Introduction

One of the earliest attempts to bring the vision of IDEs to life is described in [38, 39], where an "intelligent" blackboard automatically interprets and annotates sketches jotted down in group discussions. Related applications in narrower application areas have successfully been demonstrated by sketch-based systems for architecture [27–29] and spatial information systems [18, 31, 42, 43, 55] and the availability of pen-based computers has triggered an increase of interest in such systems.

We shall focus on approaches which handle diagrams by manipulation of symbolic entities, such as basic geometric shapes. Discussion of methods based on the manipulation of low-level picture structures, such as pixel arrays, is beyond the scope of this chapter. For a flavour of these methods the reader is referred to the treatment of Inter-Diagrammatic Reasoning by Anderson et al. in this volume and to the publications on the Bitpict system [21–23].

Research into *document understanding*, more precisely into the automatic interpretation of drawings in paper documents, marks the beginning of diagram understanding research. The primary motivation for such projects was to allow CAD systems to read older technical drawings, such as construction plans, drawn only on paper [35].

Much recent work on IDEs has focused on generic support for automatically generating an IDE from a formal high-level specification [10, 49] of a diagram notation. The goal of such systems, which we shall call Diagramming Support Systems (DSSs) in the following, is to automatise the construction of diagrammatic human-computer interfaces for any diagrammatic notation that can be sufficiently formalised.

It is interesting to ask: How is the task of diagram interpretation different in these two settings? The most obvious distinction is that document understanding is concerned with *static* diagrams while DSS, as a user interface technology, must support *dynamic diagram usage*. In a static system, a diagram is fixed once it is drawn and can be interpreted as an immutable entity without reference to its construction process. Clearly, this is true for the interpretation of printed diagrams on paper. In a dynamic system, a diagram can be changed after its construction. For a DSS this means that it has to incrementally maintain a correct interpretation under change. A second important aspect of dynamic notations is that not only the state of the diagram but also the type of change itself can convey meaning.

Another fundamental difference is that between *open systems* and *closed systems* of diagrammatic notations. A closed system is semantically self-contained, i.e. each diagram can be interpreted without referring to knowledge about the application domain or to assumptions about its usage, and, in principle, such diagrammatic notations can be formalised unambiguously. Current DSSs are mostly used with closed notations, since this greatly simplifies diagram interpretation. In contrast, an open system requires knowledge not captured in the notational system to disambiguate between conflicting interpretations. Many technical notations, such as mechanical engineering drawings, require domain knowledge for a meaningful interpretation and, therefore, many document-understanding systems have to be regarded as dealing with open systems.

Two other important examples of IDE applications are *diagrammatic query languages for spatial information systems* and *diagrammatic theorem provers*. Diagrammatic languages are widely used as visual query languages for databases. One of the most interesting types of visual query languages are sketch-based query languages for spatial information systems [18, 31, 42, 43, 55]. The basic idea behind such systems is to let the query system interpret an input sketch as an example of the spatial scenario that the user is searching for. The system needs to disambiguate the sketched query, whether by making "informed guesses" about the user's intention or by providing feedback which informs the user about the system's interpretation so that he can apply corrections.

Here we encounter another important distinction: In *unidirectional communication* a diagram is only used as input (e.g. in document understanding).

In bidirectional communication we have a true diagrammatic dialogue, where a diagrammatic notation is also used as output (e.g. in sketch-based query languages which use visual feedback to disambiguate a query). Another excellent example of bidirectional systems are IDEs for interactive theorem proving which cooperate with a mathematician to verify or falsify a conjecture. We note that in many cases diagrams in such systems may not be understood as a closed system, since it may be impossible to fix the required diagrammatic notation for examples a priori.

We will now discuss concrete examples of each of the above IDE application areas. However, before we proceed to this, it is worthwhile to analyse the subtasks which are part of a general IDE framework.

23.2 IDE Subtasks

Building IDEs is not a straightforward task. In order to do so we need a good understanding of the subtasks that an IDE has to manage and the techniques for performing these. Fortunately, these subtasks are now starting to be better understood and technology supporting the early stages is now becoming available.

Obviously, the most basic requirement is an adequate *geometric or spatial modelling* of the diagram. There is a (perhaps) surprising lack of agreement on how best to do this. Arguably the most direct way is to handle a diagram as a collection of pixels and to model geometric shapes as sets of pixels. A model of geometric relationships can then be based on intersections of these pixel sets (for the case of black and white diagrams) or more complex set operations in the case of colour diagrams. Reasoning in this fashion, however, is rather expensive. A better approach is to build complete topological calculi which are based on relationships between sets of points. The most commonly used approaches are in the spirit of Clarke's topological "calculus of individuals based on connection" [12,13] and many of them use the Region-Connection-Calculus (RCC) [15,26] directly derived from this.

An alternative approach to geometric modelling of diagrams is based on geometric entities. The diagram is broken up into basic geometric shapes such as lines, rectangles or circles, and these objects are represented symbolically. Their exact geometric properties are then captured by giving these objects real-valued attributes. Predicates or constraints expressed in terms of real-values are used to express the spatial relationships between the objects. Related ways of modelling are used in most IDEs and are exemplified in [9,32,45,46,56].

It is not uncommon to use even more abstract modelling which totally dispenses with concrete geometry, simply representing geometric relationships between diagram elements as uninterpreted symbolic relations. This direction is taken by most generic graph-grammar-based approaches to IDEs. Here a diagram is simply modelled as a (hyper) graph whose nodes represent diagram objects and whose edges represent spatial relationships between these objects (see [52] for an overview).

The second low-level subtask is *parsing*. This is the process of recognising the (usually hierarchical) syntactic structure of a diagram from a flat representation of its components. Parsing can also be regarded as the process of mapping the concrete spatial syntax of a diagram to an abstract (often relational) syntax, though it is often not clear where to place the border between abstract syntax and semantics. We note that, if the input is a pixel-based representation, parsing needs to be preceded by a *tokenising* phase, which recovers the geometric objects from the raster representation.

Diagrammatic parsing is most commonly based on multidimensional grammars. Broadly speaking there are three main approaches: variants of traditional string parsing algorithms, graph-grammar-based approaches and attributed multiset grammar-based approaches. Subsequently we will discuss an approach based on Constraint Multiset Grammars in more detail and a comprehensive survey can be found in [41]. Logical formalisms for diagrammatic parsing purposes are a relatively recent development. A detailed exposition of such a formalism which is based upon a combination of description logic with concrete geometric domains is given by Haarslev at al. elswhere in this volume.

Once the syntactic depth structure of the diagram is recovered, the next subtask is that of interpretation, i.e. building a semantic representation of the meaning of the diagram. Although conceptually very different, the same approach is often used for both parsing and interpretation. Many current IDEs use variants of grammatical specifications and in such approaches the interpretation is defined by equipping the productions with additional semantic attributes. While this is sufficient for a broad range of highly structured diagram notations that can be strictly formalised, it poses problems with *open notational system* that require outside knowledge to be interpreted. There are two ways to overcome this problem.

One approach relies on the fact that grammatical formalisms can be interpreted as a particular variant of logical formalism in a broad sense and that grammars can therefore be mapped to logical specifications. In consequence, it is possible to extend grammar formalisms with more powerful open-ended logic specification methods in a scalable fashion where and when it is required. This approach is, for example, taken in [45,46]. Another possibility is to abandon grammatical specification altogether and to switch to a different basic framework, such as to logic-based descriptions. The work presented by Haarslev et al. elswhere in this volume clearly demonstrates that this can result in a formally clean and very expressive framework. Due to the trade-off between expressiveness and effective tractability it is not yet clear whether one of these approaches is generally superior for actual IDE implementations.

The final subtasks, which close the gap to *reasoning*, are less well understood and too complex to be discussed in any detail in this short exposition. A variety of approaches have been taken to introduce reasoning into IDEs: Broadly speaking, we can try to embed the reasoning capabilities directly into the IDE framework or the IDE can be coupled with an external reasoning system. Logic-based frameworks such as GenEd probably provide the most powerful formal basis for enriching IDEs directly with reasoning capabilities. The second alternative for reasoning enabled IDEs is to let the IDE transform the diagrammatic representation into another representation that can be processed by a non-diagrammatic reasoner that interacts with the IDE. This approach is, for example, taken in some diagrammatic theorem provers [4, 5, 25].

In general, the process of *diagram transformation* appears to be central to diagrammatic reasoning: Transformation details how one diagram can syntactically and/or semantically be transformed into another diagram that directly and explicitly exhibits additional information that is either not present or only implicit and hidden in the prior diagram. Unrestricted type-0 variants of graph and multiset grammars have been used to specify diagram transformations and geometric constraint solving has proven crucial for preserving the diagram semantics during such transformations. In fact, constraint-based transformation approaches can be viewed as a heterogeneous variant combining abstract manipulation with spatial manipulation, because they usually operate simultaneously on the symbolic level and on the geometric level. As discussed in more depth in [44, 47] it proves crucial to have access to both levels.

Our discussion about diagram reasoning comes from a symbolistic point of view, in which diagrammatic reasoning is the process of transforming a diagrammatic representation to an abstract symbolic space and of performing reasoning in this conceptual space. While this is the way that virtually all existing IDEs work, it is interesting to discuss how much of the reasoning process can be performed directly on the level of the diagrammatic representation without requiring preceding interpretation of the diagram.

However, in a certain sense, viewing diagrammatic reasoning in this way makes little difference, as the same inference process is formalised, only from a different viewpoint. More precisely, we can formalise reasoning with diagrams directly, as the dynamic creation of a sequence of diagrams $\mathcal{D}_1, \mathcal{D}_2, ..., \mathcal{D}_i, ..., \mathcal{D}_n$ where $\mathcal{D}_i$ is spatial transformationed by τ_i into $\mathcal{D}_{i+1}$ and the resulting consequences can directly be read off from the final diagram D_n. The process that virtually all systems employ instead is to first use an interpretation $\mathcal{I}$ to map the initial diagram $\mathcal{D}_1$ to an abstract problem representation $\mathcal{R}_1$ and to reason in the abstract problem space. In many cases the reasoning in the abstract problem space is constructive, i.e. it works by transforming the problem representation into a new, refined problem representation. In these cases we can immediately map the abstract argument (sequence of reasoning steps) to a "true" diagrammatic argument, provided there is a mapping $\mathcal{I}^{-1}$ back from the abstract representation $\mathcal{R}_i$ to its diagrammatic representation $\mathcal{D}_i$ in each step i (see Fig. 23.1 for the case when $n = 4$). In this case it is immaterial whether we perform reasoning on the diagrammatic level or on the abstract level since they are "isomorphic".

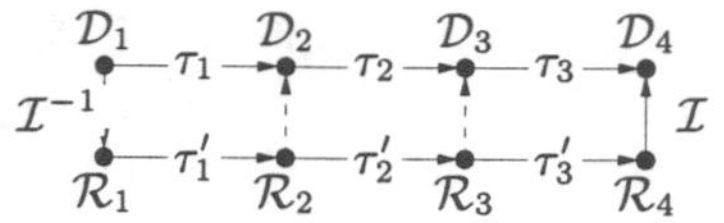

Figure 23.1. Levels of diagram transformation.

We will now turn to the description of some representative IDEs and discuss how the subtasks we have identified above are solved in these systems.

23.2.1 Document Understanding

A typical document image analysis system takes a raster image, obtained from scanning a document page containing diagrams and text, and converts this into a (moderately) high-level description of the diagram components and connections. Following [36], this process has five main phases: *data-capture and preprocessing*, in which the original document is scanned to give a raster image; *region segmentation*, in which different types of regions, for instance, diagrams, text and images, are recognised in the document; *vectorisation*, in which raster-to-vector graphics conversion is performed; *feature extraction*, in which feature points such as the point of transition from a straight line to a curve are determined; and *graphics recognition and interpretation*, in which application-specific routines are used to extract the semantic meaning of the diagram. We shall focus on this last phase, since it is the only phase which can really be called diagram understanding. The other phases are low level in nature and correspond to low-level visual processing employed by humans when determining diagram components. Indeed, computer vision and document understanding employ many of the same techniques.

Unfortunately, it is fair to say that graphics recognition and interpretation is the least understood phase of document understanding; far more attention has been paid to the difficult low-level phases. One generic approach is to perform *symbol recognition* by, for instance, matching an application-specific catalogue of graphic shapes such as rectangles or the symbol for a door or window. More high-level understanding is usually only provided in systems employing application-specific ad hoc algorithms for restricted diagram domains such as maps, CAD systems, or musical notation.

In several systems, however, although hidden behind domain-specific heuristics, common techniques based on determining layout relationships between diagram components can be glimpsed. Such systems employ grammars, production rules and semantic networks.[1] For example, a semantic network approach is employed for bar-chart recognition [57] and for symbol detection in architectural drawings [1]. In [51] block grammars are used to specify layout of the page for a particular publication format. This is used to provide intelligent segmentation and recognise components like titles, authors and figure captions. Another grammar-based system, the Diagram Understanding System [24], is notable for its genericity. It takes an attributed multiset grammatical specification of a diagram language, such as state transition diagrams, and generates a top-down parser for the language, employing spatial data structures for efficiency in parsing.

Most current image analysis systems, however, are very application specific and do not provide a generic approach to high-level diagram understanding. They are also inherently static in nature and therefore only loosely related to IDE tasks.

23.2.2 Diagramming Support

Constructing an IDE from scratch is a dauntingly difficult task. Several research efforts have therefore focused on providing generic support for building IDEs from high-level specifications. We term such systems Diagramming Support Systems (DSSs). The task of a DSS is to automatically construct an intelligent diagram editor for a given diagrammatic notation from a syntactic and semantic high-level specification of this notation. Such an editor must be capable of interpreting diagrammatic input and of translating the diagrams into other (usually symbolic) representations that are meaningful to other parts of the application. Additionally, the editor should automatically detect incorrect usage of the notation, provide semantic feedback and should generally support the drawing process as much as possible.

The most ambitious approach to this task is the construction of freeform diagram editors which allow the user to sketch a diagram without prescribing any particular way of drawing them. The underlying graphic editor interprets the diagram as it is being constructed and, in the ideal case such a system can even beautify and refine rough, approximate diagrams during their construction and detect incorrectly constructed diagrams. This approach is particularly attractive with pen-based systems since it fits well with the traditional pen-and-paper model of diagram construction. This concept, often dubbed "smart paper/intelligent pen and paper" or later "intelligent diagrams", was described in [54] and the idea of DSSs for such environments has its roots in the work of [9, 11, 28, 29]. Over the last decade several systems have at least

[1] Interestingly, in the early days of image processing and document understanding, grammatical techniques were employed in virtually all phases. However, in part because of lack of flexibility and brittleness in the presence of errors, they are no longer used in practice for low-level phases [2].

partially implemented the concept of intelligent diagrams. Notable examples include, Stretch-A-Sketch [28,29], VLCC [16], Penguins [9,10], Recopla [48,49] and GenEd [30].

A related but simpler approach to the construction of diagram editors with automatic interpretation is syntax-directed editing (see, for example, DiaGen [50] and VAS [17]). However, syntax-directed editors do not allow freestyle drawn input. The interpretation task in this context is therfore significantly easier than that in IDEs.

It is obvious that the recognition of incorrectly constructed diagrams or incorrect usage is not an issue in syntax-directed systems, since the construction of incorrect diagrams can a priori be avoided. While syntax-directed editing is already quite limiting when working with highly structured technical diagrams, such as digital circuit schemas, it seems prohibitive in settings where the usage of diagrams is more exploratory or less strict, such as in architecture and design. For these reasons, more recent versions of DiaGen are also moving towards a hybrid approach which supports freestyle drawings.

In DSSs for freeform diagrams, the ability to automatically maintain spatial constraints during direct manipulation of a diagram plays an important role, because it is often required that the semantics of a diagram is preserved, while the user modifies its appearance. Imagine, for example, an editor for electronic circuit diagrams: Clearly, when a circuit element, such as a transistor, is moved the connections should automatically follow the element, so that the circuit logic does not change. A general approach to such capabilities can be based on constraint solvers for interactive graphics [6, 33].

As an example of a DSS we will describe the Penguins system in more detail. Like most DSSs, Penguins is an editor generator which takes a high-level description of a diagrammatic language and generates a customised diagram editor for that language. The specification language is grammatical in nature and can optionally include *layout constraints* which are used in diagram beautification. Broadly speaking, the layout constraints describe in conjunction with the grammar productions what the ideal layout for a particularly structured diagram is.

The key components of Penguins and most other DSSs are a *parser generator* and a generic *diagramming environment*. The diagramming environment resembles an object-oriented graphics editor and provides standard graphic primitives, such as lines, circles, texts and arrows. In order to support freehand drawings, pen-based input is supported as an alternative input method. Pen-based sketches are analysed by a *tokeniser* which recognises basic graphical entities like lines, circles and rectangles from pen-tracking data.

The second key component, a parser generator, constructs an incremental diagram parser for the given notation from its specification. Most DSSs use multidimensional grammar formalism for this specification, and Penguins is no exception, using Constraint Multiset Grammars [40]. The parser generator constructs an incremental *diagram parser* which is interfaced with the generic diagramming environment. The two components together behave as a diagram editor which can interpret its input.

We will illustrate the specification of a diagrammatic language with the simple example of binary trees. The following grammar production from this specification specifies a binary tree with two children: The parent node P connects to two children C1 and C2 with arrows L1 and L2 respectively. Notice that the constraints between the various components of the binary tree are rather loose, allowing flexibility in diagram creation. For example, a branch C1 is considered to be the left child branch of P as long as it is placed on the lower left of P.

```
B:bintree() ::= P:node, L1:arrow,
    C1:bintree, L2:arrow, C2:bintree where
      % Constraints on the left child.
      P.down1 == L1.start & C1.root.up == L1.end
      P.down1.y <= C1.root.up.y & P.down1.x >= C1.root.down2.x
      % Constraints on the right child.}
      P.down2 == L2.start & C2.root.up == L2.end
      P.down2.y <= C2.root.up.y & P.down2.x <= C2.root.down1.x
      %Initialize attributes.
      B.root = P; B.left = C1; B.right = C2.
```

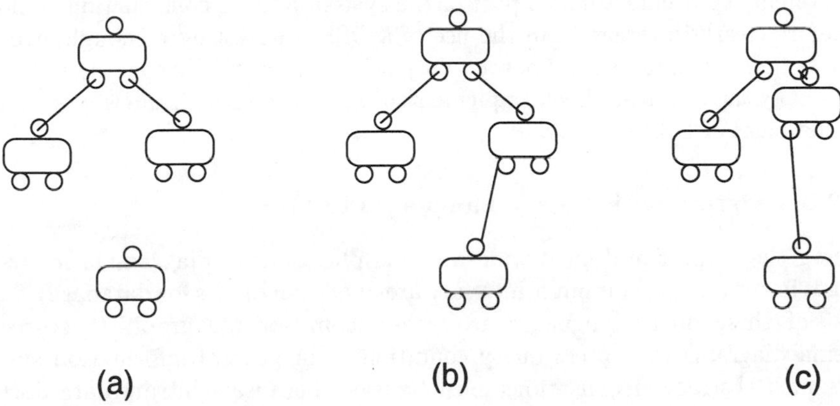

Figure 23.2. Drawing a binary tree with Penguins.

The resulting editor performs incremental parsing as each element is placed. For instance, in Fig. 23.2(a) two binary trees have been recognised: the top tree with three nodes and another tree consisting of a single node. When an arc is added, as shown in Fig. 23.2(b), the underlying parser incrementally recognises that there is now a single tree composed of the two old subtrees. The system also performs automatic correction and refinement of inaccurately drawn diagrams.

The grammar rules that are applied during the recognition of the diagram generate constraints on the spatial arrangement that are later used to preserve the semantics of the diagram when the user manipulates its appearance. For instance, if the middle node on the right-hand side is selected and moved upwards, then the parent node and the arcs follow this movement so that all edges remain connected. This is shown in Fig. 23.2(c). Finally, additional layout constraints (not shown in the above production rule) can be used for visual language-specific diagram beautification.

As in all grammar-based approaches, interpretation in the Penguins system is specified by using semantic attributes in the productions.

Geometric error correction, diagram beautification and the preservation of semantic constraints during diagram manipulation all rely heavily on constraint-solving techniques. In Penguins these are provided by the constraint-solving toolkit, QOCA [33], which is incorporated into the generic diagramming environments.

Clearly, the benefit of any DSS for the construction of a particular IDE depends on the relative complexity of the diagrammatic components in comparison to the rest of the system. In the following we will discuss two particular types of IDEs which have been constructed without DSSs. In the first case, diagrammatic theorem provers, it is obvious that the major complexity of the system is hidden in its inference component and not in its diagrammatic interface. In contrast, in the case of spatial query interfaces the diagrammatic component represents a major part of the system and its construction could therefore clearly leverage from the use of a DSS. Interestingly enough, even if these query interfaces have not actually been automatically constructed by DSSs they have at least been implemented using the same frameworks that are the basis of DSSs.

23.2.3 Diagrammatic Query Languages in GIS

Among the earliest and most widespread applications of diagrammatic languages in the human-computer interface are query languages for databases [8]. Most of these query languages are either iconic or use graph-structured schema diagrams to express query conditions. For general information systems such abstract visualisations must be used, because arbitrary data does not have any intrinsic natural visualisation. Spatial databases are an important special case, since spatial data has obvious mappings to diagrammatic representations, in which space is used to depict space.

The common idea behind diagrammatic spatial query languages is to let users draw sketched maps which depict approximations of the spatial situation they are looking for. For example, the sketch in Fig. 23.3 defines a query for border checkpoints on roads that connect two capital cities in neighboring countries.

Sketching as a method for database retrieval is also used in image databases [20, 37] but the approach there is fundamentally different, since the retrieval is not based on a symbolic interpretation of the input sketch, but on a variety of statistic similarity matches. Sketching as a way of querying spatial databases was introduced by the deductive visual query language

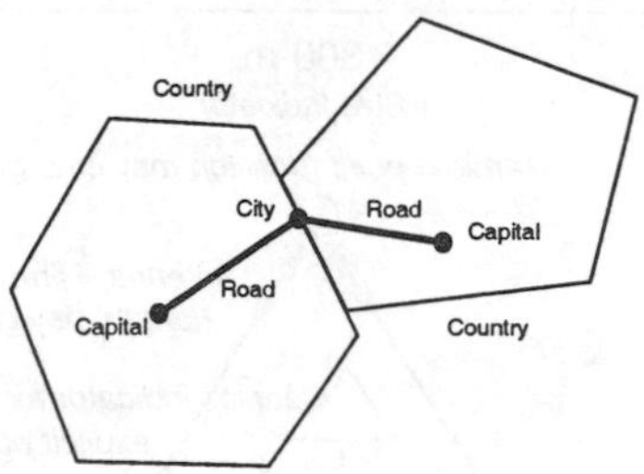

Figure 23.3. Sketch query.

Sketch [42,43]. This system analyses drawings of spatial scenarios and translates them into formulas of a logic object calculus which are processed by a back-end database. The spatial conditions in the query are derived by parsing the sketch into basic object representations and by analysing the spatial constraints between these objects in the sketch, while basic object types can be derived from the object labels.

For example, the query in Fig. 23.3 is an actual diagrammatic query corresponding to the clause

```
query(checkpoint=Ci1) <-
    Co1, Co2: country and Ci1: city and Ca1, Ca2: capital and
    Rd1, Rd2: road and B: meet(Co1.area, Co2.area) and
    Ca1 inside Co1.area and Ca2 inside Co2.area and
    Rd1 inside Co1.area and Rd2 inside Co2.area and
    Ci1 on B.extension and Rd1 touches Ci1 and Rd2 touches Ci and
    Rd1 touches Ci1 and Rd1 touches Ci2.
```

A challenging problem for a sketch-based query language is how views (derived objects) and in particular abstract symbolic information, which the database contains only in propositional form, can be expressed in the sketch. For this, Sketch introduced a method of diagrammatic deduction by allowing abstraction of a compound diagrammatic query into a new virtual object type. The spatial attributes in the query become spatial attributes of the virtual object and can be used in sketches in the same way as standard objects [43].

The concept of Spatial Query-by-sketch [18] later proposed to extend sketch interpretation with a mathematically well-founded method of dealing with the limited accuracy of sketches based on the 9-intersection model [19]. For example, from a rough sketch, it is often not clear whether it is actually intended to let two areas intersect or whether they are only meant to touch. The 9-intersection allows to define a concept of "conceptual neighbourhood" of topological relations, which can be used to relax topological query constraints.

Arguably the most advanced IDE for spatial information systems is Visco [31,55], a sketch-based query system which supports relaxation of queries as well as "don't care" spatial relations. The query language is a much enriched

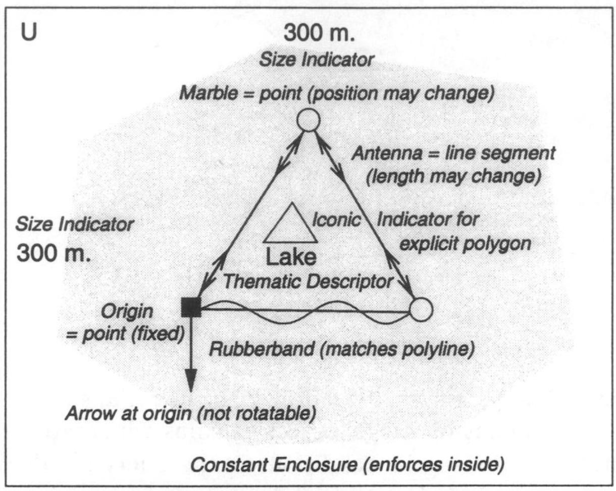

Figure 23.4. Visco elements.

version compared to those used in Sketch and Spatial Query-by-Sketch, which incorporates new primitives for spatial objects that depict not only its spatial properties, but also the way in which these can be relaxed. These primitives are based on immediate physical metaphors; for example, "rubberbands" are used for polylines that can be stretched. The available primitives are annotated in Fig. 23.4, which would be interpreted as "show all lakes of (nearly) arbitrary form and smaller than 300 × 300 m". Another more complicated query is shown in Fig. 23.5. It searches for three perfectly rectangular houses almost aligned in parallel and nearly lying on a straight line. The trade-off between expressiveness (precise but flexible queries) and conciseness (simple query language) should have become evident from these examples. Visco's semantic is based on a complete formalisation of sketched queries based on the integration of a qualitative spatial calculus [14] into description logic [7]. This is discussed in detail by Haarslev et al. elswhere in this volume.

Several important subtasks are involved in the construction and formalisation of a sketch-based query language: Firstly, the query interpreter must incorporate a parser to perform the basic analysis of the input diagram. Since sketches are practically always inaccurate, some form of "soft" approximate interpretation is required, which should be based on a mathematically well-founded geometric modelling as a second component. Finally, as a third component the query interpreter must incorporate some form of query disambiguation, which can either be done by enriching the query language, by a dialogue with the user or by "informed guessing" based on a modelling of the user's intentions.

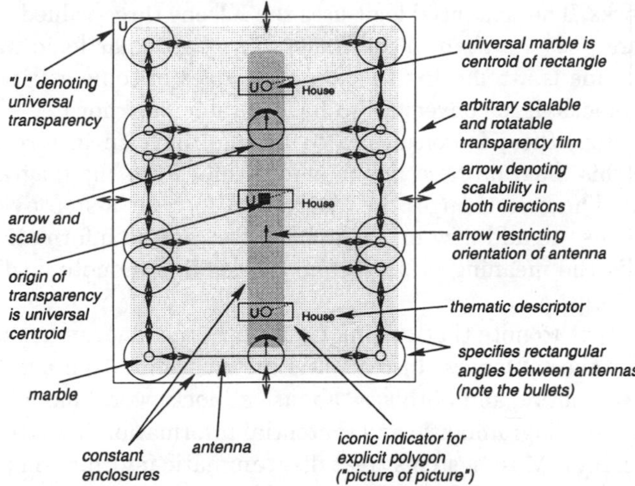

Figure 23.5. Visco query.

23.2.4 Diagrammatic Theorem Provers

IDEs for interactive theorem provers come closer to the idea of true diagrammatic "reasoning" than any other application. We shall look at two representative systems: Hyperproof [5] and Diamond [34]. Other systems are discussed in [4, 25].

Hyperproof is a logic system that combines sentential information and diagrammatic information within one proof system. The purpose of a Hyperproof is to prove or disprove a goal starting from some initial hypothesis. Both kinds of information are typically about a blocks world. While the sentential information is expressed as sentences in first-order logic, the diagrammatic information is expressed as diagrams of a blocks world. Proofs are organised in a Fitch-style natural deduction system. At every step of the proof, the information at that step will be either a first-order sentence or a diagram. The system contains the usual natural deduction rules for introduction and elimination of sentence connectives. To manage the diagrammatic information, there are rules which allow conversion between sentential information and diagrammatic information. The interaction between the diagram and the sentences is managed via logical rules. In other words, there are no operations which transfer data between the two representations.

The diagram in Hyperproof depicts a blocks world where an instructor can set up the initial diagrammatic situation. The blocks can have shape, size, demeanor (happy or sad), or name attributes. And they may share a two-place relation "likes". A block is assumed to always have a position on the grid, a shape, a size, etc.; however, it may not be specified. One kind of goal for a proof is to elicit the values of these attributes given the sentential information about the blocks. The sentential logic uses the Kleene three-valued semantics. The values are *false*, *unknown* and *true*. This particular logic was chosen because when one is attempting to show the values of some attributes for a collection of blocks, it is convenient to have a value "unknown".

At every step, there is considered to be one diagram in force. The information in this diagram descends monotonically from the diagram at the previous step. The user changes the diagram in force at a step by using the computer's mouse to add new information. Note that no information can be deleted; this is the meaning of a diagram descending monotonically from a previous diagram.

It is important to note that the information in the diagram is not treated as a model for the sentences. Instead, the information is syntactically on a par with the sentential, and both are "about" a blocks world model. No other proof system uses diagrammatic and sentential information in quite the same way as Hyperproof. Most systems treat diagrammatic information informally as annotation or as initial data whereas in Hyperproof diagrammatic information is integrated into the proof proper.

Another system we shall briefly discuss is Diamond. This is an interactive proof checker which is capable of synthesising some inductive schemes for mathematics given n-dimensional arrays of diagrammatic elements. A schematic proof which is diagrammatic is a proof that involves an inductive sequence of statements for which a particular diagram provides one instance of that sequence. An example of a schematic diagram is one showing a square array involving an integral measure for a side, say n. The array here represents an instance of n for a particular value. n is not a variable or algebraic indeterminate in the diagram; the diagram actually has n elements to a side. An example of an inherently inductive diagram is where the diagram has elements in it that represent an inductive step within the diagram, i.e., it has elements displaying the n and the $n + 1$ case where the n is a variable or algebraic indeterminate within the diagram.

The Diamond system is capable of accepting proofs with schematic diagrams of a particular kind and performing a generalisation or inductive step. This automatic generalisation step from example proofs to an invariant diagram construction method is the crux of the mathematical induction proof interactively discovered by Diamond. All proof generation in Diamond is based on the so-called ω-rule [3], which formalises inductive generalisation.

The individual diagram representations are fragments of two- and three-dimensional matrices and represent concrete cases of a particular mathematical theorem.

The diagrams are built from atomic actions and composite actions built recursively from these atomic actions. Examples of atomic actions are *rotate*, *translate*, *cut*, *join*, *project from 3D to 2D*, *remove*, *insert a segment*.

A basic proof is an initial diagram and a finite, linear sequence of diagrams that follow from the initial diagram and each via the diagrammatic actions.

Diamond attempts to match the sequences of actions such that the sequence of actions for the first proof is a prefix of the sequence of actions for the second. This guides the induction generalisation. Consider the theorem $n^2 = (2n-1) + \ldots + 3 + 1$. An example proof sequence for this diagram that Diamond can use is

The arrows indicate sequences of rules that generate each diagram from the previous diagram. Each diagram following an arrow can be seen to be an instance of the expression $2n - 1$.

It is difficult to generally classify diagrammatic proof systems, but one useful distinction that can be made is whether the diagrams denote a purely abstract (logical) domain, such as Peirce's $\alpha - \beta$-Calculus or a concrete domain such as Hyperproof's blocks world. Systems like Diamond are located somewhere between these poles, because it starts from the representation of concrete instances, but tries to bridge to abstract relations by generalising recurring patterns in the concrete depictions.

23.3 Conclusions

This chapter has reviewed several representative application areas of intelligent diagrammatic environments and has discussed some implementations of such environments. In the light of these we can take a second look at the IDE subtasks that we have identified above.

On the positive side, it appears that the lower-level tasks of modelling, parsing and interpretation are now relatively well understood. However, a number of challenging open issues remain even for these tasks.

Modelling of diagrammatic information is covered by several different frameworks which have successfully been applied for a wide range of problems, notably topological calculi, the attribute-based metric approach and the graph-based symbolic approach. Since these approaches emphasise different aspects of the problem, they are in some sense complementary. The challenge is to develop a tighter integration of these aspects, allowing metric and symbolic information and topology to be handled in a single consistent framework. A second open issue on this level is the development of tolerant geometric models that handle the inherent lack of accuracy in real diagrams. While many IDEs use ad hoc ways to compensate for this, more universal approaches to "soft" or "tolerant" geometries, mainly based on fuzzy sets, are explored in the context of qualitative reasoning and in particular in spatial information systems [53].

The phase of parsing and interpretation of diagrams is probably the best-understood part of IDEs, provided we restrict our view to static, unambiguous notations. In these cases grammar-based methods as well as logic-based approaches can be used. Both offer general frameworks for understanding diagram interpretation, but unfortunately we are still faced with a conceptual gap between the more efficient but less expressive grammar-based frameworks and the more expressive but computationally more expensive logic frameworks. However, a closer integration of these methods through mappings of grammar formalisms to logic is underway. Such mappings will allow a more modular integration of these lines of research.

One of the major unresolved challenges is the interpretation of ambiguous notations. For the disambiguation of diagrammatic information, the interpretation component typically needs to reason about the communication intent for which knowledge about the denoted domain (exterior to the notation) is needed. Current interpretation frameworks are not powerful enough to incorporate complex domain-dependent reasoning and, due to the generality of the task, it is not clear how this can be done in a single unified framework. A promising line of research is the work on open-ended approaches that can be more easily interfaced with other reasoning mechanisms.

The remaining sub-tasks of higher-level diagrammatic reasoning and transformation are much less understood and a unifying coherent theory of high-level diagrammatic reasoning is not in sight. The question of dynamicity in diagram notations appears to be central to our understanding of many questions in diagrammatic reasoning. We will need a better understanding of the *interpretation of dynamic diagram usage*, i.e. of the interpretation of changes that are made by the human user as well as of the *dynamic usage of diagrammatic notations to reason with them*, i.e. of the manipulation of diagrams to express reasoning with diagrams. The latter aspect is particularly interesting, since its investigation may help to shed light on the question if and how diagrammatic representations are essentially different from sentential representations.

Acknowledgements

The research described here was partially funded by the Australian Research Council under grant number A00103256.

References

1. Ah-Soon, C. (1998). A constraint network for symbol detection in architectural drawings. In K. Tombre and A. Chhabra (Eds), Graphics Recognition: Algorithms and systems, Vol. 1389 of LNCS. Berlin: Springer, pp. 80–90.

2. Baird, H. (1990). Industrial applications. In H. Bunke and A. Sanfeliu (Eds), Syntactic and structural pattern recognition: Theory and applications. World Scientific, pp. 369–380.
3. Baker, S., Ireland, A. and Smaill, A. (1992). On the use of the constructive omega rule within automated deduction. In A. Voronkov (Ed.), Logic programming and automated reasoning. LNAI 624. Berlin: Springer, pp. 214–225.
4. Barker-Plummer, D. and Bailin, S. (1992). Proving the diamond lemma with the GROVER theorem proving system. In AAAI symposium on reasoning with diagrammatic representations.
5. Barwise, J. and Etchemendy, J. (1996). Visual information and valid reasoning. In G. Allwein and J. Barwise (Eds), Logical reasoning with diagrams. Oxford: Oxford University Press, pp. 3–26.
6. Borning, A., Marriott, K., Stuckey, P. and Xiao, Y. (1997). Solving linear arithmetic constraints for user interface applications. In UIST'97: ACM symposium on user interface software and technology.
7. Brachman, R.J. and Schmolze, J.G. (1985). An overview of KL-ONE knowledge representation systems. Cognitive Science August:171–216.
8. Catarci, T. Costabile, M.F., Levialdi, S. and Batini, C. (1997). Visual query systems for databases: A survey. Journal of Visual Languages and Computing 8.
9. Chok, S. and Marriott, K. (1995). Automatic construction of user interfaces from constraint multiset grammars. In IEEE symposium on visual languages. Los Alamitos, CA: IEEE Computer Society Press, pp. 242–250.
10. Chok, S. and Marriott, K. (1998a). Automatic construction of intelligent diagram editors. In Proceedings of the 11th ACM symposium on user interface software and technology, pp. 185–194.
11. Chok, S. and Marriott, K. (1998b). Automatic construction of intelligent diagram editors. In Proceedings of the 11th ACM symposium on user interface software and technology.
12. Clarke, B. (1981). A calculus of individuals based on connection. Notre Dame Journal of Formal Logic 23(3):204–218.
13. Clarke, B. (1985). Individuals and points. Notre Dame Journal of Formal Logic 26(1):61–75.
14. Clementini, E., DiFelice, P. and van Oosterom, P. (1993). A small set of formal topological relationships suitable for end-user interaction. In D. Abel and B. Ooi (Eds), International symposium on advances in spatial databases (SSD'93), Singapore, June, pp. 277–295.
15. Cohn, A.G. (1997). Qualitative spatial representation and reasoning techniques. In Künstliche Intelligenz 1997 (KI'97). LNAI 1303. Springer, pp. 1–30.
16. Costagliola, G., De Lucia, A., Orefice, S. and Tortora, G. (1998). Positional grammars: Theory and practice in a visual language for interface modelling. In K. Marriott and B. Meyer (Eds), Visual language theory. New York: Springer, pp. 171–191.
17. Dinesh, T. and Üsküdarli, S. (1998). Input and output for specified visual languages. In K. Marriott and B. Meyer (Eds), Visual language theory. New York: Springer, pp. 325–351.
18. Egenhofer, M. (1996). Spatial-query-by-sketch. In IEEE sypmosium on visual languages, pp. 170–177.
19. Egenhofer, M. and Franzosa, R. (1991). Point-set topological spatial relations. International Journal of Geographical Information Systems 5(2):161–174.
20. Faloutsos, C., Barber, R., Flickner, M., Hafner, J., Niblack, W., Petrovic, D. and Equitz, W. (1994). Efficient and effective querying by image content. Journal of Intelligent Information Systems 3:231–262.

21. Furnas, G. (1990). Formal models for imaginal deduction. In Twelfth annual conference of the Cognitive Science Society. Hillsdale, NJ: Erlbaum, pp. 662–669.
22. Furnas, G. (1991). New graphical reasoning models for understanding graphical interfaces. In Proceedings of the ACM conference on human factors in computing systems (CHI). New York: ACM Press, pp. 71–78.
23. Furnas, G. (1992). Reasoning with diagrams only. In AAAI spring symposium on reasoning with diagrammatic representations, Stanford.
24. Futrelle, R. and Nikolakis, N. (1995). Efficient analysis of complex diagrams using constraint-based parsing. In Proceedings of the third international conference on document analysis and recognition (ICDAR), pp. 782–790.
25. Gelernter, H. (1963). Realization of a geometry theorem-proving machine. In E. Feigenbaum and J. Feldman (Eds), Computers and thought. New York: McGraw-Hill, pp. 134–152..
26. Gooday, J. and Cohn, A. (1996). Using spatial logic to describe visual programming languages. Artificial Intelligence Review 10:171–186.
27. Gross, M. (1994a). Recognizing and interpreting diagrams in design. In T. Catarci, M. Costabile, S. Levialdi and G. Santucci (Eds), Advanced visual interfaces (AVI). New York: ACM Press.
28. Gross, M. (1994b). Stretch-a-sketch: A dynamic diagrammer. In Proceedings of the 1994 IEEE symposium on visual languages. Los Alamitos, CA: IEEE Computer Society Press, pp. 232–238.
29. Gross, M. and Do, E.-L. (1996). Ambiguous intentions: a paper-like interface for creative design. In ACM 9th symposium on user interface software and technology (UIST). ACM Press, pp. 183–192.
30. Haarslev, V. (1998). A fully formalized theory for describing visual notations. In K. Marriott and B. Meyer (Eds), Visual language theory. New York: Springer, pp. 261–192g.
31. Haarslev, V. and Wessel, M. (1997). Querying GIS with animated spatial sketches. In IEEE sypmosium on visual languages, Capri, Italy, pp. 201–208.
32. Helm, R. and Marriott, K. (1986). Declarative graphics. In Proceedings of the 3rd international conference on logic programming, Vol. 225 of LNCS. New York: Springer, pp. 513–527.
33. Helm, R., Marriott, K., Huynh, T. and Vlissides, J. (1995). An object-oriented architecture for constraint-based graphical editing. In C. Laffra, E. Blake, V. de Mey and X. Pintado (Eds), Object-oriented programming for graphics. Berlin: Springer, pp. 217–238.
34. Jamnik, M., Bundy, A. and Green, I. (1997). Automation of diagrammatic proofs in mathematics. In B. Kokinov (Ed.), Perspectives on cognitive science, Vol. 3. New Bulgarian University, pp. 168–175.
35. Karima, M., Sadhal, K. and McNeil, T. (1995). From paper drawings to computer aided design. IEEE Computer Graphics and Applications February:24–39.
36. Kasturi, R., Raman, R., Chennubthotla, C. and O'Gorman, L. (1992). An overview of techniques for graphics recognition. In H. Baird, H. Bunke, and K. Yamamoto (Eds), Structured document image analysis. Berlin: Springer, pp. 285–324.
37. Kato, T. (1992). Cognitive view mechanism for multimedia information systems. In R. Cooper (Ed.), International workshop on interfaces to database systems. LNCS. New York: Springer.
38. Lakin, F. (1986). Spatial parsing for visual languages. In S.-K. Chang, T. Ichikawa and P.A. Ligomenides (Eds), Visual languages. New York: Plenum Press, pp. 35–85.

39. Lakin, F. (1987). Visual grammars for visual languages. In AAAI-87, 7th national conference on AI, pp. 683–688.
40. Marriott, K. (1994). Constraint multiset grammars. In IEEE symposium on visual languages. Los Alamitos, CA: IEEE Computer Society Press, pp. 118–125.
41. Marriott, K., Meyer, B. and Wittenburg, K. (1998). A survey of visual language specification and recognition. In K. Marriott and B. Meyer (Eds), Visual language theory. New York: Springer, pp. 5–85.
42. Meyer, B. (1992). Beyond icons: Towards new metaphors for visual query languages for spatial information systems. In R. Cooper (Ed.), Interfaces to database systems. Glasgow: Springer, pp. 113–135.
43. Meyer, B. (1994). Pictorial deduction in spatial information systems. In International IEEE workshop on visual languages (VL'94), St Louis, MO, October, pp. 23–30.
44. Meyer, B. (1997). Formalization of visual mathematical notations. In M. Anderson (Ed.), AAAI symposium on diagrammatic reasoning (DR-II), Boston, MA, November. AAAI Technical Report FS-97-02, AAAI Press, pp. 58–68.
45. Meyer, B. (1999). Constraint diagram reasoning. In Principles and practice of constraint programming: CP'99, Alexandria, VA, October. Berlin: Springer.
46. Meyer, B. (2000). A constraint-based framework for diagrammatic reasoning. In Applied Artificial Intelligence (special issue on constraint handling rules) 14.
47. Meyer, B. and Marriott, K. (1997). Specifying diagram animation with rewrite systems. In International workshop on theory of visual languages (TVL'97), Capri, Italy, September, pp. 85–96.
48. Meyer, B. and Zweckstetter, H. (1998). Interpretation of visual notations in the recopla editor generator. In FRVDR'98: AAAI fall symposium on formalizing reasoning with visual and diagrammatic representations, Orlando, FL, October. AAAI TR FS-98-04.
49. Meyer, B., Zweckstetter, H., Mandel, L. and Gassmann, Z. (1999). Automatic construction of intelligent diagrammatic environments. In HCI'99: 8th international conference on human-computer interaction, Munich, Germany, August. Hillsdale, NJ: Erlbaum.
50. Minas, M. and Viehstaedt, G. (1995). Diagen: A generator for diagram editors providing direct manipulation and execution of diagrams. In IEEE workshop on visual languages. Los Alamitos, CA: IEEE Computer Society Press, pp. 203–210.
51. Nagy, G., Seth, S. and Viswanathan, M. (1992). A prototype document image analysis system for technical journals. IEEE Computer 25(7):10–22.
52. Rozenberg, G. (Ed.)(1997). Handbook of graph grammars and computing by graph transformation. Singapore, World Scientific.
53. Schneider, M. (1999). Uncertainty management for spatial data in databases: Fuzzy spatial data types. In International symposium on advances in spatial databases (SSD'99), Hong Kong. LNCS 1651, pp. 330–351. LNCS 1651.
54. Weitzman, L. and Wittenburg, K. (1993). Relational grammars for interactive design. In IEEE symposium on visual languages, pp. 4–11.
55. Wessel, M. and Haarslev, V. (1998). VISCO: Bringing visual spatial querying to reality. In IEEE sypmosium on visual languages, Halifax, Canada, pp. 170–177.
56. Wittenburg, K. and Weitzman, L. (1990). Visual grammars and incremental parsing for interface languages. In IEEE symposium on visual languages. Los Alamitos, CA: IEEE Computer Society Press, pp. 111–118.
57. Yokokura, N. and Watanabe, T. (1998). Layout-based approach for extracting constructive elements of bar-charts. In K. Tombre, and A. Chhabra (Eds),

24. Towards Diagram Processing: A Diagrammatic Information System

Michael Anderson

We advocate the development of an agent capable of processing diagrammatic information directly in all its forms. In the same way that we will require intelligent agents to be conversant with natural language, we will expect them to be fluent with diagrammatic information and its processing. We present a methodology to this end, detail a diagrammatic information system that shows the merit of this line of research, and evaluate this system to motivate its future extensions.

24.1 Introduction

Of the set of behaviours that will be required of an artificially intelligent agent, a somewhat neglected member has been the ability to deal with diagrammatic information. Much attention has been paid to machine synthesis, recognition and understanding of natural language in both textual and audio forms. The understanding has been that such capabilities are required of an agent if it is expected to fully communicate with human beings and function in human environments. Much less attention has been given to machine understanding of diagrammatic information, an equally important mode of human communication. Effective capabilities in this mode will be crucial to an agent intended as a full partner in human discourse and activity. In the same way that we will require such agents to be conversant with natural language, we will expect them to be fluent with diagrammatic information and its processing. Ultimately, a machine with such capabilities will interact with a real-world environment, rife with diagrammatic information, with a higher degree of autonomy than those without such capabilities.

The main thrust of diagrammatic reasoning research to date (from an artificial intelligence perspective) has been a search for computational efficiency gains through representations, and related inference mechanisms, that analogously model a problem domain. As this has been the aim of much of the seminal work in the field (e.g. [8,9]), it is understandable that much effort has been expended in this direction. Although it is arguable that some progress has been made through this line of research, we believe that the field's most important contribution will be the development of an agent that is capable of dealing directly with diagrammatic information in all its forms.

We envision a system that takes diagrams as input, processes these diagrams, abstracting information and drawing inferences from them alone and in concert with other forms of knowledge representation, and expresses this newly gained knowledge as output in the form of text, new diagrams, actions, etc. Although the approach taken by this system will not necessarily claim cognitive plausibility, the fact that human beings do these things as a matter of course will stand as proof by existence that such a system has been fashioned. This diagram-processing system will be comprised of a number of important components. It will require a means to input diagrams such as a vision component. It will require a way to internally represent diagrams. The diagrammatic representations so acquired will require storage, as will knowledge needed to deal with these representations, necessitating some storage management component. A processing component will be required that synthesises and abstracts new knowledge from combinations of diagrammatic and other forms of knowledge representations. Various components will be required to use the new knowledge to produce desired output in a variety of situations.

Reflection on the design of these components raises a number of questions: What constitutes a diagram? In what form will diagrams be accepted as input by the system? How will diagrams be internally represented? How will knowledge be gleaned from diagrams? What is the nature and content of a priori knowledge that will be required? How will other forms of representation and inference be integrated with diagrammatic representations and inference? What is the nature of the desired output? How will this output be produced? Etc. These are hard questions with a multiplicity of answers that in themselves generate more questions. They form the parameters of the problem. Our intent is to build a test bed in which various values for these parameters can be tested, compared and contrasted, and ultimately forged into a single general-purpose diagram-processing system.

Following the methodology advocated in [1], we (1) identify the particular intelligence we seek to study, (2) define a set of telescoping restricted tasks, (3) define evaluation criteria, (4) develop a model and system for our most restricted domain, and (5) evaluate this model and system for the next level of complexity.

24.2 Topic of Study

As we have stated, we are ultimately interested in developing an agent with full diagrammatic reasoning capabilities on a par with human beings. If a picture is worth a thousand words, we would like to obviate the need for this text by being able to communicate with an agent directly via pictures or diagrams. An agent should be able to accept and understand such diagrammatic input from us and our environment as well as being able to produce such diagrams in its attempt to communicate diagrammatically representable notions to us.

24.3 Restricted Tasks Set

As this ultimate system is clearly beyond the grasp of current theory and technology, we refine it into a nested group of successively simpler tasks. Each nested level is a subset of all levels it is nested within and, further, a simplification of these containing levels. At the simplest level, we attempt to prove the merit of this line of research by developing a system that realises the goal of this level. As we succeed at one level, we will attempt to build upon this success by applying what we have learned to the next less simplified level, expanding our solution to cover new problems it entails.

Given, as the outer most level, a system with full diagrammatic reasoning capabilities, we define as a simpler subset of this ultimate system a system that has diagrammatic reasoning capabilities in some arbitrary task. Examples of such systems are map-understanding robots, scientific diagram abstracting agents, and diagram generation systems.

We choose, as the next level of simplification, to focus on diagrammatic reasoning capabilities in one particular task, namely what we term *diagrammatic information systems*: systems that allow users to pose queries concerning diagrams, seeking responses that require the system to infer information from diagrams. As a final level of simplification we constrain any given instantiation of this diagrammatic information system to be knowledgeable about particular diagram types in a particular domain.

That said, we strive to develop a core of this diagrammatic information system that remains diagram type and domain independent, capable of accepting domain-dependent diagrammatic and non-diagrammatic knowledge. In this way, each body of knowledge produces a new instantiation of the diagrammatic information system knowledgeable in the particular domain and diagram types represented by this knowledge.

24.4 Evaluation Criteria

To evaluate systems at all levels of nesting, we compare its understanding of diagrammatic information with the understanding that a human being exhibits with the same information. In this manner, we evaluate an instantiation of a diagrammatic information system in a particular domain by the number of diagram types it can handle, the variety of queries it can respond to, and the quality of responses it gives to these queries. The level of success of such a system is measured by how it approaches, matches, or exceeds the performance of a human being reasonably capable within the chosen domain.

24.5 Model and System

As the first task we have set for ourselves is an implementation of a *diagrammatic information system*, we choose one possible set of values to the parameters of this problem.

We define a *diagram* to be a *tessellation* (tiling) of a planar area such that it is completely covered by atomic two-dimensional regions or tesserae (tiles). Such a definition is broad enough to include arbitrarily small tesserae (points, at the limit), pixels, and, at the other end of the spectrum, domain-specific entities such as countries, regions, blocks, etc. Further, as this definition is not tied to any particular semantics, it is general enough to encompass all diagrams. Given the wide variety of semantic mappings employed by diagrams, a general definition that makes no semantic commitment is useful.

We sidestep a vision component by accepting bitmaps depicting diagrams as input to our system. As this is a likely output of a vision component, such a component can be appended later.

Our currently chosen approach gleans knowledge from diagrams by directly manipulating spatial representations of them. This approach is motivated by noting that, given diagrams directly input as bitmaps, any translation into another representation will require some form of direct manipulation of these bitmaps. In many cases, this translation is superfluous. Given this approach, we store input bitmaps directly with no further abstraction. This strategy not only allows us to manipulate spatial representations directly but, should the need arise, it will allow us to translate to any other representations as required.

We use, as a basis for this direct manipulation of diagrams, the theory of *inter-diagrammatic reasoning* (IDR) [3, 4]. IDR leverages the spatial and temporal coherence often exhibited by groups of related diagrams for computational purposes. Using concepts from set theory, image-processing theory, colour theory, and others, like diagrams are combined in ways that produce new like diagrams that infer information implicit in the original diagrams.

Knowledge required to process diagrams is likely to be both domain and diagram specific. Facts and rules pertinent to targeted domains are necessary, as is information germane to processing diagram types represented. We represent this knowledge both diagrammatically and non-diagrammatically, as appropriate, constraining both domains and diagram types as necessary. We achieve integration of diagrammatic and non-diagrammatic knowledge and inferencing by providing an inter-lingua abstraction that furnishes a homogeneous interface to various modes of representation and inferencing, permitting inferences to be made with heterogeneously represented knowledge.

Our output is both diagrammatic and textual, meant for direct human consumption. Although we skirt other forms of output such as action or intermediate output intended for use by some other system, there is nothing in the nature of the processing that precludes use of its product in such ways.

Our first instantiation of a diagrammatic information system is informed about *cartograms* (maps representing information as greyscale or colour shaded areas) of the United States.

Figure 24.1. Vegetation in the United States.

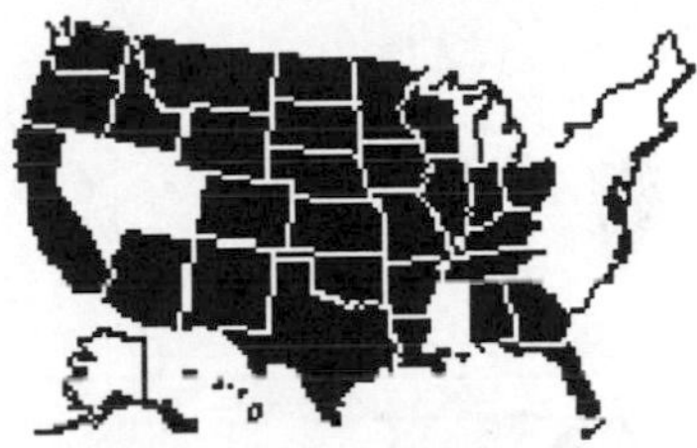

Figure 24.2. Response to query: "Which states have grassland?"

24.5.1 An Example

As an example, consider the diagram in Fig. 24.1. This is a cartogram that depicts in three levels of grey where each of the major vegetation types are situated in the United States. The darkest grey represents forest, medium grey represents grassland, and the lightest grey represents desert. Given this diagram as input to the system, as well as the semantics of the grey levels in this particular diagram, posing the query "Which states have grassland?" elicits the diagram in Fig. 24.2 as a response from the system. In this diagrammatic response, each state in which grassland exists is represented by its shape in black positioned where the state lies within the United States. We use this example to examine the implementation of this instantiation of a diagrammatic information system in further detail.

Fig. 24.1 is input to the system as a bitmap and is stored as such with no further manipulation. The system is supplied with the semantic mapping of the grey levels of the diagram to the vegetation types present. The input diagram is then parsed into three diagrams, each comprised of a single grey level. Each of these diagrams represents, then, the location of a particular vegetation type within the United States. Fig. 24.3, for example, shows the diagram resulting from this parsing that represents the locations of grassland in the United States.

Figure 24.3. Location of grassland in the United States.

Figure 24.4. Location of Nevada in the United States.

A priori diagrammatic knowledge required to respond to this query is comprised of a set of diagrams that represent the locations of each state within the United States. Fig. 24.4 is an example of such a diagram which shows the location of the state of Nevada within the United States by marking its area on the map in black. There are 50 such state diagrams.

The response to the query "Which states have grassland?" is generated by comparing each of these state diagrams with the diagram representing grassland. When a state diagram intersects the grassland diagram (both diagrams without the United States outline), the semantics of the domain dictate that that state contains grassland. All such states are then accumulated on a single diagram (with the United States outline) and presented to the user as the response to the query.

In this manner, diagrammatic responses can be generated for a wide variety of queries concerning vegetation in the United States including "Which states do not have forest?", "How many states have desert?" (Simply return a count of the state diagrams that intersect the desert diagram), "Does Rhode Island have desert?" (Simply return true if the state diagram for Rhode Island intersects the desert diagram), "Which vegetation type covers the most states?", "Do any states have both grassland and desert?", "Which states have either desert or forest?", "Do more states have grassland than desert?", "Which states have forest but not grassland?", etc.

An overview of the formalism we use to generate responses to these queries, the theory of inter-diagrammatic reasoning, follows.

24.5.2 Inter-Diagrammatic Reasoning

The theory of inter-diagrammatic reasoning [2, 3] defines diagrams as tessellations. Tesserae take their values from an I, J, K valued subtractive CMY colour scale. Intuitively, these CMY (cyan, magenta, yellow) colour scale values (denoted $v_{i,j,k}$) correspond to a discrete set of transparent colour filters where i is the cyan contribution to a filter's colour, j is the magenta contribution, and k is the yellow contribution. When overlaid, these filters combine to create other colour filters from a minimum of WHITE ($v_{0,0,0}$) to a maximum of BLACK ($v_{I-1,J-1,K-1}$). In the current work, i, j, and k are always equal, providing greyscale values from white to black only. The following unary operators, binary operators, and functions provide a set of basic tools to facilitate the process of inter-diagrammatic reasoning.

Binary operators each take two diagrams, d_1 and d_2, of equal dimension and tessellation and each return a new diagram where each tessera has a value v that is some function of the values of the two corresponding tesserae, $v_{i1,j1,k1}$ and $v_{i2,j2,k2}$, in the operands.

- OR, denoted $d_1 \vee d_2$, returns the maximum of each pair of tesserae where the maximum of two corresponding tesserae is defined as

  $$v_{\max(i1,i2),\max(j1,j2),\max(k1,k2)}$$

- AND, denoted $d_1 \wedge d_2$, returns the minimum of each pair of tesserae where the minimum of two corresponding tesserae is defined as

  $$v_{min(i1,i2),min(j1,j2),min(k1,k2)}$$

- OVERLAY, denoted $d_1 + d_2$, returns the sum of each pair of tesserae where the sum of values of corresponding tesserae is defined as

  $$v_{min(i1+i2,I-1),min(j1+j2,J-1),min(k1+k2,K-1)}$$

- PEEL, denoted $d_1 - d_2$, returns the difference of each pair of tesserae where the difference of values of corresponding tesserae is defined as

  $$v_{max(i1-i2,0),max(j1-j2,0),max(k1-k2,0)}$$

- NONNULL (NULL), denoted $NONNULL(d)$, ($NULL(d)$) is a one-place Boolean function taking a single diagram that returns TRUE if d contains any non-WHITE (all WHITE) tesserae else it returns FALSE.
- ACCUMULATE, denoted $\alpha(d, ds, o)$, is a three-place function taking an initial diagram, d, a set of diagrams of equal dimension and tessellation, ds, and the name of a binary diagrammatic operator, o, that returns a new diagram which is the accumulation of the results of successively applying o to d and each diagram in ds.
- MAP, denoted $\mu(f, ds)$, is a two-place function taking a function f and a set of diagrams of equal dimension and tessellation, ds, that returns a new set of diagrams comprised of all diagrams resulting from application of f to each diagram in ds.
- FILTER, denoted $\phi(f, ds)$, is a two-place function taking a Boolean function f and a set of diagrams of equal dimension and tessellation, ds, that returns a new set of diagrams comprised of all diagrams in ds for which f returns TRUE.

Given these inter-diagrammatic operations, the vegetation and state maps as previously described, and a null diagram (denoted $\emptyset$) standing for a diagram with all tesserae white-valued, the following more formally specifies the generation of a diagrammatic response to the query "Which states have grassland?":

$$\alpha(\emptyset, \phi(\lambda(x) NONNULL(Grassland \wedge x), State), +)$$

This (1) defines a lambda function that ANDs its parameter with the grassland diagram and returns true if the result is not null, (2) filters out diagrams from the set of state diagrams for which this lambda function does not return true (these are the state diagrams that do not intersect the grassland diagram), and (3) overlays the remaining state diagrams onto the null diagram giving the desired result. Fig. 24.5 details this example.

Responses to all of the queries suggested previously can be generated via IDR operators. As in the example, those queries requiring a diagrammatic response produce an appropriate set of diagrams which are OVERLAYed together. Those queries requiring a numeric response produce an appropriate set of diagrams and return the cardinality of it. For instance, the number of states that have grassland can be returned by taking the cardinality of the set returned by the filtering operation instead of accumulating that set upon the null diagram as is done in the example. Those queries requiring a Boolean response return the value of the NONNULL function applied to an appropriately derived diagram. For instance, a response to the query "Are there any states that have grassland?" will derive a diagram as in the example and return the result of applying the NONNULL function to it. Responses to queries seeking negative information can be derived by using the NULL function to produce an appropriate set of diagrams. For instance, a response

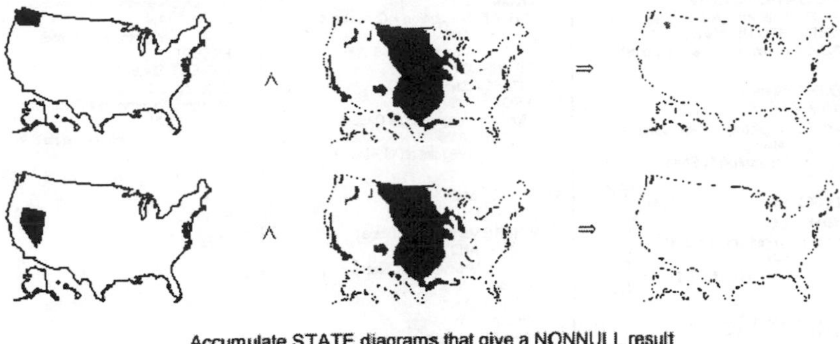
Accumulate STATE diagrams that give a NONNULL result

Figure 24.5. Generation of response to query "Which states have grassland?"

to the query "Which states do not have grassland?" can be generated by simply replacing the NONNULL function with the NULL function. Queries seeking responses to conjunctions or disjunctions need to use set intersection and set union (respectively) to produce the appropriate sets of diagrams. Responses to relational $(<,>,<=,>=,<>,=)$ queries need to compare the cardinality of each set of diagrams produced for each subquery involved.

Although IDR operators can produce responses to this wide variety of queries in this domain, it is by no means intuitive how they should be used to do so. In the following, we introduce a higher-level query language that permits a user to query diagrams more intuitively, specifying *what* they wish to know more than *how* it should be generated.

24.5.3 Diagrammatic SQL

Diagrammatic SQL (DSQL) is an extension of Structured Query Language (SQL) [5] that supports querying of diagrammatic information. Just as SQL permits users to query information in a relational database, DSQL permits a user to query information in diagrams.

We have chosen to extend SQL for use as our query language for a number of reasons. As we will show, SQL has a remarkable fit to the uses we wish to make of it. It is a reasonably intuitive language that allows specification of what data you want without having to specify exactly how to get it. It is a well-developed prepackaged technology whose use allows us to focus on more

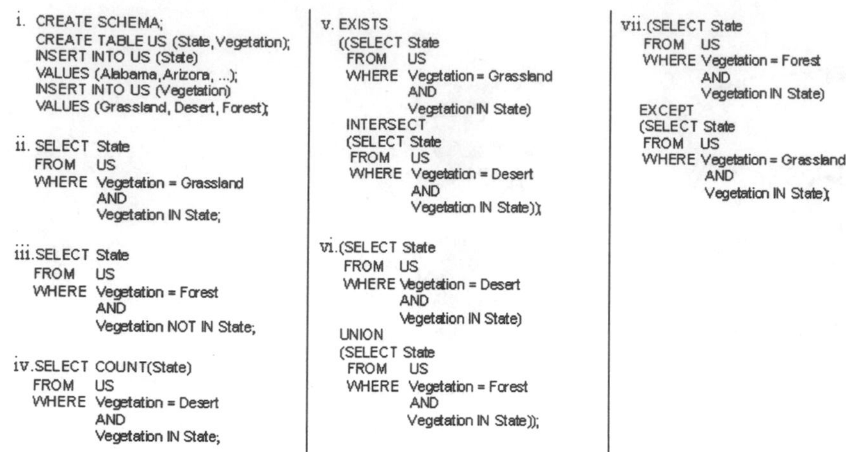

Figure 24.6. Generation of response to query "Which states have grassland?"

pressing system issues. SQL's large installed base of users provides a ready and able audience for a fully developed version of the system. The availability of immediate and imbedded modes provide means to use the system for both direct human consumption and further machine processing. The availability of natural language interfaces to SQL will allow the system to provide an even more intuitive interface for its users.

Besides providing a basis for a diagrammatic query language, a relational database that stores image data can be used by the system as a storage management component. Further, as relational databases already manage other types of data, use of one as a storage management component with a diagrammatic extension to its SQL gives the system a means to query both diagrammatic and non-diagrammatic data simultaneously. This provides a linkage between heterogeneous data allowing whole new classes of queries, for example, "What is the total population of states having desert?", "Which of the states having forest has the highest per capita income?", "What vegetation is contained by the state with the highest annual rainfall?", etc.

We have developed a grammar for a subset of DSQL that allows it to handle queries of the types previously discussed. Where SQL queries return relations, DSQL queries return sets of diagrams. These diagram sets can have their members OVERLAYed upon a null diagram for diagrammatic results or counted to return numeric results. Further, these sets can be tested for emptiness to return Boolean results or used as operands in set operations such as union, intersection, and difference. Examples of DSQL syntax and semantics follow.

DSQL Example. Fig. 24.6 shows an example data definition and sample queries in DSQL. Fig. 24.6i uses DSQL to define a schema for the diagrammatic information required by the examples presented previously. It creates a table named US that contains two diagram sets named State and Vegetation. In the current Lisp implementation, as no connection has yet been established to a relational database management system, a table is simply a list of related diagram sets. Each diagram set has inserted into it a number of diagrams appropriate to the set. In the current Lisp implementation, these are actually symbols that evaluate to diagrams.

Fig. 24.6ii is a DSQL query that represents the example query "Which states have grassland?" It has the same SELECT FROM WHERE clauses that SQL queries have and these share similar semantics with their SQL counterparts. Most often, the SQL SELECT clause specifies what attribute(s) will have values in the resulting relation. In DSQL, the SELECT clause specifies what diagram set(s) will have values returned from the query. The SQL FROM clause specifies which table(s) are involved in the query. In DSQL, the FROM clause specifies which list(s) of diagram sets are involved in the query. The SQL WHERE clause specifies which condition(s) have to be satisfied by values returned by the query. This is the same use that a WHERE clause is put to in DSQL.

The DSQL query in Fig. 24.6ii states that the set of diagrams from the diagram set State of the diagram set list US that conform to the constraints specified will be returned. The WHERE clause specifies (1) that the Vegetation diagram set of the diagram set list US is restricted to the Grassland diagram only and (2) that the diagram in the Vegetation diagram set must intersect given State diagrams. In one context, the SQL IN Boolean operator returns true if and only if the value on the left-hand side is a value in the attribute on the right-hand side. In DSQL, IN is a Boolean operator that returns true if and only if the diagrams involved intersect. When sets of diagrams are involved, as in this and following examples, the semantics of a DSQL query dictate that this intersection be tested for each member of each set. In this case, the Grassland diagram will be tested for intersection with each member of the State diagram set, in turn, allowing the query to return only those states that contain grassland. As previously detailed, the response to this query is achieved by IDR operators as:

$$\alpha(\emptyset, \phi(\lambda(x) NONNULL(Grassland \wedge x), State), +)$$

Fig. 24.6iii is a DSQL query that seeks a response to the question "Which states do not have forest?" The semantics of this query is much like the previous example. In this example, though, the Vegetation diagram set is restricted to the Forest diagram and this diagram must not intersect with a state diagram for it to be included as part of the result. The response to this query is achieved by IDR operators as:

$$\alpha(\emptyset, \phi(\lambda(x) NULL(Forest \wedge x), State), +)$$

Fig. 24.6iv is a DSQL query that seeks a response to the question "How many states have desert?" This change in mode from a diagrammatic response to a numeric response is signalled by the application of the COUNT function to the diagram set in the SELECT clause. It is realised by the following IDR formulation where cardinality is a function returning the number of members in a set:

$$cardinality(\phi(\lambda(x)NONNULL(Desert \wedge x), State))$$

Fig. 24.6v is a DSQL query that asks "Are there any states that have both grassland and desert?" The fact that a Boolean response is required is signalled by the use of the EXISTS function. In SQL, the EXISTS function tests for an empty (single attributed) relation resulting from a subquery. In DSQL, it is used to test for an empty set of diagrams resulting from any query. To produce the set to be tested using IDR operations, the set of state diagrams that have grassland is intersected (∩) with the set of state diagrams that have desert. If this resulting set is not empty ($\neg empty$), return true else return false. Following is the IDR realisation of this query:

$$\neg empty(\phi(\lambda(x)NONNULL(Grassland \wedge x), State)$$
$$\cap$$
$$(\phi(\lambda(x)NONNULL(Desert \wedge x), State))$$

Fig. 24.6vi is a DSQL query that seeks a diagrammatic response to the question "Which states have either desert or forest?" This response is generated by taking the union (∪) of the set of states that have desert and the set of states that have forest and, then, OVERLAYing them onto the null diagram. Expressed as IDR operations:

$$\alpha(\emptyset, \phi(\lambda(x)NONNULL(Desert \wedge x), State)$$
$$\cup$$
$$\phi\lambda(x)NONNULL(Forest \wedge x), State), +)$$

In a similar vein, Fig. 24.6vii is a DSQL query that seeks a diagrammatic response to the question "Which states have forest but not grassland?" This response is generated by taking the difference (−) of the set of states that have forest and the set of states that have grassland and, then, OVERLAYing them onto the null diagram. Expressed as IDR operations:

$$\alpha(\emptyset, \phi(\lambda(x)NONNULL(Forest \wedge x), State)$$
$$-$$
$$\phi(\lambda(x)NONNULL(Grassland \wedge x), State), +)$$

24.6 Evaluation of Model and System

To reiterate, our current goal is to develop an instantiation of a diagrammatic information system knowledgeable about a particular diagram type (cartograms) in a particular domain (United States) that produces quality responses to the full range of queries that would be expected of a reasonably capable human being given the same knowledge.

A subset of a DSQL grammar required to handle the range of queries exemplified in this work has been developed, a rudimentary compiler that translates this range of DSQL queries into their IDR formulations has been constructed, and the IDR operations that produce the desired output have been realised in Common Lisp. The current instantiation of the diagrammatic information system responds to an interesting range of queries posed against cartograms of the United States. This range of queries can be characterised as those whose responses are generated by various combinations of input diagrams and a priori diagrams. It is arguable that, within this range, the quality of responses equals or exceeds human capabilities with the same diagrammatic information. These are indicators of a promising line of research.

That said, there is much work yet to accomplish to realise fully the goal of even this relatively simple level. For example, not all queries that a human would be expected to answer in the example domain can be handled currently by the system. These include, for instance, queries that seek information about area or neighbourhood relations. Further, only cartograms of the same size and orientation as the a priori diagrammatic knowledge can be handled by the system. Clearly, humans are capable of handling such variations. Noise and uncertainty concerns present in real-world data have also been avoided. For instance, textual annotations on a cartogram, although helpful to a human, are noise to the current system. Ready recognition of these limitations is the product of a clearly delineated task and well-defined goal. These limitations, then, provide the focus for future work.

Queries that seek information about area provide an opportunity to integrate IDR with another theory of diagrammatic reasoning. Furnas' *BITPICT theory* [7] postulates a logic that deals with diagrams via BITPICT rule mappings that can be used to transform one diagram into another and, therefore, allow reasoning from diagrams to diagrams. A BITPICT rule is meant to convey that all instances in a diagram of the bit pattern on the left-hand side of a rule are replaced with the bit pattern on the right-hand side of that rule. As interesting as this theory is, it can be subsumed by IDR by using appropriate sets of diagrams representing the universal instantiation of BITPICT rules. With this theory irregular shapes can be normalised, allowing comparison of their relative areas. Queries such as "Does California have more desert than grassland?" can then be handled. Further, domains that seem less amenable to IDR techniques (for instance, line graphs) can be made more manageable by use of this theory (for instance, by shading the area under a curve).

Queries that seek information about neighbourhoods provide an opportunity to integrate IDR with an image-processing theory. *Mathematical morphology* [11, 12] is an image-processing technique based on shape that is used to simplify, enhance, extract or describe image data. Sets of pixels describe an *image object*. Information about the geometric structure of image objects is extracted via use of other objects called *structural elements*. Information pertaining to the size, spatial distribution, shape, connectivity, convexity, smoothness, and orientation can be obtained by transforming the original image object by various structural elements. As primitive mathematical morphological operators can be modelled by IDR operators, IDR subsumes this theory as well. One such primitive operator, *dilation*, can be intuitively viewed as an operation that adds layers to the border of a two-dimensional object. Adding a sufficiently wide layer to a state diagram, for instance, allows this modified diagram to be tested for intersection with the other state diagrams. This can produce the set of state diagrams that neighbour the original state. In combination with area-querying capabilities introduced previously, new classes of interesting queries can be handled. For example, "Which of the states surrounding Nevada have the greatest amount of forest?" is one such query that could then be handled.

Problems with real-world data can be approached using geometrical transformations. Orientation and size of cartograms can be normalised by combinations of rotation and scaling operations. The search for the required combination and parameters of operations could be guided by the user. These operations, themselves, are implementable within the theory of IDR.

In addition to the above extensions, we are developing a full DSQL grammar, a complete interpreter and compiler to translate DSQL to IDR, and support for both immediate and imbedded modes of operation. We are also planning a relational database implementation with an exploration of the heterogeneous data inference that such an implementation will allow. We are also interested in investigating the extension to DSQL of natural language interfaces for SQL. Finally, when we have satisfactorily accomplished the stated goals for this level, we will then lift constraints and focus on the set of problems introduced by exploring new diagram types in new domains.

Acknowledgements

We thank Dr Chris Armen for his insightful comments and encouragement throughout the duration of this project. This material is based upon work supported by the National Science Foundation under grant number IIS-9820368.

References

1. Allen, J. (1998). AI growing up: The changes and opportunities. AI Magazine 19(4):13–23.
2. Anderson, M. and Armen, C. (1998). Diagrammatic reasoning and colour. In Proceedings of the AAAI fall symposium on formalization of reasoning with visual and diagrammatic representations, Orlando, FL, October.
3. Anderson, M. and McCartney, R. (1995). Inter-diagrammatic reasoning. In Proceedings of the 14th international joint conference on artificial intelligence, Montreal, Canada, August.
4. Anderson, M. and McCartney, R. (1996). Diagrammatic reasoning and cases. In Proceedings of the 13th national conference on artificial intelligence, Portland, OR, August.
5. Date, C. (1989). A guide to the SQL standard, (2nd edn.). Reading, MA:Addision-Wesley.
6. Feigenbaum, E.A. and Feldman, J. (Eds.) (1963). Computers and thought. New York:McGraw-Hill.
7. Furnas, G. (1992). Reasoning with diagrams only. In N. Narayanan (Ed.), Working notes of AAAI spring symposium on reasoning with diagrammatic representations.
8. Gelernter, H. (1959). Realization of a geometry theorem proving machine. In Proceedings of an international conference on information processing,UNESCO House, pp. 273–282. (Also in [6].)
9. Larkin, J. and Simon, H. (1987). Why a diagram is (sometimes) worth ten thousand words. Cognitive Science 11:65–99.
10. Narayanan, N. (Ed.) (1992). Working notes of AAAI spring symposium on reasoning with diagrammatic representations. Menlo Park, CA:AAAI Press.
11. Serra, J. (1982). Image analysis and mathematical morphology, Vol. 1. New York:Academic Press.
12. Serra, J. (1988). Image analysis and mathematical morphology, Vol. 2. New York:Academic Press.

25. Using Diagrams to Understand Diagrams: A Case-Based Approach to Diagrammatic Reasoning

Dale E. Fish
Robert McCartney

This chapter describes work in progress on a system that "understands" a diagram of a problem situation by finding correspondences between it and similar diagrams in its case memory. The input to the system is a diagram of the problem situation consisting of some number of simple elements such as circles and lines. The goal of the system is to determine the significance of the diagram and its constituent elements by recognising the structure of the diagram and structures within it. Our approach uses high-level perception (recognition of instances of categories and relations) to detect similarities between the problem diagram and stored cases (diagrams) which are understood; the matching cases provide the basis for inferring information missing in the problem diagram. The output is the problem situation diagram annotated with descriptions of the elements and relationships between them and any inferences the system may be able to make along with their justifications.

25.1 Introduction

There are a number of problem domains where information is more easily and naturally represented using diagrams. While humans are adept at using diagrams, automated reasoning systems typically use pre-distilled codified representations. This means that in domains where diagrammatic representations are the norm, the diagrams must first be interpreted by humans. In extreme cases, this results in either simplified representations where determinations of relevancy and saliency have already been made, or representations that (attempt to) include all the information present in the original diagrams. With the former, there is little left for the system to reason about and the latter can be very difficult if not impossible given the density and complexity of information inherent even in simple diagrams. An automated system that reasons directly from diagrams precludes the necessity of spoon feeding the system-interpreted representations and retains the original information-rich diagrams. Using the actual diagrams in reasoning is especially important when there is limited domain knowledge (i.e. the first principles of a domain are unknown) or flexibility is required, both of which are characteristics of learning systems.

There has been considerable interest in the area of diagrammatic reasoning (DR) but little work using case-based reasoning (CBR) as an approach to DR. We think this may be a natural fit, reasoning with diagrams about diagrams. This work explores the possible advantages of using a store of case diagrams to interpret new diagrams.

This chapter describes work in progress on a system that "understands" a diagram of a problem situation by finding correspondences between it and similar diagrams that are already understood in its case memory. The input to the system is a diagram of the problem situation consisting of some number of simple elements such as circles and lines. The goal of the system is to determine the significance of the diagram and its constituent elements by recognising the structure of the diagram and structures within it. Our approach uses high-level perception (recognition of instances of categories and relations) to detect similarities between the problem diagram and stored cases (diagrams) which are understood; the matching cases provide the basis for inferring information missing in the problem diagram. The output is the problem diagram annotated with descriptions of the elements and relationships between them and any inferences the system may be able to make along with their justifications.

In the next section, we state the problem we are addressing and follow with a discussion of the approach we are taking. The following section elaborates on the approach by providing an example. We then discuss some other work related to our approach, some problems to be addressed and possible extensions to our work, and we conclude with a brief discussion of the contribution we believe this work provides.

25.2 Problem Statement

We are interested in looking at the utility of using diagrams we know and understand to reason about new diagrams. In general terms, the question we are dealing with is: Can diagrams be useful in understanding other diagrams? For humans working in domains where information is routinely represented diagrammatically, the answer seems to be an obvious yes. Architects represent plans with diagrams and understand these representations because of their experience in working with them. The military uses diagrams to represent tactical scenarios, sports teams use diagrams to represent plays, and directors and actors use diagrams to represent sets and stage movement. The usefulness of diagrammatic representations in these and other domains is the efficiency with which they convey information. The efficiency with which the people involved understand the diagrams is due in large part to their experience in working with them.

We are developing a CBR system that uses diagrams of situations as cases and reasons from these cases about new diagrammatically represented problem situations. We are assuming that a problem situation can be represented using some number of simple pictorial elements such as basic geometric figures, and that the identity of these elements along with their location and orientation is given, sidestepping the issue of low-level recognition. (What this really means is that the system is given instructions on how to draw the problem situation as opposed to giving it a bitmap.) What these elements represent at some basic level may also be known given a particular domain, but the roles the individual elements play in a specific diagram must be inferred by recognising the structures and relationships inherent in the diagram.

This seems to be a difficult problem because every type of structure the system is capable of recognising increases the size of the interpretation space. For example, one type of structure our system tries to recognise is groupings which are simply collections of individual elements. The basis for grouping certain elements together may be that they are near each other or they are similar in some other way. Recognising groups is beneficial because it simplifies processing to consider some number of elements collectively rather than individually. However, even if an element can be in at most one group, the number of possible groupings is given by the size of the power set of the elements. Even with a small number of elements, a small number of relationships between elements and/or groups, and a small number of possible categories to assign to the elements, the problem space expands rapidly. One role of the cases then is to focus attention. The system tries to find the same sort of structures in the problem diagram that it knows about in the cases. When similar structures and elements are found in the problem diagram, they are mapped to the structures and elements in the case that prompted their discovery and the elements in the problem diagram are inferred to have roles similar to the roles played by the case elements they map to.

The domain we have chosen is the game of football, which is a domain where diagrams are routinely used to represent plans for plays. A play is an attempt by the offence (the team with the ball) to move the ball forward. Each member of the offence must come to a set position, remaining still before the start of each play. This is called a formation. The play begins with a prearranged signal and all of the members of the team move at once, each with their own responsibility. The play diagrams used in football are compact representations of complete plans, showing the initial positions of the players and their assignments during the play. Figure 25.1 shows an example of a play diagram. This domain can be quite complicated although the diagrams themselves are very simple, using only a small set of easily recognisable symbols. The simplicity of these diagrams makes it a good domain for initial investigation (since we are more interested in a conceptual analysis of a diagram rather than low-level recognition of complex elements within a diagram) while the complexity inherent in the domain should provide fertile ground for interesting extensions. We are, of course, very interested in proving claims of generality by using this approach on other domains.

We believe a system that understands diagrams can be a useful tool in domains where diagrams are commonly used. In football, coaches sketch new plays for their offence to run and sketch the plays the opposing team's offence may use in order to prepare the defence. Each player on a team usually receives a thick playbook with all the plays a team may use. In such a domain, it would be useful to have an intelligent editor that is aware of what is being drawn. For example, the system could alert the coach if the proposed play formation is illegal, the types of players in the formation could be filled in by the system using their relative positions, and assignments could be selected from lists and drawn automatically. Coupled with a database of plays, the system would be even more useful as new plays could be created by modifying old ones. An interesting problem with stored plays would be how to allow them to be recalled by sketching their salient features. If plays could be retrieved this way, a more challenging problem would be to retrieve plays, or cases, based on similarity to a diagram of a problem situation where knowledge is limited and use the matching case or cases to infer the missing information in what we call the problem diagram. This is the problem the system described in this chapter addresses. Figure 25.2 shows a problem diagram as it would be presented to the system. The diagram consists of 11 symbols representing where the players on offence line up relative to one another for the given play. The perspective is that of the defence trying to recognise a novel offensive alignment in terms of offensive plays with which it is familiar.

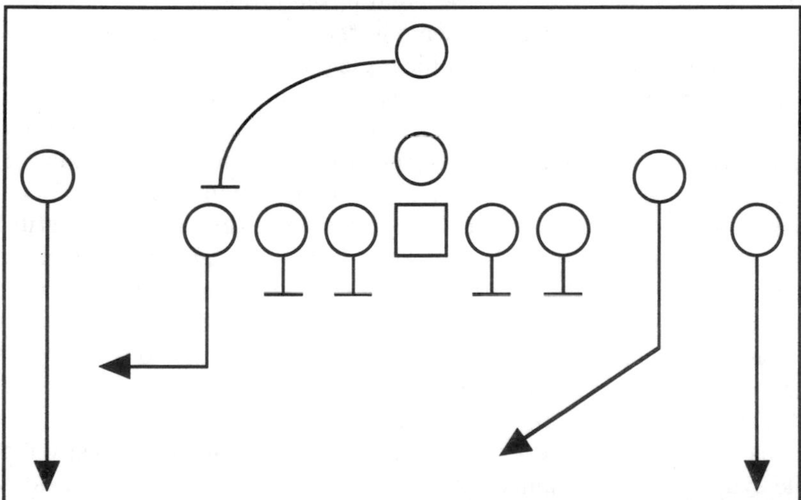

Figure 25.1. Sample play diagram showing player alignment, blocking scheme, and routes of receivers.

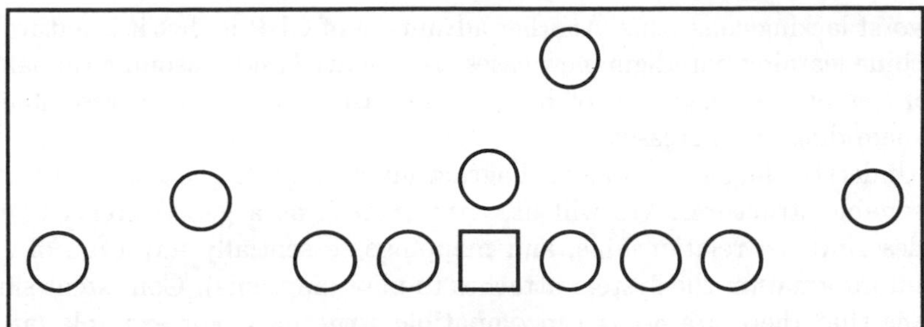

Figure 25.2. A problem diagram.

What we would like to do is take a problem diagram such as this and make good judgements about similarity between it and stored cases and find reasonable correspondences, or mappings, between the elements of two similar diagrams. We believe that this is useful since missing information about an element or structure in the problem diagram may be inferred from the information known about its corresponding element or structure in a similar case. The desired output of the system is the problem diagram annotated with labels identifying the types of the player elements, the assignments of the individual players, and the discovered structures and relationships; the mappings are included as justifications for the inferences and to indicate which groups and elements in the problem diagram are believed to share roles with which groups and elements in the case diagram. In other words, the system's goal is to infer the play from the formation.

25.3 Approach

25.3.1 High-Level Description

We are taking a CBR approach to DR. We see this as an interesting extension of the CBR paradigm, where cases and problem situations are represented diagrammatically and as a natural approach to the tough DR problem. The basic goal of this work is to develop a general approach for automated diagram understanding. By "general" we do not mean to preclude the need for domain knowledge, especially since we are using CBR, but instead that the approach itself is not wedded to a particular domain.

In any CBR system, cases are used to encapsulate specific information. This information may be specific in terms of a given domain, such as recipes in CHEF [13] or in terms of more abstract goals, such as case adaptation in DIAL [19]. The utilisation of specific knowledge (experiences in human terms) is a strength of CBR. The ability to reason from cases means that a system's domain knowledge requirement is simplified; the system does not need knowledge of first principles that are at best difficult to apprehend and at worst lacking consensus. Another advantage of CBR is that it is naturally a machine learning paradigm; new cases are acquired and reasoning (hopefully) improves as the likelihood of having cases that more closely resemble the problem diagram increases.

Understanding the problem diagram involves finding a consistent set of reasonable structures. We will use "structures" as a generic term to refer to descriptions, relationships, and mappings (essentially any type of additional information the system attaches to play diagrams). Consistent simply means that there are no two incompatible structures. For example, an element cannot have two player type descriptions attached to it, or an element in the problem diagram cannot be mapped to more than one element in a particular case. Reasonable is more difficult. For our approach, reasonableness is a measure of the quality of the mappings between structures in the problem diagram and structures in a case (i.e. how much conceptual slippage is required to make a mapping fit). For example, mapping a group of two WIDE-RECEIVER (a player category) elements to a group of two WIDE-RECEIVER elements is better than mapping a group of two WIDE-RECEIVER elements to a group of five LINEMAN (another player category) elements.

Our system uses three types of knowledge: general knowledge, general domain knowledge, and specific domain experience. The first two are represented in the system's conceptual network which includes general concepts (e.g. spatial relationships such as BEHIND and NEXT-TO) and domain concepts (e.g. player types such as QUARTERBACK and WIDE-RECEIVER). A network representation is used to encode associations among concepts. These associations enable the system to focus more on structural similarity (structures involving related concepts) and avoid the restricting specificity of superficial similarity by allowing near concepts to map to one another. Specific domain experience is represented as a collection of cases in the case memory where cases are annotated play diagrams. The conceptual network and case memory are discussed in more detail later.

An important aspect of the way our approach works is that the various types of processing are interleaved. This means all types of structure building (describing elements, identifying relations between them, and mapping structures in the problem diagram to structures in the cases) go on simultaneously. There is an initial phase where general domain knowledge is used to recognise certain distinguishing characteristics of the problem diagram to retrieve relevant cases. Cases are examples of what to look for and where to look for them, so it is desirable to use cases as soon as possible in the processing. Another advantage of this approach is scaling. In domains with more concepts, more complex relations, and complicated diagrams, finding a complete consistent description of the problem diagram would be inefficient. This interleaving of recognition processes is an idea borrowed from Tabletop [10, 15], which is discussed briefly in the section on related work.

This interleaving is accomplished by encoding each type of recognition and structure building as a separate chunk, or method, which is added to a code queue. These methods may cause other methods to be added to the queue, either because they were successful (in which case similar methods would be added) or because they require some preprocessing (in which case a different kind of method would be added). These methods are chosen to run based on priority, where priority is a function of dynamic factors, most important of which are the activation levels of the particular concepts involved. All recognition methods involve concepts, and when instances of concepts are recognised in the problem diagram the corresponding nodes in the conceptual network receive some activation which also spreads to neighbouring nodes. Methods in the code queue that recognise more highly activated concepts are chosen over methods whose corresponding nodes are less activated. To begin, the code queue is primed with several methods reflecting domain predispositions.

The system first applies general domain knowledge to the problem diagram in order to begin building descriptions of the elements of the diagram. This involves recognising elements in terms of the concepts they represent and recognising relationships between elements or other structures. The problem diagram is annotated with the structures as they are recognised and the concept nodes in the conceptual network representing the types of structures that have been recognised are activated. This activation has two effects: the system tries to recognise more of the activated types of structures (by giving priority to related methods in the code queue), and cases containing instances of the same (or similar) structures are activated.

This phase of the process does not actually finish since there may be any number of possible structures, few of which would actually be correct and or useful. Instead, the system begins to consider cases that share similarity with the problem diagram in terms of the descriptions and structures as soon as any similarities are identified. The system tries to recognise correspondences between cases and the problem diagram. A correspondence is a structure that indicates a mapping between elements or structures. Each case that is considered will have its own set of correspondences with the problem diagram. For each case, there will be a maximal consistent set of correspondences, where maximal is in terms of the quality of the mappings and the extent of the structures involved.

Recognising new elements of the problem diagram brings new cases to the foreground and causes new structures to be built. Forming correspondences causes the system to look for other elements or structures in order to form similar correspondences.

Cases are ranked based on their maximal consistent sets of correspondences with the problem diagram. These sets change as new structures and correspondences are formed, which means cases come in and out of favour. At some point, the system becomes more stable; fewer new structures are found and the ranking of the cases remains the same. When this stability reaches some threshold, the system stops. The "answer" is the current description of the problem diagram and its correspondences with the highest rated case.

25.3.2 Conceptual Network

The conceptual network consists of nodes that represent the "centre" of particular concepts connected to one another via directional links that represent the relationships between concepts. There are two main types of links: hierarchical class relationships are represented using ISA and HAS-MEMBER links; relationships described by other concepts in the conceptual network are represented by labelled links. For example, the TIGHT-END node representing a specific class of player is connected to the ELIGIBLE-RECEIVER node representing a more general class of player by an ISA link, and there is a HAS-MEMBER link connecting the ELIGIBLE-RECEIVER node to the TIGHT-END node. The nodes for the concepts LEFT-OF and RIGHT-OF are connected by a LABELLED link, where the label is the OPPOSITE node. In this way, a particular concept is represented by the node for that concept and to a lesser extent its neighbouring nodes.

The conceptual network serves multiple purposes. The first purpose it serves is providing the methods for recognising structures in the problem diagram. Each concept node contains a method or methods for recognising an instance in a diagram of the concept it represents. These methods may be entirely self-contained or may involve the recognition methods of other concepts. The recognition methods for general knowledge concepts are often self-contained. For example, the spatial relation concept NEXT-TO has a method for recognising if two elements are adjacent. Domain knowledge concepts always involve other concepts. For example, recognising the QUARTERBACK requires recognising the player element BEHIND the CENTRE, where BEHIND and CENTRE are concept nodes with their own recognition methods (BEHIND is a spatial relation and CENTRE is a player type). If the system tries to identify the QUARTERBACK , it scans the descriptions of the player objects to see if the CENTRE has been identified, meaning that a player element description has been augmented by the system identifying it as the CENTRE. If the CENTRE has been recognised, then the method for recognising the BEHIND relationship is used to identify the QUARTERBACK. If the CENTRE has not been identified, then the method for recognising it is added to the code queue along with another copy of the method for recognising the QUARTERBACK. This is one way that concepts make use of other concepts.

Given a network sort of representation, with related concepts as neighbours, we find a spreading activation approach can be used for a number of purposes. As mentioned in the high-level description, nodes are activated by successful recognition methods. The links between nodes encode a variable measure of the conceptual distance between concepts: the shorter the distance the more similar the concepts. Activation "flows" along these links so that near nodes receive some fraction of the activation level of the source node. The amount of activation that spreads is a function of the distance between the nodes. For example, if a GROUP structure of THREE elements is recognised, some activation is added to the GROUP concept and THREE concept nodes, and this activation will spread to the TWO and FOUR concept nodes, and so on. Activation levels are an indication of the importance of specific concepts in the problem diagram in terms of what has been discovered up to that point. When the activation level of a node rises above some threshold, a recognition method for that concept is added to the code queue. In this way, the spreading activation causes recognition methods similar to successful recognition methods to be added to the code queue. Since we are interested in *similar* things and not just *specific* things, spreading activation is used so that similar concepts are included.

The next purpose the conceptual network serves is that of a kind of index into case memory. This also takes advantage of the spreading activation. High-level concept nodes are connected to occurrences of themselves in the individual cases in case memory. Activating these concept nodes results in activating cases. The activation level of a particular concept node is in a sense an indication of the relevancy of that concept to the problem diagram. Cases that involve activated concepts are relevant to the problem diagram.

The conceptual network also provides a basis for mapping similar concepts onto one another. When looking for similar cases or forming correspondences between situations, we do not want to restrict the indexing or mapping to exact matches; this would put too much emphasis on superficial similarity. Instead, we would like to allow for slippage, allowing a particular concept to "remind" the system of similar concepts and similar concepts to map to each other. Of course, everything else being equal, it seems logical to prefer correspondences involving concepts that are more similar over less similar concepts. The conceptual distance between concepts provides a mechanism for preferring correspondences between closer matches.

25.3.3 Case Memory

Case memory consists of a number of play diagrams from the problem domain that have been augmented with structures the system may have formed or discovered during the same process of understanding that the problem diagram is subjected to. Figure 25.3 shows a sample case diagram of a passing play. The player elements are annotated with labels for the types of players they represent (football is a game of specialisation). For example, the element that represents the QUARTERBACK is annotated with the label QB. These labels come from the corresponding concepts in the conceptual network. There are also four group structures shown in the case diagram. These are the rectangles marked G1 - G4. The routes of the four receivers in this play are indicated by the arrows. The other lines indicate the how the rest of the team is supposed to block. There would be other structures as well, such as the spatial relationships between the groups and between some elements as well as additional information describing group compositions. These are omitted from the figure for simplicity.

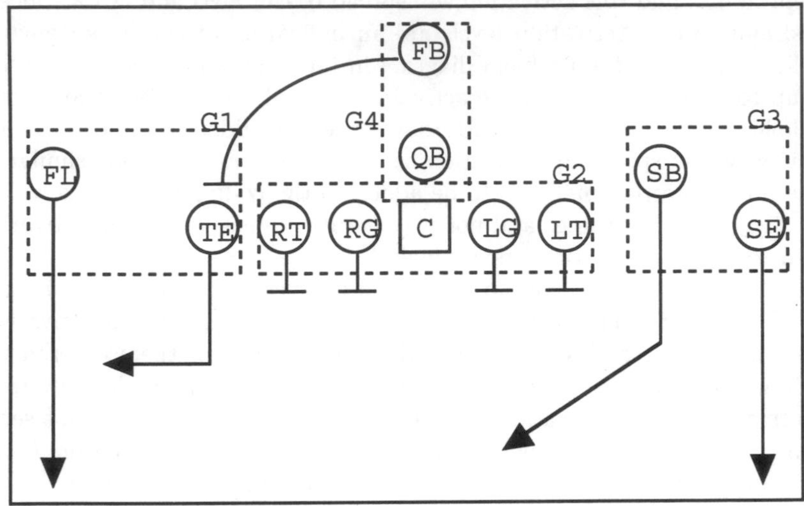

Figure 25.3. An annotated case diagram.

In considering a case, the system opens a window showing the problem diagram (upper half of the window) and the particular case (lower half of the window). The system chooses structures in the case and adds recognition methods for that type of structure to the code queue. These methods may include an argument for a specific area of the problem diagram. For example, if there is a group structure in the lower right of the case (e.g. G3 in Fig. 25.3), the system may add a method to the code queue for recognising a group in the lower right of the problem diagram. The system maps elements and structures it finds in the problem diagram to elements and structures in the case. The same element or structure cannot participate in more than one correspondence, but for any element or structure there may be several reasonable mappings, so there must be a means for preferring one correspondence over another. The quality of correspondences is evaluated according to certain predispositions (e.g. preferring to map groups similar in number and element composition). These mappings work with the conceptual network as a focusing mechanism for the system. The system looks for instances of highly activated concepts and structures involving highly activated relational concepts in the problem diagram. Concept nodes receive additional activation when their concepts participate in correspondences between cases and the problem diagram, so attention on the problem diagram is in a sense directed by its relations to cases.

Cases are scored in order to make a determination as to which cases are most relevant to the problem diagram. The score of a case is a function of the activation levels of the concepts in the case. Activated concepts that participate in structures within a case (as opposed to concepts that occur in isolation) are weighted in order to reflect the system's preference for structural similarity versus superficial similarity.

25.4 A Sample Problem

Given a problem situation described by the diagram in Fig. 25.2, the system first attempts to recognise what concepts are represented by the elements of the diagram. The rules of the game of football and certain standardisations regarding the way the game is played are used to identify the classes of players represented by the circles in the problem diagram by recognising certain spatial relationships between the elements in the diagram. For example, recognising which circle represents the QUARTERBACK involves identifying which circle stands in the relation BEHIND to the square representing the CENTRE, which in turn is recognised by identifying the only element that is a square. As the system recognises elements, it builds descriptions of them using concepts from the conceptual network. This in turn activates the corresponding concept nodes in the network, which has the effect of activating cases in case memory containing occurrences of the activated concepts. The spreading activation in the conceptual network means that cases containing similar concepts will also be activated.

At the same time, the system identifies higher-level structures in the problem diagram. These structures consist of groups of elements and relationships between elements and other structures. The system may try a number of different structures, some of which may be inconsistent with one another. For example, in forming groups, the system is predisposed to favour groups containing similar elements (i.e. elements sharing a superclass) and groups containing neighbouring elements. Since elements are not allowed to be in more than one group (unless one group is a proper subset of the other), including one grouping in the description of the problem diagram may prevent the inclusion of another. The system tries to find a consistent set of structures, favouring the set that has the most structure involving the concepts with the most activation.

As the system proceeds with building a description of the problem diagram, it begins to consider cases, preferring cases with similar structures found in the problem diagram. Considering cases means finding correspondences between structures or elements in the case and the problem diagram. These processes are interleaved. This means that in considering a certain case, if a number of good correspondences have been identified but some structure in the case has no corresponding structure in the problem diagram, the system will look for such a structure. The idea is to avoid trying to completely describe the problem diagram and instead let partial mappings with similar cases indicate what to look for.

Figs. 25.4 and 25.5 are possible results of the system, showing correspondences between the problem diagram (the upper half of each figure) and two cases. For each of the two cases, other correspondences are possible but these may be preferred by the system for a variety of reasons. For example, in Fig. 25.4 the correspondences marked M1 and M4 suggest a degree of diagonal symmetry between the diagrams. The system prefers this mapping over the mirror symmetry mapping (G1 to G1 and G3 to G3) because there is more similarity in the make-up of the groups mapped as shown.

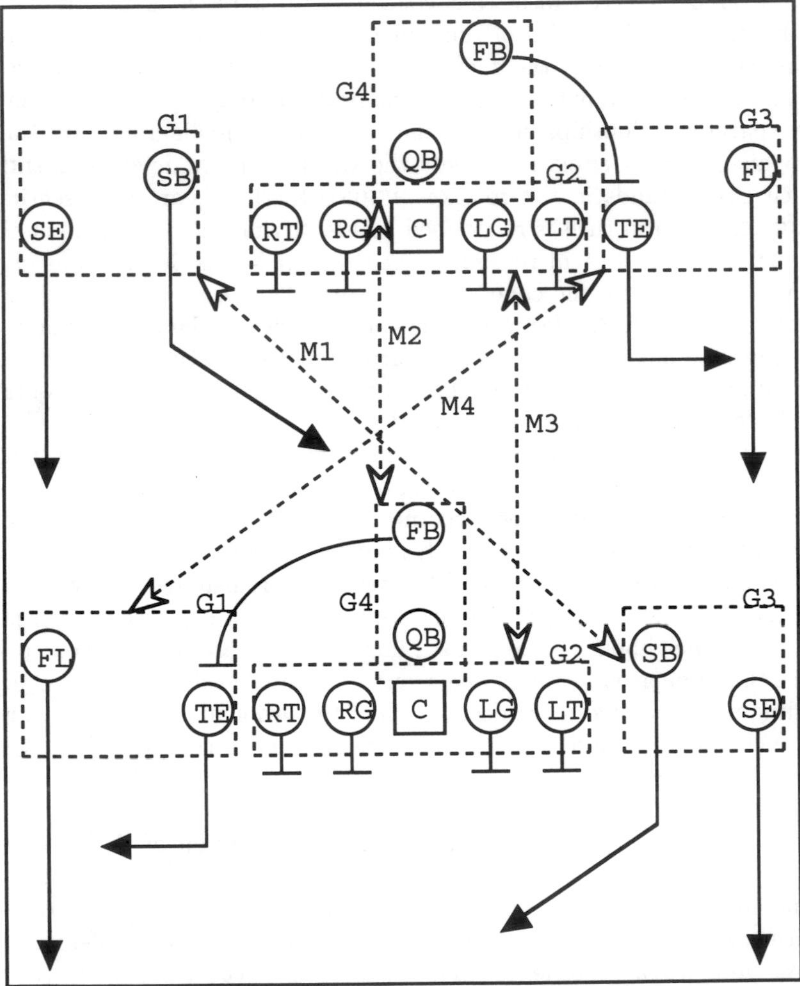

Figure 25.4. Correspondences between a case and the problem diagram.

As mentioned, building descriptions and structures in the problem diagram are interleaved with forming correspondences with similar cases. The correspondences help focus the system's attention by suggesting where to look and what to look for in the problem diagram. For example, if the grouping G1 in the problem diagram in Fig. 25.4 had not been recognised (and therefore correspondence M1 had not been found), the existence of the unmapped group G3 in the case prompts the system to look for a group of two elements of type SLOT-BACK and SPLIT-END. Similarly, the correspondence marked M4 prompts the system to look for such a group on the opposite side of the formation by activating the diagonal-symmetry concept, which is a particular type of mapping. The group G1 is then formed since it has the right number of elements, the types are identical to the elements of group G4 in the case, and it is where it should be. The correspondence M1 can then be recognised.

Individual elements within groups are mapped also. These mappings are omitted from the figures for simplicity. When an element in the problem diagram maps to an element in a case, the elements are inferred to have the same assignment. This is where the pass routes and blocking assignments in the problem diagram in Figs. 25.4 and 25.5 come from. For example, in Fig. 25.4, the tight-end labelled TE in the problem diagram maps to the tight-end in the case. This player is assigned to run a type of route called a "quick out", which means run about five yards straight ahead and then head for the near sideline. The system recognises types of assignments (these are encoded as concepts in the network) and knows that how a particular assignment looks depends on which side a particular player is located. This is why the route of the tight-end in the problem diagram looks flipped; he is assigned the same type of route, not the same bitmap.

The system considers multiple cases simultaneously, concentrating on the most promising cases. Rating a case involves scoring the structures within the case and its correspondences with the problem diagram. Structures involving more concepts with more activation are scored higher than structures with fewer concepts having less activation, and correspondences involving structures with higher scores are scored higher than correspondences involving structures with lower scores or correspondences between isolated elements. Also, concepts participating in structures and correspondences receive additional activation. This helps bias the system toward structural similarity as opposed to superficial similarity.

The case in Fig. 25.4 scores higher than the one in Fig. 25.5. All the groups mapped to each other in Fig. 25.4 have the same number and type of elements. In addition, five elements in the problem diagram are recognised as representing concepts that are subclasses of the ELIGIBLE-RECEIVER concept. Thus the node for this concept is highly activated, which contributes to the high score for the case in Fig. 25.4 since it contains five instances of concepts with ISA relationships to ELIGIBLE-RECEIVER. The player formation in

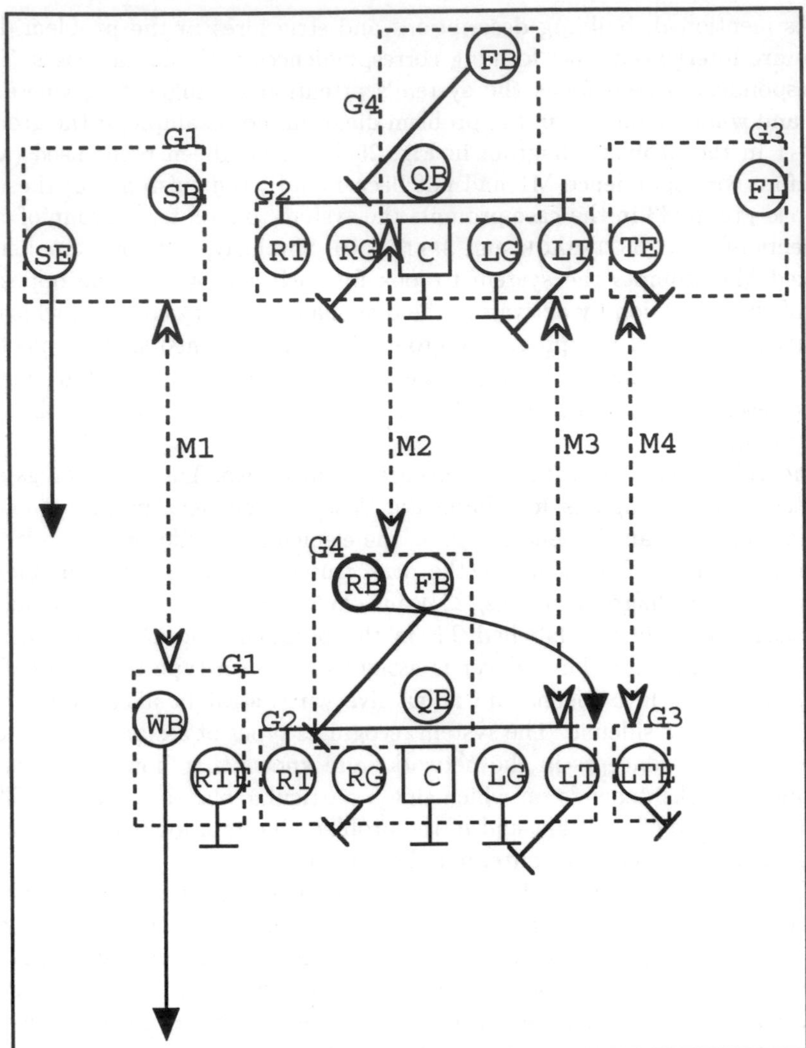

Figure 25.5. Correspondences between another case and the problem diagram.

the case in Fig. 25.5 is too different from the problem diagram to map all the player elements reasonably. For example the FLANKER labelled FL in the problem diagram is recognised as this type of player only because of its location relative to the rest of the formation, not because of any correspondence to an element in the case. Without a mapping, there is no way to infer his assignment.

Eventually the system begins to stabilise. This means that fewer new descriptions, relationships, and correspondences are found, the maximal consistent set of descriptions in the problem diagram remains the same, and the ordering of the cases remains the same. When this stabilisation reaches some threshold (or activity falls below some threshold) the system is done. The "answer" is the annotated problem diagram. Its correspondences with the highest rated case are included to show why certain inferences were made.

25.5 Related Work

Tabletop is a system developed by French [10] that models analogical reasoning. It is presented with a source and a target and attempts to identify analogous structures between them. Tabletop's problem is a table set (sometimes haphazardly) for two; an object on one side of the table is "touched" and Tabletop must decide what the analogous object on the other side of the table would be. It is included here mainly because we borrowed some important aspects of it for our approach such as using a conceptual network to model associations and focus system resources on activated concepts and the idea of interleaving building representations and forming correspondences. It is also included here because, although it is not purported to be a diagrammatic reasoning system, it could easily be one, and it is a motivation for our work. We originally thought of our work as an attempt to use CBR to do the sort of high-level perception that Tabletop does, but it soon became obvious that what we were doing was DR. Tabletop's knowledge is represented in its conceptual network or "slipnet" and it makes no use of a case memory.

POLYA [22] is a CBR system that constructs proofs for high school geometry problems. The diagram of the problem POLYA is to solve is used to provide features that are used as indices to plans in case memory. POLYA interleaves the execution of two types of plans: plans that extract more features from the diagram (search plans) and plans that actually write the proof. This is a very interesting piece of work and shares a lot of similarities with our approach, especially the idea that features discovered in the problem diagram are used to prompt the activation of methods to look for certain other features, instead of trying to describe the diagram completely. An important difference is that the search plans are not diagrams. They are methods for finding more features based on the features that have already been identified. Using actual diagrams similar to the problem diagrams to guide processing may make our approach better suited to weak theory domains and it may be more general. Nevertheless, another important difference is that POLYA is working, so we will have to wait to see if these claims are valid.

Anderson and McCartney [1, 2] describe an efficient use of stored diagrams applicable to a number of domains. A syntax and semantics of interdiagrammatic reasoning is developed along with operators that allow retrieval of cases using diagrammatic matching. Our approach is different in that retrieval is based on similarity of concepts and structures identified in the diagrams (i.e. what the diagrams represent) as opposed to similarity of planar tessellations.

COACH [6] is a CBR system that works in the football domain. The emphasis of the work is plan creation. COACH generates new plays by recognising the bug in a failed play and applying abstract strategies for modifying plans which are indexed by generalised descriptions of plan failures. Cases in COACH are primarily a starting point; planning from scratch is expensive and difficult in weak-theory domains. The differences between COACH and our approach are the system goals (planning versus recognition) and COACH does not use diagrams.

25.6 Future Work

We are in the process of implementing a system using the approach described in this chapter that applies this approach to limited aspects of the football domain. We have implemented small test versions of the main components of the system but have not yet put the pieces together. The conceptual network consists of a player type hierarchy and several relational concepts. A handful of the recognition methods are implemented and limited annotating of the problem diagram works. The case memory consists of only a few test cases with relatively little similarity between them.

We expect there are a number of problems that will have to be addressed if our approach is to work. First is how to choose initially which cases to apply to the problem diagram. The problem here is that all football play diagrams share substantial superficial similarity, and the main reason behind using cases for recognition was to avoid the difficulty of trying to analyse the deeper structure of the problem diagram without the focus cases provide. This problem will be exacerbated with a large case base. One approach is to use abstractions of the top-level structures in the cases and retrieve cases based on what types of similar structures are found in the problem diagram. Another approach might be that it may not matter that much which cases are initially applied if cases are indexed by the specific structures they contain. For example, a group may be found in response to some case being activated; as the internal description of that group is fleshed out, a case or cases with a similarly described group may be activated.

It will still be difficult to determine exactly how to prefer certain cases over others. There are competing factors such as the number of structures mapped, the quality of the individual mappings, and inter-structure relationships (e.g. symmetry). Real results should give us a better idea of which set of metrics are useful and how they should be weighted.

Another difficulty is how to know when to stop looking. How do you know when the best set of correspondences you have at some point is as good as it is going to get, or that further recognising and mapping will not yield any significant improvements? One way to handle this may be to think of the approach as an anytime algorithm and the answer is the case-to-problem diagram mappings in order of ranking. This of course is inadequate if the goal is something specific like "identify the primary receiver" since the primary receiver elements in the various cases being considered may not be mapped yet to any element in the problem diagram. The approach we have in mind is to have the system quit when activity falls below some threshold (e.g. when the frequency of finding new structures drops off) and the rankings have stabilised. When we get some results we will be able to judge the effectiveness of this approach.

There is also the problem of context. Each case that is considered will involve different concepts or similar concepts to different degrees. For example, considering one case may prompt activation of the diagonal-symmetry concept, while another may prompt activation of the mirror-symmetry concept. We would like to be able to switch contexts without any adverse influence. One approach might be for each case to have its own private concept activation levels. However, it seems that it would still be desirable to be able to have cases respond to some degree to global activation levels so that successes may be repeated. This is a very interesting problem. Note for example that this context switching can occur with a single case: the elements of a group in the problem diagram may map diagonally to the elements of the corresponding group in the case, while the rest of the diagram maps straight on.

The evaluation of the effectiveness of our approach will require that we also test other domains. One interesting extension that would broaden the range of applicable domains would be to consider sequences of diagrams representing movement as the problem situation. Cases would also be sequences where each case represents an episode. The problem would be to use the cases to predict subsequent diagrams in the problem sequence and modify predictions dynamically.

25.7 Conclusions

This paper describes an approach to diagrammatic reasoning that uses high-level perception to recognise concepts and relationships in diagrams and make inferences by recognising correspondences with similar diagrams. This work is largely unimplemented at present. Evaluation of the effectiveness of our approach will follow an analysis of the performance of the working system. An assessment of its utility will require application of our approach to a variety of domains.

References

1. Anderson, M. and McCartney, R. (1996). Diagrammatic reasoning and cases. In proceedings of the 13th national conference on artificial intelligence. Menlo Park, CA: AAAI Press, pp. 1004-1009.

2. Anderson, M. and McCartney, R. (1995). Inter-diagrammatic reasoning. In Proceedings of the 14th international joint conference on artificial intelligence. San Mateo, CA: Morgan Kaufmann, pp. 878-884.
3. AFC Association (1995). Football coaching strategies. Champaign, IL: Human Kinetics.
4. Chandrasekaran, B., Glasgow, J. and Narayanan, N.H. (Eds) (1995). Diagrammatic reasoning: Cognitive and computational perspectives. Menlo Park, CA: AAAI Press.
5. Chandrasekaran, B., Narayanan, N.H. and Iwasaki, Y. (1993). Reasoning with diagrammatic representations: A report on the spring symposium. AI Magazine 14:49-56.
6. Collins, G.C. (1989). Plan Creation. In C.K. Riesbeck and R.C. Schank (Eds), Inside case-based reasoning, Hillsdale, NJ: Erlbaum, pp. 249-290.
7. Dreayer, B. (1994). Teach me sports: Football. Santa Monica, CA: General Publishing Group.
8. Estes, W.K. (1994). Classification and cognition. New York: Oxford University Press.
9. Flores, T. (1993). Coaching football. Lincoln Wood, IL:Masters Press.
10. French, R.M. (1995). The subtlety of sameness: A theory and computer model of analogy-making. Cambridge, MA: MIT Press.
11. Gentner, D. and Stevens, A.L. (Eds) (1983). Mental models. Hillsdale, NJ: Erlbaum.
12. Goel, V. (1996). Sketches of thought. Cambridge, MA: MIT Press.
13. Hammond, K.J. (1986). CHEF: A model of case-based planning. In Proceedings of AAAI-86. Los Altos, CA: Morgan Kaufman, pp. 267-271.
14. Hawkes, D.D. (1995). Football's best offensive playbook. Champaign, IL: Human Kinetics.
15. Hofstadter, D. (1995). Fluid concepts and creative analogies: Computer models of the fundamental mechanisms of thought. New York: BasicBooks.
16. Kolodner, J. (1993). Case-based reasoning. San Mateo, CA: Morgan Kaufmann.
17. Kolodner, J.L. (1984). Retrieval and organizational strategies in conceptual memory: A computer model. Hillsdale, NJ: Erlbaum.
18. Kolodner, J.L. and Riesbeck, C.K.(Eds) (1986). Experience, memory, and reasoning. Hillsdale, NJ: Erlbaum.
19. Leake, D.B., Kinley, A. and Wilson, D. (1997). Case-based similarity assessment: Estimating adaptability from experience. In Proceedings of the 14th national conference on artificial intelligence. Menlo Park, CA: AAAI Press/MIT Press, pp. 674-679.
20. Lockhead, G.R. (1992). On identifying things: A case for context. In B. Burns (Ed.), Percepts, concepts and categories: The representation and processing of information. Amsterdam: Elsevier, pp. 381-410.
21. McCartney, R. (1993). Episodic cases and real-time performance in a case-based planning system. Expert Systems with Applications 6:9-22.
22. McDougal, T.F. and Hammond, K.J. (1995). Using diagrammatic features to index plans for geometry theorem-proving. In B. Chandrasekaran, J. Glasgow and N.H. Narayanan (Eds), Diagrammatic reasoning: Cognitive and computational perspectives. Menlo Park, CA: AAAI Press, pp. 691-709.
23. Millikan, R.G. (1984). Language, thought, and other biological categories. Cambridge, MA: MIT Press.
24. Mitchell, M. (1993). Analogy-making as perception: A computer model. Cambridge, MA: MIT Press.

25. Nahinsky, I.D. (1992). Episodic components of concept learning and representation. In B. Burns (Ed.), Percepts, concepts and categories: The representation and processing of information. Amsterdam: Elsevier, pp. 381-410.
26. Novick, L.R. (1988). Analogical transfer: Processes and individual differences. In D.H. Helman (Ed.), Analogical reasoning: Perspectives of artificial intelligence, cognitive science, and philosophy. Dordrecht: Kluwer, pp. 125-145.

26. Interpreting the Engineer's Sketch: A Picture is Worth a Thousand Constraints

Thomas F. Stahovich

This chapter describes a program called SKETCHIT that transforms a single sketch of a mechanical device into multiple families of new designs. To "interpret" a sketch, the program first determines how the sketched device should have worked. The program then derives constraints on the geometry to ensure the device works that way. The program is based on qualitative configuration space (qc-space), a novel representation that captures mechanical behaviour while abstracting away the particular geometry used to depict this behaviour. The program employs a paradigm of abstraction and resynthesis: it abstracts the initial sketch into qc-space then maps from qc-space to new geometries.

26.1 Introduction

Drawings have always been an important tool for engineers, with the sketch on a napkin an important and traditional means of thought and communication. Yet to date, CAD software has been at best a drafting tool, producing carefully drawn pictures, but neither understanding them the way people do, nor capable of accepting as input an informal sketch of the sort engineers commonly create.

We are working to change that by developing a program that can read, understand, and use sketches of mechanical devices like the one shown in Fig. 26.1. Our program, called SKETCHIT, is capable of taking a single stylised sketch of a mechanical device and generalising it to produce multiple new designs.[1]

Engineering sketches, by their very nature, are inaccurate descriptions of a device. Taken literally, the geometry in Fig. 26.1, for example, may not actually produce the desired behaviour. Nevertheless, a skilled engineer is able to see how a roughly sketched device was supposed to have worked and

[1] Portions of this chapter are reprinted, with permission, from Generating Multiple New Designs from a Sketch, Stahovich, T. F., Davis, R., and Shrobe, H., in *Proceedings of the Thirteenth National Conference on Artificial Intelligence*, pp 1022–1029, copyright © 1996, American Association for Artificial Intelligence.

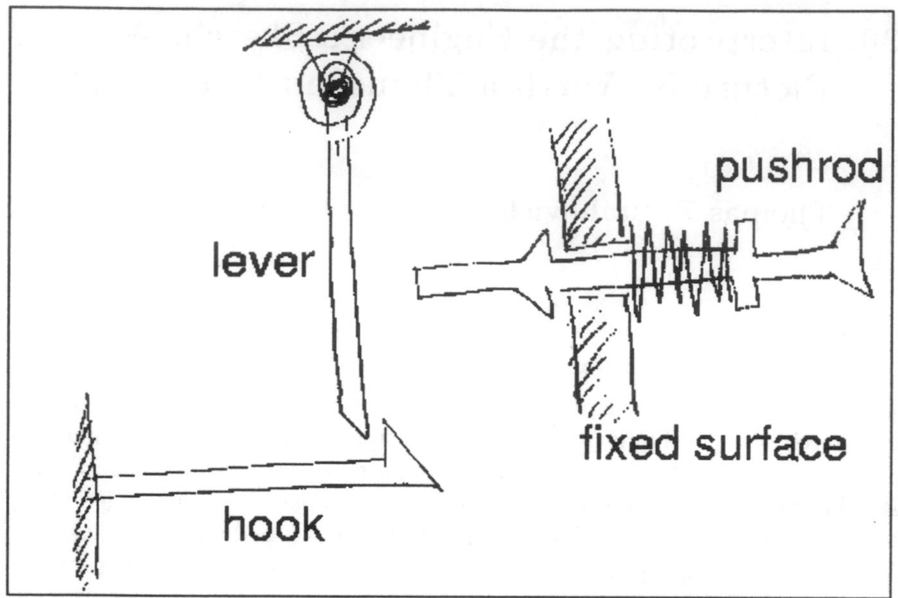

Figure 26.1. A sketch of a circuit breaker.

hence what the geometry should have been. In effect, achieving the correct behaviour places constraints on the device's geometry and understanding these constraints allows the engineer to overcome the inaccuracies in the sketch. Thus, interpreting an engineering sketch is as much about understanding physics as it is about understanding geometry. Our software directly embodies this observation: to interpret the geometry of a sketch, the program first identifies what behaviours the parts should provide. The program then derives constraints to ensure the geometry actually produces these behaviours.

To identify the behaviours of the individual parts of a device, the program transforms the sketch into a novel representation we call qualitative configuration space (qc-space). Qc-space captures the behaviour of the original design while abstracting away the particular geometry used to suggest that behaviour. If the sketch as drawn does not produce the desired behaviour, the program adjusts the qc-space until it does. The program then uses a library of geometric interactions to transform each identified behaviour into new geometry with constraints ensuring that behaviour. The constraints define a family of geometries that all produce a particular kind of behaviour. Thus, as the program transforms the qc-space back into geometry, it transforms the initial sketch into a family of designs. Because the library may contain multiple implementations for a particular kind of behaviour, the program is capable of generating multiple families of new designs. The program represents each new family with what we call a behaviour-ensuring parametric

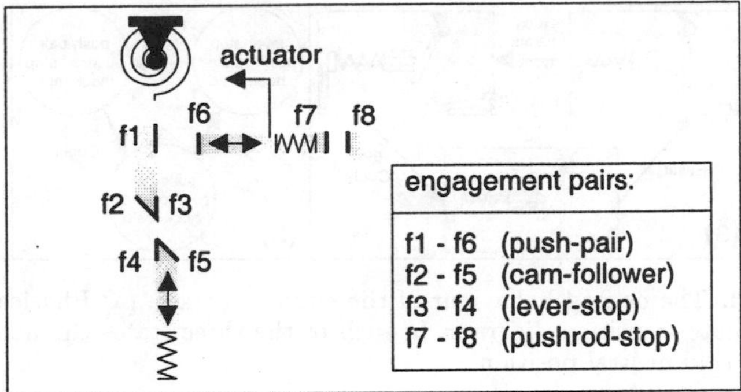

Figure 26.2. Sketch as actually input to the program. Engagement faces are bold lines. The actuator represents the reset motion imparted by the user.

model ("BEP-Model"): a parametric model augmented with constraints that ensure the geometry produces the desired behaviour.[2]

We use the design of a circuit breaker (Fig. 26.1) to illustrate the program in operation. In normal use, current flows from the lever to the hook; current overload causes the bimetallic hook to heat and bend, releasing the lever and interrupting the current flow. After the hook cools, pressing and releasing the pushrod resets the device.

SKETCHIT takes as input a stylised sketch of a device and a state transition diagram describing the desired overall behaviour of the device. The latter provides guidance in identifying what behaviours the individual parts of the device should provide.

The designer describes the circuit breaker to SKETCHIT with the stylised sketch shown in Fig. 26.2, using line segments for part faces and icons for springs, joints, and actuators. SKETCHIT is concerned only with the *functional geometry*, i.e., the faces where parts meet and through which force and motion are transmitted (lines f1–f8). The designer's task is thus to indicate which pairs of faces are intended to engage each other. Consideration of the connective geometry (the surfaces that connect the functional geometry to make complete solids) is put off until later in the design process.

The designer describes the desired overall behaviour of the circuit breaker with the state transition diagram in Fig. 26.3. Each node in the diagram is a list of the pairs of faces that are engaged and the springs that are relaxed. The arcs are the external inputs that drive the device. This particular state transition diagram describes how the circuit breaker should behave in the face of heating and cooling the hook and pressing the reset pushrod.

Figure 26.4 shows a portion of one of the BEP-Models that SKETCHIT derives in this case. The top of the figure shows the parameters that define the sloped face on the lever (f2) and the sloped face on the hook (f5). The

[2] A parametric model is a geometric model in which the shapes are controlled by a set of parameters.

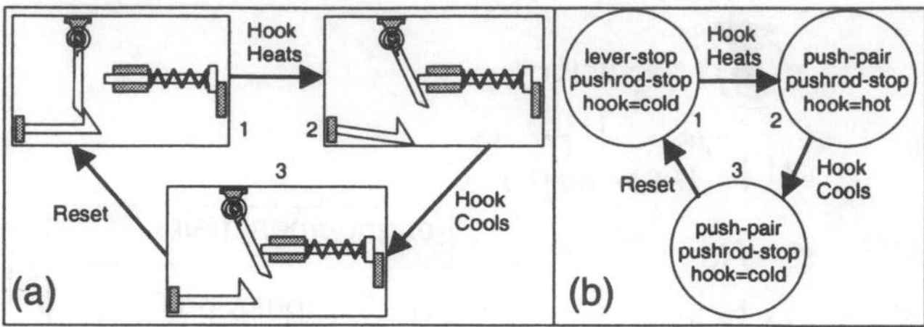

Figure 26.3. The desired behaviour of the circuit breaker. (a) Physical interpretation. (b) State transition diagram. In each of the three states the hook is either at its hot or cold neutral position.

bottom shows the constraints that ensure this pair of faces plays its role in achieving the overall desired behaviour: i.e., moving the lever clockwise pushes the hook down until the lever moves past the point of the hook, whereupon the hook springs back to its rest position. As one example of how the constraints enforce the desired behaviour, the ninth equation, `0 > R14/TAN(PSI17) + H2_12/SIN(PSI17)`, constrains the geometry so that the contact point on face f2 never moves tangent to face f5. This in turn ensures that when the two faces are engaged, clockwise rotation of the lever always increases the deflection of the hook.

The parameter values shown at the top of Fig. 26.4 are solutions to the constraints of the BEP-Model, hence this particular geometry provides the desired behaviour. These specific values were obtained using the numerical constraint solver in DesignView, a commercial CAD program. Using this constraint solver (or a similar one), we can easily explore the family of designs defined by this BEP-Model. Figure 26.5, for example, shows another solution to this BEP-Model. Because these parameter values satisfy the BEP-Model, even this rather unusual geometry provides the desired behaviour. As this example illustrates, the family of designs defined by a BEP-Model includes a wide range of design solutions, many of which would not be obtained with conventional design approaches.

Figures 26.4 and 26.5 show members of just one of the families of designs that the program produces for the circuit breaker. SKETCHIT produces other families of designs (i.e., other BEP-Models) by selecting different implementations for the pairs of interacting faces and different motion types (rotation or translation) for the components. Figure 26.6 shows an example of selecting different implementations for the pairs of interacting faces: In the original implementation of the cam-follower engagement pair, the motion of face f2 is roughly perpendicular to the motion of face f5; in the new design of Fig. 26.6, the motions are parallel. Figure 26.7 shows a design obtained by selecting a new motion type for the lever: In the original design the lever rotates, here it translates.

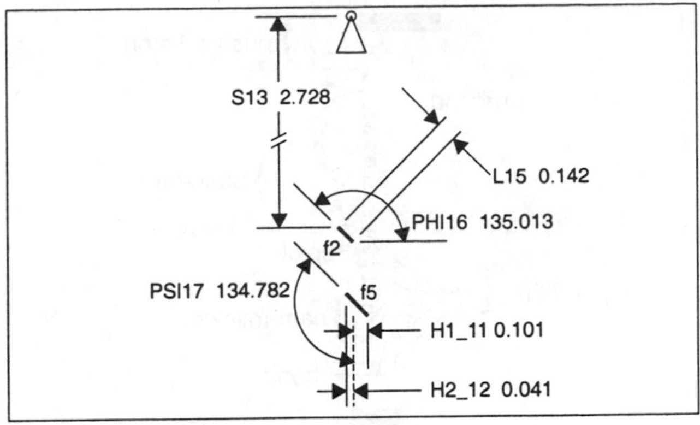


```
H1_11 > 0       H2_12 > 0        S13 > H1_11       L15 > 0
PHI16 > 90      PHI16 < 180      PSI17 > 90        PSI17 < 180
0 > R14/TAN(PSI17) + H2_12/SIN(PSI17)
R14 = SQRT(S13^2 + L15^2 - 2*S13*L15*COS(PHI16))
```

Figure 26.4. Output from the program (a BEP-Model). Top: the parametric model; the decimal number next to each parameter is the current value of that parameter. Bottom: the constraints on the parameters. For clarity, only the parameters and constraints for faces f2 and f5 are shown.

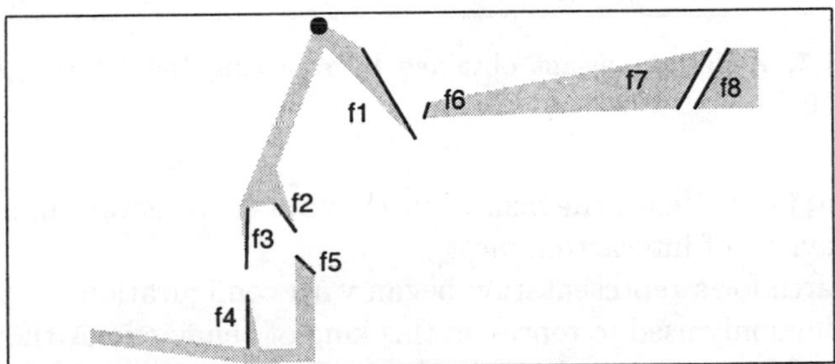

Figure 26.5. Another solution to the BEP-Model of Fig. 26.4. Shading indicates how the faces might be connected to flesh out the components. This solution shows that neither the pair of faces at the end of the lever nor the pair of faces at the end of the hook need be contiguous.

26.2 Representation: QC-Space

SKETCHIT's approach to its task is to use a representation that captures the behaviour suggested by the sketch while abstracting away the particular geometry used to depict this behaviour. This allows the program to generalise the initial design by selecting new geometries that provide the same behaviours.

For the class of devices that SKETCHIT is concerned with, the overall behaviour is achieved through a sequence of interactions between pairs of

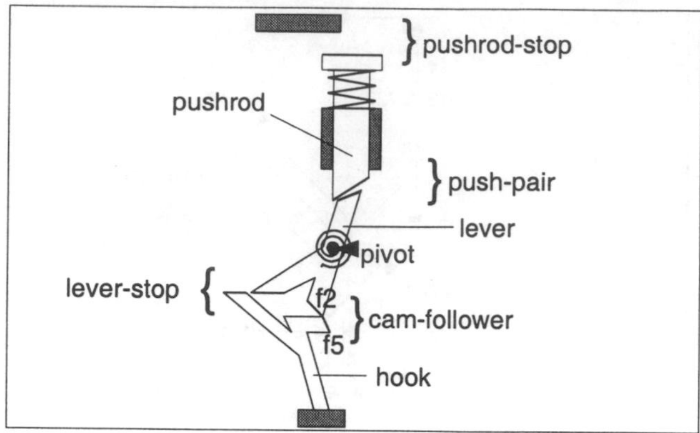

Figure 26.6. A design variant obtained by using different implementations for the engagement faces. In the position shown, the pushrod is pressed so that the hook is just on the verge of latching the lever.

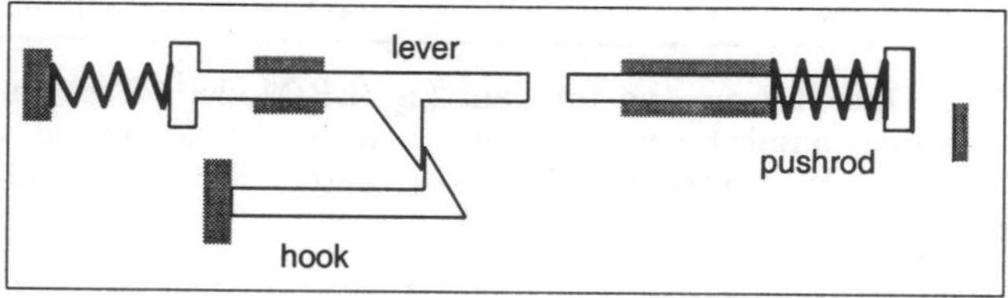

Figure 26.7. A design variant obtained by replacing the rotating lever with a translating part.

engagement faces. Hence the behaviour that our representation must capture is the behaviour of interacting faces.

Our search for a representation began with configuration space (c-space), which is commonly used to represent this kind of behaviour. Although c-space is capable of representing the behaviours we are interested in, it does not adequately abstract away their geometric implementations. We discovered that abstracting c-space into a qualitative form produces the desired effect; hence we call SKETCHIT's behavioural representation "qualitative configuration space" (qc-space).

This section begins with a brief description of c-space, then describes how we abstract c-space to produce qc-space.

26.2.1 C-Space

Consider the rotor and slider in Fig. 26.8. If the angle of the rotor U_R and the position of the slider U_S are as shown, the faces on the two bodies will touch. These values of U_R and U_S are termed a *configuration* of the bodies in which the faces touch, and can be represented as a point in the plane, called a configuration space plane (cs-plane).

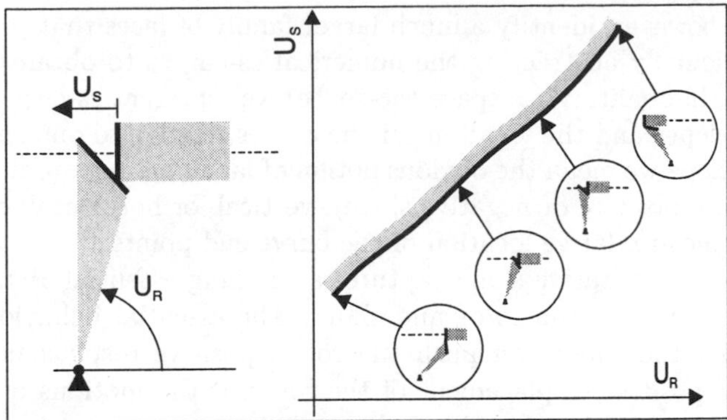

Figure 26.8. Left: A rotor and slider. The slider translates horizontally. The interacting faces are shown with bold lines. Right: The c-space. The inset figures show the configuration of the rotor and slider for selected points on the cs-curve.

If we determine all of the configurations of the bodies in which the faces touch and plot the corresponding points in the cs-plane (Fig. 26.8), we get a curve, called a configuration space curve (cs-curve). The shaded region "behind" the curve indicates blocked space: configurations in which one body would penetrate the other. The unshaded region "in front" of the curve represents free space: configurations in which the faces do not touch.

The axes of a c-space are the position parameters of the bodies; the dimension of the c-space for a set of bodies is the number of degrees of freedom of the set. To simplify geometric reasoning in c-space, we assume that devices are fixed-axis. That is, we assume that each body either translates along a fixed axis or rotates about a fixed axis. Hence in our world the c-space for a pair of bodies will always be a plane (a cs-plane) and the boundary between blocked and free space will always be a curve (a cs-curve).[3] However, even in this world a device may be composed of many fixed-axis bodies, hence the c-space for the device as a whole can be of dimension greater than two. The individual cs-planes are orthogonal projections of the multidimensional c-space of the overall device.

26.2.2 Abstracting to QC-Space

C-space is already an abstraction of the original geometry. For example, any pair of faces that produces the cs-curve in Fig. 26.8 will produce the same behaviour (i.e., the same dynamics) as the original pair of faces. Thus, each cs-curve represents a family of interacting faces that all produce the same behaviour.

[3] The c-space for a pair of fixed-axis bodies will always be two-dimensional. However, it is possible for the c-space to be a cylinder or torus rather than a plane. See Section 26.3.2 for details.

We can, however, identify a much larger family of faces that produce the same behaviour by abstracting the numerical cs-curves to obtain a qualitative c-space. In qualitative c-space (qc-space) we represent cs-curves by their qualitative slopes and the locations of the curves relative to one another. By qualitative slope we mean the obvious notion of labelling monotonic curves as diagonal (with positive or negative slope), vertical, or horizontal; by relative location we mean relative location of the curve end points.[4]

To see how qualitative slope captures something essential about the behaviour, we return to the rotor and slider. The essential behaviour of this device is that the slider can push the rotor: positive displacement of the slider causes positive displacement of the rotor. If the motions of the rotor and slider are to be related in this fashion, their cs-curve must be a diagonal curve with positive slope. Conversely, any geometry that maps to a diagonal curve with positive slope will produce the same kind of pushing behaviour as the original design.

There are eight types of qualitative cs-curves, shown in Fig. 26.11. Diagonal curves always correspond to pushing behaviour; vertical and horizontal curves correspond to what we call "stop behaviour", in which the extent of motion of one part is limited by the position of another.

The key, more general insight here is that *for monotonic cs-curves, the qualitative slopes and the relative locations completely determine the first-order dynamics of the device.* By first-order dynamics we mean the dynamic behaviour obtained when the motion is assumed to be inertia-free and the collisions are assumed to be inelastic and frictionless.[5] The consequence of this general insight is that qc-space captures *all* of the relevant physics of the overall device, and hence serves as a design space for behaviour. It is a particularly convenient design space because it has only two properties: qualitative slope and relative location.

Another important feature of qc-space is that it is constructed from a very small number of building blocks, viz., the different types of qcs-curves in Fig. 26.11. As a consequence we can easily map from qc-space back to geometric implementation using precomputed implementations for each of the building blocks. We show how to do this in Section 26.3.2.

---

[4] We restrict qcs-curves to be monotonic to facilitate qualitative simulation of a qc-space.

[5] "Inertia-free" refers to the circumstance in which the inertia terms in the equations of motion are negligible compared to the other terms. One important property of inertia-free motion is that there are no oscillations. This set of physical assumptions is also called quasi-statics.

26.3 The SKETCHIT System

Figure 26.9 shows a flow chart of the SKETCHIT system with its two main processes: "Behaviour Extraction" and "Constraint and Geometry Synthesis". These processes are discussed in the sections that follow.

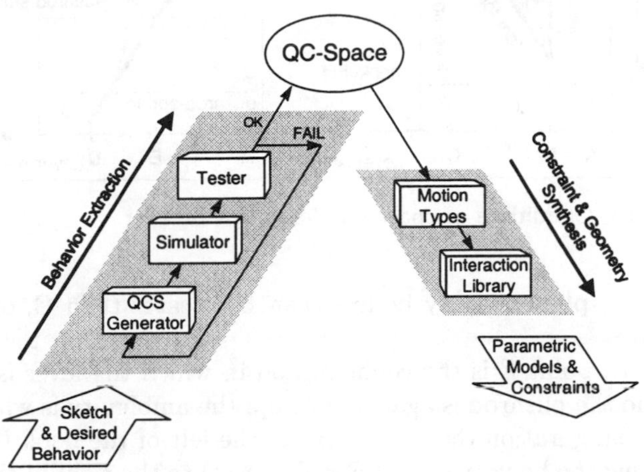

Figure 26.9. Overview of SKETCHIT system.

26.3.1 Behaviour Extraction Process

SKETCHIT uses generate and test to abstract the initial design into one or more working qc-spaces, i.e., qc-spaces that provide the behaviour specified in the state transition diagram.

The generator produces multiple candidate qc-spaces from the sketch, each of which is a possible interpretation of the sketch. The simulator computes each candidate's overall behaviour (i.e., the aggregate behaviour of all of the individual interactions), which the tester then compares to the desired behaviour.

The generator begins by computing the numerical c-space of the sketch. It then abstracts each numerical cs-curve into a qcs-curve, i.e., a curve with qualitative slope and relative location.

As with any abstraction process, moving from specific numerical curves to qualitative curves can introduce ambiguities. For example, in the candidate qc-space in Fig. 26.10 there is ambiguity in the relative location of the abscissa value (E) for the intersection between the push-pair curve and the pushrod-stop curve. This value is not ordered with respect to B and C, the abscissa values of the end points of the lever-stop and cam-follower curves in the

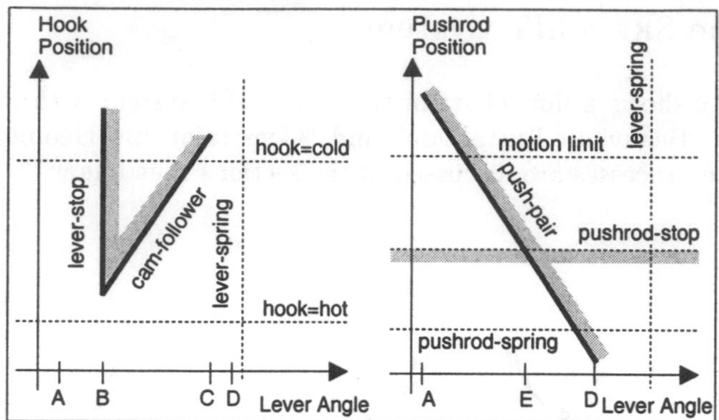

Figure 26.10. Candidate qc-space for the circuit breaker.

hook-lever qcs-plane: E may be less than B, greater than C, or between B and C.[6]

Physically, point E is the configuration in which the lever is against the pushrod and the pushrod is against its stop; the ambiguity is whether in this particular configuration the lever is (a) to the left of the hook (i.e., E < B), (b) contacting the hook (i.e., B < E < C), or (c) to the right of the hook (i.e., C < E). When the generator encounters this kind of ambiguity, it enumerates all possible interpretations, passing each of them to the simulator.

The relative locations of these points are not ambiguous in the original, numerical c-space. Nevertheless, SKETCHIT computes all possible relative locations, rather than taking the actual locations directly from the numerical c-space. One reason for this is that it offers one means of generalising the design: The original locations may be just one of the possible working designs; the program may be able to find others by enumerating and testing all of the possible relative locations.

A second reason the program enumerates and tests all possible relative locations is because this enables the program to compensate for flaws in the original sketch. These flaws arise from interactions that are individually correct, but whose global arrangement is incorrect. For example, in Fig. 26.2 the interaction between the lever and hook, the interaction between the pushrod and the lever, and the interaction between the pushrod and its stop may all be individually correct, but the pushrod-stop may be sketched too far to the left, so that the lever always remains to the left of the hook (i.e., the global arrangement of these three interactions prevents the lever from actually interacting with the hook). By enumerating possible locations for the intersection between the pushrod-stop and push-pair qcs-curves, SKETCHIT will correct this flaw in the original sketch.

[6] We do not consider the case where E = B or E = C.

Currently, the candidate qc-spaces the generator produces are possible interpretations of the ambiguities inherent in the abstraction.[7] The simulator and tester determine which of these interpretations produce the desired behaviour. SKETCHIT employs an innovative qualitative simulator designed to minimise branching of the simulation. See [13,15,17] for a detailed presentation of the simulator which computes the motion of the parts of a device as a trajectory through qc-space (this process is itself a diagrammatic reasoning task).

26.3.2 Constraint and Geometry Synthesis

In the synthesis process, the program turns each of the working qc-spaces into multiple families of new designs. Each family is represented by a BEP-Model.

Qc-space abstracts away both the motion type of each part and the geometry of each pair of interacting faces. Hence there are two steps to the synthesis process: selecting a motion type for each part and selecting a geometry for each pair of engagement faces.

Selecting Motion Type. SKETCHIT is free to select a new motion type for each part because qc-space abstracts away this property. More precisely, qc-space abstracts away the motion type of parts that translate and parts that rotate less than a full revolution.[8]

Changing translating parts to rotating ones, and vice versa, permits SKETCHIT to generate a rich assortment of new designs.

Selecting Geometries. The general task of translating from *c-space* to geometry is intractable [1]. However, *qc-space* is carefully designed to be constructed from a small number of basic building blocks, 40 in all. The origin of 32 of these can be seen by examining Fig. 26.11: there are four choices of qualitative slope; for each qualitative slope there are two choices for blocked space; and the qc-space axes q_1 and q_2 can represent either rotation or translation. The other eight building blocks represent interactions of rotating or translating bodies with stationary bodies.

Because there are only a small number of basic building blocks, we were able to construct a library of implementations for each building block. To translate a qc-space to geometry, the program selects an entry from the library for each of the qcs-curves.

[7] We are also working on a more sophisticated generator that considers a wider range of candidates in order to correct more serious flaws in the original sketch. See Section 26.5.

[8] Qc-space cannot abstract away the motion type of parts that rotate more than a full revolution because the topology of the qc-space for such parts is different: If one of a pair of parts rotates through full revolutions, its motion will be 2π periodic, and what was a plane in qc-space will become a cylinder. (If both of the bodies rotate through full revolutions the qc-space becomes a torus.) Hence, if a pairwise qc-space is a cylinder or torus, the design must employ rotating parts (one for a cylinder, two for a torus) rather than translating ones.

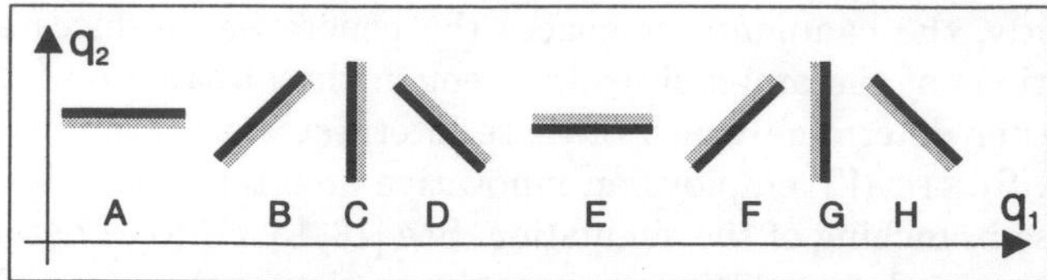

Figure 26.11. The eight types of qcs-curves. For drawing convenience, diagonal qcs-curves are shown as straight line-segments; they can have any shape as long as they are monotonic.

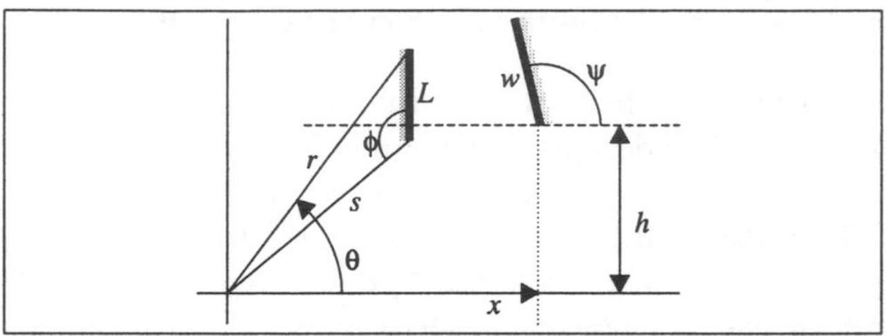

Figure 26.12. Implementation for qcs-curve F. The two faces are shown as bold lines. The rotating face rotates about the origin; the translating face translates horizontally. θ is the angle of the rotor and x (measured positive to the *left*) is the position of the slider.

Each library entry contains a pair of parameterised faces and a set of constraints that ensure that the faces implement a monotonic cs-curve of the desired slope, with the desired choice of blocked space. Each library entry also contains algebraic expressions for the end point coordinates of the cs-curve.

For example, Fig. 26.12 shows a library entry for qcs-curve F in Fig. 26.11, for the case in which q_1 is rotation and q_2 is translation. For the corresponding qcs-curve to be monotonic, have the correct slope, and have blocked space on the correct side, the following 10 constraints must be satisfied:

$$
\begin{array}{lll}
w > 0 & L > 0 & h > 0 \\
s < h & r > h & \pi/2 < \phi \leq \pi \\
\psi > 0 & \psi < \arcsin(h/r) + \pi/2 & r = (s^2 + L^2 - 2sL\cos(\phi))^{1/2} \\
\multicolumn{3}{l}{\arccos(h/r) + \arccos(\frac{L^2+r^2-s^2}{2Lr}) < \pi/2}
\end{array}
$$

The end point coordinates of the cs-curve are:

$$\theta_1 = \arcsin(h/r) \; x_1 = -r\cos(\theta_1) \; \theta_2 = \pi - \arcsin(h/r) \; x_2 = -r\cos(\theta_2)$$

Figure 26.13 shows a second way to generate qcs-curve F, using the constraints:

$$
\begin{array}{lll}
h_1 > 0 & h_2 > 0 & s > h_1 \\
L > 0 & \pi/2 < \phi < \pi & \pi/2 < \psi < \pi \\
\multicolumn{2}{l}{0 > r/\tan(\psi) + h_2/\sin(\psi)} & r = (s^2 + L^2 - 2sL\cos(\phi))^{1/2}
\end{array}
$$

The end point coordinates of this cs-curve are:

$$\theta_1 = -\arcsin(h_2/r) \qquad x_1 = -r\cos(\theta_1) + h_2/\tan(\psi)$$
$$\theta_2 = \arcsin(h_1/s) + \arccos(\tfrac{s^2+r^2-L^2}{2sr}) \qquad x_2 = -s\cos(\arcsin(h_1/s)) - h_1/\tan(\psi)$$

In the first of these designs the motion of the slider is approximately parallel to the motion of the rotor, while in the second the motion of the slider is approximately perpendicular to the motion of the rotor.[9] The two designs thus represent qualitatively different implementations for the same qcs-curve.

To generate a BEP-Model for the sketch, our program selects from the library an implementation for each qcs-curve. For each selection it creates new instances of the parameters and transforms the coordinate systems to match those used by the actual components. The relative locations of the qcs-curves in the qc-space are turned into constraints on the end points of the qcs-curves. Finally, the program assembles the parametric geometry fragments and constraints of the library selections to produce the parametric geometry and constraints of the BEP-Model.

Our library contains geometries that use flat faces,[10] although we have begun work on using circular faces. We have at least one library entry for each of the 40 kinds of interactions. We are continuing to generate new entries.

SKETCHIT is able to produce different BEP-Models (i.e., different families of designs) by selecting different library entries for a given qcs-curve. For example, Fig. 26.5 shows a solution to the BEP-Model SKETCHIT generates by selecting the library entry in Fig. 26.13 for the cam-follower qcs-curve. Figure 26.6 shows a solution to a different BEP-Model SKETCHIT generates by selecting the library entry in Fig. 26.12 for the cam-follower. As these examples illustrate, the program can generate a wide variety of solutions by selecting different library entries.

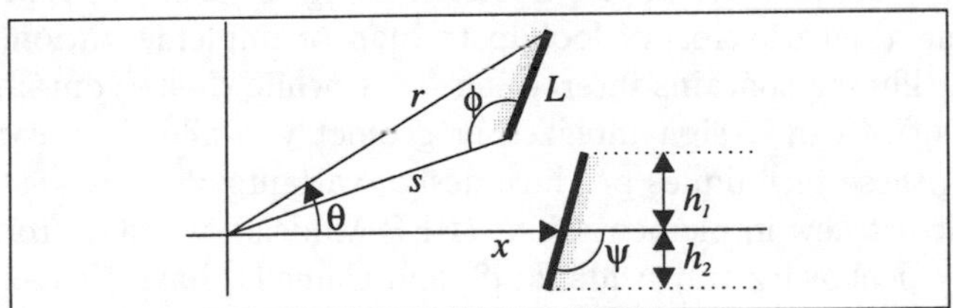

Figure 26.13. Another implementation for qcs-curve F. The two faces are shown as bold lines. The rotating face rotates about the origin; the translating face translates horizontally. θ is the angle of the rotor and x (measured positive to the *left*) is the position of the slider.

[9] The first design is a cam with offset follower; the second is a cam with centred follower.

[10] Circular faces are used when rotors act as stops.

26.4 Related Work

There is little previous work in sketch understanding. Narayanan et al. [11] use a diagram of a device to reason about its behaviour, but they use a pre-parsed description of the behaviours of each component while we reason directly from the geometry of the interacting faces.

Faltings [4] suggests that a sketch is not a single qualitative model but rather represents a family of precise models. He demonstrates that taking a sketch as a qualitative metric diagram it is possible to compute the "kinematic topology" (an abstraction of the "place vocabulary" [3]). The kinematic topology may contain ambiguities suggesting behaviours that may be obtained by modifying the geometry. Methods for determining which modifications will yield these other behaviours is an open issue.

Our work is closely related to work in design automation. Our techniques can be viewed as a natural complement to the bond graph techniques of the sort developed in [20]. Our techniques are useful for computing geometry that provides a specified behaviour, but because of the inertia-free assumption employed by our simulator our techniques are effectively blind to energy flow. Bond graph techniques, on the other hand, explicitly represent energy flow but are incapable of representing geometry.

Our techniques focus on the geometry of devices that have time-varying engagements (i.e., variable kinematic topology). Therefore, our techniques are complementary to the well-known design techniques for fixed topology mechanisms, such as the gear train and linkage design techniques in [2].

There has been a lot of recent interest in automating the design of fixed topology devices. A common task is the synthesis of a device which transforms a specified input motion to a specified output motion [10,19,21]. For the most part, these techniques synthesise a design using an abstract representation of behaviour, then use library lookup to map to implementation. However, because our library contains interacting faces, while theirs contain complete components, we can design interacting geometry, while they cannot. Like SKETCHIT, these techniques produce design variants.

To construct new implementations (BEP-Models), we map from qc-space to geometry. Joskowicz and Addanki [8] and Caine [1] have also explored the problem of mapping between c-space and geometry. They obtain a geometry that maps to a desired c-space by using numerical techniques to directly modify the shapes of parts. However, we map from qc-space to geometry using library lookup.

Our work is similar in spirit to research exploring the mapping from shape to behaviour. [9] uses kinematic tolerance space (an extension of c-space) to examine how variations in the shapes of parts affect their kinematic behaviour. Their task is to determine how a variation in shape affects behaviour; ours is to determine what constraints on shape are sufficient to ensured the desired behaviour. [5] examines how much a single geometric parameter can change, with all others held constant, without changing the place vocabulary (topology of c-space). Their task is to determine how much a given parameter can change without altering the current behaviour; ours is to determine the constraints on all the parameters sufficient to obtain a desired behaviour.

More similar to our task is that of Faltings and Sun [6]. They describe a system that modifies user selected parameters until there is a change in the place vocabulary, and hence a change in behaviour. Then, just as we do, they use qualitative simulation to determine if the resulting behaviour matches the desired behaviour. They modify c-space by modifying geometry; we modify qc-space directly. They do a form of generalisation by generating constraints capturing how the current geometry implements the place vocabulary; we generalise further by constructing constraints that define new geometries. Finally, our tool is intended to generate design variants while theirs is not.

Our work (see also [13, 15, 17]) builds upon the research in qualitative simulation, particularly the work in [3], [7], and [12]. Our techniques for computing motion are similar to the constraint propagation techniques in [18].

26.5 Future Work

As Section 26.3.1 described, the current SKETCHIT system can repair a limited range of flaws in the original sketch. We are continuing to work on techniques for repairing more serious kinds of flaws.

Because there are only two properties in qc-space that matter (the relative locations and the qualitative slopes of the qcs-curves), to repair a sketch, even one with serious flaws, the task is to find the correct relative locations and qualitative slopes for the qcs-curves.

We can do this using the same generate and test paradigm described earlier, but for realistic designs this search space is still far too large. We are exploring several ways to minimise search, such as debugging rules that examine *why* a particular qc-space fails to produce the correct behaviour, based on its topology. The desired behaviour of a mechanical device can be described by a path through its qc-space, hence the topology of the qc-space can have a strong influence on whether the desired path (and the desired behaviour) is easy, or even possible. For example, the qc-space may contain a funnel-like topology that "traps" the device, preventing it from traversing the desired path. If we can diagnose these kinds of failures, we may be able to generate a new qc-space by judicious repair of the current one.

We are also working to expand the class of devices that SKETCHIT can handle. Currently, our techniques are restricted to fixed-axis devices. Although this constitutes a significant portion of the variable topology devices used in actual practice (see [12]), we would like to extend our techniques to handle particular kinds of non-fixed-axis devices. We are currently working with a commonly occurring class of devices in which a pair of parts has three degrees of freedom (rather than two) but the qc-space is still tractable.

26.6 Conclusion

We have developed a computer program capable of transforming a stylised sketch of a mechanical device into multiple families of new designs. To "interpret" a sketch, the program first identifies what behaviours the parts should provide. The program then derives constraints on the geometry to ensure it produces these behaviours. In effect, the program uses physical reasoning to understand the geometry. We have used the program to design a range of devices that includes a circuit breaker and a yoke and rotor mechanism.

One reason this work is important is that sketches are ubiquitous in design. They are a convenient and efficient way to both capture and communicate design information. By working directly from a sketch, SKETCHIT takes us one step closer to CAD tools that speak the engineer's natural language.

Acknowledgements

Support for this project was provided by the Advanced Research Projects Agency of the Department of Defense under Office of Naval Research contract N00014-91-J-4038.

References

1. Caine, M.E. (1993). The design of shape from motion constraints. Technical Report 1425, MIT AI Lab.
2. Erdman, A.G. and Sandor, G.N. (1984). Mechanism design: Analysis and synthesis, Vol. 1. Englewood Cliffs, NJ: Prentice-Hall.
3. Faltings, B. (1990). Qualitative kinematics in mechanisms. Artificial Intelligence 44:89–119.
4. Faltings, B. (1992). Qualitative models in conceptual design: A case study. In Reasoning with diagrammatic representations, papers from the 1992 spring symposium, technical report SS-92-02. AAAI Press, pp. 69–74.
5. Faltings, B. (1992). A symbolic approach to qualitative kinematics. Artificial Intelligence 56:139–170.
6. Faltings, B. and Sun, K. (1996). Faming: Supporting innovative mechanism shape design. Computer-Aided Design 28:207–215.
7. Forbus, K.D., Nielsen, P. and Faltings, B. (1991). Qualitative spatial reasoning: The clock project. Technical Report 9, Northwestern University, Institute for the Learning Sciences.
8. Joskowicz, L. and Addanki, S. (1988). From kinematics to shape: An approach to innovative design. In Proceedings AAAI-88, pp. 347–352.
9. Joskowicz, L., Sacks, E. and Srinivasan, V. (1995). Kinematic tolerance analysis. In 3rd ACM symposium on solid modeling and applications.
10. Kota, S. and Chiou, S.-J. (1992). Conceptual design of mechanisms based on computational synthesis and simulation of kinematic building blocks. Research in Engineering Design 4:75–87.

11. Narayanan, N.H., Suwa, M. and Motoda, H. (1994). How things appear to work: Predicting behaviors from device diagrams. In Proceedings AAAI-94, pp. 1161–1167.
12. Sacks, E. and Joskowicz, L. (1993). Automated modeling and kinematic simulation of mechanisms. Computer-Aided Design 25(2):106–118.
13. Stahovich, T.F. (1996). SKETCHIT: A sketch interpretation tool for conceptual mechanical design. Technical Report 1573, MIT AI Lab., March.
14. Stahovich, T.F., Davis, R. and Shrobe, H. (1996). Generating multiple new designs from a sketch. In Proceedings of the thirteenth national conference on artificial intelligence, pp. 1022–1029.
15. Stahovich, T.F., Davis, R. and Shrobe, H. (1997). Qualitative rigid body mechanics. In Proceedings of the fourteenth national conference on artificial intelligence.
16. Stahovich, T.F., Davis, R. and Shrobe, H. (1998). Generating multiple new designs from a sketch. Artificial Intelligence 104(1–2):211–264.
17. Stahovich, T.F., Davis, R. and Shrobe, H. (2000). Qualitative rigid-body mechanics. Artificial Intelligence 119(1–2):19–60.
18. Stallman, R.M. and Sussman, G.J. (1976). Forward reasoning and dependency-directed backtracking in a system for computer-aided circuit analysis. Technical Report Memo 380, MIT AI Lab., September.
19. Subramanian, D. and Wang, C.-S. (1993). Kinematic synthesis with configuration spaces. In 7th international workshop on qualitative reasoning about physical systems, pp. 228–239.
20. Ulrich, K.T. (1988). Computation and pre-parametric design. Technical Report 1043, MIT AI Lab.
21. Welch, R.V. and Dixon, J.R. (1994). Guiding conceptual design through behavioral reasoning. Research in Engineering Design 6:169–188.

27. Diagramming Research Designs

Herman J. Adèr

In the social sciences different graphical methods have been developed to represent research designs. The chapter starts with a review of some of these methods: figures often turn out to be ambiguous, in particular where the interpretation of the arrows between nodes is concerned. Furthermore, rendering the placement in time of the various operations stays mostly implicit, while indication of strategy is not depicted graphically. A *functional notation* is sketched which allows the drawing of well-defined diagrams using a graphical node-and-arrow language. An alternative to the representation of an experimental design is given first and, at the end, a diagram of an actual survey study. Finally, the consequences for various fields are discussed. This kind of diagramming could be beneficially used in methodological consultation and education. Modelling meta data and meta analysis are other potential application fields.

27.1 Introduction

What may be the reason to represent methodological concepts in a diagram? It is not difficult to think of situations in which a reliable and easy-to-use formal representation could be beneficial. For instance, since *planning* largely determines the research design and thus the data analysis and the interpretation of the results, an unambiguous representation of the research plan could help to avoid design flaws.

An example is the design of a protocol for a clinical trial in medicine. A diagram could help to spot ambiguous points in the whole procedure during the trial and specify it in complete detail. With the representation of methodological knowledge two concepts play a key role: *context* and *time*. Of course, the notion of context is important in many other fields. What makes this case a special one is that we can indicate what the context consists of. Apart from ideas and notions from applied statistics, many elements of the context of a research design originate from the research field (the subject matter field) in which the basic research question was posed.

As to the notion of *time*, for many research designs the temporal aspect is essential, be it to study variability over time (in reproducibility studies and studies on responsiveness) or to study development of a phenomenon over time. This may be on a per person basis as occurs in a repeated measurement design or in a more epidemiological sense in a cohort study. Correspondingly, a host of statistical methods is available to handle the resulting data. The analysis of repeated measurements, Trend analysis, Time series analysis, Longitudinal analysis, and the analysis of survival data are all well-studied statistical techniques, specially devised to analyse temporal data. We will touch upon this issue in the section on the Representation of Time. Since these notions are so important, both Context and Time should be expressible in any representation system that tries to represent methodological knowledge.

In an attempt to apply semantic concepts to the analysis of multimedia systems and multimodal communication systems Stenning and Inder give a nice exposition of different families of representation systems [7]. In this chapter, we restrict ourselves to a particular kind of diagram, called *planar graphs*. These are graphs placed in the two-dimensional plane that have as their building stones *nodes* and *edges* between nodes.

In Section 27.2 different ways to represent statistical and methodological concepts are reviewed. We adhere to the way of drawing graphs that is customary to the field they are taken from and discuss the intrinsic meaning of the diagrams and eventual ambiguities that may occur. In later sections, we use *functional notation*. A succinct description of this notation is given in the section by that name.

Generally speaking, diagrams are supposed to be self-explaining and easy to comprehend and implicitly it is assumed that all elements shown as well as their interrelations are well defined. It is not always realised that the formal correspondence between a diagram and its meaning may be no trivial matter at all. One of the aims of this chapter is to clarify where and when graphical representation methods can be usefully applied in research design without leaving room for ambiguity.

27.2 Approaches in the Social Sciences and Medicine

Several approaches have been developed in the realm of social science research, but we restrict discussion to graphical methods to represent experimental design and path diagrams as occur in structural equation modelling.

Experimental Design. Figure 27.1 shows the way Cook and Campbell [3] indicate the designs of (quasi)-experiments. Picture (a) indicates the basic experiment: observation followed by treatment, followed by another observation. In (b) a control group is present, on which the same observations are done, but no treatment is given. The dashed line indicates that the experimental groups are not randomly formed. In picture (c) the precondition is

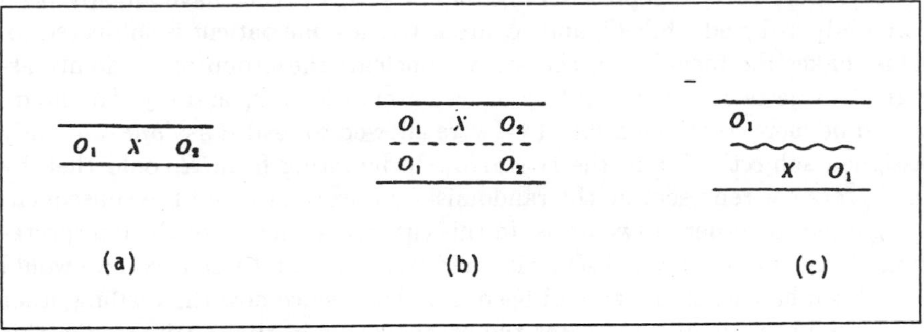

Figure 27.1. Notation to indicate quasi-experimental designs. O: Observation, X: Treatment. (a) Observation, treatment, observation (b) Non-randomised design with a control group (c) Design in which the precondition is observed in a different (but comparable) population.

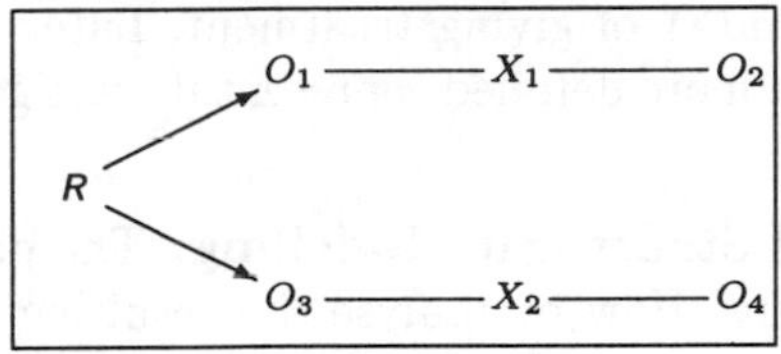

Figure 27.2. Experimental design: before-after two-group design. R: Randomly assigned subjects; O_i: Observation; X_i: Treatment (original caption).

observed in a different (but comparable) population. Note that the figure differs from other figures in this section in that it is not built from nodes with interconnected edges. A modern graphical representation of an experiment is given in Figure 27.2 [6]. This representation gives an impression of the general structure of a randomised clinical trial (see [6] for several alternative designs). After randomisation, subjects are assigned to either one of two treatment arms.

Both Figure 27.1 and 27.2 are ambiguous (although they may be effective if benevolently interpreted). In particular, in Figure 27.1, information on the structure of the experiment is intermingled with information on experimental units: in Figure 27.1 dashed and wavy lines indicate population characteristics, while the figure itself describes the temporal arrangement of the observations.

In Figure 27.2, R represents a group of subjects that have already been randomly assigned while O_i and X_i are activities *one* patient is subjected to. This makes the meaning of the arrows unclear: the group of randomly assigned subjects cannot possibly *produce* observations O_1 and O_3. The figure would be more comprehensible if we were allowed to read R as "*one* randomly assigned subject". R and the two arrows originating from R could then be interpreted as representing the randomisation procedure and the subsequent assignment to either of two arms. In this case we are left with the interpretation of the links between observation and treatment and *vice versa*: we would have been happier if arrows had been used here, since now the reading order in the figure implicitly indicates the arrangement in time. Admittedly, this is also indicated by the indices of observations and treatments: odd indices indicate the first administration while even ones indicate the second administration. But when we compare this to Figure 27.1, in which comparable observations have the same index, it is no longer clear whether O_1 and O_3 indicate the same kind of observation (this also holds for O_2 and O_4).

To end this niggling discussion, note that the nodes in this figure do not refer to objects but rather to activities, R being the act of randomisation, O that of observing and X of giving treatment. Later on, in the section on Functional Notation, a more detailed representation is given using Functional notation.

Path Diagrams and Structural Modelling. The previous two examples concern research designs. How to analyse the resulting data does not immediately follow from the figures. In this paragraph we mention diagramming methods that directly relate to a specific statistical analysis technique. Often, variables collected during an experiment or a survey can be divided in two distinct categories: *dependent* variables which commonly represent the measurements of interest of the study, and *independent* variables (also called "covariates", "predictors", "effect modifiers" or "confounders" depending on the field of research and main focus of the study) which in some way are assumed to influence the dependent variables.

Regression models are often used to statistically analyse this kind of configuration. Wright [8] introduced so-called "path diagrams", to visualise regression models. In Figure 27.3 an example is given of a path diagram. The objects that occur in this particular figure are variables Z_i and Y and the error term ε. The ρ_{ij}s labelling the bidirectional arrows are product-moment correlation coefficients indicating the strength of association between covariates Z_i. The path coefficients p_{Y_i} and p_{Y_ε} indicate the strength of the influences of the covariates and unsystematic error on the dependent. Thus, if we have to formulate what meaning should be attached to the arrows in the figure, we could read the bidirectional arrows as "is associated with" and the one directional arrows as "is influencing".

Wright's work led to the introduction of "Structural Equation Modelling" (SEM by abbreviation) in which similar graphical representations are used.

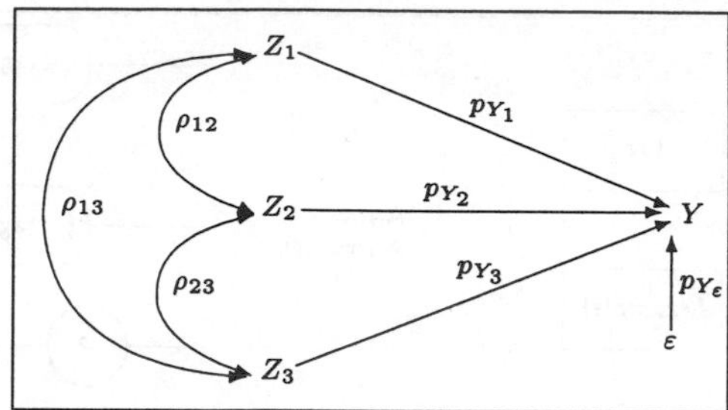

Figure 27.3. Example of a path diagram of a regression model. Y: dependent variable; z_i: independent variables (predictors); ρ_{ij}: correlation coefficients; p_{Y_i}, p_{Y_ε}: path coefficients.

Initial interest in the technique was raised when some authors tried to assess causal relationships and used structural equation modelling for it. This direction is nowadays called "causal modelling".

SEM offers a wide variety of modelling possibilities: confirmatory factor analysis, regression analysis and perhaps most importantly, models with latent parameters, all allowing for an easy graphical representation. It is a way of analysing in which both methodological and statistical considerations play an important part. As such, this field has much potential when it comes to the formulation of the basics of research methodology and the corresponding graphical representation.

27.2.1 What Should an Ideal Graphical Representation Look Like?

In Adèr [1, (Ch. 5)] many aspects are mentioned that would make a formalism to express methodological concepts more useful. Although the formulation there is more general than for graphical representations alone, it fully applies to graphical representations:

- *Recognisability.* The representation should look familiar for those who use it. Since statisticians and methodologists have sufficient knowledge of mathematics, mathematical indications and formalisation methods may be used.
- *Rough descriptions.* The representation should allow to refrain from full detail. Rough indications should be possible that can later on be specified in a more detailed way.
- *Handling complex pieces of information.* The previous point entails that it should be possible to label larger pieces of information of high complexity. No further specification of the complex internal structure should be needed.

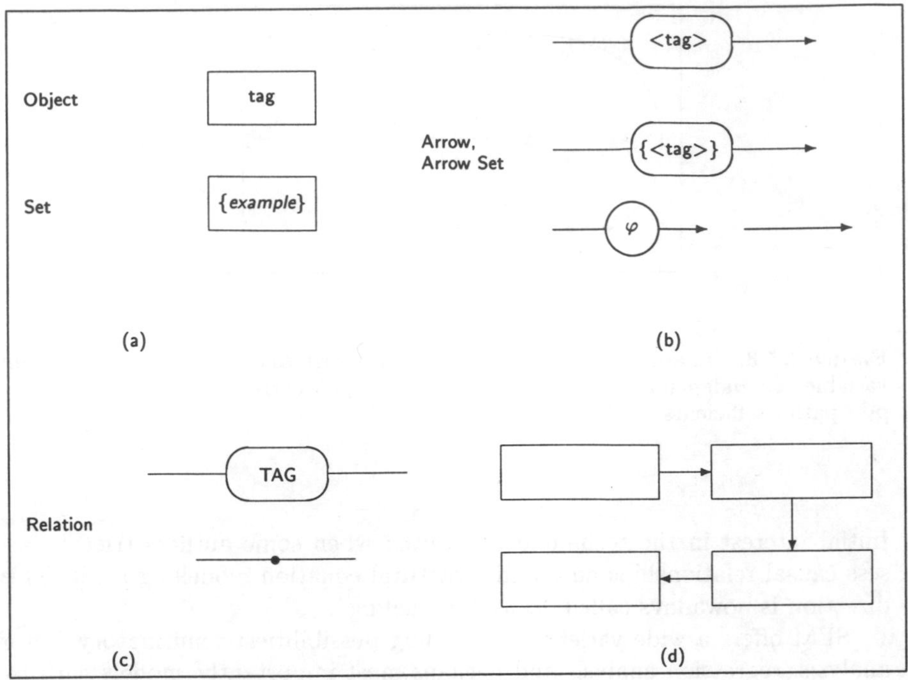

Figure 27.4. Basic elements of F-graphs. (a) Objects and Sets; (b) Arrows and arrow sets; (c) Relations; (d) Indication of time dependence

- *Extensibility.* It must be simple to adapt or extent the graphical representation to a specific research domain, although the basic elements and operations should stay the same.

In the introduction we emphasised the need to represent the context dependence. We also argued that being able to express the time dependence of notions was essential. Closely related and maybe more familiar to AI researchers is the need to represent strategic knowledge. And similarly, any representation should allow us to indicate logical inference processes.

27.3 Functional Notation

In Adèr [1, Ch. 5] so-called *Functional notation* (or *F-notation* for short) is developed. It has a well-defined graphical variant. The graphs drawn under this regime are called *F-graphs.* The principles of these F-graphs are briefly described in this section. For a more principled discussion the reader is referred to the original source.

Two basic points of the definitional framework of F-notation should be particularly mentioned. First, *Sets are assumed to be finite.* This is no severe restriction and it has to do with the finite nature of graphical representations. Secondly, *Arrows having as a domain a set of arrows and arrows having as image an arrow are allowed.* Why this is useful will become clear in the sequel.

Basic Elements of F-Graphs. Figure 27.4(a) shows how objects and sets are represented. An object is indicated by a box with a "tag" (label) placed inside. A set is indicated by a box with curly brackets ({}) around the tag. The tag in this case is called an *example*, an element of the set to which the other elements are alike. Objects are always boxed.

Figure 27.4(b) shows various representations of arrows. Like an object, an arrow may be equipped with a tag placed in an oval or a circle. Tags of arrows are surrounded by angle brackets: `<tag>`. These tags can be used to indicate the operation the arrow represents ("produce", "go to next" and so forth). To indicate a set of arrows (or: an *Arrow Set*), curly brackets are used as in set denotations. Since a function corresponds to a special arrow set, functions may be indicated either this way or by an arrow with a Greek letter tag. Arrows can also be untagged.

Figure 27.4(c) gives examples of the way a relation is indicated. When it is untagged, the link is marked with a dot, needed to distinguish it from parts of arrows. Tags of relations are in capitals. To indicate that an object is time dependent it is provided with an index indicating the time point to which it is related (see Fig. 27.4(d)). Thus, only distinct time points can be indicated. When objects occur at the same time point, they have the same index. In the figure, arrows indicate the transition from one time point to the next. It is very well possible that more than one set of time points is used in a figure. To indicate this a different set of indices may be used, for instance instead of numbers in circles we often use boxed numbers.

Figure 27.5 shows how different logical connections are represented. The top diagram shows an object that has a predicate attached to it as an attribute. Predicates also function in implications which are simply indicated by an arrow with a $\Longrightarrow$ tag.

The bottom diagram of Fig. 27.5 gives the representation of an action that is followed by one of two alternative actions, dependent on the value of two conditions that exclude each other. In this case the arrow set consists of two arrows, labelled `Cond` and `Not Cond`. A $\vee$-symbol indicates that a choice has to be made from the arrows in the arrow set. The arrow set together with the $\vee$-operation is called a *Chooser*. When no ambiguity can arise the $\vee$-symbol is left out.

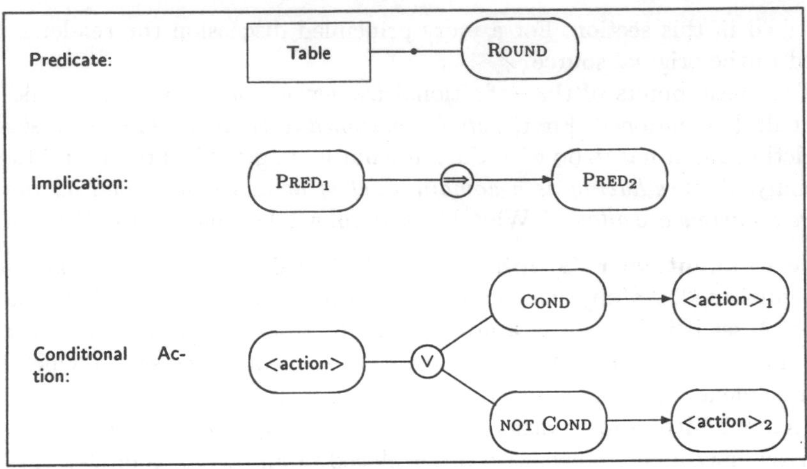

Figure 27.5. Predicates, implications and choosers.

27.3.1 Example: Design of a Clinical Trial

As an example, in Fig. 27.6 an alternative representation of a clinical trial is given (compare Fig. 27.2). The upper panel of the figure represents the inclusion procedure. According to inclusion and exclusion criteria {crit.} patients or subjects are included in or excluded from the trial. $< ic >$ is a chooser that indicates that informed consent is asked. Since the inclusion is the first step, an index $\boxed{1}$ is placed in the oval that indicates the randomisation process. The next panel indicates the randomisation process. Like $< ic >$, $< r >$ is a (tagged) chooser: It assigns an included patient to either one of two mutual exclusive alternative treatment arms, A_1 or A_2. The two lower panels of the figure indicate the study design for one patient participating in either one of the two treatment arms of the trial. As before, the small indices in boxes indicate subsequent points in time.

Two things should be noted: First, all four panels consist of labelled ovals since each one indicates an activity. Secondly, all these figures describe the situation *for one subject only.* This is typical for a randomised clinical trial, since in this type of design patients come in one after another and are randomised. Thus, time points usually differ for each subject when considered in "absolute" time. If we need to express this, each time index can be provided with a subject index to indicate that the time sets are different. But in most cases, it is assumed that the time sets of different patients can be identified. This at least is assumed in most analysis techniques. (Note that, in contrast, in survey research random sampling takes place beforehand.)

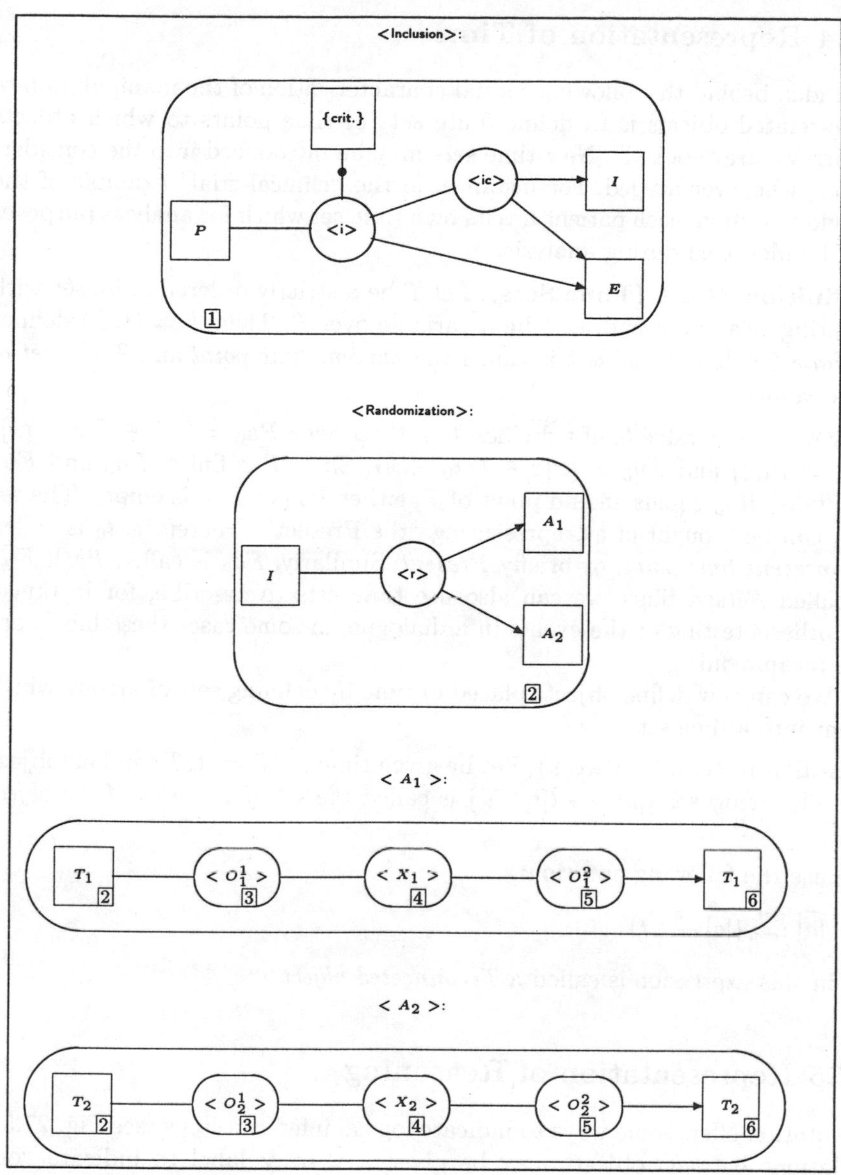

Figure 27.6. F-graph of a clinical trial design. P: Patient. <i>: Inclusion Procedure. <ic>: Ask for Informed Consent. I, E: Included or excluded patient. <r>: randomisation Procedure. A_1, A_2: first and second treatment arm. T_i: Subject assigned to treatment group. O_i: Observation in group i. X_i: Treatment in group i.

27.4 Representation of Time

The idea behind the following formal characterisation of the manipulation of time-related objects is to define finite sets of time points to which objects or arrows are "hooked". New time sets may be introduced into the considerations whenever needed. For instance, in the "clinical trial" example of the previous section, each patient has its own time set which for analysis purposes can be identified during analysis.

Definition 27.4.1 (Time Sets). Let T be a strictly ordered finite set with ordering relation $\prec$ and let $\mathbf{t}$ be a variable over T. Then $\mathcal{T} = (\mathbf{t}, T)$ defines a *Time Set.* The variable $\mathbf{t}$ is called the *current time point* and T *the set of time points.*

Any particular value t_0 of $\mathbf{t}$ divides $\mathcal{T}$ in three sets: $Pa_0 = \{x|x \in T, x \prec t_0\}$, $Pr_0 = \{[t_0]\}$ and $Fu_0 = \{x|x \in T, t_0 \prec x\}$. Since T is finite, Pa_0 and Fu_0 are finite. If t_0 equals an end point of T, either Pa_0 or Fu_0 is empty. The set Pr_0 can be thought of as representing "the Present". Therefore, t_0 is called *the present time point*, or briefly *Present.* Similarly, Pa_0 is called *Past*, Fu_0 is called *Future.* Since we can also use time sets to describe, for instance, hypothesis testing or the events in a dialogue, in some cases these labels are less meaningful.

We can now define objects placed in time by defining sets of arrows which point into a time set.

Definition 27.4.2 (States). Let be given time set $\mathcal{T} = (\mathbf{t}, T)$ and an object $[a]$. The arrow set $\{[a] \longrightarrow t | t \in T\}$ is called *the set of* $\mathcal{T}$*-states of the object* $[a]$.

We use the following notation:

$$\widehat{[a]} \overset{\text{def}}{::=} \{[a] \longrightarrow t\}$$

$\widehat{[a]}$ in this expression is called a $\mathcal{T}$*-connected object.*

27.5 Representation of Reasoning

F-notation offers some ways to indicate logical inference steps (see Fig. 27.5): Relations between objects may be given a $\vee$ or & label to indicate "or" and "and" relationships. Similarly, by labelling a Chooser, an arrow set can be provided with a logical operator that indicates which arrows should be applied. Thus, Reasoning by graphical means is well possible and eventually instructive, but to express complicated logic relationships the language of logic is preferable. It is specifically developed for this and provides efficient and parsimonious ways to express whatever complicated train of thought. A lot of research has been done, and is still going on, on the applicability of different logics for knowledge representation. Many aspects and possibilities are discussed in [5].

Inference Mechanisms. How may inference steps playing a role in Reasoning on methodological matters be formally handled? To keep things simple, let us assume that logical inference in this field obeys "at most" first-order predicate logic.

Logical inference on a set of statements $\mathcal{S}$ may then be described using rules of the form `if` T_1 `then` T_2 (or as implications of the form $T_1 \Rightarrow T_2$). In this, T_1 and T_2 are sets of predicates that can be attached to $\mathcal{S}$.

Such rules indicates that if the *antecedent* T_1 is `true`, then it follows that the *consequent* T_2 is also `true`.

Using the implication operation repeatedly, a series of inference steps can be indicated:

$$T_1 \Longrightarrow T_2 \Longrightarrow T_3 \ldots$$

If special sets of predicates T_{start} and T_{goal} are defined, the inference process aims to find a series of steps that can connect these two.

In view of definition 27.4.2, it seems natural to consider $\widehat{\{T_i\}}$, i.e to make $\{T_i\}$ $\mathcal{T}$-connected. This corresponds to the idea that inferential reasoning proceeds in steps, put one after another, leading to some pre-defined target statement.

The time points of $\mathcal{T}$ then correspond to the states $\{T_i\}$ assumes during the inference process.

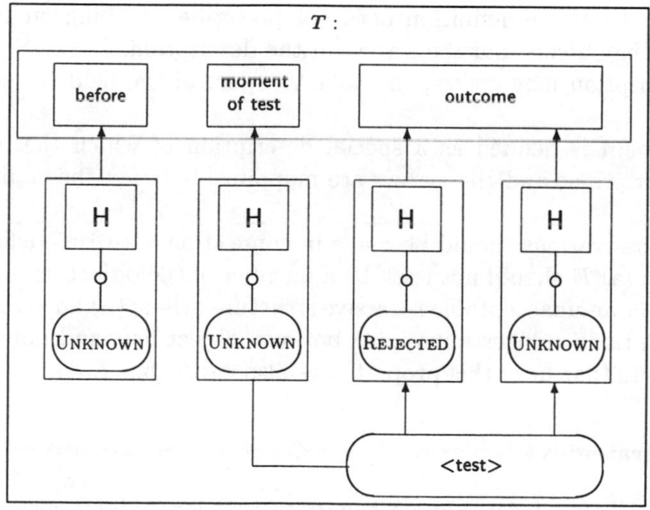

Figure 27.7. Hypothesis testing.

Example: Hypothesis Testing. An example of a short chain of inference steps occurs in hypothesis testing: The underlying set of statements $\{H\}$ corresponds to the hypotheses formulated. Attributes that are chosen from the set {`Unknown`, `Rejected`} may be attached to each H (see Fig. 27.7).

The testing procedure `<test>` consist in a Chooser that either maps on a rejected or an unknown hypothesis. The time set $\mathcal{T}$ that is used here contains three points: T = {`before, moment of test, outcome`}. Hypotheses that, after the test, have still an attribute `Unknown` cannot be rejected; in the other case they are falsified.

27.6 Descriptions and Domain

The line of thought in this section is as follows. We start off by assuming that in any field of interest notions and reasoning about notions can be formulated in terms of arrows and objects. We then show that to represent concepts like "Strategy" or "Context" boils down to finding the proper set of arrows that can indicate strategic steps or the contextual inclusion.

We first introduce two new notions, *Description* and *Domain*.

Definition 27.6.1 (Description). A *Description* S is a pair (O, A) consisting of a set of objects O and a set of arrows A. O is called the *Object List* and A the *Arrow List* of S.

Note that the above definition does not prescribe anything on the specific nature of the objects and the arrows in the description.

A description may correspond to any aspect of the field we want to describe.

A *Domain* is defined as a special description of which the objects are again descriptions and the arrows are mappings between the composing descriptions.

Two observations should be made in connection with this definition of a domain D: (a) D should not itself be a member of the object list of any of its members, or an unacceptable recursive structure arises; (b) an arrow between two descriptions induces a mapping between object lists and between arrow lists. Arrows that have this property are also called *functors*.

27.6.1 Strategies

Dube and Weiss [4] describe an expert system for maintenance of message trunks in a telephone network. In their article, strategic knowledge is represented by means of a so-called "State Transition Model". States of a system are represented as nodes in a directed graph. Obviously, this approach is quite general and may be easily transmitted to, for instance, the planning of the data analysis phase of a research project.

Figure 27.8 gives an example (it is constructed after Fig. 4 of Dube and Weiss). Each node may refer to a complicated procedure in which several steps are taken. For instance, S_{start} and S_{goal} may stand for "data cleaning" and "report writing", respectively. Likewise, the choice between subsequent nodes (here indicated by a chooser) may involve complicated considerations.

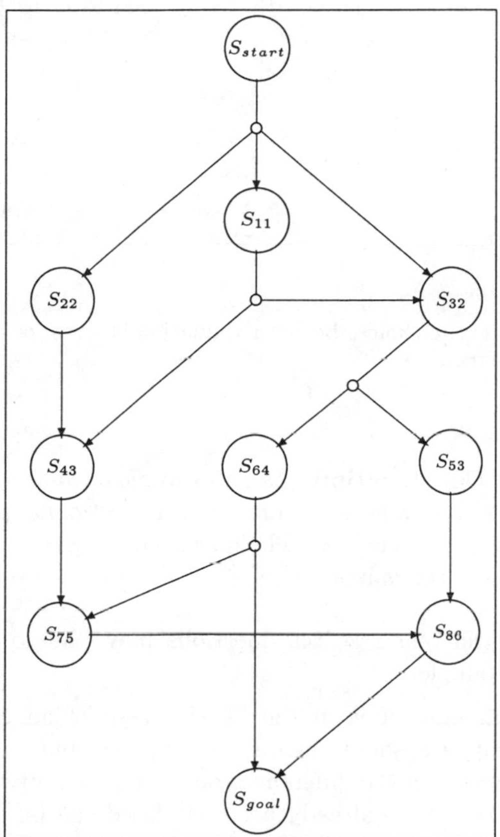

Figure 27.8. Representation of strategy as a transition network. The goal S_{goal} may be reached via different paths. The second index indicates the time point.

A strategy is now defined as a path in this graph from start to goal. Some alternative strategies are $S_{start}, S_{11}, S_{43}, S_{75}, S_{86}, S_{goal}$ and $S_{start}, S_{32}, S_{64}, S_{goal}$. (Note that the second index indicates the time point at which the activity takes place.)

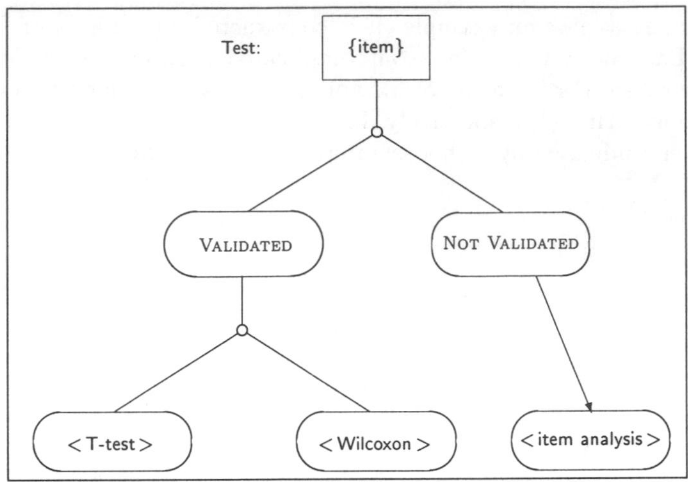

Figure 27.9. Strategic choices between validation of a test or using test scores as a measurement instrument.

Example: Test Construction. As an example of some steps in a methodological analysis strategy, let us assume that a questionnaire has been administered in two groups of subjects. The questionnaire consists of several tests, some of which have been validated before, while others have been constructed anew.

In this case, different research questions may lead to different analysis strategies. For example:

- when the main interest is in the development of an new measurement instrument, emphasis should be on item analysis; but
- when the interest is in the differences between subgroups those tests of the questionnaire that have already been validated can be compared using a T-test or a non-parametric test.

In the graphical representation given in Fig. 27.9, a particular test (`Test`) consists of items ({item}). The test is item analysed in case a new instrument has to be developed. When there are only validated items, the groups can be compared using a test for mean differences.

Note that the figure does not express anything about the actual data analysis procedure, which in both cases may involve elaborate methodological considerations.

27.6.2 Context

We start this section with an example (see Fig. 27.10).

To arrive at a statistical model, a research question is reformulated in methodological terms and the specification of a statistical model is based on that. In the process, the context of each step is in the context of the previous ones:

- The research question is formulated in terms of the research domain.
- The methodological reformulation has this question as its context and thus also the research domain.
- The context of the statistical formulation and the subsequent data analysis includes both methodological and research domain considerations.
- And finally, to interpret the results, all of the above contexts play a more or less important role.

Generally speaking, any (methodological) structure S_0 receives a particular meaning stemming from its context. The question how to represent this embedding is a question after the best way to *import* elements of the context C_{S_0} into the structure we want to describe. To do that, it is necessary to first be able to describe both S_0 and its context in some elementary way.

Let us (optimistically) assume that aspects of any domain (in our case: a research domain) may be described using objects and arrows, so that we can think of it in the way we defined it before: as a finite collection of descriptions, each determined by an object list and an arrow list. For convenience, assume that S_0 can be specified as a description (otherwise, we have only to look for a description of which S_0 is a member). The idea of the following definition is that when S_0 is one of the descriptions of a domain $\mathcal{M}$, then its context is the complement relative to $\mathcal{M}$.

Definition 27.6.2 (Context). The *context* C_{S_0} of a description S_0 in domain $\mathcal{M}$ is the union of the other descriptions S in $\mathcal{M}$:

$$C_{S_0} = \bigcup_{S \in \mathcal{M} \setminus S_0} S \tag{27.1}$$

Thus, descriptions constitute each other's context. Since C_{S_0} consists in a union of descriptions, we may consider the unions of the object lists and arrow lists, too. Let's call them $O_{C_{S_0}}$ and $A_{C_{S_0}}$, respectively.

The operation of importing elements from D_{S_0} now consists in finding a proper mapping of $(O_{C_{S_0}}, A_{C_{S_0}})$ into (O_{S_0}, A_{S_0}). (Note that when the context of S_0 is not particularly influential, only a few elements of $O_{D_{S_0}}$ need to be imported.)

In this way the relatively vague notion of context may be represented in a strictly formal way. In Fig. 27.10 these mappings are indicated by the arrows ρ and ι (ρ for reformulation; ι for interpretation.)

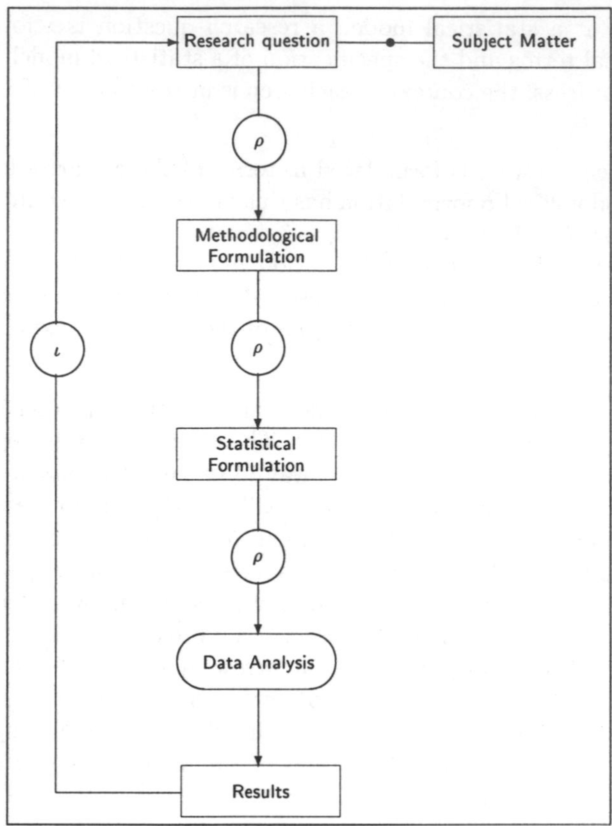

Figure 27.10. Subject matter knowledge, research question, methodological and statistical formulation form each others contexts. ρ: Reformulation; ι: Interpretation.

27.6.3 Example: Inclusion, Sampling and Analysis procedure in a case–control study

This example concerns the design and data collection of a case–control study by Bleiker et al. [2] using data from a cohort study conducted in a middle-sized Dutch town. The aim of the study is described as follows:

> The self-report of personality traits after diagnosis of breast cancer may have been influenced by the distress raised by this life-threatening disease. The aim of the study is to investigate to what extent the diagnosis of breast cancer, and its consequences, influences the self-assessment of personality traits by patients.

Table 27.1. Population and sample (Nijmegen)

Invited (1989–1990)	30,000
Screened (1989–1990)	16,465
Completed SAQ-N (1989–1990)	9,705
Completed SAQ-N for the second time (1990)	822

In two subsequent years (1989 and 1990) all female inhabitants of Nijmegen (a middle-sized town in the eastern part of the Netherlands) were invited to participate in a breast cancer screening and asked to complete the *Self Assessment Questionnaire-Nijmegen* (SAQ-N by abbreviation). About $1\frac{1}{2}$ years after the first administration, a subsample of 822 women completed the questionnaire for a second time. Among them women, diagnosed for breast cancer who received surgery and radio therapy during 1989–1992. Table 27.1 gives an overview of the procedure and the number of women involved in each step.

Figure 27.11 visualises the procedure in F-notation. Figure 27.12 gives more detail on the screening procedure.

Note that for the inclusion criterion <`woman`> and the criterion for the age of the subjects are implicitly represented by a set of arrows. The age criterion produces the initial sample. In both the first and the second screening round, cases are diagnosed. Only data of Ca_1 (cases of the first screening) are included in the rest of the study, since the SAQ-N is administered twice to them. However, (part of) the data of the first screening round (Q_1) may be used to do, for instance, an initial item analysis.

`Ca`, `Oth` and `Q` are output of the `screening` procedure. Although there are several relations between these "objects", these are not indicated here since they are not relevant to the study design. Furthermore, `screening` associates a woman with a "filled in questionnaire" and it indicates if she is a case or not. Note that <`diagnosis`> is a Chooser.

The screening procedure is only meaningful if substituted in Fig. 27.11. Thus the sampling procedure as depicted there constitutes its context. The formal inclusion of the description of the screening procedure into the sampling procedure involves the use of a functor that appropriately maps the parameters `woman`, `Case`, `Other`, `Q` and the <`SAQ-N`> operation (administration of the `SAQ-N`). The screening procedure, the administration of the SAQ-N and the output of the diagnosis are are all time-related. Only two time points are needed, corresponding to the two screenings.

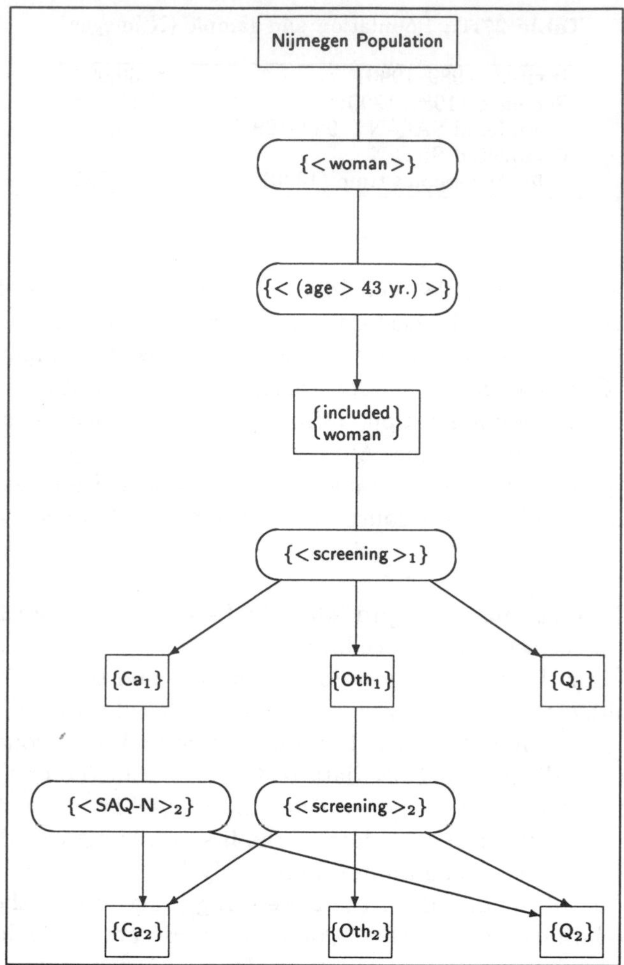

Figure 27.11. Overview of the inclusion criteria and sampling procedure of the Nijmegen study. Q: Filled in questionnaire; Ca: Case; Oth: Other woman.

27.7 Discussion

We discussed several examples of graphical representations of study designs as are commonly used in the social sciences and medicine. Often basic elements are objects and edges between objects. As we saw in the discussion of Fig. 27.2, sometimes the nodes do not indicate static entities (objects) but *activities* (like "randomise" or "observe"). Since directed edges (arrows) are commonly used to indicate activities, it follows that the distinction be-

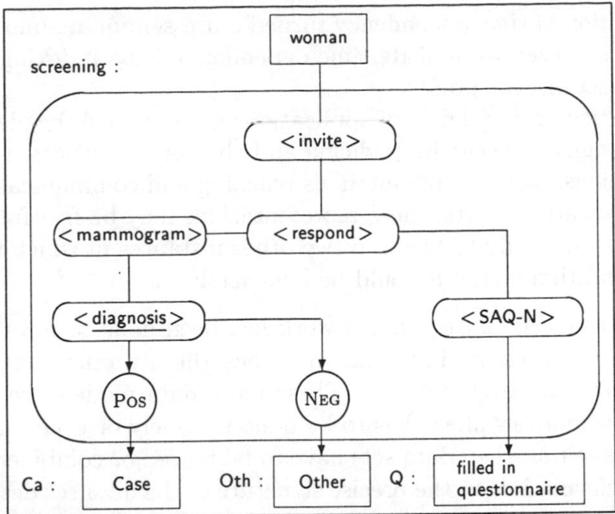

Figure 27.12. The screening procedure in the Nijmegen study.

tween objects and arrows is not strict. The ambiguity of the figures seems to be mostly caused by the *meaning* the reader has to attach to these arrows. Sometimes they simply mean "produces", "go to the next step" or "pass on information", but sometimes completely different interpretations are needed like "influences" or "is dependent on". Usually, this meaning is not explicitly given with the figure and, for various reasons, assumed self-evident.

From a methodological point of view, path diagrams as used in structural equation modelling come closest to what a notation should look like, since they suggest a kind of causality that is relevant for the formulation of research design. However, descriptions of time dependency are not altogether adequate. Nor is it self-evident how to indicate context dependency.

We introduced a particular diagramming technique based on Functional Notation. The basic elements are objects and arrows. Although some elements are special to this technique like the indication of choices and the explicit indication of time points, there is much similarity to other diagramming techniques. F-graphs allow to include concepts that are only roughly defined (see for instance Fig. 27.10). It is also possible to include notions that have a complex internal structure immaterial for the enunciation of the moment that are fully specified afterwards (see Fig. 27.12, which is a precise specification of the screening procedure that appears in the arrow sets of Figure 27.11.)

The diagrams in this chapter may serve as a demonstration of the "Extensibility" requirement mentioned in the section on the features of an ideal graphical represention: They offer a representation in a variety of situations, from "Hypothesis testing" in Fig. 27.7 to the strategies represented in Fig.

27.8. Indication of time dependency turned out essential in almost all figures. The solution chosen to indicate time dependency is by indexing referring to an implicit set of time points.

As a particular useful field of application we mentioned the planning phase of study designs as occur in medicine and the social sciences. Obviously, in other situations, like research methods teaching and communicating research designs in research papers, these representations may be fruitfully applied.

We want especially to mention two other instances in which these graphical representation methods could be beneficial:

- *Meta data models.* Lately, much work has been done to develop methods to formally represent data that describes the structure and conceptual meaning of a research data set. These meta data models are useful when part of the analysis phase has to be done routinely or even automatically, and are essential when data sets have to be reused or combined, since they contain information on the precise structure of the data set described. This is an area where the (graphical) methods presented in this chapter could be fruitfully applied to represent the formal structure.
- *Meta research.* In this field, data often results from written research reports. In many cases, it is difficult to demonstrate that procedures applied in the studies are comparable. If precise scripts of the experiments or trials were available, a responsible combination of results could be made based on appropriate statistical procedures. (In contrast to research planning and meta data modelling, as yet not much work has been done in this area at the point of formalisation.)

References

1. Adèr, H.J. (1995). Methodological knowledge: Notation and implementation in expert systems. Phd thesis, University of Amsterdam.
2. Eveline Bleiker, E., Van der Ploeg, H.M., Adèr, H.J., Van Daal, W.A.J. and Hendriks, J.H.C.L. (1995). Personality traits of women with breast cancer: Before and after diagnosis. Psychological Reports 76:1139–1146.
3. Cook, T.D. and Campbell, D.T. (1979). Quasi-experimentation: Design and analysis issues for field settings. Chicago: RandMcNally.
4. Dube, R. and Weiss, S.M. (1989). A state transition model for building rule-based systems. International Journal of Expert Systems 2(3/4):291–329.
5. Genesereth, M.R. and Nillson, N.J. (1987). Logical foundations of artificial intelligence. Los Altos, CA: Morgan Kaufmann.
6. Judd, C.M., Smith, E.R. and Kidder, L.H. (1991). Research methods in social relations. (6th edn). New York: Holt, Rinehart & Winston.
7. Stenning, K. and Inder, R. (1995). Applying semantic concepts to analyzing media and modalities. In J. Glasgow, N.H. Narayanan, and B. Chandrasekaran (Eds), Diagrammatic reasoning: Cognitive and computational perspectives, Ch. 10. Menlo Park, CA: AAAI Press /MIT Press,
8. Sewel Wright, S. (1934). The method of path coefficients. Annals of Mathematical Statistics 5:161–215.

28. How to Build a (Quite General) Linguistic Diagram Editor

Jo Calder

In linguistics, diagrams have played a crucial rule in conveying key concepts, analyses and fragments of theories. We describe a design for an editor, *Thistle*, which allows the construction and manipulation of a wide variety of linguistic (and other) diagrams and a general method for attaching semantics to such diagrams. Our key insight is that, if we treat diagrams as conforming to some grammar, we can then make use of grammatical insight from the field of formal language theory in analysing problems and designing systems. Our design represents a generalisation of all systems proposed in the computational linguistic literature of which we are aware. We discus some technical issues and offer an illustrative range of applications for this design. Some applications give rise to intriguing results in the field of educational resources. The current implementation permits instances of the editor for a wide variety of linguistic theories. We give some complete specifications for significant classes of diagrams. Overall, we argue, "grammars for diagrams" is a perspective which we expect to be fruitful in the coming years.

28.1 Introduction

We propose a novel design, *Thistle*, for an editor for diagrams representing various kinds of linguistic information. We argue for the compromises we suggest as a trade-off between generality and usability. We demonstrate the latter property through a wide range of applications. For the sake of exposition, this chapter focuses on the relatively high-level design of the editor and does not exhaustively cover the details of our current implementation. For this, the reference documentation [5] can be consulted. All of the classes of diagram described or mentioned here are available as on-line demonstrations.

28.2 Motivation

Within linguistics and computational linguistics, diagrams play a crucial role in representing the content of theories (the use of trees to define inclusion hierarchies, for example), in standing as informal demonstrations of the truth of particular claims and, therefore, in sharing ideas with the community as a whole. Popular graphical devices include trees, attribute-value matrices (*AVMs*), e.g. [16], and conventions such as those used in Discourse Representation Theory (DRT) [12]. It has been clear for a number of years that the linguistic community would benefit from a general-purpose "diagram editor" allowing users to construct and manipulate diagrams. A large range of uses exists for such a program, including the debugging of existing grammars, the construction and delivery of teaching and drilling materials and the production of diagrams for publication in some media or other. From a more general perspective, such a system offers a way of defining and interacting with documents with complex structure.

Why hasn't the community produced such an obviously desirable program? First, change has been a characteristic of the technical devices used in many branches of linguistics, and it seems in principle impossible to predict which graphical conventions will gain currency in linguistic discourse and publications. Moreover, if diagrams vary in unpredictable ways, there might seem to be no hope of providing a uniform interface for the user. A consequence of these factors is that the implementation and maintenance costs of such a program appear unacceptably high, perhaps unquantifiably so.

There seem to be two responses to this situation. One response, as seen in the tree editors described in [15] and [4] and in the feature structure editor in [13], is to fix a relatively small amount of graphical devices and restrict the operations defined over, and potential combinations of, those devices (perhaps to the extent that only operations which don't violate consistency with respect to a particular grammar are allowed).

An alternative response is to aim for the generality of the kind seen in the general field of diagram editing and visual programming, of which [21], other papers from that source, and [14] are good examples. And, of course, constructing diagrams by hand in a generic drawing package represents a common, but *in extremis* measure. There are several reasons why, for our purposes, generality is a disadvantage.

First, generality in this context typically goes along with complexity in the mathematical objects to be depicted, often requiring the use of sophisticated layout algorithms, cf. [2]. Second, there is a corresponding complexity in the specification of diagrams. That complexity may require arbitrary computation to be performed and therefore demand the power of an unrestricted programming language to describe that computation. Finally, it seems to be an assumption of such approaches that the well-formedness of a diagram should equate with the consistency of the interpretation of that diagram in the domain represented by the diagram. See [17] for a clear statement of this position. This is much too strong a requirement in many cases of interest to us: one may wish to construct an inconsistent AVM, for example, precisely to verify that some other processor correctly detects the inconsistency. One may also wish to construct diagrams in formalisms which are undecidable, for example formulae in first- or higher-order logics. In that situation, it cannot make sense to ask an editor to enforce consistency. In the end, in an appropriately general system, it should be possible to decide on a case-by-case basis whether consistency with respect to the domain in question should be enforced. We discuss in the next section how the design we present here obviates these problems, and allows the inexpensive and portable implementation of an appropriately general editor.

28.3 Design

We provide in this section a high-level specification of the editor. Details of implementation are given later.

28.3.1 Assumptions

We make two basic assumptions. First, the well-formedness of diagrams is stated in terms of a context-free grammar. This point will be illustrated below. Such an assumption is entirely in accord with practice in the areas of the specification of syntax and semantics of linguistic and semantic formalisms, including the graphical conventions used by such formalisms. Second, there is a small set of graphical primitives to state the layout of diagrams and a means for labelling parts of diagrams. Our context-free assumption above means that, generally, we can require the layout problem to be deterministic for each proper subpart of a diagram and thus for diagrams as a whole, as well.

Graphical Primitives. Our current design makes use of three kinds of primitives. *Leaf* elements which specify the typeface in which to set a sequence of characters, for example `plain`, `italic`, etc. (a total of seven). *Shape* primitives surround a single figure, for example with brackets of various kinds or with a box or circle and so on (a total of five). *Layout* primitives arrange one or more figures into larger diagrams, and these provide for vertical, horizontal, tree and array layouts (six primitives).[1] So, leaf primitives are fully specified by a series of characters; layout primitives take one or more operands each of which may be any of the primitives; shape primitives require a single operand.[2] These primitives are selected on the grounds of generality, while preserving the property that layout is deterministic.

[1] In general, these primitives may take options to control details of layout, for example the selection of smaller or larger fonts, or alignment within layouts. In examples here, these options have been suppressed for clarity. Similarly, primitives for controlling the appearance of branches and horizontal and vertical padding are not described here. Available tree layouts include the "standard" vertical orientation commonly used in linguistic presentations, and horizontally disposed dendrograms (aka cladograms).

[2] Fig 28.4 is an example of a tree described fully using some of these primitives.

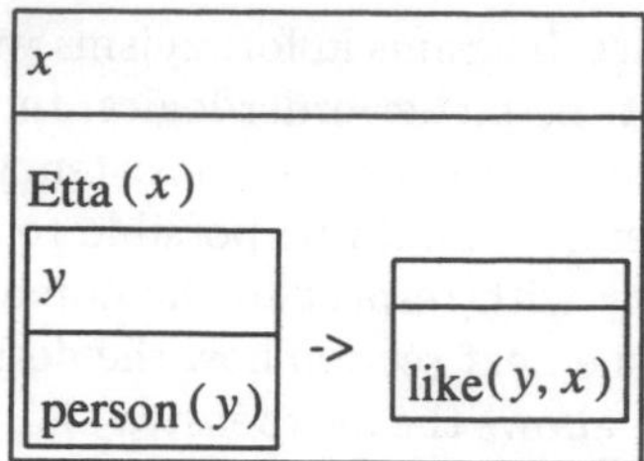

Figure 28.1. An example discourse representation structure.

Specifying Diagrams. In addition to specifying layout, we also need to indicate when a type of diagram has variable subparts, and what types of diagram may appear in those subparts. Let's take a particular example, namely the class of *discourse representation structures* (*DRS*; see e.g. [12]). An illustrative diagram is shown in Fig 28.1. The diagram can be taken as a representation of the meaning of the English sentence *Anyone who knows Etta likes her.* This class of diagrams shows a number of relevant properties. First, the diagrams are hierarchically organised. Second, organisation is also recursive. Third, the kinds of subparts that may appear at a chosen location in a diagram are dependent on, but not determined by, the immediate context in which the subpart is to appear.

In the case of DRSs, we wish to say that a *drs* consists of a *universe*, which is a collection of *referents*, and its *conditions*. The universe appears above the line of each box. Each of the conditions may be *atomic*, an *implication* or of still other types. As a point of terminology, where any number of diagrams may appear in a particular location, we will say that the diagrams that may occur there represent a *repeating type.*

Each of the elements in italic above indicates the type of a particular subpart of a larger diagram, and constitute a context-free rule relating a diagram and its subparts. In the abstract (i.e. ignoring details of layout) and with the usual interpretation of Kleene star,[3] we end up with the following characterisation:[4]

$$\text{drs} \rightarrow \text{referent}^* \quad \text{condition}^* \tag{28.1}$$

In order for the content of a diagram to be interpretable, we allow the subparts of a diagram to be named, for example (and again in the abstract):

$$\text{drs} \rightarrow \text{universe:referent}^* \quad \text{conditions:condition}^* \tag{28.2}$$

The names of subparts must be unique within any one type of diagram. All that remains is for such specifications to include layout information. A

[3] That is, zero or more subparts of the given type may appear.
[4] Details of the concrete syntax our prototype adopts are given below.

possible specification would then be as follows, where square brackets delimit sequences of specifications, and `hbox` and `vbox` provide horizontal and vertical dispositions.

```
drs → box(vbox([hbox(universe:referent*),
                line,
                vbox(conditions:condition*)]))
```
(28.3)

We term a collection of such statements, intended to characterise a particular kind of diagram, a *diagram class specification.*

In some cases, of which an example is the treatment of conditions in a DRS shown above, more than one type of diagram may appear in some position in a diagram. Above we saw that a condition in a DRS may be realised as an atomic condition or an implication. In this situation, one may specify a "union" of diagram types. Overall (and ignoring labels), a grammar of diagrams allows two kinds of production rules:

$$M \rightarrow C_1 \ldots C_m, m \geq 1 \\ N \rightarrow C_1 | \ldots | C_n, n > 1 \tag{28.4}$$

where M and N are non-terminal symbols and the rewrite for any non-terminal is unique. C is a non-terminal or terminal symbol. The first states that a diagram of type M consists exactly of subdiagrams of types $C_1 \ldots C_m$. The second, a diagram *union*, states that diagram types $C_1 \ldots C_n$ are alternative ways of realising a diagram of type N. It is clear that any context-free grammar can be rewritten so as to fall within this class. As we will see below, this choice of normal form contributes greatly to the simplicity of the editor's user interface.

In the case where a subpart of a diagram requires only a sequence of characters to complete it, the class specification may give a regular expression to describe patterns of characters that may appear. (See for example [23].)

The labelling of subparts of a diagram allows the content of a diagram to be represented in terms of sets of paths through the diagram. Let T be the set of types used in a diagram class, V the set of variable names, N the set of natural numbers, and S the set of all sequences over the character set in use. A path is then a sequence of elements of one of the following forms (where $t \in T$, $v \in V$, $n \in N$, and $s \in S$):

$$vt \quad vnt \quad vs \tag{28.5}$$

The first picks out the subpart v of a diagram and assigns a diagram type t to it. The second references the nth diagram in a repeating context and assigns a type to it. So, a path such as

```
conditions 1 implication left
```
(28.6)

refers to the LHS DRS in an implication which appears as the (say) second element in the conditions of a DRS. Similarly

```
universe 0 id "x"
```
(28.7)

identifies the content of the first referent in a DRS's universe. A further example is given in Section 28.6.

Ultimately, this type of specification is interestingly reminiscent of proposals for "rule-to-rule" semantics, for example in [9], where the interpretation (and in our case that can be taken to mean "graphical interpretation") of a structure is given in terms of a function of its subparts. More practically, one effect of the restriction to context-free rules is that it is extremely easy to generate an SGML DTD document type definition (DTD), as in [10] for the content of a particular class of diagrams. This at once provides a validator for data that the editor may be expected to display and a means of specifying stream-based communication protocols between the editor and other applications. Needless to say, the existence of a declarative specification of diagram types goes a long way towards avoiding the problem of obsolescence. In our implementation, SGML is used as the "persistence format" for users' data.

28.3.2 User Interface

One of the most obvious benefits of the above assumptions is that the range of possible actions a user may perform on a diagram is extremely limited, regardless of how complex a class of diagrams is. In general, the actions of the user consist only of selecting a subpart of a diagram and choosing one of the diagram types allowed at that point or of performing some other action on the selected subpart. The grammar is then used to constrain the range of possible types at any one location. The only "structure-based" editors we are aware of with comparable generality are those, such as `psgml` described in [19], which interpret an SGML DTD to determine allowable material in a context-dependent way.[5]

The virtues of this simplicity should be obvious, but are worth stating. First, for educational purposes, users unfamiliar with some class of diagrams are explicitly guided through possible choices, in a way which provides immediate feedback on the consequence of choices. Second, this form of interaction is efficient. Effectively, the user provides all and only that information required to fully specify a diagram. Finally, there will be a corresponding simplicity in the relationship of the editor with a back-end processor controlling the operations of the editor for the purpose of animating operations over diagrams.

28.3.3 Limitations

There are substantial restrictions in the design we propose. There are many classes of diagrams used in linguistics which are more complex than trees, for

[5] See Section 28.7 for some further comments.

example autosegmental diagrams, cf. [3], state transition diagrams, as used in finite state morphology, or the networks of Systemic Functional Grammar. (For the latter two cases, see, e.g., [22, (pp. 240, 293)].) In order to support the construction of diagrams in those particular areas, more complex systems are inevitably required. Our proposal is not intended to be so general, for precisely the reasons and benefits discussed above.

On the other hand, there are other limitations closer to home. A natural operation over attributes in an AVM is to order them (and their values) in some way. Similarly, an AVM editor might allow type constraints as discussed in [6] to be automatically verified. One might build such information into a diagram specification (and it may be feasible in some cases to do so automatically). These limitations stem from the essential part of our design which separates clearly the graphical conventions at use in some class of diagrams from the interpretation of the content of diagrams. Under that view, if one requires some formally equivalent, but graphically different representation of some information, it makes sense for the determination of equivalence to be made by a processor dedicated to a particular formalism. In other words, issues to do with the interpretation of a diagram are not to be decided by the editor. It is our experience that the benefits fully justify this distinction.

28.4 Applications

This system has been used to deliver drilling materials to undergraduates studying syntactic trees and a simplified form of DRT.[6] In [7], it is demonstrated that viewing dynamic diagrams (perhaps with an accompanying discussion by one or more people) enhances performance markedly on tasks such as syntactic category labelling and tree construction. This enhancement is seen even when the grammar rules and categories are novel, and is (most intriguingly) still marked if no verbal explanation of the diagrams is provided.

Students with no prior knowledge of the use of trees to characterise the organisation of sentences were divided into five groups. Some of them watched videos derived by screen capture of Thistle editor sessions (*ES*). The groups were presented with the following material:

- No intervention, *NI*: not presented with any material.
- Linear text, *LT*: presented with a prose description of the use of trees in syntactic description.
- Diagram only, *DO*: ES only.
- Dialogue, *DL*: ES, LT and dialogs between a tutor and a student about the task in hand.
- Discourse, *DI*: ES, LT and monologue descriptions of actions by a tutor.

[6] Fig 28.3 shows an editor based on the relevant class of diagrams. Examples of the materials discussed below can be seen at:
`http://www.cogsci.ed.ac.uk/~rcox/drs_vicars_bak/drs_intro.html`

Pre- and post-tests were administered involving the tasks of naming syntactic categories, labelling categories in trees and construction of trees. Relative to NI, all other groups significantly improved their performance between pre- and post-tests. In analysing the effect on performance in different tasks, it was found that, in syntactic tree construction, ES groups performed better than the LT-only group (at very close to significance, $p = 0.53$). As the authors note, viewing the editor sessions alone (i.e. without accompanying commentary *or* access to prose descriptions) is "surprisingly effective" in teaching this domain.

In other applications, we have provided an interface to a locally developed tokenisation engine. This tool provides a graphical interface to complex rules. Off-the-shelf technology, in the form of an SGML processor [20], provides a simple mapping to the format required by the tokeniser. We have developed (on the basis of [18]) the treatment of diagrams in [16], used to construct Fig. 28.2. The system has been used to provide Web-based visualisation tools

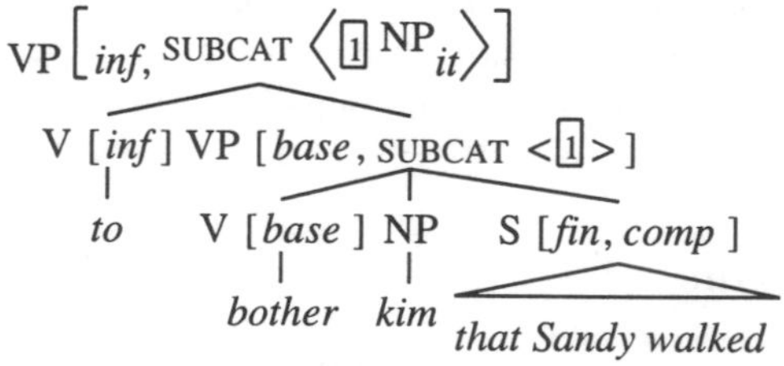

Figure 28.2. From Pollard and Sag [16, (p. 152)]. © 1994 by the University of Chicago Press. All rights reserved. Published 1994.

for the major corpus of dialogues described in [1]. It also forms the basis of the *Interarbora* service[7] for Web-based delivery of formatted trees. Other classes of diagrams for which we have provided comprehensive grammars include trees with unlimited branching and multipart node labels, categorial derivations in alternative styles, metrical trees and cluster diagrams.

There are many other kinds of applications which can be envisaged for such a system. Here we mention just a few. The "derivation checkers" or tree editors of [4] and [15] can be viewed as a mode in which each action by a user is verified for consistency with respect to a grammar. Recasting that mode within the context of delaying systems for the interpretation of constraint-based formalisms (e.g. [8]) would provide a debugger in which the grammar writer could perform an instantiation and view the results, perhaps in an animated fashion. On the other hand, the "off-line" construction of trees would provide a way of querying tree banks in a more perspicuous way than via the manual construction of a query in some query language.

[7] `http://www.ltg.ed.ac.uk/~jo/interarbora/`

28.5 Implementation

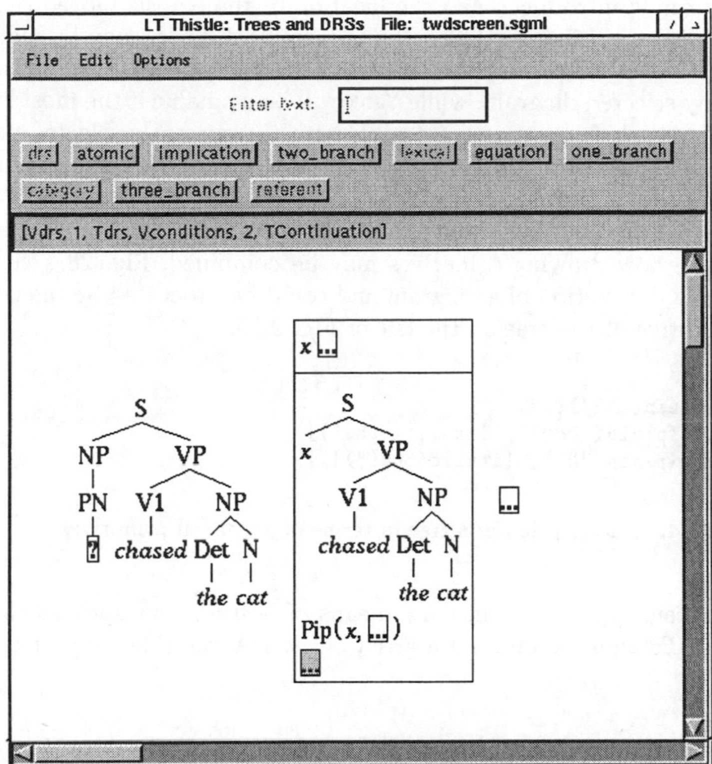

Figure 28.3. Screen capture of a tree and a DRS constructed using the editor. The selected point of structure is the ellipsis immediately below the word "*Pip*". The shaded words towards the top of the window represent types which are not available at that location. They are: "drs", "lexical", "category" and "referent". The line immediately above the diagrams indicates the path to the selected location.

Figure 28.3 above is a screen capture of an editor instance using a diagram class specification very much like that given in Section 28.6.1.[8] In this implementation, a box containing an ellipsis indicates a position permitting one or more occurrence of a diagram type or types, a box containing a question mark indicates a location allowing a single occurrence of the available types, and a question mark on its own indicates a location where characters may appear. In the state shown in the figure, the lowest ellipsis (i.e. the one

[8] The system as described here has been implemented in Java, and pages including instances of a variety of different editors are available for downloading on-line. See `http://www.ltg.ed.ac.uk/software/thistle/demos/index.html`

immediately below "Pip") is *selected.* The state of the buttons labelled by diagram type names reflects the choice open to the user at that position in structure. On instantiating a diagram at a location marked by an ellipsis, a new diagram is introduced and the location of the ellipsis moved rightward or downward according to the enclosing layout.[9] Ellipses may be hidden (or revealed) by choosing the option Show .... The operation Kill allows the deletion of any selected diagram, while Yank will be available if the most recently deleted material is of a type compatible with the currently selected position. Other operations include preparing a printable form of the image or a DTD for the class of diagrams.

We use a function-like syntax to indicate the primitives and their operands. To indicate how drawing primitives may be combined, Fig. 28.4 illustrates the use of a description of a diagram and could be processed by the editor to draw a subtree of the tree on the left of Fig. 28.3.

```
tree(plain("NP"),
  [tree(plain("Det"), [italic("the")]),
   tree(plain("N"), [italic("cat")])])
```

Figure 28.4. A description of a tree in terms of graphical primitives.

A diagram type is specified by means of a statement such as shown in Fig. 28.5. (Further examples are given below.) A variable subpart of a dia-

```
diagram_spec(drs, box(vbox([hbox(var(universe, [referent])),
                            line(),
                            vbox(var(conditions, [condition])) ])))
```

Figure 28.5. Concrete syntax for diagram specification.

gram is indicated by the syntax `var(`*`name`*`, `*`type`*`)`. That is, a diagram of the stated type may appear in this position and be referred to by the stated name. The use of square brackets, as in both uses of `var` above, is equivalent to the Kleene star in the abstract formulation shown previously, i.e. any number of diagrams of that type may occur at the position in question. As a further illustration, consider the definitions shown in Fig. 28.6. As their names suggest, the first of these limits the daughters of a tree to two, while the second allows any number of daughters. The last line illustrates the concrete syntax for diagram unions.

---

[9] There is also an operation Insert which inserts an ellipsis to the left or above the current selection.

```
diagram_spec(two_branch, tree([var(mother, category),
                               var(left, leaf_or_tree),
                               var(right, leaf_or_tree)])
diagram_spec(arbitrary_tree, tree([var(mother, category),
                                   var(daughter, [leaf_or_tree])]))

diagram_union(tree_top, [one_branch, two_branch])
```

Figure 28.6. Some example tree specifications.

```
diagram_union(leaf_or_tree, [one_branch, two_branch, three_branch,
                             lexical, referent])

diagram_spec(one_branch, tree([var(mother, category),
                               var(daughter, leaf_or_tree)]))

diagram_spec(category, plain(var(Name, Text)))
diagram_spec(lexical, italic(var(Lex, Text)))

diagram_union(condition, [atomic, equation, implication,
                          tree_top])

diagram_spec(atomic, hbox([separator(plain(""))],
                          [plain(var(relation, Text)),
                           bracket([delimiter(round)],
                           hbox([separator(plain(", "))],
                                [var(referents, [referent])]))]))
```

Figure 28.7. Part of a diagram specification for the diagram editor shown in Fig. 28.3. The symbol `Text` is equivalent to the regular expression `".*"`, i.e. any sequence of characters is allowed.

28.6 Extended Examples

28.6.1 Specification for a DRS and Tree Editor

The specifications shown in Figs. 28.6 and 28.7 provide an almost complete specification for an editor like that shown in Fig. 28.3, permitting the editing of trees with limited branching and DRSs. It has been simplified by the omission of some options controlling details of alignment and of the definitions of the diagram types `three_branch`, `implication`, `referent` and `equation`. The definition of the diagram union `tree_top` needs to be extended from that given in Fig. 28.6. Also not included is the statement of diagram types allowed at the outermost level and their layout. The options to `hbox` and `bracket` control the separator used within the horizontal box and the shape of bracket respectively.

28.6.2 Binary Trees

Diagram Class. The specification shown in Fig. 28.8 characterises a class of binary trees whose leaf nodes are abstract symbols. It is fully commented. This specification is chosen to illustrate several points. First, it shows how the graphic primitives may be used to construct graphical objects which function as "terminal symbols" in the specification. See the definition of `square` and `circle`. Second, the content of non-leaf node names is restricted by regular expression. Third, unions may contain references to other unions: `leaf_or_tree` references `leaf`, so allowing types to be grouped into classes in the most appropriate and convenient way. Fig. 28.9 shows a diagram constructed using this class.

```
% Identify this class of diagrams (optional)
diagram_class("binary trees (twd example)", "1.0")

% Name the derived document type definition (optional)
dtd_name("btreetwd.dtd")

% allow any number of trees at the top most level
diagram_spec(top, hbox([align(top)], var(tree, [binary])))

% The class of all elements that can appear within a tree
diagram_union(leaf_or_tree, [binary, leaf])

% A leaf node is any sequence of alphabetics
diagram_union(leaf, [square, circle])

% A square is a box around some space.  A circle likewise.
diagram_spec(square, box(space()))
diagram_spec(circle, box([shape(circle)], space()))

% A non-terminal node name is any nonempty sequence
% of alphanumerics.
diagram_spec(node, plain(var(node, "[a-zA-Z0-9]+")))

% binary trees have three subparts, the mother,
% left and right elements
diagram_spec(binary, tree([var(mother, node),
                           var(left, leaf_or_tree),
                           var(right, leaf_or_tree)]))
```

Figure 28.8. A complete specification for a class of binary trees.

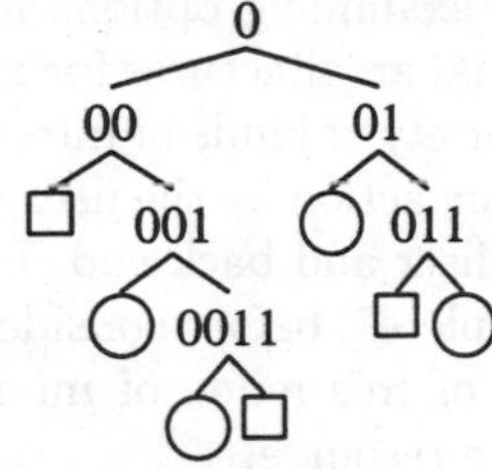


```
tree 0 binary left binary left square
tree 0 binary left binary mother node node "00"
tree 0 binary left binary right binary left circle
tree 0 binary left binary right binary mother node node "001"
tree 0 binary left binary right binary right binary left circle
tree 0 binary left binary right binary right binary right square
```

Figure 28.9. An example binary tree (left) from the class given by the specification in Fig. 28.8 and characterised in part by the set of paths shown on the right.

28.7 Conclusions, and Current and Future Work

We have presented a design for a linguistic diagram editor which, although limited in the range of graphics it permits, nevertheless provides a configurable system of substantial benefit to a wide class of users. An implementation is available, and already in use for a wide range of applications.

In our recent work, we have extended the system to allow sequences of diagrams to be constructed and viewed. We have also shown that grammatical inference can be exploited to broaden the range of operations permitted over diagrams. In particular, suppose we have constructed a diagram, and that a subdiagram *d* of type *t* appears at some location within that diagram. If it is the case that

1. some other type t' of diagram distinct from *t* can appear at *d*'s location in the diagram and
2. a diagram of type *t* is permitted to appear as a subpart within the other type t'

then we can invoke the operation of *adjunction*. This operation replaces the diagram *d* with a new diagram d' of type t' and inserts *d* at some location within d'. Adjunction here is the exact analogue of the operation used within *tree adjoining grammars* [11]. This point serves as a reminder that, once we adopt a grammatical perspective in the study of diagrams, many results of formal language theory can be used in the analysis of diagram systems. As a further point, it is known that the set of paths (as in Section 28.3.1) generated by a context-free grammar forms a regular language, and that more expressive languages can be generated by constraining sets of paths by more powerful grammars. This is one route we are keen to explore.

In current work, we are examining options for a back end processor for diagrams. A range of potential architectures for interaction are under consideration. We expect that a variety of kinds of interaction will be necessary. For example, in some settings, any action by the user may require a response from a combined system of visualiser and back end. In other settings, it is necessary that a diagram is "complete", before consideration by a back end makes sense. An obvious instance of this mode of interaction is where a diagram represents a query over some resource.

There are clear connections between the work we report here and "WYSIWYG" editors, and likewise between our declarative approach to the specification of diagrams and the now prevalent distinction of *structure* and *styling*. In practical terms, of course, this area is now the focus of much work, as the recent growth of XML and the stylesheet language XSL demonstrate.[10] We can use Thistle to provide visualisation for XML documents, and are considering ways of generalising this to cover some of the same ground as the W3C's *Amaya* browser/editor.

Evaluation of the educational usefulness of the system continues. In the future, we expect to provide diagram specifications for still other formalisms, and an interface allowing the dynamic control of the editor by other programs. We anticipate that the restriction to context-free organisation of diagrams will be acceptable for many purposes. On the other hand, extensions to the system to broaden the range of the diagram types would make the system more useful still and, in future work, we are keen to examine strategies which involve the semi-automatic layout of complex diagrams.

Acknowledgements

The research described here was funded in part by the Economic and Social Research Council Grant L127 25 1023 and by the Engineering and Physical Sciences Research Council grant GR/L21952. I would like to thank Richard Tobin, Ewan Klein and the Thinking with Diagrams '98 reviewers for critical comments on earlier proposals and presentations. I alone am responsible for errors that remain.

References

1. Anderson, A.H., Bader, M., Bard, E.G., Boyle, E.H, Doherty, G.M., Garrod, S.C., Isard, S.D., Kowtko, J.C., McAllister, J.M., Miller, J., Sotillo, C.F., Thompson, H.S. and Weinert, R. (1991). The HCRC map task corpus. Language and Speech 34(4):351–366.

[10] See the documents from the World Wide Web Consortium at `http://www.w3.org`

2. di Battista, G., Eades, P., Tamassia, R., and Tollis, I. G. (1994). Algorithms for drawing graphs: An annotated bibliography Computational Geometry Theory and Applications 4:235–282.
3. Bird, S. and Klein. E. (1990). Phonological events. Journal of Linguistics 26:33–56.
4. Calder, J. (1993) Graphical interaction with constraint-based grammars. In Proceedings of the third pacific rim conference on computational linguistics, Vancouver, 22–24 April, pp:160–169.
5. Calder, J. (1998). Thistle: Diagram display engines and editors. Technical report TR-97, Human Communication Research Centre, University of Edinburgh, July.
6. Carpenter, B. (1992). The logic of typed feature structures. Cambridge Tracts in Theoretical Computer Science. Cambridge,UK: University Press.
7. Cox, R., McKendree, J., Tobin, R., Lee, J. and Mayes, T. (1999) Vicarious learning from dialogue and discourse: A controlled comparison. Instructional Science 27:431–458.
8. Dörre, J. and Dorna, M. (1993). CUF: A Formalism for linguistic knowledge representation. In J. Dörre (Ed.), Computational aspects of constraint-based linguistics description. ILLC/Department of Philosophy, University of Amsterdam.
9. Gazdar, G., Klein, E., Pullum, G. and Sag, I.A. (1985). Generalized phrase structure grammar, Oxford: Basil Blackwell.
10. Goldfarb, C.F. (1990). The SGML handbook. Oxford: Clarendon Press.
11. Joshi, A.K., Vijay-Shanker, K. and Weir, D.J. (1991) The convergence of mildly context-sensitive grammatical formalisms. In P. Sells, S.M. Shieber and T. Wasow (Eds), Foundational issues in natural language processing. Cambridge, MA: MIT Press.
12. Kamp, H and Reyle, U. (1993). From discourse to logic. Dordrecht: Kluwer.
13. Kiefer, B. and Fettig, T. (1995). Fegramed: An interactive graphics editor for feature structures. Research Report RR-95-06, Universität des Saarlandes, Saarbrücken.
14. Myers, B.A., Giuse, D., Dannenberg, R.B., Vander Zanden, B., Kosbie, D., Pervin, E., Mickish, A., and Marchal, P. (1990). Garnet: Comprehensive support for graphical, highly-interactive user interfaces. IEEE Computer 23(11):71–85.
15. Paroubek, P., Schabes, Y. and Joshi, A.K. (1992). XTAG: A graphical workbench for developing tree adjoining grammars. In Proceedings of the third conference on applied natural language processing, Trento, Italy, 31 March–3 April, pp: 216–223.
16. Pollard, C.and Sag, I.A. (1994). Head-driven phrase structure grammar. CSLI: Stanford and University of Chicago Press: Chicago and London.
17. Serrano, J.A. (1997). The use of semantic constraints on diagram editors. In Proceedings of VL'95, 11th international IEEE symposium on visual languages, Darmstadt, Germany, 5–6 September.
18. Smithers, G. (1997). A diagram editor specification for head-driven phrase structure grammar. Unpublished dissertation, Department of Linguistics, University of Edinburgh.
19. Staflin, L. (1996). PSGML, a major mode for SGML documents. See http://www.lysator.liu.se/projects/about_psgml.html.
20. Thompson, H.S. and McKelvie, D. (1996). A software architecture for SGML annotation. In SGML Europe. Graphical Communications Association: Alexandria, VA.
21. Viehstaedt, G. and Minas, M. (1995). Generating editors for direct manipulation of diagrams. In B. Blumenthal, J. Gornostaev and C. Unger (Eds),

Proceedings of the 5th international conference on human-computer interaction (EWHCI'95), Moscow, Russia. LNCS 1015. Berlin: Springer, pp: 17–25.

22. Winograd, T. (1983) Language as a cognitive process. Vol. I: Syntax. Reading, MA: Addison-Wesley.
23. Wall, L. and Schwartz, R.L. (1991). Programming in Perl. Sebastopol, CA: O'Reilly.

29. AVOW Diagrams: A Novel Representational System for Understanding Electricity

Peter C-H. Cheng

AVOW diagrams are a novel representational system for electricity. This chapter describes how the representation comprehensively covers many aspects that are essential to the full conceptualisation of this domain. AVOW diagrams provide a relatively homogeneous characterisation of diverse electrical components and give a uniform analysis of the different modes of operation of simple and complex circuits. AVOW diagrams support the different perspectives that can be taken of electricity, including interpretation as formal abstract constraint-based laws, as concrete material attributes, and as causal relations. Examples are presented to illustrate the benefits of AVOW diagrams over the conventional algebraic characterisations. The practical limits of this diagrammatic representation are also discussed.

29.1 Introduction

The Representational Analysis and Design Project is investigating the fundamental role that external representations have in complex cognitive tasks such as problem solving, knowledge acquisition, discovery and learning [1,2,6,11]. The representational system chosen can greatly affect the ease with which a task is performed and it will substantially determine the particular methods that are used to complete the task. The overall goal of the project is to develop principles for the design of effective representational systems [6,11]. The project has used a variety of approaches to study what makes effective representations, including: invention of new representational systems for particular domains [8,10]; computational modelling of complex cognition with different representations [1]; empirical studies of human reasoning and learning with different representations [2–4]; analysis of the role of representation in scientific discovery [1]; development and evaluation of software tools and systems that exploit the new representations to support complex tasks [2,5]. The work reported in this chapter exemplifies the first of these approaches with the introduction of a novel diagrammatic representation for a complex domain – AVOW diagrams for electricity. This representational system was invented by the author to make electricity easier to understand and learn.

Good understanding of basic electricity requires different types of knowledge, on several levels of abstraction, covering various different perspectives (e.g., [12–14]. There are basic laws of electricity governing the measurable properties of individual components of circuits. There are a host of components; resistors, switches, diodes, rheostats, capacitors, batteries, insulators, wires and so forth. They are assembled into networks which have behaviours dependent on both the configuration of the network and the specific properties of particular components. Circuits can operate in different modes with different behaviours and when they malfunction the behaviours can be radically altered. Some circuits have simple constraints governing the relations among properties that are relatively easy to remember, but for complex circuits network laws and theorems have to be used to analyse the operation of the circuit. On a long time scale a circuit may be treated as a steady state system, but over a short time the same circuit may have transient behaviours that require separate analysis. For some purposes it is useful to consider electricity in terms of formal relations of abstract constraints among physical quantities, which are conventionally encoded using algebra. For other purposes it is convenient to take a more material or concrete perspective, treating electricity as a fluid pumped through channels, which is usually described in natural language. At a microscopic level electricity is the flow of electrons whose individual passage around a circuit may be imagined. The electrical properties of individual components may be considered in terms of the bulk properties and physical dimensions of the components.

AVOW diagrams constitute an effective representational system for understanding electricity, because it encompasses all of these different aspects and perspectives. The following four sections present illustrative examples of how AVOW diagrams achieve this. (See [8] for a full account.) The penultimate section discusses the limitations of AVOW diagrams. The conclusion summarises how AVOW diagrams have been used in the present project to study the nature of effective representations.

29.2 Basic Properties and Relations for Components

AVOW (Amps, Volts, Ohms and Watts) stands for the units of the four basic electrical properties; current, voltage (potential difference, electromotive force), resistance and power. Each component, load or resistor in a circuit is represented in an AVOW diagram by a rectangle or AVOW box (Fig. 29.1). Table 29.1 shows how the features of AVOW boxes represent electrical properties. An AVOW box encodes both Ohm's law and the power law, the basic relations in electricity. Ohm's law states that $V = I \times r$; or in terms of an AVOW box, $height = width \times gradient$. The power law states $P = V \times I$; or $area = height \times width$. A particular drawing of an AVOW box represents a load with a particular set of values, as in the case of Fig. 29.1 (see Table 29.1).

Table 29.1. AVOW box features representing electrical properties of resistors

Property (units)	Symbol	AVOW box representation	Example: Figure 29.1
Current (amps)	I	*width*	$I = 2$
Voltage (volts)	V	*height*	$V = 3$
Resistance (ohms)	r	*gradient of the diagonal*	$r = V/I = 3/2$
Power (watts)	P	*area*	$P = 3 \times 2 = 6$

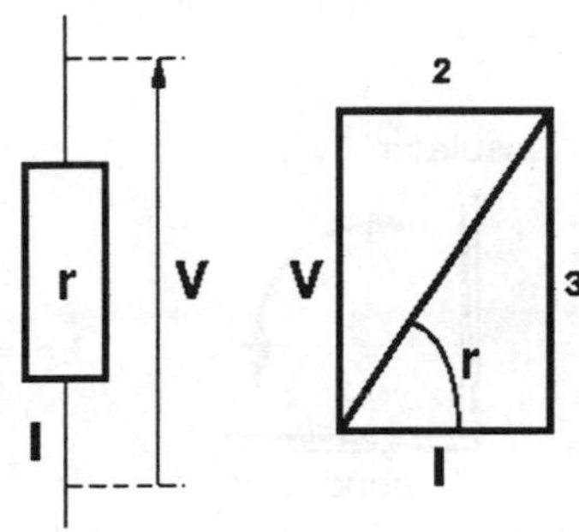

Figure 29.1. AVOW box for a load.

Notice that for constant resistance the voltage and the current are in proportion to each other, as implied by Ohm's law, as shown in Fig. 29.2. With V, I or P fixed, other similar relations also hold. The power is a function of the square of the voltage, as implied by both laws taken together (i.e., $P = V^2/r$).

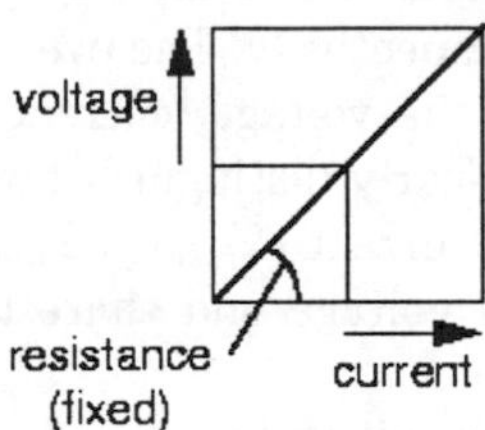

Figure 29.2. Relation among properties.

AVOW boxes provide a neat conceptualisation of conductors and insulators, which are extreme idealised cases of resistors with zero or infinite resistance, respectively. These extreme cases are represented by AVOW boxes that are simply horizontal or vertical line segments, because the gradient must be zero or infinite for these values of resistance, respectively. These cases provide useful common conceptualisations of other things. A switch may be viewed as a special AVOW box that flips back and forth between a horizontal and vertical line segment as it is turned on and off, as shown in Fig. 29.3. Similarly, diodes can be represented in the same way as they have nearly zero or infinite resistance depending on the sign of the applied voltage.

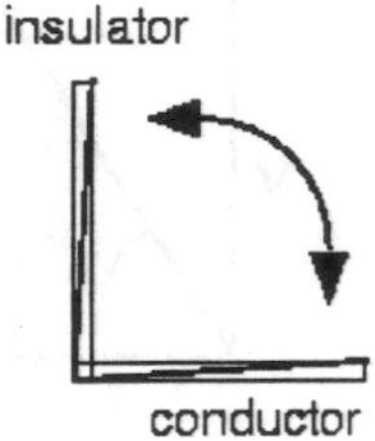

Figure 29.3. Switch.

29.3 Network Analysis

AVOW boxes may be composed or assembled to model configurations of components in networks. Figures 29.4 and 29.5 show how composite AVOW diagrams model resistors in parallel or series, by placing AVOW boxes side by side or by stacking them, respectively. The overall height, width, area and gradient of the diagonal give the voltage, current, power and resistance of the networks. The diagrams clearly distinguish how parallel networks share the voltage and distribute the current ($V_{total} = V_a = V_b, I_{total} = I_a + I_b$) and series networks distribute the voltage and share the current (i.e., $V_{total} = V_a + V_b, I_{total} = I_a = I_b$).

Something that is counter-intuitive for many learners of electricity is why the overall resistance of two parallel resistors is less than that of either resistor alone, despite the fact that in some sense the two have been "added" together in the network (and as seen in the unusual formula for parallel resistors $1/r_{total} = 1/r_a + 1/r_b$). Figure 29.4 explains why this is the case as two AVOW boxes side by side for a parallel network have a greater overall width

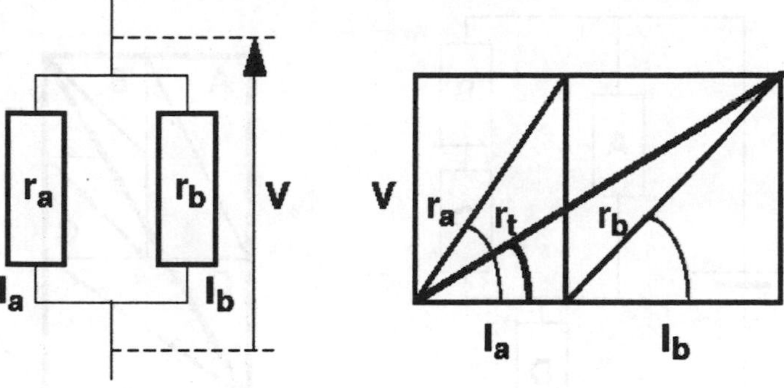

Figure 29.4. Parallel resistors.

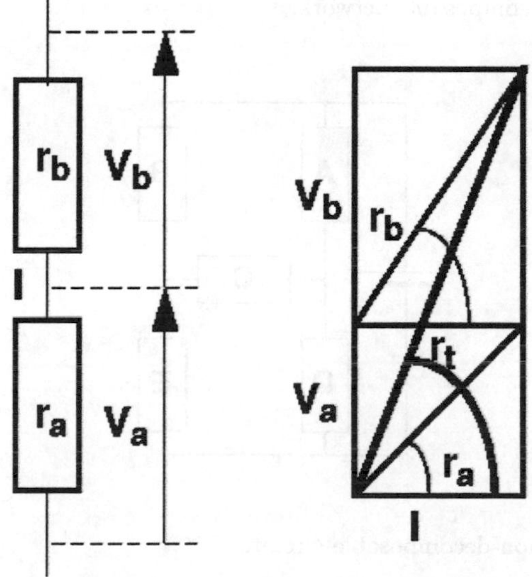

Figure 29.5. Resistors in series.

(total current), so the gradient of the complete diagram (resistance) must be less than the gradient (resistance) of either box alone.

Complex networks with parallel and series sub-networks nested within each other can be modelled recursively by applying the composition rules. A well-formed composite AVOW diagram must be a rectangle that is completely filled with AVOW boxes, with no overlaps or gaps, to model voltage and current conservation correctly. Figure 29.6 shows such a circuit and a corresponding composite AVOW diagram. This type of decompositional analysis

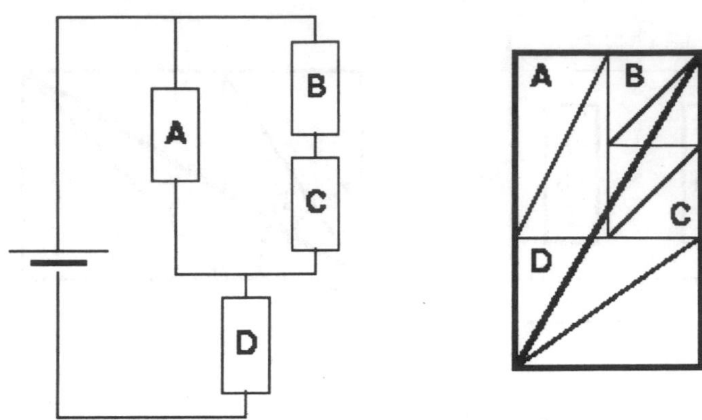

Figure 29.6. A decomposable network.

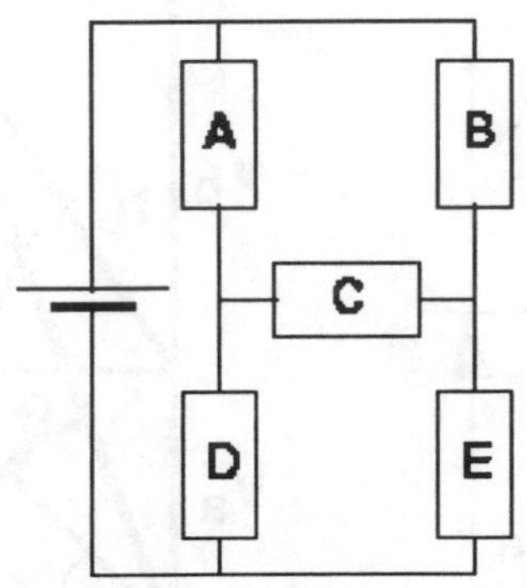

Figure 29.7. An non-decomposable circuit.

mirrors the algebraic approach to the analysis of circuits where complex circuits are gradually reduced to simpler circuits by calculating the resistance of sets of parallel and series loads and replacing them with a single resistor with an equal value. This technique fails for the algebraic representation with circuits that have components that are not simply in series or parallel, such as Fig. 29.7. Is C in parallel or in series with A? However, AVOW diagrams can still be drawn for such circuits (Figs. 29.8a and 29.8b). The circular symmetry corresponds to the symmetry of the loads surrounding resistor C. Depending on the particular values of the loads, C can be in parallel with A (Fig. 29.8a) or C in series with A (Fig. 29.8b). As usual the overall resistance of the network is the gradient of the diagonal. To find this resistance using algebra requires the solution of five simultaneous equations set up using Kirchhoff's laws, which are considered next.

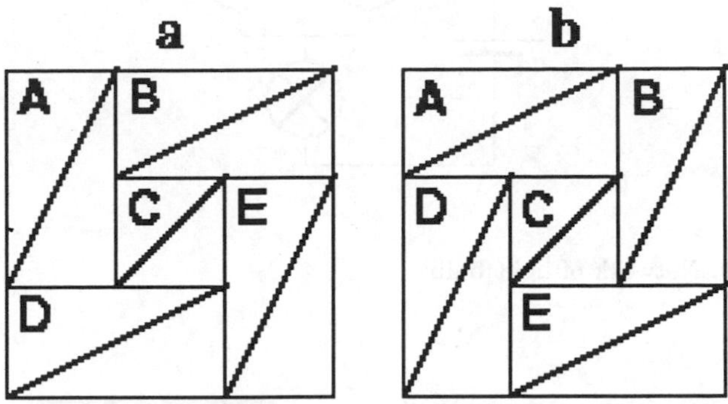

Figure 29.8. AVOW diagrams for Fig. 29.7.

Kirchhoff's laws are essential for analysing complex networks in the conventional algebraic approach to electricity, because they encode the conservation of current and voltage within networks. The compositional constraints of AVOW boxes also encode Kirchhoff's laws and the diagrams provide a simple visualisation of these laws. Kirchhoff's first law states that the total current flowing towards a node is equal to the current flowing away from it. A node is any connection between two or more branches or components in a network. In terms of the AVOW diagrams, these are represented by horizontal line segments between boxes (e.g., between D and A–C in Fig. 29.6). For all such unbroken line segments the sum of the width of the boxes above the line equals the sum of the width beneath the line (e.g., $I_A + I_C = I_D$). Kirchhoff's second law states that the sum of the products of current and resistance for each part of a closed circuit equals the total voltage of the source. In terms of the AVOW diagrams this may be interpreted as the total height of any sequence of vertically contiguous AVOW boxes will equal the height of the overall box for the complete circuit, which is also the voltage of the source. In Fig. 29.6 the stack of AVOW boxes B–C–D and A–D are two such sequences.

Insulators such as the circuit board on which the components are mounted, and conductors including the wires connecting loads, both have a natural place in composite AVOW diagrams. They are represented by vertical and horizontal line segments, respectively.

Common faults in circuits are shorts or breaks when something accidentally becomes a conductor or an insulator, respectively. Because insulators and conductors are represented in AVOW diagrams, analysing circuit be-

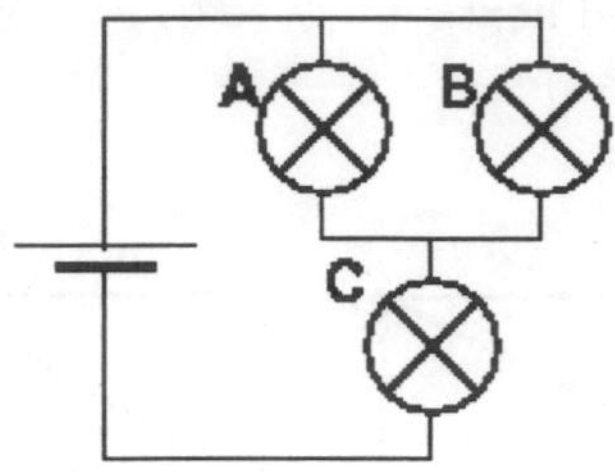

Figure 29.9. Network of light bulbs.

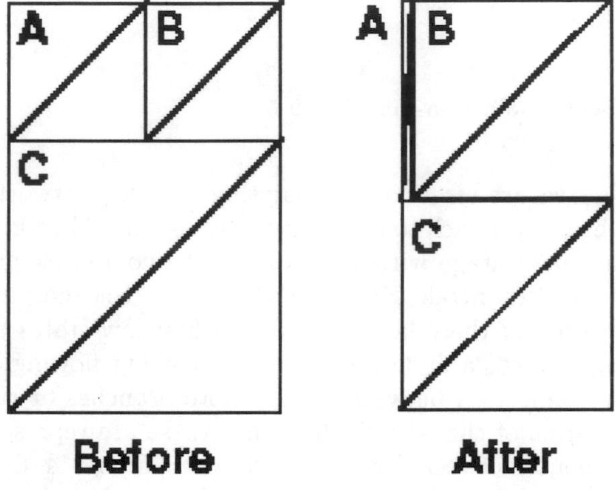

Figure 29.10. Normal and faulty operation.

haviour under fault conditions is often relatively simple because the overall topology of the AVOW diagram usually does not have to be revised. In comparison under the algebraic analysis different formulas have to be written for different circuit configurations on the assumption that the faulty component can be ignored. For example, what happens to the relative brightness of the three identical bulbs in the circuit in Fig. 29.9 when bulb A blows? The before and after AVOW diagrams are shown in Fig. 29.10; note the equivalent gradients in the "before" AVOW diagram. The AVOW box for bulb A appears as an insulator in the "after" diagram, while bulbs B and C are resized to compensate.

29.4 Power Sources and Internal Resistance

Power sources, such as batteries or mains supplies, are represented by a simple rectangle like an AVOW box, but with no diagonal. For a constant voltage source the height is fixed and for the rare cases of constant current sources the width is given. For a closed circuit with just one ideal power supply the rectangle for the source is identical to the perimeter of the composite AVOW diagram for the circuit.

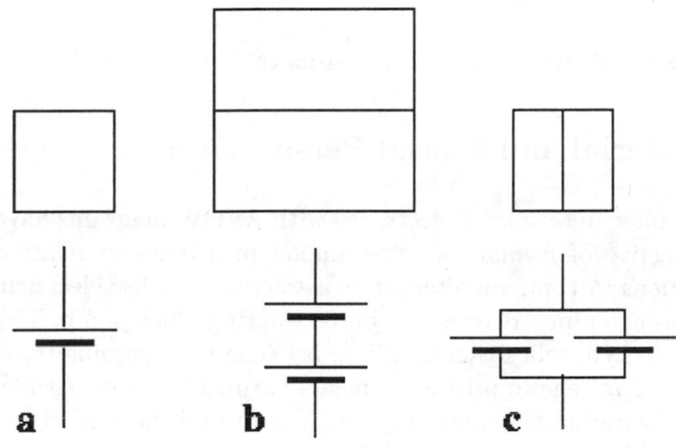

Figure 29.11. Arrangements of batteries.

Fig. 29.11 shows various configurations of identical batteries, all connected to networks with unit resistance. Figure 29.11(a) is a single battery. Like AVOW boxes for loads, the boxes for sources can be stacked when batteries are connected in series or placed side by side when batteries are in parallel, as depicted in Figs. 29.11(b) and 29.11(c). Stacking source boxes increases the voltage, but batteries in parallel do not increase the current. The current is determined by the overall resistance of the network to which the batteries are connected, so each battery has to deliver less current.

Real batteries have some internal resistance. This is normally modelled by adding a resistor directly in series with the battery, which is treated as an ideal source. Figure 29.12 shows a single load across a real battery with an extra symbol for the internal resistance. The corresponding AVOW diagram is beside it.

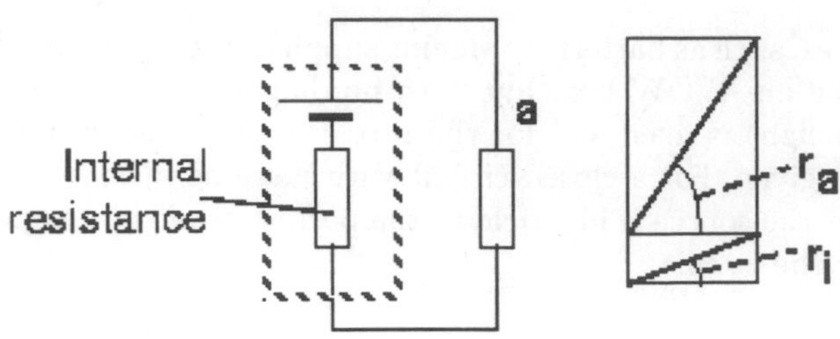

Figure 29.12. Treatment of internal resistance.

29.5 Material and Causal Perspectives

So far the interpretations of electricity with AVOW diagrams have been from the perspective of formal relations among properties in relation to circuit configurations. A common alternative is to conceptualise electricity as a fluid flowing through pipes, driven by a battery acting like a pump. The analogy of electricity to hydraulics is compelling, because the mapping from current to flow rate and voltage to pressure appears natural. However, students' mental models of hydraulic systems are typically poor and the correct mathematical models of hydraulic systems are different from the laws governing electricity. For instance, relative heights and velocities are important factors in fluid dynamics but what do they map to in electricity?

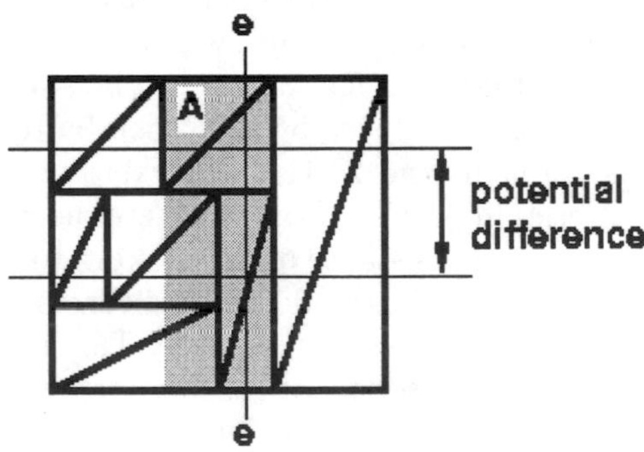

Figure 29.13. Material interpretation.

AVOW diagrams can be used to support a material interpretation. Current can be imagined as flowing down the diagram, with the flow rate through a particular component being in proportion to the width of its AVOW box. In Fig. 29.13 the flow originally through resistor A, shown by the shading, is seen to split into two streams, which pass through successive components. In the limit, a vertical line down an AVOW diagram can be read as the path of a particular electron through the circuit; for example, the line e–e in Fig. 29.13 shows an electron travelling through A and one other resistor.

A causal interpretation is sustained by knowing that any horizontal cut through an AVOW diagram is a line of equal voltage, an equi-potential, Fig. 29.13. The force, or potential difference, to drive the flow between two points in a circuit is given by the vertical distance between the points. When the voltage is constant, increasing the resistance will reduce the flow of current. Winding up the resistance will require a greater voltage if the current is to remain constant.

29.6 Limits of AVOW Diagrams

AVOW diagrams give a fairly broad coverage of electricity encompassing many types of components and circuit configurations at different levels and from alternative perspectives. However, there are limits to the convenient use of AVOW diagrams for modelling electricity. This section describes some of the cases where AVOW diagrams are not such an effective representational system.

All the circuits considered so far have been two-dimensional in the sense that the circuit diagram can be drawn without any wires crossing. To model a "three-dimensional" circuit AVOW boxes for certain loads must be cut at the overall diagram boundary and "wrapped around" to the opposite side. Such AVOW diagrams are in effect cylinders (see [8] for a full explanation).

Some circuits have multiple power sources that are simply in series or parallel with each other, as in Fig. 29.11. Others have batteries distributed throughout the circuit. AVOW diagrams can model such circuits but the rules of composition must be amended. AVOW boxes may overlap and the overall diagram need not be a rectangle. The simplicity of AVOW diagrams for basic circuits is lost, which again makes AVOW diagrams cumbersome to use. However, the way the composition rules have to be extended does provide a consistent interpretation and good visualisation of what is occurring in such complex circuits [8].

With the algebraic approach any sense of the topology of the circuit reflected in the form of the equations is obscure at best and quickly lost in any manipulations of the formulas. The potential for direct mappings between circuit topology and the spatial layout of AVOW boxes in a composite diagram is one of the distinctive features of AVOW diagrams. However, to effectively support this mapping with more complex circuits, the circuit has to be drawn with the components arranged so that current is flowing downwards through the diagram, to match the general interpretation of AVOW diagrams. However, the direction of current may not be known until after the AVOW diagram has been drawn, so there will be a need to iteratively modify successive versions of the circuit diagram and AVOW diagram. Figures 29.8(a) and 29.8(b) illustrate how the layout of the AVOW diagram may change even though the circuit topology is constant, when the direction of the current in resistor C is reversed. Interactive computer environments that permit the direct manipulation of circuit and AVOW diagrams on screen can help alleviate this problem. The prototype system for learning about electricity built by Cheng [5,6] has such capabilities.

Finally, the electrical signal travels at nearly the speed of light through circuits, but electrons drift along more slowly, and in the direction opposite to the conventional direction of the flow of current. AVOW diagrams do not provide any direct visualisation of these particular facts.

29.7 Conclusion

AVOW diagrams are a novel representational system for understanding electricity. The depth and breadth of the conceptualisation that AVOW diagrams provide have been examined (further details and examples can be found in [8]). The potential of AVOW diagrams is being investigated in various ways. A prototype computer-based discovery learning environment has been built to explore the possible design of software that exploits the representation [5,6]. The psychological processes of learning with AVOW diagrams have begun to be investigated [4,7]. An evaluation of novices learning about electricity has just been completed, in which students using AVOW diagrams gained a much deeper understanding than those using a conventional algebraic approach [9]. For example, students taught with AVOW diagrams were able to construct diagrams like Fig. 29.11 given Fig. 29.10, and many were to solve even more complex problems.

AVOW diagrams constitute an example of Law Encoding Diagrams, LEDs, a general class of representational systems that have some interesting properties that are being investigated in the Representational Analysis and Design Project [10,11]. By contrasting the effects of conventional representations (e.g., algebra) and LEDs on complex tasks the project has discovered various desirable systemic properties of effective representational systems, which may be taken as representational design principles. These principles are being evaluated by using them to design new LEDs for other domains [10] and by investigating how well the new representations support problem solving and learning.

Acknowledgements

The research is funded by the UK Economic and Social Research Council through the Centre for Research in Development, Instruction and Training. Thanks must go to members of the Centre who have supported and encouraged me in this work.

References

1. Cheng, P.C.H. (1996a). Scientific discovery with law encoding diagrams. Creativity Research Journal 9(2,3):145–162.
2. Cheng, P.C.H. (1996b). Law encoding diagrams for instructional systems. Journal of Artificial Intelligence in Education 7(1):33–74.
3. Cheng, P.C.H. (1996c). Learning qualitative relations in physics with law encoding diagrams. In G.W. Cottrell (Ed.), Proceedings of the eighteenth annual conference of the cognitive science society. Hillsdale, NJ: Erlbaum, pp. 512–517.
4. Cheng, P.C.H. (1998a). A framework for scientific reasoning with law encoding diagrams: Analyzing protocols to assess its utility. In M.A. Gernsbacher and S.J. Derry (Eds), Proceedings of the twentieth annual conference of the cognitive science society. Hillsdale, NJ: Erlbaum, pp. 232–235.
5. Cheng, P.C.H. (1998b). Some reasons why learning science is hard: Can computer based law encoding diagrams make it easier? In B.P. Goettl, H.M. Halff, C. Redfield, and V. Shute (Eds), Intelligent tutoring systems. Berlin:Springer, pp. 96–105:
6. Cheng, P.C.H. (1999a). Interactive law encoding diagrams for learning and instruction. Learning and instruction 9(4):309–326.
7. Cheng, P.C.H. (1999b). Networks of law encoding diagrams for understanding science. European journal of psychology of education 14(2):167–184.
8. Cheng, P.C.H. (1999c). AVOW diagrams: A representational system for modelling electricity (Technical no. 63). ESRC Centre for Research in Development, Instruction and Training, University of Nottingham.
9. Cheng, P.C.H. (1999d). Electrifying representations for learning: An evaluation of AVOW diagrams for electricity. (Technical no. 64). ESRC Centre for Research in Development, Instruction and Training, University of Nottingham.
10. Cheng, P.C.H. (1999e). Representational analysis and design: What makes an effective representation for learning probability theory? (Technical no. 65). ESRC Centre for Research in Development, Instruction and Training, University of Nottingham.
11. Cheng, P.C.H. (1999f). Unlocking conceptual learning in mathematics and science with effective representational systems. Computers in education 33(2–3):109–130.
12. Duncan, T. (1973). Advance physics: Material and mechanics. London: John Murray.
13. Hewitt, P.G. (1992). Conceptual physics (7th edn). New York: Harper Collins.
14. Hughes, E. (1977). Electrical technology (5th edn). London: Longman.

30. AsbruView: Capturing Complex, Time-Oriented Plans – Beyond Flow Charts

Robert Kosara
Silvia Miksch
Yuval Shahar
Peter Johnson

Flow charts are one of the standard means of representing actions or algorithms in many domains. However, applying flow charts in dynamically changing environments, like clinical treatment planning, reveals their limitations. Flow charts do not include the temporal dimension in their design, do not allow complex paths through many components, and scale very badly. These are only some of the requirements for a means of communicating clinical therapy plans. As an alternative, a plan-representation language called *Asbru* was designed, which overcomes all the limitations of flow charts. It is, however, impossible for a domain expert to work with Asbru directly. Therefore, a visualisation called *AsbruView* is presented here, which uses three-dimensional diagrams and metaphors – running tracks and traffic signs – to make the parts of Asbru easily understandable and usable. Even very complex clinical plans are easy to understand with AsbruView.

30.1 Introduction: Clinical Protocols

We are motivated by the demands in medical, real-world environments: improving the quality of health care through increased awareness of proper disease management techniques. Health maintenance organizations (HMOs) are urged to increase productivity and simultaneously reduce costs without adversely affecting the quality of patient care. One step towards this aim is the implementation of commonly accepted and standardised health care procedures. Treatment planning from scratch typically is not necessary, as general clinical procedures exist which should guide the medical staff. These procedures are called *clinical practice guidelines and protocols.* A guideline can be defined as "all methods that identify actions that are to be performed and that specify conditions that govern when it is appropriate to perform them" [18]. A clinical protocol is a more detailed version of a clinical guideline and refers to a class of therapeutic interventions.[1] Such protocols are used for utilisation review, for improving quality assurance, for reducing variation in clinical practice, for guiding data collection, for better interpretation and management of the patient's status, and for activating alerts and reminders, for improving decision support [18].

The first section sketches the problem area, namely the design of clinical therapy and treatment plans. Authoring of such plans is a non-trivial task; therefore, a means of representing such plans appropriately, a language called *Asbru*, is introduced in the second section. Asbru, however, is impossible to use for people working in the domain of therapy planning, and thus a means of representing Asbru plans in an understandable way had to be found. This visualisation of Asbru is called *AsbruView*, and is discussed in detail in the third section. Flow charts are most commonly used in the medical domain. A short introduction to flow charts is given in the fourth section. Why we didn't choose flow charts, however, is made clear in the fifth, final, section.

30.2 The Problem Area: Authoring Clinical Protocols

In most cases, physicians and other medical staff need not invent therapy or treatment plans for their patients anew every time. The medical staff can fall back on predefined protocols. Such plans are usually represented in free text (i.e., a natural language), in decision tables, or in flow charts. Authoring plans, however, is a non-trivial task. Part of its complexity stems from the inappropriate means to represent plans in order to communicate and reason about them.

Difficulties to author a treatment plan arise on various levels. Problems are also caused by the different purposes the plans or protocols are used for. The structure and composition of a treatment plan are quite manifold. Many variables must be accounted for and many different conditions must be taken into consideration. However, the instructions (the overall plan) must still remain readable, understandable, and lucid. In a therapy plan, a goal needs to be achieved in a certain time. The way or path to this goal is not always obvious and can be achieved following various paths. It must even be possible to perform actions that are not part of the actions' set, but still follow the underlying intentions of a treatment plan. This means, physicians must be able to not adhere to protocols, believing their actions to be closer to the intentions of the protocol design. Hence, a treatment plan needs to capture the intentions too, allowing to continue a particular plan even when the performed actions vary.

[1] In computer science, clinical protocols can be a interpreted as plans. Therefore, we are using the expressions "protocol" and "plan" interchangeably.

The plans' intentions can be used for different purposes, like critiquing. Does applying the plan really lead to fulfilling its intentions? Can the plan be applied under different conditions, and if so, which are those? These questions are not only needed for the selection of the correct plan for a certain patient, but also for improving the plans, and thus the patient's treatment.

In medicine, new ways of solving problems are being discovered day by day, new side effects are found, etc. So treatment plans change very often, as new treatments and new conditions are added, while others may be changed or even removed entirely. This leads to the problem that after the development of a clinical plan has been finished it may already be out of date.

Regarding these points, it is not surprising that clinical protocols are often vague, incomplete and sometimes even contradictory.

30.2.1 Representing Plans: Historical Synopsis

Clinical protocols or plans can be seen as procedures or algorithms, which need to be executed depending on health conditions of a patient and within a particular time span. An appropriate modelling language and visualisation (diagrams) capturing all different features are needed.

On the one hand, in computer science (particularly in the research of programming languages and software engineering), different approaches were introduced to capture such procedures, known as algorithms or programs. It started in the early years of programming. Some milestones: in 1947 the first graphical representations of algorithms were designed, called flow charts [6]; after a long discussion about styles of programming (e.g., *goto*-less/*goto*-free programming), "structured programming" and block structured diagrams were introduced [16]; followed by different modelling techniques, such as petri nets, graphical (visual) programming, and object-oriented programming [5]. Additionally, different computer-oriented knowledge interchange languages (e.g., KIF [4]), ontologies [7], and plan representation languages (e.g., PROPEL language [11]) were discussed to represent domain-specific procedural knowledge. In general, these representations have significant limitations and are not applicable in dynamically changing environments, like medical domains (e.g., they assume instantaneous actions and effects; they neglect that actions often are continuous (durative) and might have delayed effects and temporally extended goals; they overlook that unobservable underlying processes determine the observable state of the world). A more detailed review is given in [14].

On the other hand, scientists in medicine and medical informatics have recognised the importance of protocol-based care to ensure a high quality of treatment. An important approach was the definition of the Arden syntax [8], which encodes situation–action rules. This syntax has significant limitations too: it currently supports only atomic data types, lacks a defined semantics for making temporal comparisons or for performing data abstraction, and provides no way to represent clinical guidelines that are more complex than individual situation–action rules [15]. Therefore, the Arden syntax is not applicable for our purposes.

Other (more recent) examples are the *GuideLine Interchange Format* (GLIF) [17], the PROforma project [2], etc. A complete survey of existing methods is beyond the scope of this chapter, however.

30.2.2 Which Features do we Need?

First, we need a plan *representation*, which (1) captures a hierarchical decomposition of plans, (2) is expressive with respect to temporal annotations, plan's intentions and effects, and (3) has a rich set of sequential, concurrent, and cyclical operators. Thus, it should enable designers to express complex procedures in a manner similar to a real programming language (although typically on a higher level of abstraction), but in a more appropriate and useful way.

Second, we need a plan *visualisation* which is able to capture: (1) hierarchical decomposition of plans (which are uniformly represented in a plan-specification library); (2) time-oriented plans; (3) sequential, concurrent, and cyclical execution of plans; (4) continuous (durative) states, actions, and effects; (5) intentions considered as high-level goals; and (6) conditions, that need to hold at particular plan steps. Additionally, all different time-oriented components of skeletal plans should be visualised in an easy to understand way. The domain experts, such as physicians, should understand the basic idea of skeletal plans. However, the domain experts do not need to be familiar with the syntax of skeletal plans (clinical protocols) to author them.

30.3 The Plan Representation Language Asbru: A First Solution

A common way to overcome the limitations mentioned in Section 30.2.1 is the representation of procedural knowledge as a library of skeletal plans. Skeletal plans are plan schemata at various levels of detail that capture the essence of procedures, but leave room for execution time flexibility in the achievement of particular goals [3]. However, the basic concepts of skeletal plans are not sufficient in the medical domain, either.

Asbru[2] reflects all the described complexity in a language using a LISP-like syntax. In Asbru, the following parts of a plan can be specified: *preferences*, *intentions*, *conditions*, *effects*, and *plan body (actions)*.

Preferences constrain the applicability of a plan (e.g., select-criteria: exact-fit, roughly-fit) and express a kind of behaviour of the plan (e.g., kind of strategy: aggressive or normal).

[2] Asbru is part of a larger project, called *Asgaard*. In Norse mythology, *Asbru* (or *Bifrost*) was the bridge to *Asgaard*, the home of the gods.

Intentions are high-level goals that should be reached by a plan, or maintained or avoided during its execution. These intentions are very important not only for selecting the right plan, but also for critiquing treatment plans as part of the ever ongoing process of improving the treatment. This makes intentions one of the key parts of Asbru.

Conditions need to hold in order for a plan to be ***started***, ***suspended***, ***reactivated***, ***aborted***, or ***completed***. Two different kinds of conditions (called preconditions) exist, which must be true in order for a plan to be started: filter-preconditions cannot be achieved by treatment (e.g., subject is female); setup-preconditions can. After a plan has been started, it can be suspended (interrupted) until either the restart condition is true (whereupon it is continued at the point where it was suspended) or it has to be aborted. If a plan is aborted, it has failed to reach its goals. If a plan completes, it has reached its goals, and the next plan in the sequence is to be executed.

Effects describe the relationship between plan arguments and measurable parameters by means of mathematical functions. A probability of occurrence is also given.

The *Plan Body (Actions)* contains plans or actions that are to be performed if the preconditions hold. A plan is composed of other plans, which must be performed in sequence, in any order, in parallel, or periodically (as long as a condition holds, a maximum number of times, and with a minimum interval between retries).

A plan is decomposed into sub-plans until a non-decomposable plan – called an action – is found. This is called a ***semantic stop condition*** for the decomposition of plans. All the sub-plans consist of the same components as the plan, namely preferences, intentions, conditions, effects, and the plan body itself. An in-depth discussion of Asbru can be found in [14].

30.3.1 Asbru Syntax

Plans in Asbru are written like in a programming language, as text that follows a very strict syntax. An example for a plan in Asbru syntax is given in Fig. 30.1.

30.4 Asbru for Users: AsbruView

Its LISP-like syntax makes Asbru easy to understand for people familiar with programming languages, but physicians usually do not have degrees in computer science. So a way of visualising Asbru had to be found that would make plans easy to understand, while making all of Asbru's features available.

These graphics are impossible to draw by hand, but the use of a computer opens new possibilities, like metaphor graphics and the use of colours.

```
(PLAN controlled-ventilation
  (PREFERENCES (SELECT-METHOD BEST-FIT))
  (INTENTION:INTERMEDIATE-STATE
                          (MAINTAIN STATE(BG) NORMAL controlled-ventilation *))
  (INTENTION:INTERMEDIATE-ACTION (MAINTAIN STATE(RESPIRATOR-SETTING) LOW
                                                 controlled-ventilation *))
  (SETUP-PRECONDITIONS (PIP (<= 30) I-RDS *now*)
    (BG available I-RDS [[_, _], [_, _], [1 MIN,_] (ACTIVATED initial-phase-1#)]))
  (ACTIVATED-CONDITIONS AUTOMATIC)
  (ABORT-CONDITIONS ACTIVATED
    (OR (PIP (> 30) controlled-ventilation [[_, _], [_, _], [30 SEC, _], *self*])
      (RATE(BG) TOO-STEEP controlled-ventilation
                                  [[_, _], [_, _], [30 SEC,_], *self*])))
  (SAMPLING-FREQUENCY 10 SEC))
  (COMPLETE-CONDITIONS
    (FiO2 (<= 50) controlled-ventilation [[_, _], [_, _], [180 MIN, _], *self*])
    (PIP (<= 23) controlled-ventilation  [[_, _], [_, _], [180 MIN, _], *self*])
    (f (<= 60) controlled-ventilation [[_, _], [_, _], [180 MIN, _], *self*])
    (state(patient) (NOT DYSPNEIC) controlled-ventilation
                                 [[_, _], [_, _], [180 MIN, _], *self*]))
    (STATE(BG) (OR NORMAL ABOVE-NORMAL) controlled-ventilation
                                  [[_, _], [_, _], [180 MIN,_], *self*])
    (SAMPLING-FREQUENCY 10 MIN))

  (DO-ALL-SEQUENTIALLY
    (one-of-increase-decrease-ventilation)
    (observing))
)
```

Figure 30.1. An example of Asbru code (part of a clinical treatment protocol for infants' respiratory distress syndrome (I-RDS)).

30.4.1 AsbruView Basics

AsbruView consists of two views: Topological View and Temporal View [9,10]. In this chapter, only Topological View and time annotations (which are a part of Temporal View) will be covered.

The basic part of Asbru and AsbruView is the plan. Inspired by examples in [19, 20], a rather unusual metaphor was found to represent these plans: running tracks (see Figs 30.2 – 30.6). Additionally, traffic signs and other symbols from the world of traffic control were taken to symbolise certain concepts [13].

We are using three-dimensional objects. The width represents the time axis, the depth represents plans on the same level of decomposition, and the height represents the decomposition of plans into sub-plans: The sub-plans of a plan appear on top of it (i.e., the base of the diagram is the root plan; see Fig. 30.2).

In Topological View, the time axis is only a topological quality, not an exact axis – it does not have a scale. Exact temporal dimensions are expressed in time annotation glyphs (see Section 30.4.4).

Plans can be reused as sub-plans of several plans. To make reused plans easier to spot, colours are used as a "fourth dimension": a plan is drawn using the same colour wherever it appears.

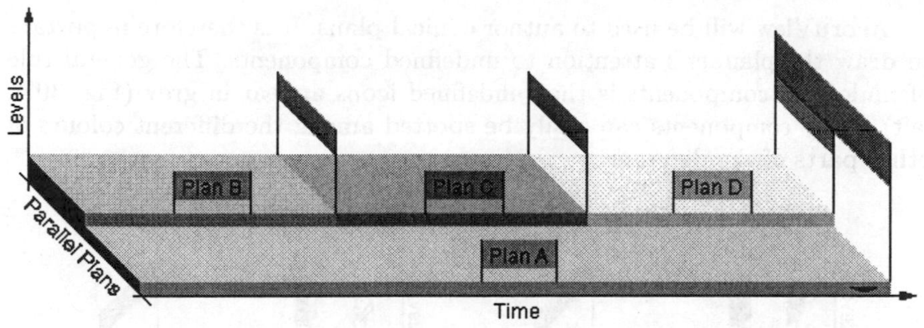

Figure 30.2. The decomposition of plans into sub-plans, and how sequential plans are represented.

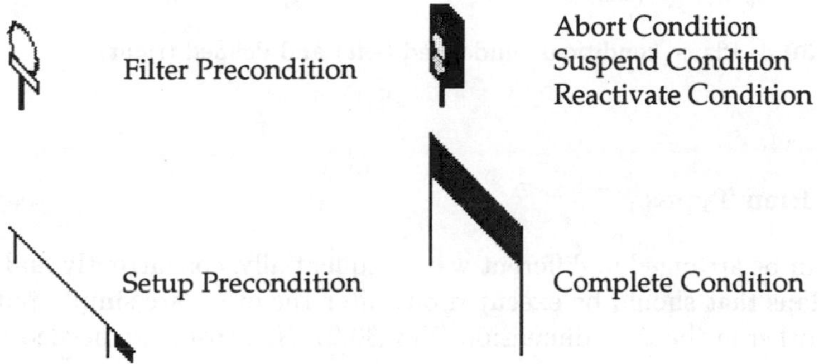

Figure 30.3. Metaphor graphics used for conditions in AsbruView.

30.4.2 Conditions

A patient is considered to be allowed to enter the running track when all preconditions become true for her. These preconditions are represented by a 'no entrance with exceptions' sign for the *filter-preconditions* (Fig. 30.3), and a barrier for the *setup-preconditions.* The characteristics of the two preconditions are reflected in the metaphors: the *setup-precondition* can be achieved by performing some action, so the barrier opens. The *filter-precondition* cannot be achieved, just as the 'exceptions' sign cannot be changed to gather access.

As long as no conditions become true that suspend the plan (the corresponding condition is symbolised by the yellow light of the traffic light), the patient 'runs' (i.e., the plan's actions are performed). When the *suspend-condition* becomes true, the runner must wait for the green light (*reactivate-condition*). Should the *abort-condition* (represented by the red light) become true, the runner is stopped and not allowed to continue. When the plan is completed (i.e., the *complete-condition*, represented by the finishing flag, has become true), she has reached the finishing line — and can proceed to the next plan.

AsbruView will be used to author clinical plans. It is therefore important to draw the planner's attention to undefined components. The general rule of undefined components is that undefined icons appear in grey (Fig. 30.4, left). Grey components can easily be spotted among the different colours of other parts of the diagram.

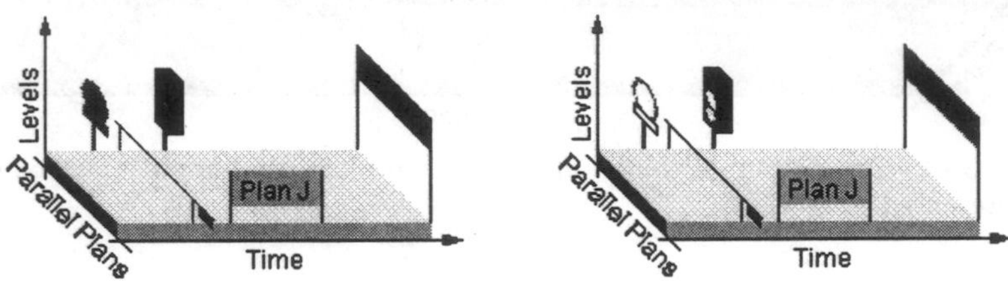

Figure 30.4. Plans' conditions: undefined (left) and defined (right).

30.4.3 Plan Types

Plans can be arranged in different ways: sequentially, concurrently and cyclically. Plans that should be executed one after the other are simply put next to each other in the time dimension (Fig. 30.2). No arrows are needed to link the plans.

Following Asbru keywords, plans that should be executed in parallel are called *all-together* or *some-together* plans (depending on whether all or only a subset of them must be executed). If the order of execution is not known at design time, the plans are called *all-any-order* and *some-any-order* plans, respectively.

Plans that should be executed in parallel are put alongside each other, so that more than one running track exists for a certain 'length' of time (see Fig. 30.5). To execute plans in parallel means that these plans should be started at the same time and performed in parallel. However, these plans may complete at different time points, depending on their *complete-conditions.*

Additional symbols are needed to visualise the other operators. If not all plans of a set have to be executed, but some can be omitted, these optional plans are covered with a question mark texture (Fig. 30.6). Compulsory plans are painted with plain colour (the set of compulsory plans is called the *continuation-condition*).

An *any-order* plan is drawn with a depression that plans can be put into once their order has been decided on. The sub-plans of such a plan are drawn so as to prejudge their temporal order as little as possible (Fig. 30.6).

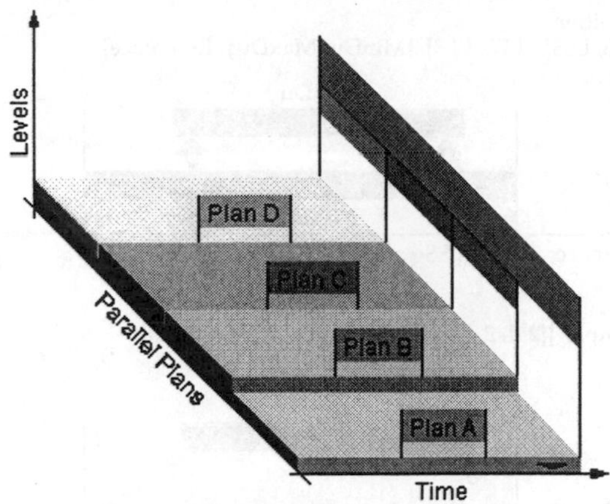

Figure 30.5. The representation of parallel plans.

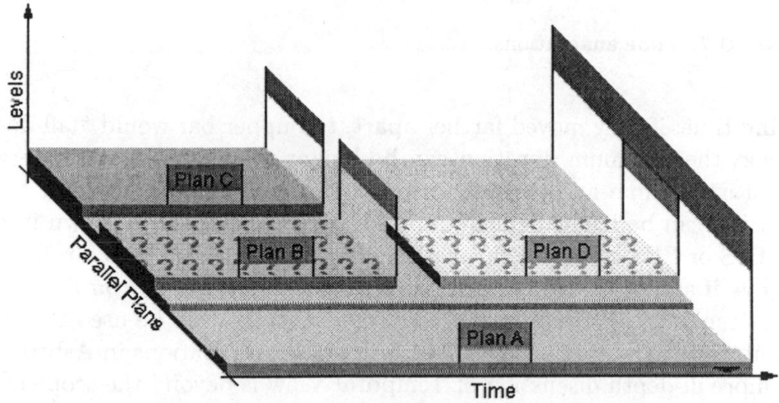

Figure 30.6. Some of the plans are to be executed in any order. Plans that do not have to complete are marked with a question mark texture.

30.4.4 Time Annotations

In Asbru, the amount of time a plan takes to execute is not necessarily known beforehand. A plan's (or condition's) temporal extent can be specified in Temporal View in terms of Time Annotation glyphs (Fig. 30.7). A glyph (also known as "Chernoff face" [1]) is a shape whose features reflect properties of the represented data – usually many properties in a single shape.

The representation used for these time annotations is a more abstract one. Two horizontal bars, one above the other, show the minimum and maximum duration (*MinDu* and *MaxDu*) of a plan. The earliest and latest starting (*ESS* and *LSS*) and finishing shifts (*EFS* and *LFS*) are labelled, and two diamonds support the upper bar (Fig. 30.7). This is meant to indicate that the minimum duration is dependent on the latest starting time and the earliest

Definition:
[[ESS, LSS], [EFS, LFS], [MinDu, MaxDu], Reference]

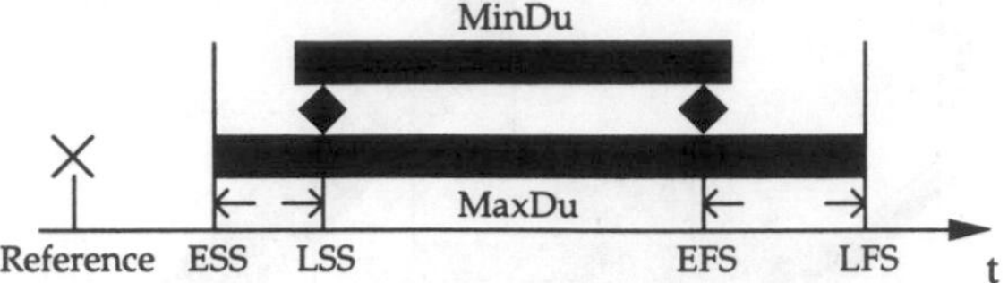

Example: [[2 d, 3 d], [_, 11 d], [6 d, _], Diagnosis]

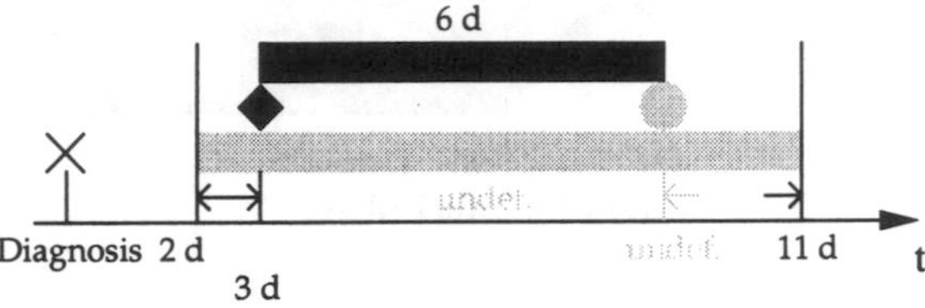

Figure 30.7. Time annotations.

finishing time: if they moved farther apart, the upper bar would "fall down". However, the minimum duration can be longer than the difference between latest starting time and earliest finishing time. Any set of parts of a time annotation can be left undefined; an undefined component is drawn in grey. If the LSS or EFS are not defined, they change their shape into a roll, which can move if any of the other parts are defined (Fig. 30.7, bottom).

So, despite its abstract nature, this representation can be used to explain the contraints as well as the possibilities of time annotations in Asbru.

A more in-depth discussion of Temporal View is beyond the scope of this chapter, but can be found in [9, 10].

30.5 What are Flow Charts?

A flow chart is a diagram for representing algorithms and showing the decision structure of a program. The basic elements of a flow chart are little boxes and arrows. The arrows connect the different boxes and sketch the control flow of statements (actions). These connectors can be used to sequence particular statements, to loop over one or more statements, or to go to another path of statements (*goto* statement). The different shapes of boxes denote the various kinds of statements (e.g., circles are start and stop buttons, little rectangles are commands or actions; diamonds are decisions; advanced rectangles are input and output data). Boxes can be numbered to refer to other diagrams that refine their contents. When introduced in [6], the little boxes and their contents served as a high-level language, grouping the inscrutable machine language statements into clusters of significance.

When talking about flow charts, we refer to the "classic" flow charts defined by [6] and standardised in DIN 66001, not one of the many extensions (e.g., [12,16]). These "classic" flow charts have a number of drawbacks, which may be of minor or no importance at all in other areas, but which make them impossible to use for our problem domain.

30.6 Why not Flow Charts?

30.6.1 Time

Clinical treatment involves one very important variable that is not accounted for in flow charts: time. It can be of vital importance not only to treat the patient right, but also to apply the treatment on time, or a certain number of times (with a certain interval in between). Time also plays an important role for the conditions that must be met in order to make use of a plan. A certain observable may not trigger a plan simply because of its value, but because of its behaviour over time (e.g., rising or falling at a certain pace, or for a certain time).

The progress of time is simply not visible in flow charts; but in AsbruView, it is a part of the design.

30.6.2 Scalability and Object Orientation

In Asbru, therapies are broken up into small, manageable parts. Once a part is defined, it can be used as a building block for more complicated plans, which can themselves be used as building blocks, etc.

In order to use these building blocks, one simply includes them in the plan (as an alternative, for example). The conditions for applying the plan are part of the "building block" plan itself, and thus need not be redefined in every plan that uses it.

Fig. 30.8 shows a flow chart of an Asbru plan. It is easy to see how much simpler an AsbruView diagram is to understand, and how complicated a flow chart would become if it consisted of many plans.

When a patient is treated, a virtually unlimited number of complications may occur. In such a case, an Asbru plan is aborted, and the appropriate plan is sought to deal with the complication. It is impossible to add conditions and links for every complication when using flow charts. So another means has to be used, which shows how ineffective and insufficient flow charts are for this domain.

This certain "object orientation" for conditions has another dimension as well: Plans inherit the using plan's preferences, intentions, conditions, and effects – but not their actions. This increases the possible modularity of the plans and reduces redundancy.

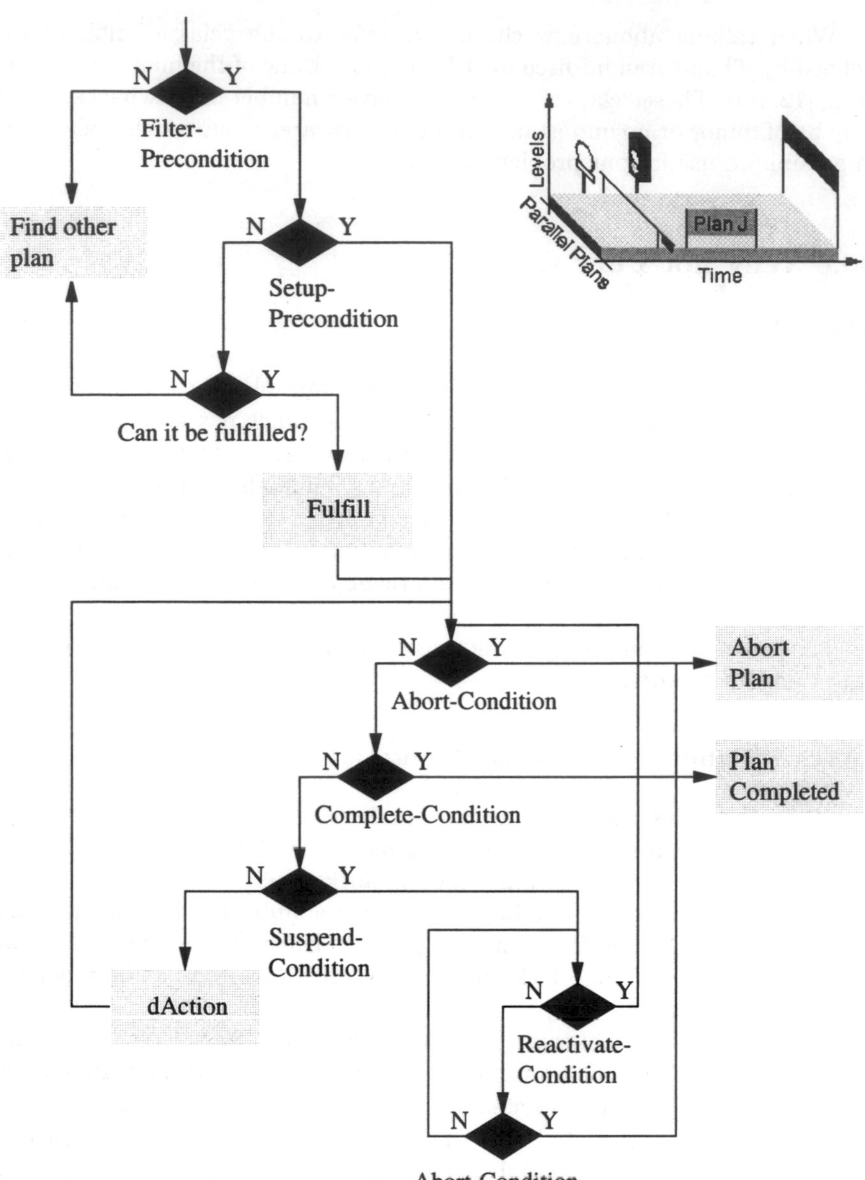

Figure 30.8. Asbru's conditions in a flow chart. Compare the flow chart to the simple symbols used in the inset (and Fig. 30.4).

30.6.3 Complex Constructs

Flow charts don't support different kinds of decisions, either. There is only one "if condition A is true then goto α else goto β" construct. For practical purposes, it is impossible to create a "do some of (α, β, γ)", or a "do α, β, and γ in any order", even if one could build more complex parts from the available primitives. Some of these constructs are possible, but make the resulting flow charts impossible to read – which clearly contradicts the concept of flow charts.

30.6.4 Computer-Aided Protocol Design

With most flow charting tools, one gets very little support when changing parts of a chart. Suppose there are three plans, A, B and C, in that order. After we have drawn the arrows accordingly, we realise that C should go in between A and B. So we have to delete the arrows from A to B and from B to C, move C between A and B, and then connect A to C and C to B.

AsbruView supports such changes: Simply take plan C and drop it on the edge between A and B – it will be inserted in the correct spot.

Conditions that have yet to be defined are displayed in grey. This is an important help for the plan designer to keep track of what has been done already and what still needs to be done.

30.6.5 Miscellaneous

As mentioned, plans need not only be applied, but their usefulness needs to be analysed in order to improve the quality of treatment. Flow charts provide no means of including a plan's intentions – AsbruView does. This is, of course, also true for the different kinds of conditions (that cannot be further differentiated with flow charts), preferences, and effects.

Real-world clinical plans tend to consist of a huge number of actions and conditions and to be of a quite complicated structure. Therefore, more complicated plans are difficult if not impossible to oversee, when drawn using flow charts. Additionally, flow charting tools typically provide no means for splitting a larger plan into parts – especially in an "incremental" manner, i.e., the decision to split the plan (and how) is made *after* the design has been started.

30.7 Conclusion and Future Plans

We have described the various problems authors of clinical plans face. Flow charts are a very commonly used representation for clinical plans. However, flow charts are not the appropriate means for representing such complex, time-oriented plans. We presented three-dimensional diagrams, called AsbruView, which overcome these limitations. The proposed diagrams utilise the metaphors of running tracks and traffic control, to visualise such clinical plans in an easy to understand way. The benefits of AsbruView are: (1) to handle all temporal dimensions of plans, conditions, intentions, and effects; (2) to cope with all possible, as well as unpredictable, orders of plan execution; (3) to deal with all the exception conditions that might arise; and (4) to deal with domain-specific features, like plans' intentions.

Our aim is to use AsbruView during the design and the execution phase. Therefore, we will adapt AsbruView to be used to author a protocol during the design phase as well as to visualise the performed protocols during the execution phase in a user-appropriate and task-specific way.

Acknowledgements

The authors wish to thank Georg Duftschmid, Johannes Gärtner, Klaus Hammermüller, Werner Horn, Franz Paky, and Christian Popow for helpful comments and discussions; and Leonore Neuwirth for her help with getting hold of an important paper. This work is part of the Asgaard Project, which is supported by *Fonds zur Förderung der wissenschaftlichen Forschung* (Austrian Science Fund) grant P12797-INF.

References

1. Chernoff, H. (1973). The use of faces to represent points in k-dimensional space graphically. Journal of the American Statistical Association 68:361–368.
2. Fox, J., Johns, N. and Rahmanzadeh, A. (1997). Protocols for medical procedures and therapies: A provisional description of the PROforma language and Tools. In E. Keravnou, C. Garbay, et al. (Eds), Proceedings of the 6th conference on artificial intelligence in medicine Europe (AIME-97), Grenoble, France, 23–26 March. Berlin: Springer, pp. 21–38.
3. Friedland, P.E. and Iwasaki, Y. (1985). The concept and implementation of skeletal plans. Journal of Automated Reasoning 1(2):161–208.
4. Genesereth, M.R. and Fikes, R.E. (1982). Knowledge interchange format, version 3.0 reference manual. Tech.report logic-92-1, Computer Science Department, Stanford University.
5. Goldberg, A. and Rubin, K.S. (1995). Succeeding with objects: Decision frameworks for project management. Reading, MA: Addison-Wesley.
6. Goldstine, H. and von Neumann, J. (1947). Planning and coding problems for an electronic computing instrument, Part ii, Vol.1. US Army Ordinance Department. Reprinted in von Neumann, J. (1963). Collected works, Vol. V. New York: McMillan, pp. 80–151.
7. Guarina, N. and Giaretta, P. (1995). Ontologies and knowledge bases. In N.J.I. Mars (Ed.), Towards very large knowledge base. Amsterdam: IOS Press.
8. Hripcsak, G., Ludemann, P., Pryor, T.A., Wigertz, O.B. and Clayton, P.D. (1994). Rationale for the Arden syntax. Computers and Biomedical Research 27:291–324.

9. Kosara, R. (1999). Metaphors of movement: A user interface for manipulating time-oriented, skeletal plans. Master's thesis, Vienna University of Technology, Vienna, Austria, May.
10. Kosara, R. and Miksch, S. (1999). Visualization techniques for time-oriented, skeletal plans in medical therapy planning. In W. Horn, Y. Shahar, G. Lindberg, S. Andreassen and J. Wyatt (Eds), Proceedings of the joint European conference on artificial intelligence in medicine and medical decision making (AIMDM'99), Berlin: Springer, pp. 291–300, Aalborg, Denmark.
11. Levinson, R. (1995). A general programming language for unified planning and control. Artificial Intelligence 76(1–2):319–275.
12. Martin, J. (1973). The "natural" set of basic control structures. SIGPLAN Notices 8(12):5–14.
13. Miksch, S., Kosara, R., Shahar, Y. and Johnson, P. (1998). Asbruview: Visualization of time-oriented, skeletal plans. In Proceedings of the 4th international conference on artificial intelligence planning systems 1998 (AIPS-98), Pittsburgh PA, Menlo Park, CA: AAAI Press.
14. Miksch, S., Shahar, Y. and Johnson, P. (1997). Asbru: A task-specific, intention-based, and time-oriented language for representing skeletal plans. In Proceedings of the 7th workshop on knowledge engineering: Methods and languages (KEML-97). Milton Keynes, UK: Open University.
15. Musen, M.A., Gennari, J.H., Eriksson, H., Tu, S.W. and Puerta, R.A. (1995). PROTÉGÉ-II: A computer support for development of intelligent systems from libraries of components. In Proceedings of the 8th world congress on medical informatics (MEDINFO-95), pp. 766–770.
16. Nassi, I. and Shneiderman, B. (1973). Flowchart techniques for structured programming. SIGPLAN Notices 8(8):12–26.
17. Ohno-Machado, L., Gennari, J.H., Murphy, S., Jain, N.L., Tu, S.W., Oliver, D.E., Pattison-Gordon, E., Greenes, R.A., Shortliffe, E.H. and Barnett, G.O. (1998). The GuideLine Interchange Format: A model for representing guidelines. Journal of the American Medical Informatics Association 5(4):357–372.
18. Pattison-Gordon, E., Cimino, J.J., Hripcsak, G., Tu, S.W., Gennari, J.H., Jain, N.L. and Greenes, R.A. (1996). Requirements of a sharable guideline representation for computer applications. Report no. smi-96-0628, Stanford University.
19. Tufte, E.R. (1990). Envisioning information. Cheshire, CT: Graphics Press.
20. Tufte, E.R. (1997). Visual explanations. Cheshire, CT: Graphics Press.

31. Acting with Diagrams: How to Plan Strategies in Two Case Studies

Nadine Lucas

Nathalie Cousin-Rittemard

In this chapter we present two case studies on how diagrams were used to communicate across different academic fields: linguistics and mechanics. The problem we faced was to define a global structure from data usually considered at a lower scale. The definition of implicit rules in previous representations of phenomena was the first stage. Reasoning on measure and scale led to reuse of conventional reference systems. Changes in data disposition on the sheet of paper illustrate the intermediate step of the perceptive and intellectual construct.

31.1 Introduction

What can a mechanicist talk about with a linguist? Reasoning, diagrams, and communication. How would you draw the flow of a fluid in a rotating machine and how would you draw the unfolding of discourse?

As was stated by Tufte [10], the problem in visual representations is "escaping flatland".

31.2 What is the problem?

The problems we are tackling are complex: fluid mechanics and rhetorics. Although they seem wide apart, both subjects involve the notion of structure. The structures can be naively thought of as a three-dimensional approach, which is obviously linked to the phenomenon itself in the case of physics: the world is supposed to have three spatial dimensions. In the case of a text, we start from the perception of symbols arranged in the flatland: a sheet of paper. We immediately select, say, a sentence, and rearrange the "linear word order" into a syntactical structure, usually drawn as a tree. We thus gain an extra item of information: hierarchy, symbolised by top and bottom of the diagram.

In rhetorics and fluid mechanics, the structure definition is implicit to the subject. In case of linguistics, the structure is assimilated to a sentence.

As for fluid mechanics, a structure may be viewed: as a locally spatial set of the positive and negative parts of the temporal fluctuation with respect to a temporal average of a variable in the case of unsteady flows; and as the global set of analytical (self-similar) zones in the case of steady flows.

Our objective is to yield a global description of phenomena or/and a description of a global phenomenon. Furthermore, we would like to describe the coexistence and the dynamics of phenomena occurring in different spatial, temporal and spatiotemporal scales.

Hence, we act with diagrams. The diagrams become an investigation support: aids to intellectual activity, objects we can reflect upon and criticise. Eventually the diagrams carry out the concepts to be transmitted.

It soon becomes clear that diagrams need not be realistic. Ultimately, a diagram should enhance the understanding of the observed phenomena. In fact, as pointed out by Bachelard [2], in science we can only see through concepts which give some sort of coherence to our perception of the world. Hence, the implicit rules of diagram construction become important: What is explained and what is left out?

31.3 The matter

In our view, the definition of the object to be observed is important: its nature (direct image or indirect image of the "observable", and static or dynamical object) and its acquired representations. In the diagrams we use, we have to see through many choices. The scientific object under attention is not necessarily a direct representation of observations; it is often a selection or a combination of data. The inherited properties of the object comprise the choice of reference system, definition of variables, and choice of scale for time and space. We have to understand what was the problem, how it was drawn, and how it was solved. But sometimes, and for both academic subjects, we would like something more.

We distinguish the data: what is physically perceived from the observed object, on one hand, and what is assumed to be an indirect image of the real world on the other hand. Our data are not rough data. We have assumed that the previously defined structures are local structures and that a larger-scale analysis can be conducted. Hence the fruits of our work are eventually structures but considering another scale.

31.4 The Problem of Point of View

Having distinguished the objects, we then consider the necessity of making a distinction between *physical perception* (observation stage) and the *mental perception* (judgement stage). Henri Poincaré's advocation of the no-use of the verb *perceive* [7] has pointed out the confusion between the sensitive and the jugement senses of the term perception. For example, when we physically perceive a geometric distance to be smaller than another, the distance often is greater than the other one. But when we qualitatively "judge", comparing the order between the same distances, according to conveniently adopted conventions, we often state that one distance is smaller than the other one.

As first-hand observation is the basis for scientific work, discriminating between *physical* and *mental perception* helped us state the conventional value of the objects we consider and to use physical perception as a sharpened tool.

Two main questions then arise.

Firstly, the problem of the point of view: what do we "physically perceive" and what can we "mentally perceive"?

Hence, how do we state the observer's point of view on a phenomenon which is at least three-dimensional? Do we try to stick to a two-dimensional view, a collection of views or do we switch to a three-dimensional view? In which frame do we choose to express our view on the scrutinised object?

Lastly, the question of measure is directly related to the problem of envisioning and sketching phenomena: which measure? Furthermore, calculus seems to induce the related requirement of defining the well-suited range of operations leading to a meaningful result, within a defined frame and for the defined set of datas.

31.5 Shift and Solve

31.5.1 Linguistics

Starting from the most common diagrams used in linguistics, Fig. 31.1 shows how syntactical trees or stemmas expand a segment considered to be "linear": the sentence. The observer defines a tree structure, here in a drastically simplified Chomsky grammar, and elements which are functional segments. By convention, the window through which language is observed frames a sentence, made of phrases. These segments are in turn broken down into subsegments until the ultimate step of analysis, which is the word (a graphical unit). Thus the nodes in the tree implicitly represent the layers of hierarchy provided for in the description. The tree can develop two branches at each layer, as in the case of Chomsky grammar, or more, as in the case of Tesnière. The *linear order* seen on the horizontal axis is related to the *functional order*, to be read on the vertical axis. But the horizontal and vertical axes are not drawn, and the implications of the choices in drawing are not often discussed as such. For instance, the nodes usually are more numerous at the deepest layer, and their number regularly decreases until the last node. But in some cases there is a solitary node at one intermediate layer, and the assumption that a solitary node means "completion of the structure" must be reassessed.

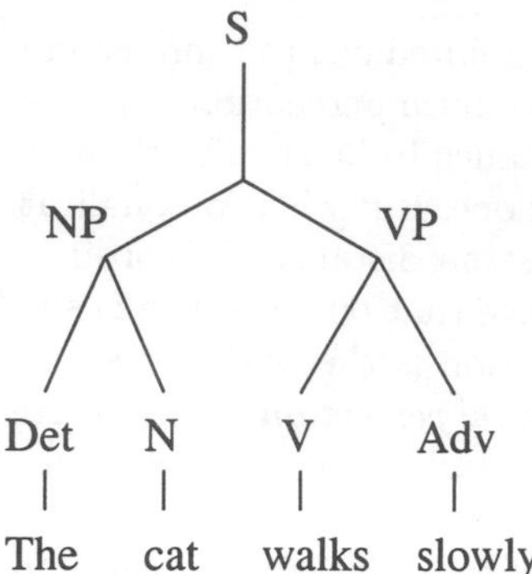

Figure 31.1. Simplified syntactical tree.

The box description is used by another school of linguists, the distributionalists [4]. The usual frame of observation, called the *resulting form*, is the sentence unit. When a text is described, the *grain* is different [5]. Figure 31.2 and the following show a short text arranged in boxes. Here the structure, the *resulting form*, is instantiated by a paragraph (as defined by typography) and its elements are the sentences.

The disposition on the sheet of paper is made along rules, according to the linguistic perspective on the data (the text). In Fig. 31.2, the layout shows the thematic development. The choice is to clearly distinguish the *Theme* and the *Rheme*. The sub-diagrams in the margins indicate the method of arrangement of the elements: the order in which to read the sentences. The top slab, corresponding to the *Theme*, is to be read in pairs of sentences. The bottom slab, corresponding to the *Rheme*, is to be read sentence after sentence as forming a square. This Pythagorean representation was chosen among the many squares illustrated in [6]. The last element is out of the slabs and consequently described as having a closure function.

In such a diagram, the chronological order of appearance of the sentences is broken, and is replaced by a spatial order enhancing the interpretation of the thematic role of the subsets in the global structure. Moreover, as the rules to draw figures depend explicitly on the number of elements, the link between geometrical figures and rhetorical figures is established and can be discussed.

Figure 31.3 shows the same short text arranged in a different perspective: macro-syntax. Each element is symbolised as a box and not as a line. Each numbered square box, drawn in a dotted line, contains a sentence defined as a graphical unit (with a) and b) for the fifth sentence split by the two-dot punctuation mark). The black rounded boxes delineate the functional groups and are related by arrows indicating the syntactical relations. The aim of the figure is to show the dependency relations between the groups of sentences. The thickness of lines indicate the successive inclusions. The disposition rules

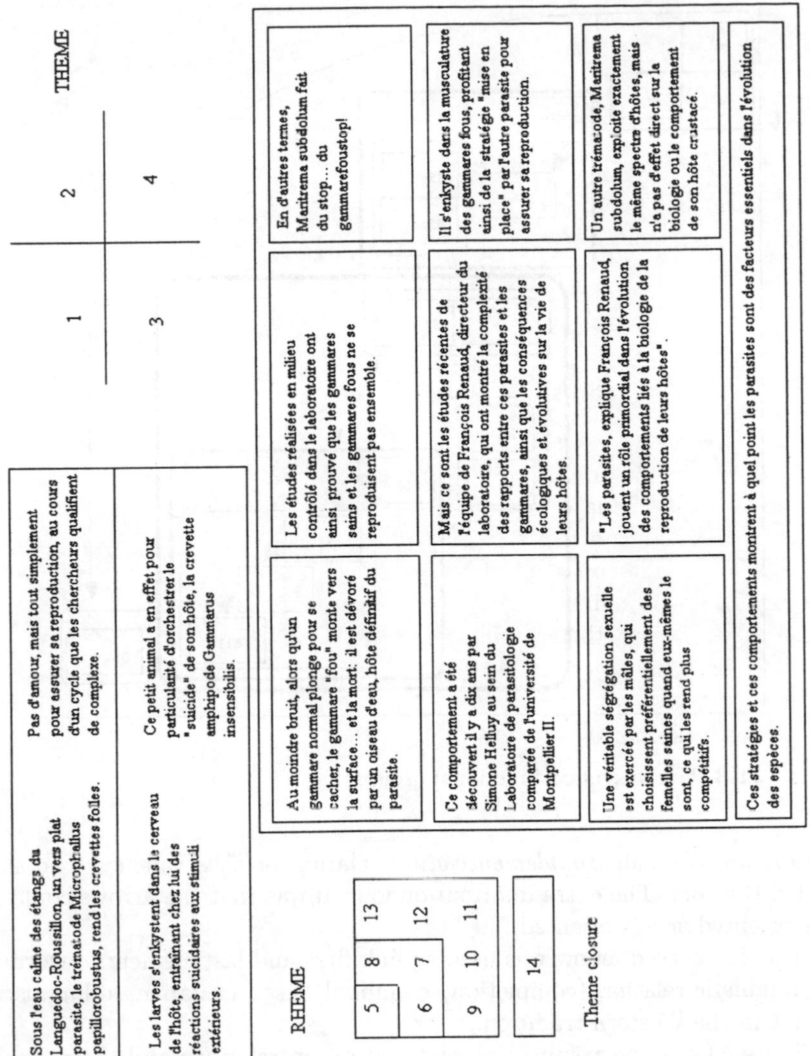

Figure 31.2. Text in boxes: thematic development.

are conventions. For instance, a shared convention in the linguistic community using sets is to draw two boxes in a row to symbolise "co-ordination" while two boxes, one on top of the other, symbolise "sub-ordination".

The diagonal layout convention for the relation between two functional groups is personal. At the first stages of the study, it was simply felt more convenient to draw arrows between large boxes placed on a diagonal. In fact, as the convention "one box on top of another" had a conventional meaning, it was felt necessary to adopt a different drawing convention for a different

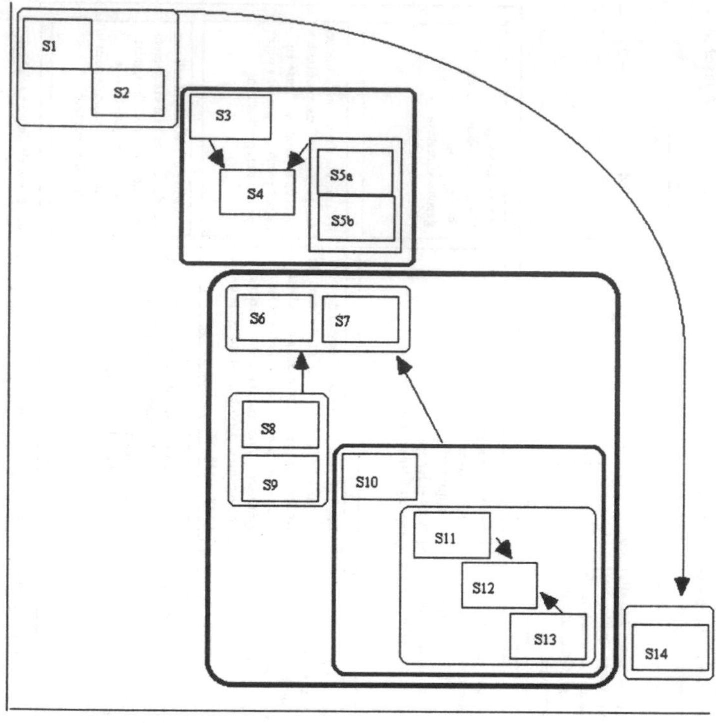

Figure 31.3. Macro-syntax: functional groups.

relation we may call *complementisation* relation or case valency. The shift fulfilled this aim. The extra information gained was that the arrows could be right oriented or left oriented.

Last, the curved arrow was used to link first and last elements according to a linguistic relation (completion) commonly used in Japanese linguistics but not in the Western tradition.

Figure 31.4 is the result of an elaboration on relations and their spatial rendering on the sheet of paper. The perspective in the analysis is again syntax but, with an additional information on the intermediate level of analysis relating to discourse. Instead of being regularly placed below the previous one and shifted to the right, the rounded black boxes (the functional groups) are sometimes placed below the previous one and shifted to the left. This is seen for the box grouping sentences S11–S13, which is placed left of the S10 box. The last sentence S14 is also placed leftwards. The change in progression reflects changes in morpho-syntactical forms, including pronominal sentence, and use of long-range anaphorical terms such as *autre* (other) and *même* (same). There is another specification to be stated (the choice between the right and the left shift), which allows a better differentiation of criteria.

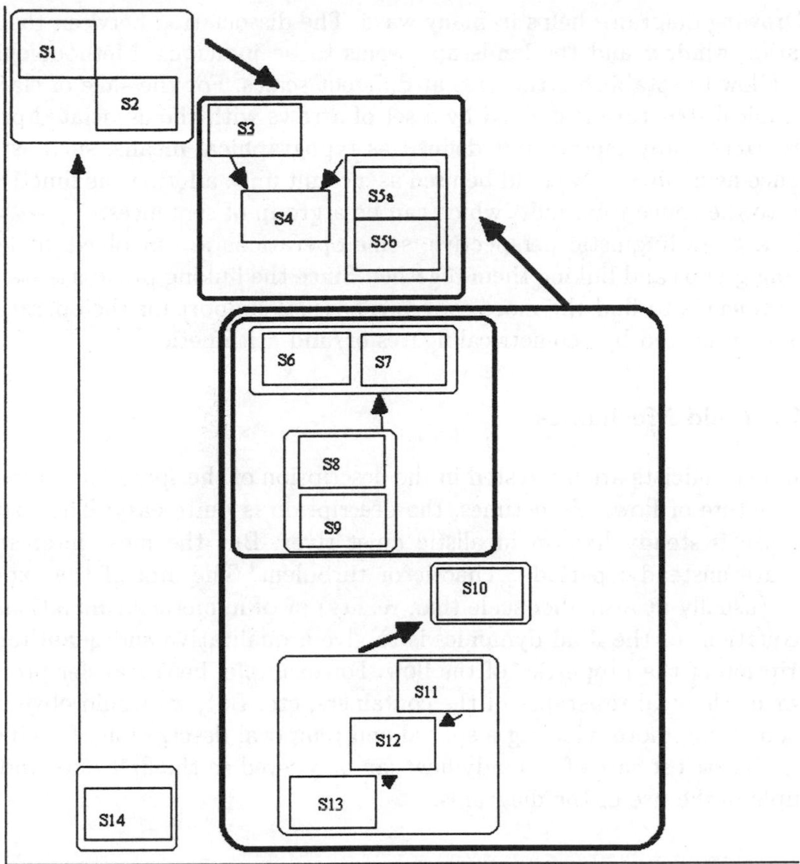

Figure 31.4. Macro-syntax of discourse.

The diagram as a whole can be read on the horizontal axis with the sheet of paper held vertically, as is usual. Then, the *local* relations between smaller black boxes can be read as complementisation, co-ordination, etc. But when the sheet of paper is held horizontally, the *macro*-relations become more conspicuous: the three larger black boxes reflect the same *structure* (with a centre indicated by the arrows) as the *local* ones. The analysis scale is differentiated by the way the paper is held, or by the reading angle of the diagram. The curvature effects (successive right and left shifts) are now read with the sheet of paper held horizontally as showing an upward movement with the apex marked by box S10 and a downwards movement. Both movements show an adjunct sentence (S10 citative and S14 evaluative) as their borderline. This allows a more detailed account of the completion concept than the curved arrow (see Fig. 31.3), which disappeared as such and was replaced by a straight arrow (see Fig. 31.4).

Drawing diagrams helps in many ways. The dissociation between the observation window and the landscape seems to be judicious. Methodological tools allow to establish structures at different scales. For the sake of clarity, a canonical structure is defined by a set of arrows with the associated properties. Hence, any metric unit defined as typographical means, such as the sentence here, obviously could be used as a count unit, alloting the functional value to the conceptual unit, which can be a group of sentences.

In a given linguistic perspective, some operations are involved: at least forming groups and linking them together. Since the linking process is stated, the problem is to find the more convenient visual support for the operations being transmitted by geometrical figures or/and arithmetic.

31.5.2 Fluid Mechanics

Fluid mechanicists are interested in the description of the spatial and temporal structure of flows. Sometimes, the description is quite easy: The spatial structure is steady like an idealistic quiet river. But the most interesting flows are unsteady: periodic, chaotic or turbulent. The aim of the experiments (usually at a smaller scale than reality) or of numerical simulations of the equations of the fluid dynamics is to give a qualitative and quantitative description of the properties of the flow. For example, heat transfer properties or mechanical resistance of the containers, etc. But, we would obviously like something more: yielding a spatial and temporal description of the flow. Then, the easiest case of a steady flow can be viewed as the first outstanding example of the use of the diagrams.

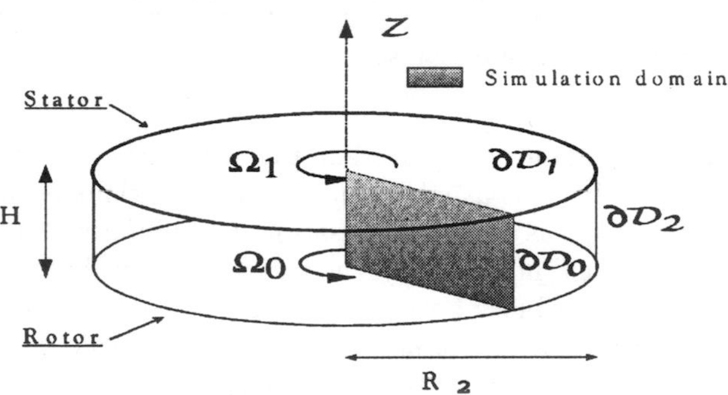

Figure 31.5. Two-disk system.

The two disk systems flows are characteristic of turbomachinery flows. The physical domain is a fluid domain between two parallel coaxial rotating disks closed by a vertical end-wall (see Fig. 31.5). On one hand, Fig. 31.5 is a descriptive sketch of something real, something which could be found in everyday life: a schematised concrete object. On the other hand, the diagram shows that we assume that the flow is axi-symmetric, which means that the simulation domain is a "slice" of fluid (drawn in grey). Figure 31.5 is then characterised by the coexistence of mathematical abstraction (the simulation domain) included in reality (the physical domain).

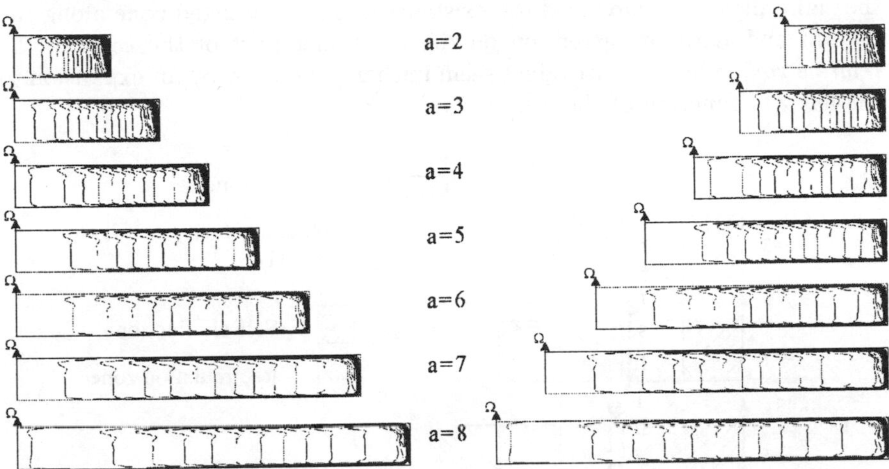

Figure 31.6. a (left) , b (right). Equally spaced isovalues of stream function, for aspect ratio in a range of a = 2,3,...,8. The isovalues vary from 0 at boundary down to $-6.7\ 10^3$.

Figure 31.6 is highly abstract. It gives a computerised representation of the set of data observed in order to describe the steady flows as a function of the cavity size, or more properly as a function of aspect ratio: the ratio between the radius and the axial clearance ($a = R_2/H$). The aspect ratio, as the geometrical global parameter of the cavity, is one of three global parameters characterising the flow in the cavity. The two other global parameters are fixed in this study (see [3]).

Figure 31.6 highlights the importance of the data layout in the flatland. In order to allow comparison, say, the "natural" and usual layout of the data is Fig. 31.6a (at left) and a new but convenient layout of the data is Fig. 31.6b (at right). For both frames the same equally spaced isovalues[1] of a characteristic function are drawn as a function of the aspect ratio from top to bottom (a = 3, 4,..., 8). The left boundary of the sections is the axis, the right boundary is the vertical end-wall and the top and bottom boundary are the disks.

The sections of Fig. 31.6a are "naturally" left-aligned with the axis of rotation, which is the "natural" point of view in the sense of pyramidal piling

[1] In common life, an example of isovalues lines are the lines joining points of the same altitude on a map for walkers. Here the isovalue lines are drawn for a characteristic function of flow, so that they are also trajectories.

of the cavities (only by imagination), considering increasing sizes. Conversely, the sections of the Fig. 31.6b are conveniently right-aligned with the vertical end-wall, which is unusual but highlights the existence of an ***absolute zone*** in the neighbourhood where the isovalues seem unchanged as a function of the aspect ratio. The division in zones became more conspicuous in Fig. 31.6b: the right alignment enhanced the existence of the unchanged zone along the margin and attracted attention on the remainder part of the section, the *relative zone*, where the isovalues seem unchanged except by an expansion to the left as a function of the aspect ratio.

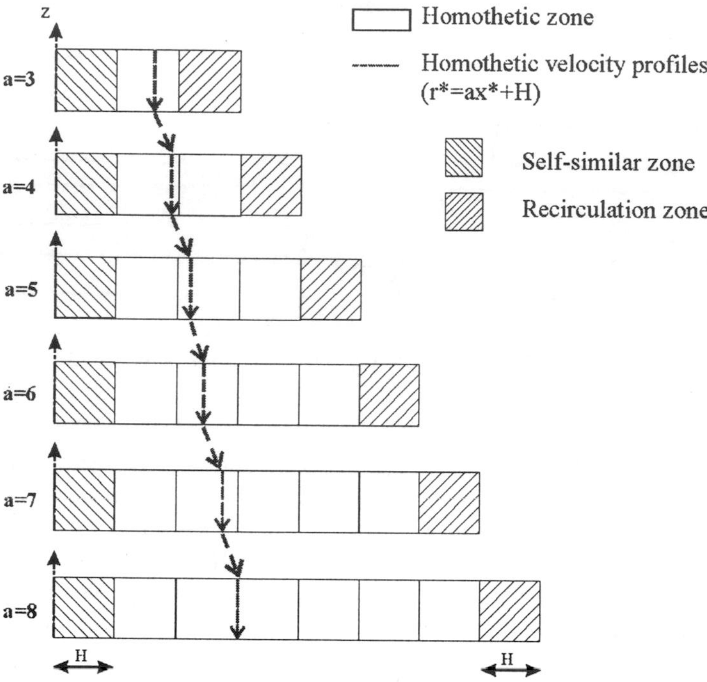

Figure 31.7. Steady flow as a function of the aspect ratio.

This leads to a novel interpretation of the diagram in terms of self-similar zone, homothetic zone and recirculation zone. Figure 31.7 eventually shows the three zones defining the global structure of the flow as a function of the global geometrical aspect ratio parameter: the two zones previously defined and extra information on the structure of the *relative zone*. Figure 31.7 is left-aligned in order to underline the "expansion law" in relation to the end of the self-similar zone (left hatched zone). The "expansion law" is underlined by the superposition as a function of the aspect ratio (red dot arrows) of the vertical velocity profiles (red dot lines). Another point of view on the global structure of the steady flow as a function of the aspect ratio is given in Fig.

31.8 with the pyramidal piling of the sections. Although the final results are not yet published, self-similitude could be hypothesised. A more abstract and clearer mathematical view on the phenomenon is given.

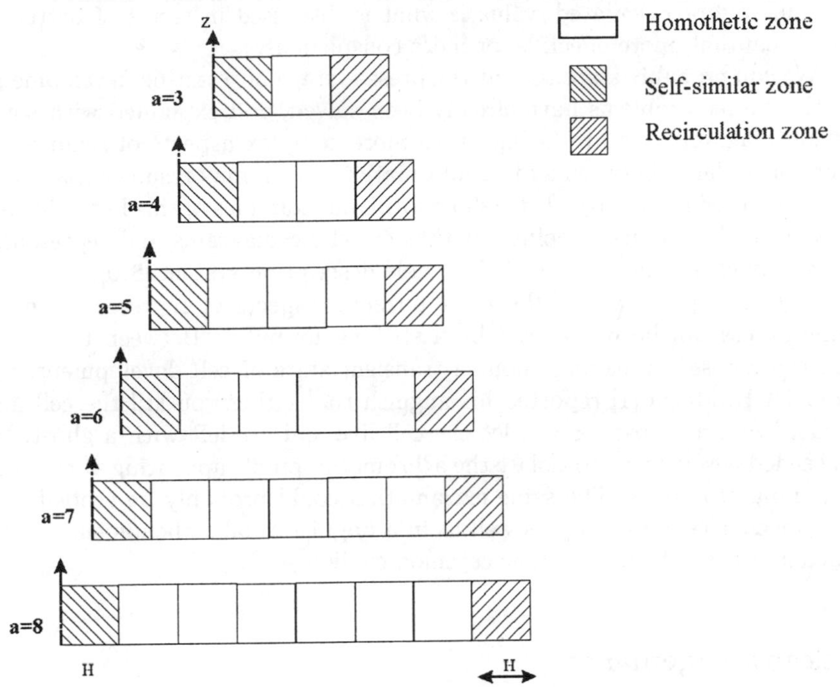

Figure 31.8. Steady flow as a function of the aspect ratio.

31.6 Conclusion

We have evoked the same epistemological problems in our respective academic fields: data disposition, orientation, system of reference, local and global structure, from a point of view which can be considered to be "static" and internal with respect to the diagram.

We have also considered shifting diagrams, rotation of the page, using information given by a diagram to produce the next one. This is more external to the matter and closer to the way we not only think with diagrams but also act with them.

We presented some stages of our reflection on structure problems by showing intermediate re-representations, leading to reperception of the global structure abstracted from a collection of local units.

The rules of the game can be more or less sophisticated; they describe the set of relations the observer wishes to use, and to teach. Interpretation of the diagram is always necessary, because what man is interested in is "functional value" or "exchange value". While measure is what is objectively represented, manipulated or calculated, value is what is discussed in terms of better or more beautiful, more effective, or more conspicuous.

We can probably state that at the present stage of scientific development, all the simple problems have already been solved, and explained with some simple diagrams. We are facing up to more complex aspects of reality and therefore balance between adding information, which may result in maziness, or reorganized knowledge: Tufte shows brilliant examples of this last solution. Stewart and Golubitsky point out that enriching diagrams and representations may cause the collapse of the whole description system [8,9].

In this respect, one of the most puzzling aspects of representation we came across can be introduced by a striking formula: "Between the ghost and the corpse". Speaking about a transient state of cell development, the biologist H. Mazia [1] reported his anguish that either you kill the cell and are left with a corpse, or you let the cell live and are left with a ghost. He succeeded nevertheless to isolate the achromatic spindle appearing at the very beginning of mitosis. The same exclamation could probably be echoed in a fluid mechanics laboratory as well as in a linguistics lab. Phenomena taking place in a flash of time are now common challenges.

Acknowledgements

Cited material is reprinted from Dagneaux, P. 1997, L'enigme des crevettes folles, *le journal du CNRS*, (July/August 1997, no. 91–92), p.22 with kind permission.

References

1. Atlan, H. (1979). Entre le cristal et la fumée: Essai sur l'organisation du vivant. Paris: Seuil.
2. Bachelard, G. (1938). La formation de l'esprit scientifique (pocket edn 1996). Paris: J. Vrin.
3. Cousin-Rittemard N. (1996). Contribution à l'étude des instabilités des écoulements axisymétriques en cavité interdisques de type rotor-stator. Doctoral dissertation, University of Paris 6, Paris.
4. Hockett, C.F. (1958). A course in modern linguistics. New York: Macmillan.
5. Lucas, N. (1993) Syntaxe du paragraphe dans les articles scientifiques en japonais et en français. In Moirand et al. (Eds), Parcours linguistiques de discours spécialisés. Berne: Peter Lang.
6. Nelsen, R.B. (1993). Proofs without words: Exercices in visual thinking. Washington: Mathematical Association of America.
7. Poincaré, H. (1902). La science et l'hypothèse. Paris: Flammarion.

8. Stewart, I. (1994). Visions géométriques. Paris: Belin.
9. Stewart, I. and Golubitsky, M. (1992). Fearful Symmetry: is God a geometer? New York: Penguin.
10. Tufte, E.R. (1990). Envisioning information. Cheshire, CT: Graphics Press.

32. Specifying Diagram Languages by Means of Hypergraph Grammars

Mark Minas

For working with diagrams on a computer screen we need diagram editors, i.e., graphical editors specialised in the specific diagram language. In order to create such a diagram editor in a methodical way, a formal representation of each diagram and of the whole diagram language is required. This chapter describes continued work on how to specify a wide range of diagram languages in terms of a hypergraph model together with hypergraph grammars. The specification of a diagram language can serve as input for an automated generator which creates a diagram editor for the specified diagram language. Editors support syntax-directed editing as well as free-hand editing of the diagrams, which are internally represented by hypergraphs. For free-hand editing, a hypergraph parser is used to obtain the diagrams' syntactic structure and to distinguish correct diagrams or diagram parts from incorrect ones.

32.1 Kinds of Diagrams

Diagrams are a powerful means for representing complex situations since they directly support the visualisation of multidimensional relationships and thus human perception. Diagrams are established in nearly every field. Independent of their specific field of application, however, three main building principles of diagrams can be identified where each diagram class may incorporate more than one of these principles: Diagrams like *graphs* express relationships between node-like, visual items by connecting them using lines or arcs. Typical examples are finite automata, Entity-Relationship diagrams, flow charts, Petri nets, or connection diagrams in electrical engineering. Instead, *recursively* built diagrams consist of a few primitive building blocks which can be grouped together in order to create new, more complex building blocks. An example are Nassi–Shneiderman diagrams which build larger structures by horizontally or vertically attaching smaller structures. State charts are graph-like as well as recursive: States are connected by arcs, and each state can consist of a state chart again. Finally, diagrams may use general *spatial relationships* like touching, overlapping, intersecting etc. for expressing complex situations. Typical examples are Euler diagrams for expressing sets and operations on sets, and VEX as a visual λ-calculus which uses circles containing or touching other circles, which may be connected by lines and arrows, too, in order to express abstraction, application, and bindings [2].

This chapter describes an approach for specifying diagrams using any of the described building principles, i.e., any of the spatial building principles. The key issue of this approach is the opportunity to use such a specification as an input for a tool. The tool automatically creates a graphical environment which serves as an editor for creating and modifying diagrams of the specified diagram class.

32.2 DiaGen

The work described in this chapter is continued work on DiaGen, a framework together with a generator for creating graphical editors for a specific diagram class from a formal specification. So far DiaGen has considered diagram languages consisting of graph-like and recursively built diagrams only. This chapter shows that the concepts used in DiaGen – hypergraphs as an internal diagram model and hypergraph grammars – are also a simple means for specifying diagram languages which use general spatial relationships.

DiaGen as described in [6, 8, 12] consists of an editor framework and a generator. A formal specification of a diagram class serves as input for the generator which creates custom components that build – together with the framework – a graphical editor customised for the specified diagram class. Main features supported by this approach are:

1. Diagrams are internally represented by hypergraphs (see next section); a diagram class is thus a hypergraph language together with a mapping from hypergraphs to their visual representation as diagrams. A context-free hypergraph grammar is used to describe the hypergraph language. For diagrams which cannot be described by a context-free grammar, additional graph transformations serve as a means to create context-sensitive shares.
2. Nodes and hyperedges carry attributes, and each grammar production is augmented by layout constraints on attributes accessible in the production. A constraint-solver provides automatic, user-adjustable layout of diagrams.
3. Diagrams can be edited in a syntax-directed way by using transformations on derivation trees for their context-free share and on hypergraphs described in the specification. To hide those details from the user, interactions of the user and the editor are described by certain interaction automata, thus offering editing diagrams by direct manipulation.
4. Free-hand editing is also supported. The user can arbitrarily add, delete, move, or modify parts of the diagram as with a drawing tool. The underlying hypergraph model is modified accordingly; a hypergraph parser

distinguishes correct diagrams from incorrect ones by keeping the underlying hypergraph's syntactic meta-structure up to date. Free-hand editing with parser support relaxes the need to specify a full set of transformations on diagrams for syntax-directed editing since free-hand editing can be used for (yet) unspecified diagram operations. Therefore, this editing mode enhances usability of editors and also makes rapid prototyping of diagram editors possible because – as an extreme case – specification of diagram operations can be omitted completely.

32.3 Hypergraph Representation of Diagrams

Hypergraphs have proved to be an intuitive means for internally representing diagrams [6–8, 12]. A hypergraph is a generalisation of a graph, in which edges are *hyperedges* [1]. Each hyperedge has a type and a (fixed) number of tentacles. Each tentacle is connected to a node. We say the hyperedge *visits* these nodes. The familiar directed graph can be seen as a hypergraph in which each hyperedge has two tentacles: *source* and *target* of the edge.

32.3.1 Graph-Like and Recursively Built Diagrams

For graph-like or recursively built diagrams, there is a direct representation of diagrams by hypergraphs. Figure 32.1 shows a simple Nassi–Shneiderman diagram and its hypergraph model: Nodes in the hypergraph stand for points in the plane; hyperedges are diagram elements whose position is given by the nodes being visited by the hyperedge. Each hyperedge tentacle represents an area of a diagram element that can be attached to others. In the case of Nassi–Shneiderman diagrams each corner of a diagram element is such an attachment area, which has to be represented by a tentacle being connected to a node.

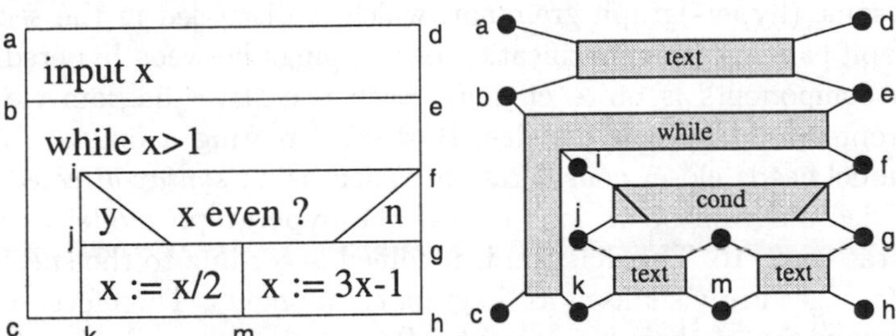

Figure 32.1. A Nassi–Shneiderman diagram and its hypergraph representation. Nodes are depicted by black dots, hyperedges by shaded areas which are connected to visited nodes. Letters a . . . m do not belong to the diagram. They rather illustrate which point of the diagram is represented by which node.

Finite automata are another diagram class which are easily modelled by hypergraphs. Figure 32.2 shows a sample finite automaton for the regular language $(ab + aab)^*$ and its hypergraph model. States are modelled by hyperedges with only one tentacle which represents the borderline of the state's circle. Transitions are represented by hyperedges with three tentacles: the two end points of the arrow (source s and target t in Fig. 32.2) and the arrow line (description d in Fig. 32.2). The arrow line tentacle is needed for representing transition labels. Each label is represented by a hyperedge with one tentacle which is connected to the same node as the arrow line tentacle. Please note that a node may represent an area of the plane instead of a single point as in the Nassi–Shneiderman example.

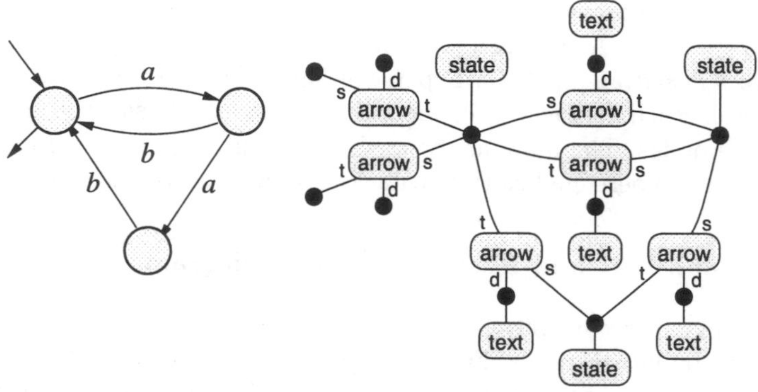

Figure 32.2. A finite automaton for the regular language $(ab + aab)^*$ and its hypergraph representation. Nodes are depicted by black dots, hyperedges by shaded areas that are connected to visited nodes.

A specification of a diagram class based on such a hypergraph model consists of a mapping between hyperedges and their diagram counterparts and a description of all valid hypergraph models. The latter is usually specified by some (hyper-)graph grammar, which is discussed in the section on hypergraph parsing. The specification of mappings between hyperedges and diagram components is quite obvious: Each primitive diagram element is simply represented by a hyperedge. However, creating a diagram's hypergraph model needs closer consideration: When using *syntax-directed editing*, pre-specified diagram operations modify the hypergraph model as the primary data structure. The diagram is modified according to the specification how hyperedges are visualised as diagram components. Therefore, diagrams behave like views of their hypergraphs. However, when providing *free-hand editing*, diagrams are modified directly, and the hypergraph representation has to be adjusted accordingly. The crucial operation of these adjustments is how hyperedges that represent spatially related diagram components are connected by visiting the same nodes. For graph-like diagram classes such as the finite automata example or recursively built diagram classes such as the Nassi–Shneiderman example, this task is easily fulfilled: Each diagram component is represented by a hyperedge which already visits an appropriate number of nodes for the component's "sensitive" areas of the plane. Whenever diagrams are edited such that sensitive areas overlap, the corresponding nodes get unified; whenever overlapping areas get separated, the previously unified nodes get separated again. For diagrams with general spatial relationships as building principle, however, using an internal hypergraph model and keeping it synchronised with a diagram, which is modified by free-hand editing, is a more complicated task. This is considered next.

32.3.2 Diagrams Using General Spatial Relationships

When diagrams use general spatial relationships like *containing*, *intersecting*, or *touching*, it is not sufficient to represent only diagram components by hyperedges; the hypergraph also has to contain hyperedges that make the significant relationships explicit. However, the hypergraph still has to model its represented diagram in a way which makes it easy to incrementally adjust the hypergraph when editing the diagram.

We consider VEX [2] as an example of a diagram language which primarily uses spatial relationships, but also recursive and graph-like features for expressing λ-expressions: in VEX each variable identifier is represented by an empty circle that is connected by a line to a so-called *root node*. A root node is again an empty circle with one or more lines touching it, leading to all identifiers representing the same variable. A root node may either be internally tangential to another circle – it then represents a parameter of a λ-abstraction – or it is not included by any other circle; it then denotes a free variable. A circle representing a λ-abstraction contains its parameter circle and a VEX (sub-)diagram as its body. An application of two expressions is depicted by two externally tangential circles with an arrow at the tangent point. The head of the arrow lies inside the argument circle. Figure 32.3 shows VEX expressions for $(\lambda x.x)y$ and $\lambda x.(xx)$.

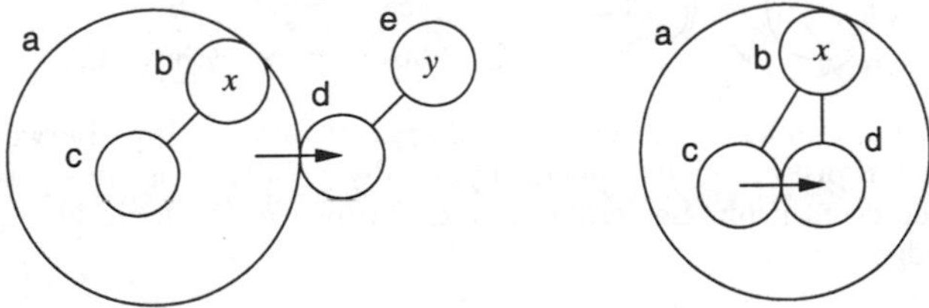

Figure 32.3. Two VEX expressions for $(\lambda x.x)y$ and $\lambda x.(xx)$. Letters a ... e do not belong to the diagram. They are used for referencing circles only.

VEX diagrams consist of circles, lines, and arrows, which are represented by corresponding hyperedges. For simplicity we omit labelling text. Therefore, each diagram element has two attachment areas: circles have their borderline and their area; lines as well as arrows have their end points as attachment areas where they connect to other diagram elements. Plain edges can be used for representation; hyperedges visiting more than two nodes are not required: The edges representing arrows and lines simply connect nodes representing the corresponding end points. Circles are also represented by (directed) edges connecting two nodes: the source node ("borderline node") of the edge represents the borderline of the circle, the target ("area node") of the edge its inner area (Fig. 32.4).

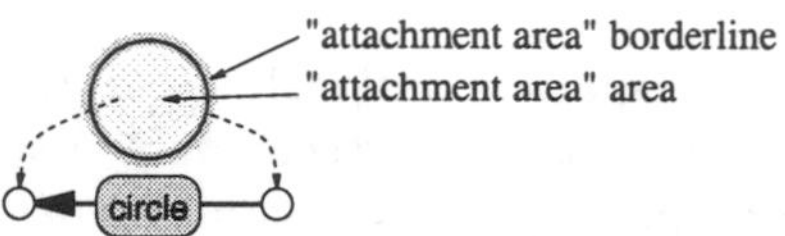

Figure 32.4. Hyperedge representation of a circle.

VEX's main spatial relationships relate two circles which may be internally or externally tangential, or they relate arrows with circles where arrow head and tail lie inside of two circles. The latter situation is represented similarly to connecting two circles by a line: the arrow's hyperedge simply connects the circle nodes which represent the circles' areas. However, a situation where one circle is contained in another one cannot be described by simply unifying their area nodes. It would not be clear which circle is the inner one. Furthermore, if there were a third circle contained in the outer one, we would have to unify its area node with the other area node, too, losing all the information of how the second and the third circle are related (Fig. 32.5).

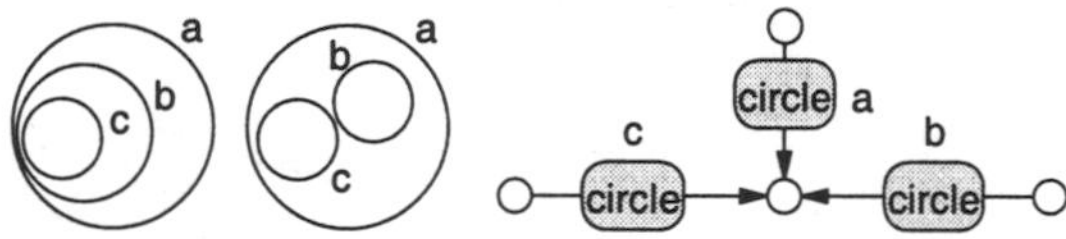

Figure 32.5. Inadequate representation of spatial relationships between circles. If inclusion is not represented by special hyperedges, both diagrams have the same hypergraph representation. Letters a, b, and c indicate which circle corresponds to which hyperedge.

In order not to lose information by representing a diagram by an internal hypergraph, additional hyperedge types *touch* and *inside* are used. The first one denotes an (undirected) hyperedge[1] connecting the borderline nodes of tangential circles, the latter one denotes a directed edge from the area node of a circle to the area node of another circle which contains the first one. We do not need a direct representation of circles being internally or externally tangential. This is expressed by a *touch* edge between the border nodes of both circles and an *inside* edge between the area nodes if the circles are internally tangential and without such an *inside* edge if they are externally tangential. In order to make detection of invalid diagrams possible on the level of their internal diagrams, we also need an additional *intersect* edge which connects the border nodes of intersecting circles. The existence of such an edge is an indicator for an invalid VEX diagram. Figure 32.6 shows the corresponding hypergraph representations for the VEX diagrams of Fig. 32.3. *Circle* hyperedges are labelled with the same letters as the corresponding circles in Fig. 32.3.

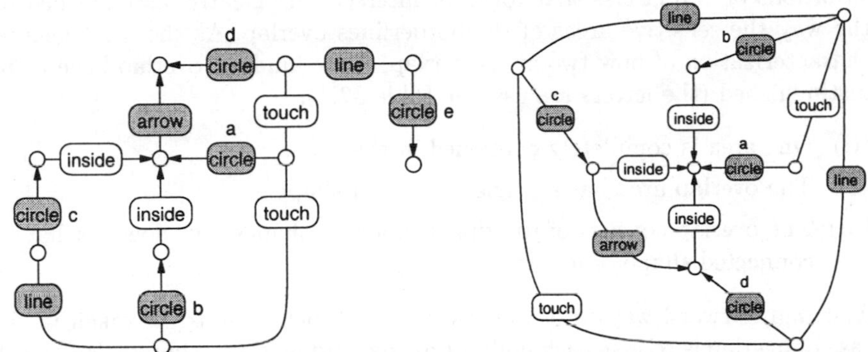

Figure 32.6. Hypergraph representations of the VEX diagrams of Fig. 32.3. Edges for diagram components have a grey label, spatial relationship edges have a white one. Letters a ... e indicate which circles in Fig. 32.3 are represented by which hyperedge.

When editing a diagram in free-hand mode, the hypergraph has to be adjusted accordingly. For lines that connect two circles, VEX follows the graph-like building principle. In this case, for each of the line's end points, its node gets unified with the borderline node of the corresponding circle if their sensitive areas overlap. Application arrows are similarly handled; they connect the corresponding circles' area nodes when their sensitive areas overlap. However, *touch* and *inside* edges have to be created automatically when necessary, and they have to be removed again when the circles are no longer related. Therefore, we have the following general situation.

[1] An undirected edge can easily be represented by two directed edges which connect the same nodes, but with opposite direction.

Primitive diagram components are represented by hyperedges. In the beginning, when they are created, each of these hyperedges is added to the already existing hypergraph with a set of new nodes. The new hyperedge is connected to these nodes. Each node represents a sensitive area of the plane. As soon as sensitive areas of two nodes overlap, one of three actions has to be taken which has to be revoked as soon as the original sensitive areas no longer overlap:

1. Either, the corresponding nodes get unified. This connects the new hyperedge to the existing hypergraph. The shape of the sensitive area of the unified node has to be reconsidered then.
2. Or the corresponding nodes get connected by a hyperedge representing the spatial relationship between the corresponding diagram components.
3. Or nothing happens.

The selected operation may also depend on the kind of overlap, e.g., the situations of two circles that touch or intersect are clearly distinguished by the way the sensitive areas of the borderlines overlap. As the most general characterization of how two areas overlap, three kinds of overlap have to be distinguished (the letters are used in Table 32.1):

(C) One area is completely contained in the other.

(S) The overlap area has a single connected shape.

(M) The overlap consists of multiple connected shapes, but does not have a connected shape of its own.

A straightforward way to define the actions which have to be taken in the case of overlap is to give each node a type depending on the hyperedge, which visits the node, and to create a table, which describes the action to be taken if two nodes of given type overlap and the overlap is of a certain kind.

We use VEX as an example again. We have three types of nodes: Circle nodes represent the circles's *area* or *border*, and arrows as well as lines have nodes of type *end point.* Table 32.1 shows all the actions to be taken when sensitive areas of different nodes overlap. Rows and columns of the table represent the different node types. Each column is separated into three subcolumns according to the kind of overlap. The table's fields contain the edges which have to be added to the hypergraph or "unify" if the corresponding nodes have to be unified; the table then contains the new types of unified nodes as underlined names, too.

With such a table it is now straightforward to have a general editor framework which can be customised to an editor for a specific diagram class that keeps an internal hypergraph model up to date. So far, the specification consists of a set of primitive diagram elements – for VEX we need circles, lines, and arrows – and their hyperedge representations. For each of these components, sensitive areas have to be defined where they can have "contact" with

Table 32.1. Action to be taken when the sensitive areas of two nodes overlap.

node type →	*area*			*border*			*end point*		
↓ overlap →	C	S	M	C	S	M	C	S	M
area	*inside*						unify area		
border					*touch*	*inter-sect*	unify border	unify border	
end point	unify area			unify border	unify border				

other diagram components. These areas are represented by nodes of the hypergraph. Furthermore, each of the sensitive areas has to have a specific type where different areas may have the same type. Finally, the specification has to contain a table like Table 32.1 which describes how to connect hypergraph parts when sensitive areas of different nodes overlap.

In order to really create an editor for the specified diagram class, it has to be aware of the diagram class's syntax, i.e., we need a means to check the correctness of the hypergraph model by some hypergraph parser which simultaneously can extract additional information, e.g., an abstract syntax representation, for further processing.

32.4 Hypergraph Parsing

We describe diagrams in terms of their hypergraphs. In order to define a diagram language in terms of a hypergraph language, ***hypergraph grammars*** are an appropriate means. In [6] *context-free* hypergraph grammars with optional *embeddings* have been used to define the class of valid hypergraphs and thus the class of valid diagrams. A hypergraph parser was presented to check whether hypergraphs are valid with respect to the specified grammar. In this section, the class of hypergraph grammars is further extended in order to cover an even wider range of diagram classes.

A hypergraph grammar HG consists of a set T of terminal hyperedge types and a set N of non-terminal ones with both sets being disjoint. Furthermore, there is a finite set of hypergraph productions,[2] and a starting hypergraph $S \in Graph_{N \cup T}$, where $Graph_{N \cup T}$ denotes the set of all hypergraphs containing only hyperedges of types in $N \cup T$. Each production $p = (L \leftarrow G \rightarrow R)$ consists of three hypergraphs $L, G, R \in Graph_{N \cup T}$ and two hypergraph morphisms $G \rightarrow L$ and $G \rightarrow R$, where G ("gluing hypergraph") is a sub-hypergraph of L ("left-hand side", LHS) and R ("right-hand side", RHS). The morphisms unambiguously define the mapping from G into L or R. A hypergraph production p is applied to a hypergraph H by finding

[2] In the following, we use the terms *production* and *rule* as synonyms.

L as a subgraph (*redex*) of H, removing the subgraph $L \setminus G$ from H and merging in $R \setminus G$ instead, where the gluing hypergraph G and the mapping (hypergraph morphism) $G \to R$ exactly describe how the RHS has to fit in. The resulting hypergraph H' is called ***derived*** from H, $H \to H'$, in one step. The hypergraph language $L(HG)$ is the set of all terminal hypergraphs $H \in Graph_T$ which can be derived from the starting hypergraph S in a finite number of steps, $S \overset{*}{\to} H$.

A special case of such hypergraph grammars are context-free hypergraph grammars which only contain context-free productions, where L consists of a single hyperedge of non-terminal type, G being the set of all nodes of L, and $R \in Graph_{N \cup T}$ being an arbitrary hypergraph containing the nodes of G.

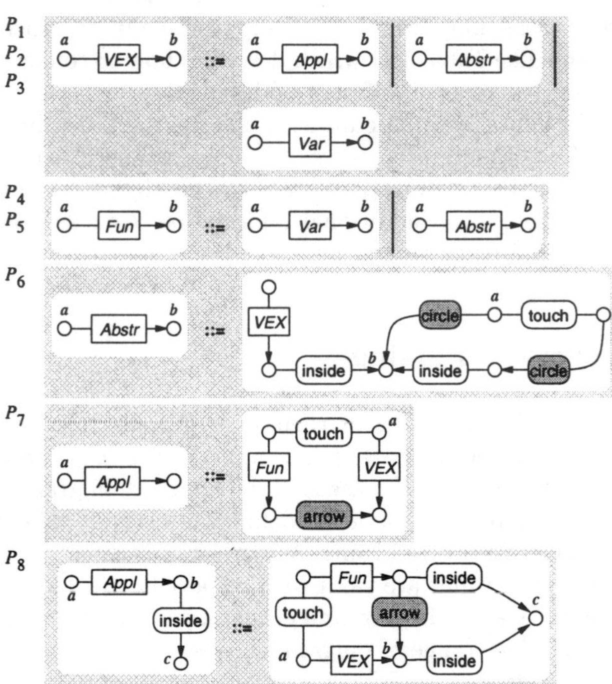

Figure 32.7. Some selected hypergraph productions of the VEX grammar. Edges with oval labels are terminal edges; rectangular labels depict non-terminal edges.

For VEX and the hypergraph model described earlier, there does not exist any context-free hypergraph grammar. However, a grammar according to the general hypergraph grammar definition is easily created. Figure 32.7 shows a selection of eight productions in a more readable representation which contains only LHS and RHS of a production. The gluing hypergraph is implicitly defined by the labelled nodes and the hyperedges which are contained in the LHS and the RHS at the same time. The complete grammar consists of 15 productions [11]. $P_1, \ldots, P_6$ are actually context-free ones, but P_8 is not. P_8's gluing hypergraph consists of the nodes a, b, and c and of the *inside* hyperedge. P_7 is somehow special since it must not be applied if P_8 is applicable. Application of P_7 is then prohibited because only one of the LHS nodes is in the gluing hypergraph. If the host hypergraph has an additional edge visiting the node corresponding to the unlabelled one, removing $L \setminus G$ would yield an inconsistent hypergraph since this node would be removed, leaving the additional edge without a visited node.[3] Figure 32.8 shows a sample derivation using P_1, P_8, and P_3.

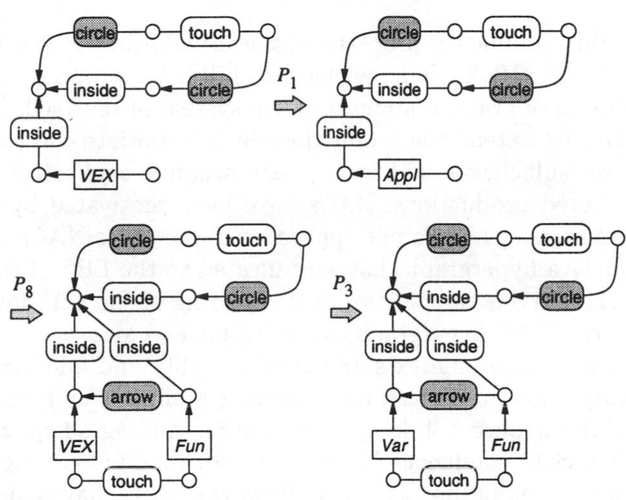

Figure 32.8. A sample derivation using the VEX grammar.

For a diagram editor with free-hand editing, a parsing algorithm has to reconstruct a derivation from the starting hypergraph to the internal hypergraph model of the current diagram. The diagram is a valid diagram if the parser succeeds, otherwise it is not. For context-free hypergraph grammars, this problem is more or less efficiently decidable [6]. For the general grammar, this problem is no longer decidable. For certain restricted graph grammars[4] such as the *layered graph grammars* [9] there are special parsing algorithms; however, they are quite inefficient. Recently, the class of *reserved graph grammars* (RGGs) [17] has been proposed, which allows for a straightforward way to parse hypergraphs.[5] The derivation for a hypergraph H is reconstructed by

[3] In graph transformation theory, this condition, which prohibits application of graph productions, is called *gluing* or *contact condition.*

[4] In the literature, mainly graphs as the simple form of hypergraphs are used. However, most results for graphs also apply to hypergraphs.

[5] *Reserved graph grammars* have been introduced by Zhang and Zhang for special kinds of graph which are actually hypergraphs.

exchanging LHS and RHS of each production ("reversed productions") and to start a derivation at H using reversed productions until the derivation stops. If the resulting hypergraph is the starting hypergraph, H is valid. For RGGs, parsing is always terminating since there is a well-founded ordering on hypergraphs decreasing for each derivation step during parsing [17]. Furthermore, the resulting system of reversed productions is confluent, i.e., for all derivations $H \xrightarrow{*} H_1$ and $H \xrightarrow{*} H_2$, there is always a hypergraph H' with $H_1 \xrightarrow{*} H'$ and $H_2 \xrightarrow{*} H'$. Therefore we can conclude that the simple RGG parser finds a derivation from any hypergraph H to the starting hypergraph if and only if there exists such a derivation. The parser cannot run into a dead end.

However, this confluence property is the crucial property which is frequently, e.g., for our VEX grammar, hard to fulfil. We propose a simple but effective extension of graph grammars: if the system of reversed productions is not confluent, we extend the productions by appropriate context and/or – if this is not yet sufficient – add appropriate *negative application conditions* (NACs) to affected productions. NACs have been motivated by [5]: a production with matching LHS is not applicable if one of its NACs is satisfied. A NAC is simply a hypergraph that is connected to the LHS of the reversed production. The NAC is satisfied for an embedding of the LHS into the host hypergraph if the NAC hypergraph can be embedded also.

Of course, additional contexts and NACs modify and add further information to hypergraph grammars; the original grammar is no longer the only description of the hypergraph language's syntax. Adding adequate contexts and NACs that yield confluence but that do not modify the original grammar's language might be not too easy. However, tools may assist the user here.

This chapter does not present a fully fledged set of reversed productions for VEX, owing to its size. For details see [11]. Instead we give an example of why we need additional contexts and NACs for VEX. Figure 32.9 shows a hypergraph for which the reversed productions of P_3 as well as P_4 are applicable. Actually, reversed P_3 must not be applied since the variable has to act as a function ("*Fun*"). The situation when reversed P_3 may be applied is easily distinguished from the situation when reversed P_4 may be applied by considering the context: if an *arrow* edge is present as indicated, reversed P_4 has to be applied, otherwise P_3. Figure 32.9 shows the appropriately extended reversed productions.

32.5 Related Work

In the field of frameworks for visual language environments based on diagrams there are several related approaches; the most closely related ones are VLCC [3], the visual environment [10], and the VisPro toolset based on reserved graph grammars [17].

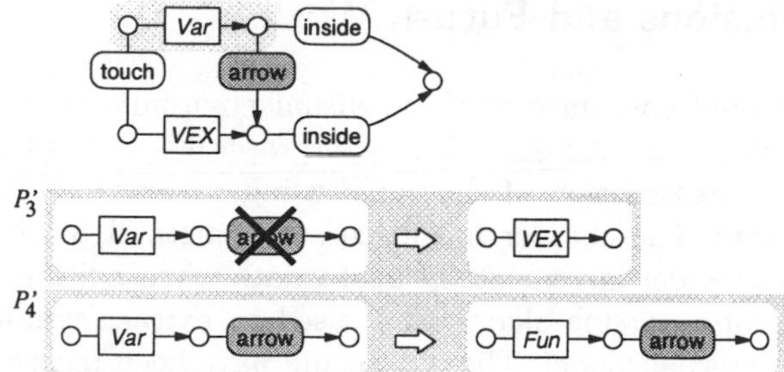

Figure 32.9. Hypergraph with overlapping redices – indicated by the greyish subgraph – and appropriately extended reversed P_3 and P_4 (see Fig. 32.7) with negative application condition (indicated by "X") or additional context.

In contrast to the three building principles of diagrams presented in this chapter, Costagliola et al. distinguish two main syntactic models for visual languages. The *connection-based* one corresponds to our graph-like principle; the *geometric-based* one covers our recursive principle as well as the one with general spatial relationships. They use an object-oriented hierarchy for representing diagrams according to their syntactic models instead of a uniform representation as in our work. For connecting visual components, their VLCC system uses sensitive areas which are defined for each diagram component, as in DiaGen.

The approach by Rekers and Schürr actually uses two kinds of graph as internal representations of diagrams: the *spatial relationship graph* (SRG) abstracts from the physical diagram layout and represents higher-level spatial relationships. Additionally, an *abstract syntax graph* (ASG) is kept uptodate with the SRG representing the diagram's logical structure. Two different but connected context-sensitive graph grammars are used in order to define the syntax of SRGs and ASGs. Free-hand editing of diagrams is planned to modify the first graph; syntax-directed editing is going to modify the second. The other graph is modified accordingly. DiaGen uses only one internal hypergraph which corresponds to their SRG. However, DiaGen uses hypergraphs which makes specification more "natural" and easier, we believe.

Finally, Zhang and Zhang's VisPro system depends on reserved graph grammars which allow for efficient parsing even of context-sensitive graph grammars. The hypergraph parser presented in this chapter actually uses the same parser and extends it by additional negative application conditions which yield a wider range of (hyper-)graph languages to which it can be applied.

32.6 Conclusions and Future Work

We have presented continued work on visual environments for graphically working with diagrams. An extended hypergraph model has been presented which allows representation of diagrams using various building principles in a uniform way. The chapter has discussed a method to keep this internal hypergraph model incrementally up to date when editing the diagram. A new hypergraph parsing algorithm is used to extract syntactic information from the internal model. These concepts have been incorporated into a framework for visual environments (DiaGen, http://www2.informatik.uni-erlangen.de/DiaGen).

Further work also consists of research into how to connect a visual environment as considered here with other systems, e.g., for diagrammatic reasoning. For this we need suitable representations of (abstract) syntax and semantics of diagrams as described in [4]. The translation into such a description can be specified by extending the hypergraph grammar by attributes and attribute evaluation rules similar to compiler generators like Yacc. Other work deals with another similarity with common compilers. Efficiency of parsing might be improved by a two-stage approach: Instead of parsing the internal hypergraph model directly, lexical analysis first tries to recognise larger patterns within the model and creates a reduced hypergraph model, which is then more easily and efficiently parsed.

References

1. Berge, C. (1989). Hypergraphs. Amsterdam: North-Holland.
2. Citrin, W., Hall, R. and Zorn, B. (1995). Programming with visual expressions. In [14], pp. 294–301.
3. Costagliola, G., Lucia, A.D., Orefice, S. and Tortora, G. (1997). A framework of syntactic models for the implementation of visual languages. In [16], pp. 58–65.
4. Erwig, M. (1997). Semantics of visual languages. In [16], pp. 304–311.
5. Habel, A., Heckel, R. and Taentzer, G. (1996). Graph grammars with negative application conditions. Fundamenta Informaticae 26(3,4).
6. Minas, M. (1997). Diagram editing with hypergraph parser support. In [16], pp. 230–237.
7. Minas, M. and Shklar, L. (1996). A high-level visual language for generating web structures. In [15], p. 248f.
8. Minas, M. and Viehstaedt, G. (1995). DiaGen: A generator for diagram editors providing direct manipulation and execution of diagrams. In [14], pp. 203–210.
9. Rekers, J. and Schürr, A. (1995). A graph grammar approach to graphical parsing. In [14], pp. 195–202.
10. Rekers, J. and Schürr, A. (1996). A graph based framework for the implementation of visual environments. In [15], pp. 148–155.
11. Stingl, B. (1998). Specifying animated diagram classes by graph transformation systems. Diploma thesis, University of Erlangen, Germany.
12. Viehstaedt, G. and Minas, M. (1994). Interaction in really graphical user interfaces. In [13], pp. 270–277.

13. VL'94 (1994). Proceedings of the 1994 IEEE symposium on visual languages, St Louis, MO. Los Alamitos, CA: IEEE Computer Society Press.
14. VL'95 (1995). Proceedings of the 1995 IEEE symposium on visual languages, Darmstadt, Germany. Los Alamitos, CA: IEEE Computer Society Press.
15. VL'96 (1996). Proceedings of the 1996 IEEE symposium on visual languages, Boulder, CO. Los Alamitos, CA: IEEE Computer Society Press.
16. VL'97 (1997). Proceedings of the 1997 IEEE symposium on visual languages, Capri, Italy. Los Alamitos, CA: IEEE Computer Society Press.
17. Zhang, D-Q. and Zhang, K. (1997). Reserved graph grammar: A specification tool for diagrammatic VPLs. In [16], pp. 288–295.

Index